Applied and Numerical Harmonic Analysis

Signal Analysis and Prediction

A. Procházka
J. Uhlíř
P.J.W. Rayner
N.G. Kingsbury
Editors

Birkhäuser
Boston • Basel • Berlin

Ales Procházka
Institute of Chemical Technology
Prague, Czech Republic

Jan Uhlíř
Czech Technical University
Prague, Czech Republic

P.W.J. Rayner
N.G. Kingsbury
University of Cambridge
England, United Kingdom

Library of Congress Cataloging-in-Publication Data

Signal analysis and prediction / [edited by] Procházka ... [et al.]
p. cm. -- (Applied and numerical harmonic analysis)
Includes bibliographical references and index.
ISBN 978-1-4612-7273-1 (Basel)
1. Signal processing--Statistical methods. 2. Prediction theory.
I. Procházka, Ales, 1948- . II. Series.
TK5102.9.S533 1998 98-4740
621.382'23--dc21 CIP

Printed on acid-free paper

Birkhäuser 

ISBN 978-1-4612-7273-1

9 8 7 6 5 4 3 2 1

Contents

List of Contributors

Søren Vang Andersen, Aalborg University, Institute of Electronic Systems, Center for Personkommunikation (CPK), Fredrik Bajers Vej 7, DK-9220 Aalborg, Denmark
E-mail: sva@cpk.auc.dk

Matthias Arnold, Friedrich Schiller University Jena, Medical Faculty, Institute of Medical Statistics, Computer Sciences and Documentation, Jahnstraße 3, 07740 Jena, Germany
E-mail: iia@imsid.uni-jena.de

Jens Baltersee, Integrated Systems for Signal Processing, Aachen University of Technology, Templergraben 55, D-52056 Aachen, Germany
E-mail: balterse@ert.rwth-aachen.de

Harm J.W. Belt, Philips Research Laboratories, Prof. Holstlaan 4, NL-5656 AA Eindhoven, Netherlands
E-mail: Belt@natlab.research.philips.com

Svetlana Borovkova, Shell Research and Technology Centre, Amsterdam, P.O.Box 38000, 1030 BN Amsterdam, The Netherlands
E-mail: Svetlana.A.Borovkova@opc.shell.com

Maja Bračič, University of Ljubljana, Faculty of Electrical Engineering, Tržaška 25, 1000 Ljubljana, Slovenia
E-mail: maja@osc.fe.uni-lj.si

Albertus C. den Brinker, Faculteit der Elektrotechniek, Technische Universiteit Eindhoven, P.O. Box 513, NL-5600 MB Eindhoven, Netherlands
E-mail: A.C.d.Brinker@ele.tue.nl

Robert Burton, Oregon State University, Department of Mathematics, Kidder Hall 368, Corvallis, OR, 97331 USA
E-mail: burton@math.orst.edu

Catharina Carlemalm, Division of Automatic Control, Department of Signals, Sensors and Systems, Royal Institute of Technology, SE-100 44 Stockholm, Sweden
E-mail: catharina.carlemalm@s3.kth.se

Andrea Cavallaro, University of Trieste - DEEI, Via A. Valerio 10, 34127 Trieste, Italy

Jonathon Chambers, Signal Processing Section, Department of Electrical Engineering, Imperial College of Science, Technology and Medicine, Exhibition Road, London, SW7 2BT, United Kingdom
E-mail: j.chambers@ic.ac.uk

Georges A. Darbellay, Inst. of Information Theory and Automation, Academy of Sciences, Pod vodárenskou věží 4, 182 08 Prague 8, Czech Republic
E-mail: dbe@utia.cas.cz

Herold Dehling, University of Groningen, Department of Mathematics, P.O.Box 800, 9700 AV Groningen, The Netherlands
E-mail: dehling@math.rug.nl

Elias N. Demiris, University of Patras, School of Engineering, Department of Computer Engineering and Informatics, Patras 26500, Greece
E-mail: Demiris@ceid.upatras.gr

Domingo Docampo, Departamento de Tecnologías de las Comunicaciones, E.T.S.I. Telecomunicación, Campus Universitario, 36200 Vigo, Spain
E-mail: ddocampo@tsc.uvigo.es

Andrei Doncescu, University of La Rochelle, 17000 La Rochelle France
E-mail: andrei@mail.univ-lr.fr

Juan Francisco Fariña, Departamento de Tecnologías de las Comunicaciones, E.P.S.I. Telecomunicación, Univ. Carlos III de Madrid, 28911 Leganés (Madrid), Spain
E-mail: jfarina@tsc.uc3m.es

Christian Faye, Equipe Traitement des Images et Signaux, 6 Avenue du Ponceau, 95014 Cergy-Pontoise, France

Bill Fitzgerald, University of Cambridge, Department of Engineering, Trumpington Street, Cambridge, CB2 1PZ, United Kingdom
E-mail: wjf@eng.cam.ac.uk

Jean-Paul Gourret, University of La Rochelle, 17000 La Rochelle France
E-mail: jean-paul.gourret@mail.univ-lr.fr

Peter M. Grant, Signals and Systems Group, Department of Electrical Engineering, Univ. of Edinburgh, Edinburgh, EH9 3JL, United Kingdom
E-mail: Peter.Grant@ee.ed.ac.uk

Dominique Guégan, ENSAE-CREST, Timbre J120, 3 av. P. Larousse 92245 Malakoff Cedex France
E-mail: guegan@ensae.fr

Roland Günther, Faculty of Mathematics and Informatics, Institute of Stochastics, Ernst-Abbe-Platz 1-3, 07740 Jena, Germany
E-mail guenther@minet.uni-jena.de

Fredrik Gustafsson, Division of Automatic Control, Department of Electrical Engineering, Linköping University, SE-581 83 Linköping, Sweden
E-mail: fredrik@isy.liu.se

John M. Hannah, Signals and Systems Group, Department of Electrical Engineering, Univ. of Edinburgh, Edinburgh, EH9 3JL, United Kingdom

Egon Hansen, Aalborg University, Institute of Electronic Systems, Center for Personkommunikation (CPK), Fredrik Bajers Vej 7, DK-9220 Aalborg, Denmark
E-mail: eh@cpk.auc.dk

Hynek Hermansky, Oregon Graduate Institute of Science & Technology, Department of Electrical and Computer Engineering, P.O. Box 91000, Portland, OR 97291, U.S.A.
E-mail: hynek@ece.ogi.edu

Ulrich Heute, University of Kiel, Institute for Network and System Theory, Kaiserstr. 2, 24143 Kiel, Germany
E-mail: uh@techfak.uni-kiel.d400.de

Abdulnasir Hossen, Applied Science University, Computer Science and Information Systems Department, 11931 Amman, Jordan

Marcelo Iribarren, University of Concepción, Dept. of Electrical Engineering, P. O. Box 53-C, Concepción, Chile
E-mail: miribarr@manet.die.udec.cl

Søren Holdt Jensen, Aalborg University, Institute of Electronic Systems, Center for Personkommunikation (CPK), Fredrik Bajers Vej 7, DK-9220 Aalborg, Denmark
E-mail: shj@cpk.auc.dk

Nick Kingsbury University of Cambridge, Department of Engineering, Trumpington Street, Cambridge, CB2 1PZ, United Kingdom
E-mail: ngk@eng.cam.ac.uk

Tomáš Kreisinger, Czech Technical University, Faculty of Electrical Engineering, Technická 2, 166 27 Prague 6, Czech Republic
E-mail: kreising@feld.cvut.cz

Gernot Kubin, Institute of Communications and High-Frequency Engineering, Vienna Univ. of Technology, Gusshausstrasse 25/389, A–1040 Vienna, Austria
E-mail: G.Kubin@ieee.org

Jyh-Ming Kuo, Kaohsiung Polytechnic Institute, Taiwan

Anthony N. Lasenby, MRAO, Cavendish Laboratory, Madingley Road, Cambridge, CB3 OHE, United Kingdom
E-mail: anthony@mrao.cam.ac.uk

Joan Lasenby, University of Cambridge, Department of Engineering, Trumpington Street, Cambridge, CB2 1PZ, United Kingdom
E-mail: jl@eng.cam.ac.uk

Spiridon D. Likothanassis, University of Patras, School of Engineering, Department of Computer Engineering and Informatics, Patras 26500, Greece and Computer Technology Institute (C.T.I.), Patras 26100, Greece
E-mail: Likothan@cti.gr

Man Lin, Linköping University, Department of Computer Science, 581 83 Linköping, Sweden
E-mail: linma@ida.liu.se

Lennart Ljung, Department of Electrical Engineering, Linköping University, S-581 83 Linköping, Sweden
E-mail: Ljung@isy.liu.se

Julian Magarey Centre for Sensor Signal and Information Processing, Technology Park Adelaide, The Levels, SA5095, Australia
E-mail: jfam@cssip.edu.au

Aleksej Makarov, Swiss Federal Institute of Technology, Micro-Processor and Interface Laboratory, 1015 Lausanne, Switzerland
E-mail: Aleksej.Makarov@epfl.ch

Danilo Mandic, Signal Processing Section, Department of Electrical Engineering, Imperial College of Science, Technology and Medicine, Exhibition Road, London, SW7 2BT, United Kingdom
E-mail: d.mandic@ic.ac.uk

Stefano Marsi, University of Trieste - DEEI, Via A. Valerio 10, 34127 Trieste, Italy
E-mail: marsi@ipl.univ.trieste.it

Cesar San Martin, University of Concepción, Dept. of Electrical Engineering, P. O. Box 53-C, Concepción, Chile
E-mail: csanmart@manet.die.udec.cl

Andrew C. McCormick, University of Strathclyde, Department of Electronic and Electrical Engineering, 204 George Street, Glasgow, G1 1XW, United Kingdom

Ludovic Mercier, CREST, Laboratoire de statistique, Timbre J340, 3 av. P. Larousse 92245 Malakoff Cedex France
E-mail: mercier@ensae.fr

Martina Mudrová, Institute of Chemical Technology, Department of Computing and Control Engineering, Technická 1905, 166 28 Prague 6, Czech Republic
E-mail: Martina.Mudrova@vscht.cz

Bernard Mulgrew, Signals and Systems Group, Department of Electrical Engineering, Univ. of Edinburgh, Edinburgh, EH9 3JL, United Kingdom
E-mail: mb@ee.ed.ac.uk

Asoke K. Nandi, Signal Processing Division, University of Strathclyde, Department of Electronic and Electrical Engineering, 204 George Street, Glasgow, G1 1XW, United Kingdom
E-mail: asoke@eee.strath.ac.uk

André Neubauer, Siemens AG, Microelectronics Design Center,
Wacholderstraße 7, 40489 Düsseldorf, Germany
E-mail: neubauer@ezmd.hl.siemens.de

David E. Newland University of Cambridge, Department of Engineering,
Trumpington Street, Cambridge, CB2 1PZ, United Kingdom
E-mail: den@eng.cam.ac.uk

José M. Páez-Borrallo, Dpto. de Señales, Sistemas y Radiocomunicaciones,
ETS Ing. Telecomunicación Universidad Politécnica de Madrid,
Ciudad Universitaria s/n, 28040 Madrid, Spain
E-mail: paez@gaps.ssr.upm.es

Iván A. Pérez-Álvarez, Dpto. de Señales y Comunicaciones, Univ. de Las
Palmas de Gran Canaria, Campus, Univ. de Tafira, 35017 Las Palmas, Spain
E-mail: ivan@masdache.teleco.ulpgc.es

Petr Pollák, Czech Technical University, Faculty of Electrical Engineering,
Technická 2, 166 27 Prague 6, Czech Republic
E-mail: pollak@feld.cvut.cz

Jose C. Principe, Computational NeuroEngineering Laboratory,
University of Florida, Gainesville, Florida, FL 32611, U.S.A.
E-mail: principe@cnel.ufl.edu

Aleš Procházka, Institute of Chemical Technology, Department of Computing
and Control Engineering, Technická 1905, 166 28 Prague 6, Czech Republic
E-mail: A.Prochazka@ieee.org

Anthony Quinn, Department of Electronic and Electrical Engineering,
University of Dublin, Trinity College, Dublin 2, Ireland
E-mail: aquinn@ee.tcd.ie

Peter Rayner, University of Cambridge, Department of Engineering,
Trumpington Street, Cambridge, CB2 1PZ, United Kingdom
E-mail: pjwr@eng.cam.ac.uk

Pedro Saavedra, U. of Concepción, Dept. of Mechanical Engineering,
P. O. Box 53-C, Concepción, Chile

Yoo-Sok Saw, Centre for Communications Research, Dept. of Electrical and
Electronic Engineering, Univ. of Bristol, Bristol, BS8 1UB, United Kingdom
E-mail: Yoo-Sok.Saw@bristol.ac.uk

M. Reza Serafat, Inst. for Network & System Theory, University Kiel, Germany
E-mail: res@techfak.uni-kiel.d400.de

Giovanni L. Sicuranza, University of Trieste - DEEI,
Via A. Valerio 10, 34127 Trieste, Italy
E-mail: sicuranza@univ.trieste.it

Pavel Sovka, Czech Technical University, Faculty of Electrical Engineering,
Technická 2, 166 27 Prague 6, Czech Republic
E-mail: sovka@feld.cvut.cz

Aneta Stefanovska, University of Ljubljana, Faculty of Electrical Engineering, Tržaška 25, 1000 Ljubljana, Slovenia
E-mail: aneta@osc.fe.uni-lj.si

Martin Štorek, Institute of Chemical Technology, Department of Computing and Control Engineering, Technická 1905, 166 28 Prague 6, Czech Republic
E-mail: Martin.Storek@vscht.cz

Gilles Thonet, Swiss Federal Institute of Technology, Signal Processing Laboratory, 1015 Lausanne, Switzerland
E-mail: Gilles.Thonet@epfl.ch

Jan Uhlíř, Czech Technical University, Faculty of Electrical Engineering, Technická 2, 166 27 Prague 6, Czech Republic
E-mail: uhlir@feld.cvut.cz

Ludong Wang, Hughes Network Systems, Germantown, MD 20876, U.S.A.

Rodolphe Weber, Laboratoire d'Electronique, Signaux et Images, Université d'Orléans, 45067 Orléans, France
E.mail: weber@lesi.univ-orleans.fr

Herbert Witte, Institute of Medical Statistics, Computer Sciences and Documentation, Jahnstraße 3, 07740 Jena, Germany
E-mail iew@imsid.uni-jena.de

Tianruo Yang, Linköping University, Department of Computer Science, 581 83 Linköping, Sweden
E-mail: tiaya@ida.liu.se

Santiago Zazo-Bello, Universidad Alfonso X El Sabio, Villanueva de la Cañada, 28691 Madrid, Spain
E-mail: szazo@uax.es

Preface

Methods of signal analysis represent a broad research topic with applications in many disciplines, including engineering, technology, biomedicine, seismography, econometrics, and many others based upon the processing of observed variables. Even though these applications are widely different, the mathematical background behind them is similar and includes the use of the discrete Fourier transform and z-transform for signal analysis, and both linear and non-linear methods for signal identification, modelling, prediction, segmentation, and classification. These methods are in many cases closely related to optimization problems, statistical methods, and artificial neural networks.

This book incorporates a collection of research papers based upon selected contributions presented at the First European Conference on Signal Analysis and Prediction (ECSAP-97) in Prague, Czech Republic, held June 24-27, 1997 at the Strahov Monastery. Even though the Conference was intended as a European Conference, at first initiated by the European Association for Signal Processing (EURASIP), it was very gratifying that it also drew significant support from other important scientific societies, including the IEE, Signal Processing Society of IEEE, and the Acoustical Society of America. The organizing committee was pleased that the response from the academic community to participate at this Conference was very large; 128 summaries written by 242 authors from 36 countries were received. In addition, the Conference qualified under the Continuing Professional Development Scheme to provide PD units for participants and contributors.

This edited collection of 35 chapters provides extended versions of many of the invited lectures as well as selected regular contributions. These chapters present fundamental methods, with the chapters based on the invited lectures including significant tutorial material and references for further study. The book is organized in four parts:

PART I. SIGNAL ANALYSIS covers fundamental areas of the mathematical background of the initial stage of signal processing, including methods of time-frequency and time-scale signal analysis, wavelet transforms, methods of higher order statistics, and the use of geometric algebra in signal analysis. These methods are applied for both one-dimensional and multi-dimensional signal processing. A number of real applications are presented.

PART II. SIGNAL IDENTIFICATION AND MODELLING includes both linear and non-linear methods of signal and system identification, their mathematical description and modelling. Selected methods analyse autoregressive models with time-varying parameters, and particular attention is given to the Bayesian approach in signal processing.

PART III. SIGNAL PREDICTION presents selected methods of this extensively studied area of signal processing, including the study of signal predictability, non-linear dynamic modelling, fuzzy systems, the use of artificial neural networks, and adaptive systems. This section also includes contributions on smoothing techniques and chaotic time series.

PART IV. SPEECH AND BIOMEDICAL SIGNAL PROCESSING covers two basic areas. The first studies methods of speech processing and includes modelling, detection, coding, and recognition, especially in noisy environments. The second application is biomedical signal processing and classification.

We believe that the chapters presented here represent a balanced selection of theoretical and applied papers covering fundamental topics from the discipline of signal analysis and prediction.

Acknowledgements

It is a great pleasure for us to express our gratitude to all contributors who submitted their papers for ECSAP-97 and to all 69 members of the International Scientific Committee for their valuable comments and for their careful reviews of the extended summaries. The results of three anonymous reviews for each paper gave the editors a basis for suggesting the enlargement of selected contributions. We would like to express our deep thanks to all the authors for the careful preparation of their chapters and for their helpful cooperation during the complicated process of compiling this book. We are very grateful also to all our colleagues and research students for their help in the final stage of preparation of this book, especially to Magdaléna Kolínová for her careful correction of all the TEX files, to Martina Mudrová for her transformation of graphic formats, to Martin Štorek, Jaroslav Stříbrský, and Roman Laštovka for their help with computer processing of files submitted through the Internet, and to many other individuals who contributed to the scientific level and formal appearance of this book. We would like to express our thanks also to the institutes that enabled us to complete this project, namely to the Engineering Department of the University of Cambridge in the United Kingdom, to the Institute of Chemical Technology in Prague, and to the Faculty of Electrical Engineering of the Czech Technical University in Prague.

Aleš Procházka
Jan Uhlíř
Prague, Czech Republic

Peter Rayner
Nick Kingsbury
Cambridge, United Kingdom

INTERNATIONAL SCIENTIFIC COMMITTEE

Part I

SIGNAL ANALYSIS

1

Time-Frequency and Time-Scale Signal Analysis by Harmonic Wavelets

David E. Newland[1]

ABSTRACT
New details of the theory of harmonic wavelets are described and provide the basis for computational algorithms designed to compute high-definition time-frequency maps. Examples of the computation of phase using the complex harmonic wavelet and methods of signal segmentation based on amplitude and phase are described.

1.1 Introduction

The author of this paper is interested in analysing vibration records. The objective of vibration analysis is usually to learn about the physical process of vibration excitation, or about the response properties of the system that is excited, or about both these features. Because of the well-known dynamic properties of linear systems in the frequency domain, and because most of the vibration world can be modelled by linear equations, frequency decomposition is the standard method of analysis. The FFT algorithm for digital Fourier analysis makes this a simple and elegant method of decomposition. But its application is restricted, in theory, to steady-state, deterministic vibratory conditions or to the response to stationary random excitation when there are long record lengths.

For transient conditions, when frequency content may be changing rapidly with time, the analysis is more complicated. This paper refers first to two well-known methods for the frequency analysis of transient records: the short-time Fourier transform method (STFT) and the Wigner-Ville method. The properties and limitations of these methods for vibration analysis are reviewed briefly to set the scene for wavelet methods and, in particular, for the harmonic wavelet method which has been found to be a successful basis for the time-frequency analysis of transient records.

For this chapter, the term 'wavelet' is used in a general sense. Harmonic wavelets and their variants are not necessarily self-similar at different scales and then they do not form a set of constant Q filters. One of their advantages is that their bandwidths at different scales (frequencies) may be chosen arbitrarily (see Newland [8, 13, 14]). The harmonic wavelet transform has a close relationship to the STFT but there

[1] University of Cambridge, Department of Engineering, Trumpington St., Cambridge, CB2 1PZ, United Kingdom, E-mail: den@eng.cam.ac.uk

are essential differences between the two methods that will be discussed later.

Harmonic wavelets have simple formulations in the frequency domain. However modifications to harmonic wavelet theory to allow wavelets to be weighted by a frequency window function are described below and appear to offer practical advantages. A disadvantage is that the frequency band occupied by an individual wavelet becomes less precise as a result of windowing. Also a wavelet's definition requires more information than a single (centre) frequency. The centre frequency must be accompanied by a bandwidth figure; together these define the wavelet's scale. When bandwidth and centre frequency remain in a fixed proportion, a single scale number completely defines the wavelet. But for practical applications it has been found that wavelets with a proportional bandwidth which is variable may be more effective. This amounts to using a mixed family of non-orthogonal wavelets: a mathematician's nightmare but an engineer's delight!

Signal decomposition into wavelets is carried out by correlation. In the author's case, a vibration signal has to be correlated with a chosen wavelet at a particular time. Perfect reconstruction is possible when a family of orthogonal wavelets is used. A discrete signal with N terms then yields N wavelet coefficients. The computational efficiency of wavelet algorithms for signal decomposition is important but, for vibration analysis, improved time-frequency resolution can be obtained by using non-orthogonal wavelets and computing more than N wavelet coefficients for an N-term signal. Later, a series of practical examples will illustrate this conclusion.

1.2 Short-Time Fourier Transforms

The FFT computes Fourier coefficients for a chosen record length. A (discrete) real signal of N terms generates N complex Fourier coefficients, half of which are complex conjugates of the others. There are $N/2$ cosine components and $N/2$ sine components which together recover the signal. The process of recovery is inefficient because it involves a great deal of cancellation between different harmonics, each of which extends over the full record length at constant amplitude. Each Fourier coefficient is the result of correlating the corresponding harmonic, at constant amplitude and phase, with the whole length of record.

To obtain information on how frequency composition changes with time, a short data window is used so that the full record is partitioned into short, weighted segments each of which is analysed separately. Fourier coefficients are computed for the data in the window (only). The window is then moved to a new position and this process repeated. Adjacent windows may overlap and it is generally necessary to make them overlap to obtain satisfactory definition in the resulting time-frequency map. If the width of the window covers a time span T, the frequency spacing between harmonics is $1/T$ (see, for example, Newland [7]). Using a short data window means that $1/T$ is large and so precise frequency definition is impossible. This fundamental uncertainty is unavoidable. High resolution in time and frequency cannot be obtained simultaneously.

For each short length of data (each window position), the FFT generates harmonics which are equally-spaced in frequency. Their spacing is $1/T$ for all the harmonics, so that each frequency coefficient covers the same frequency band of width $1/T$. This is a feature of the STFT: all frequency coefficients give results averaged over the same (constant) bandwidth $1/T$. To obtain adequate discrimination at low frequencies may require a long window length T. Then time accuracy is lost. Also high frequencies may have closer discrimination than needed and long computation times may be needed which produce mainly surplus frequency data. Although the spacing between spectral lines can be reduced by adding zeros to the recorded signal, this serves to interpolate between the previously computed results but does not improve frequency precision which is limited by the uncertainty principle (see, for example, Newland [13]).

1.3 Wigner-Ville Method

An alternative approach is based on the idea that, in theory, an instantaneous correlation function can be computed. The notion is based on the concept of an ensemble of sample functions which together describe a random process. Suppose that it is possible to calculate an instantaneous correlation function

$$R_{xx}(\tau, t) = E[x(t - \tau/2)x(t + \tau/2)] \tag{1.1}$$

at time t. The operator E denotes the average of a statistical ensemble, which it has to be assumed is available for analysis. The Fourier transform of (1.1) gives

$$S_{xx}(\omega, t) = \frac{1}{2\pi} \int_{-\infty}^{\infty} R_{xx}(\tau, t) \exp(-i\omega\tau)\, d\tau \tag{1.2}$$

where $S_{xx}(\omega, t)$ is a function of time t as well as of the frequency parameter ω. (The position of the $1/2\pi$ is that customary in random vibration analysis; it differs from normal signal processing practice.) This instantaneous spectral density function provides information on how the mean-square of the ensemble is distributed over frequency and time. But in practice it is never possible to compute the ensemble average correlation function $R_{xx}(\tau, t)$. Instead the ensemble averaging operation is omitted and (1.1) substituted into (1.2) without the averaging operation, to give

$$\Phi_{xx}(\omega, t) = \frac{1}{2\pi} \int_{-\infty}^{\infty} x(t - \tau/2)x(t + \tau/2) \exp(-i\omega\tau)\, d\tau \tag{1.3}$$

where $\Phi_{xx}(\omega, t)$ is a new function of t and ω, called the Wigner-Ville distribution of $x(t)$ [16, 17].

A detailed review of the properties of the function $\Phi_{xx}(\omega, t)$ is given in the book by Cohen [1]. Although the integral (1.3) is centered at time t, it is an infinite integral and so covers the character of $x(t)$ far away from the chosen time t. Therefore it does not describe only the local behaviour of $x(t)$. This is the same fundamental problem that limits the STFT. Because of the continuing nature of harmonic waves, it is impossible to have a truly local spectral density. There are two other disadvantages. One is that $\Phi_{xx}(\omega, t)$ is not inherently positive, so its interpretation can be difficult. The second is that $\Phi_{xx}(\omega, t)$ introduces confusing cross terms when signals with multi-frequency components are analysed (see the discussion in Mark [4]). These drawbacks can be overcome, at least in part, but at the expense of greatly increased computation time [13] and it has been claimed that better time-frequency maps can be generated from Cohen-Posch distributions (a modified form of Wigner-Ville distribution) than by the STFT [3]. The final word has not been written on this competition, but the harmonic wavelet transform provides an alternative method with the computational economy of the STFT and map definition properties similar to those achievable with the Cohen-Posch method.

1.4 Harmonic Wavelet Calculations

In essence the wavelet approach is the same as the STFT except that any basis function can be used (only harmonic functions of constant amplitude and phase are used by the STFT). The input signal $f(t)$ is correlated with this basis (or test) function, $w(t)$. Because $w(t)$ is localised and generally has harmonic characteristics, it is called a wavelet. Any waveform may be used for the wavelet, provided that it is localised at a particular time (or position). The wavelet coefficient $a(t)$ is defined by the correlation equation

$$a(t) = \int_{-\infty}^{\infty} f(\tau) w^*(\tau - t)\, d\tau. \tag{1.4}$$

Knowledge of $a(t)$ provides information about the structure of $f(t)$ and its relationship to the shape of the analysing wavelet $w(t)$. Because we are allowing for all the quantities to be complex, $w^*(t)$ the complex conjugate of $w(t)$ appears in (1.4) (see ref. [6]). When $f(\tau)$ correlates with $w^*(\tau - t)$, then $a(t)$ will be large; when they do not correlate, $a(t)$ will be small. More information is obtained if the process is repeated with a different wavelet function, $w_1(t)$.

Any basis functions $w(t)$ can be used. In this paper only one particular class of wavelet will be used: generalised harmonic wavelets [8]. These have important advantages for vibration analysis when detailed time-frequency maps are required and they are easy to use, as will be demonstrated below. One advantage is the simplicity of harmonic wavelets in the frequency domain, and we begin by converting (1.4) from the time domain by taking Fourier transforms of both sides of (1.4) which becomes the well-known frequency equation (see for example ref. [5], ch. 9)

$$A(\omega) = F(\omega) W^*(\omega) \tag{1.5}$$

where the quantities shown with capital letters in (1.5) are the Fourier transforms of the corresponding quantities shown by lower-case symbols in (1.4), so that

$$A(\omega) = \frac{1}{2\pi} \int_{-\infty}^{\infty} a(t) \exp(-i\omega t)\, dt \tag{1.6}$$

$$F(\omega) = \frac{1}{2\pi} \int_{-\infty}^{\infty} f(t) \exp(-i\omega t)\, dt \tag{1.7}$$

and

$$W(\omega) = \int_{-\infty}^{\infty} w(t) \exp(-i\omega t)\, dt. \tag{1.8}$$

Therefore instead of computing the integral in (1.4), the same result may be achieved by multiplying the Fourier transforms in (1.5).

In order to make practical calculations, these Fourier transforms are computed by applying the FFT algorithm. It is a simple operation to compute the FFT of

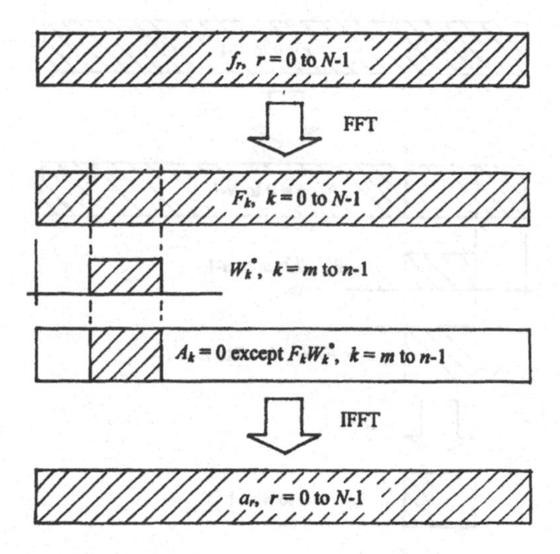

FIGURE 1.1. FFT algorithm to compute harmonic wavelet coefficients for wavelets in the frequency band $m2\pi \leq \omega < n2\pi$ (for a record of unit length)

the input signal and it is then necessary to multiply this, term by term, with the complex conjugate of the FFT of the wavelet to be used. For the complex harmonic wavelet [6, 8], the Fourier transform $W(\omega)$ is especially simple. It is zero everywhere except in a finite band of frequencies defined by

$$m2\pi \leq \omega < n2\pi \tag{1.9}$$

where $m < n$ and need not be integers. Within this band it has the constant (real) value

$$W(\omega) = 1/(n-m)2\pi. \tag{1.10}$$

Because harmonic wavelets are complex, their Fourier transform is one-sided, so that $W(\omega)$ remains zero for all negative frequencies.

A practical calculation is illustrated diagrammatically in Fig. 1.1.

For an input signal $f(t)$ which is sampled N times to give the sequence $f_0, f_1, f_2, \ldots f_{N-1}$, this calculation produces N complex wavelet coefficients. They are the result of making the correlation calculation in (1.4) for the centre of the wavelet $w_{m,n}(t)$ at each of the N integer positions that correspond to the sampled values f_0, f_1, f_2 etc. When there are N sampled values in the domain $0 \leq t < 1$, each wavelet is spaced $1/N$ from the next.

1.5 Coefficients for an Orthogonal Set

In the theory of harmonic wavelets [6, 8], it is shown that a complete set of orthogonal wavelets can be assembled by spacing wavelets in all frequency bands along the real axis $0 \leq \omega < \infty$. Those in the band $m2\pi$ to $n2\pi$ are spaced $1/(n-m)$ apart in the time domain, instead of $1/N$ apart as in Fig. 1.1.

Consider a sequence with $N = 16$ terms and take $m = 3$, $n = 7$ (for a discrete algorithm it is convenient to use integers). If the time axis covers the domain

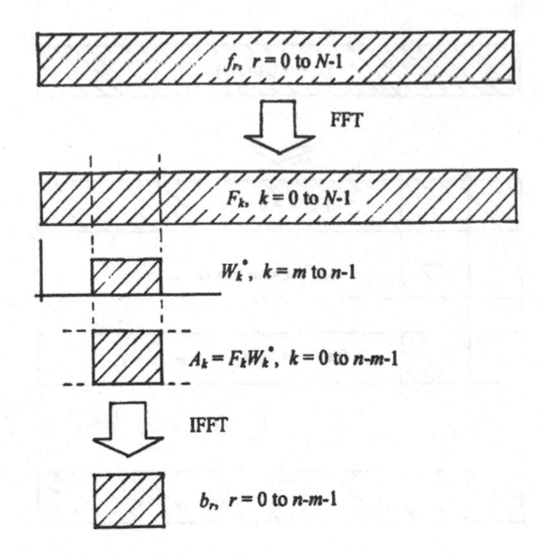

FIGURE 1.2. Modified version of the FFT algorithm in Fig. 1.1 which computes a smaller set of (orthogonal) harmonic wavelet coefficients

$0 \leq t < 1$, the corresponding frequency axis covers the domain $0 \leq \omega < \pi$ (see, for example [7]). The harmonic wavelets defined by $m = 3$, $n = 7$ occupy a frequency band $3\pi/8$ to $7\pi/8$. The orthogonal set of 16 wavelets has $n - m = 4$ wavelets in this band and they are spaced 1/4 apart along the time axis, so that there are 4 wavelets at positions 0, 1/4, 1/2 and 3/4. Hence, instead of computing 16 coefficients for wavelets (in the chosen frequency band) at positions 0, 1/16, 1/8, 3/16 etc., only 4 wavelet coefficients need to be calculated at 0, 1/4, 1/2, 3/4 in order to generate enough information for the full orthogonal set.

This reduced number of coefficients can be computed by the algorithm shown in Fig. 1.2 [6, 8]. To distinguish them from those calculated according to Fig. 1.1, these are identified as $b_0, b_1, b_2, b_3, \ldots, b_{n-m-1}$.

Harmonic wavelet coefficients can be calculated either by the algorithm illustrated in Fig. 1.1, or by the algorithm in Fig. 1.2. Fig. 1.1 calculates N wavelet coefficients for every frequency band chosen, while Fig. 1.2 calculates only enough wavelet coefficients to construct an orthogonal set of wavelets. Fig. 1.2 is an optimum calculation because only enough wavelet coefficients are calculated to permit perfect reconstruction of the input signal from the output (wavelet) coefficients. Fig. 1.1 calculates redundant information in this respect by producing an excess of wavelet coefficients. However the extra coefficients may be extremely useful in obtaining a detailed picture of the frequency characteristics of the input signal.

1.6 Relation between Orthogonal and Non-Orthogonal Coefficients

The coefficients produced by the algorithm in Fig. 1.1 are related to those calculated according to Fig. 1.2. From Fig. 1.1, there are N coefficients a_0 to a_{N-1}. From Fig. 1.2 there are $n - m$ coefficients b_0 to b_{n-m-1}. The a's and b's are related as follows.

According to Fig. 1.1,

$$a_r = \sum_{k=0}^{N-1} A_k \exp(i2\pi kr/N), \quad r = 0, 1, 2, \ldots, N-1 \tag{1.11}$$

which, since $A_k = 0$ for all r except $m \leq r \leq n-1$, reduces to

$$a_r = \sum_{k=m}^{n-1} A_k \exp(i2\pi kr/N), \quad r = 0, 1, 2, \ldots, N-1. \tag{1.12}$$

In comparison, the wavelet coefficients calculated by the algorithm in Fig. 1.2 are

$$b_s = \sum_{j=0}^{n-m-1} A_{j+m} \exp(i2\pi js/(n-m)), \quad s = 0, 1, 2, \ldots, n-m-1. \tag{1.13}$$

Replacing $(j+m)$ by k in (1.13) gives

$$\begin{aligned} b_s &= \sum_{k=m}^{n-1} A_k \exp(i2\pi(k-m)s/(n-m)) \\ &= \exp(-i2\pi ms/(n-m)) \sum_{k=m}^{n-1} A_k \exp(i2\pi ks/(n-m)), \\ s &= 0, 1, 2, \ldots, n-m-1. \end{aligned} \tag{1.14}$$

Now compare the summation in (1.12) with that in (1.14). They each have the same number of terms. Let the index r be related to the index s by

$$r = sN/(n-m) = sc \tag{1.15}$$

where

$$c = N/(n-m) \tag{1.16}$$

for $s = 0, 1, 2, \ldots, n-m-1$. The two summations will be the same if $r = 0, c, 2c, \ldots, N-c$ when the coefficients fall into corresponding pairs a_0, b_0; a_c, b_1; a_{2c}, b_2, etc.

The phase difference between coefficients in the two sets is given by the exponential term $exp(-i2\pi ms/(N/c))$ in (1.14) and will be zero if

$$mc/N = \textit{any integer}, \tag{1.17}$$

including zero. Consider an example. Let $N = 100, m = 3, n = 7$. Then, from (1.16), $c = 25$. The corresponding pairs are a_0, b_0; a_{25}, b_1; a_{50}, b_2, a_{75}, b_3. But mc/N is not an integer, so there are phase differences between corresponding coefficients.

If $N = 128, m = 8, n = 16$, then $c = 16$ and $mc/N = 1$ and, in this example, there is no phase difference between corresponding coefficients a_0, b_0; a_{16}, b_1; a_{32}, b_2; a_{48}, b_3; a_{64}, b_4; $\ldots, a_{112}, b_7$. So (1.14) computes some of the coefficients in the longer series (1.12), although generally with modified phase.

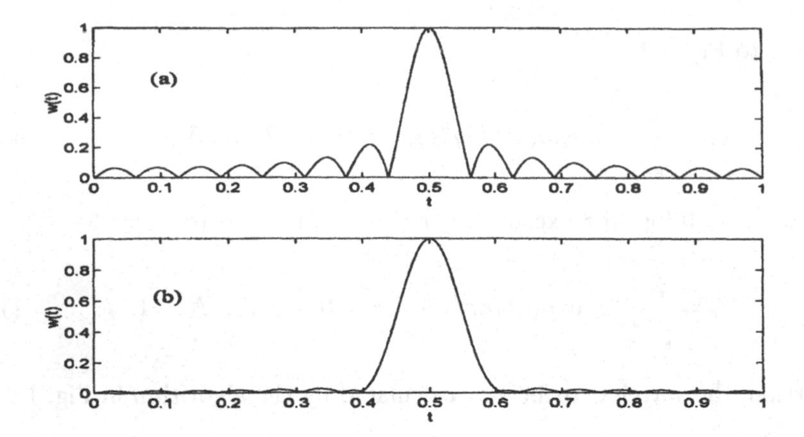

FIGURE 1.3. Comparison between the moduli of (a) a harmonic wavelet (which has a basic 'boxcar' window) and (b) the same wavelet after windowing with a Gaussian weighting function whose 'standard deviation' is 1/4 of the boxcar width

1.7 Computational Algorithm

In order to compute the *magnitudes* of the wavelet coefficients defined by the frequency domain equation (1.5), we can (a) compute the inverse FFT shown in Fig. 1.1, or (b) compute the shortened inverse FFT shown in Fig. 1.2, or (c) compute the inverse FFT of any length series which consists of all zeros except for the sequence taken from Fig. 1.2. If $N = 16$, $n - m = 4$, it can be seen that (a) gives a sequence 16 terms; (b) gives a shortened sequence of 4 terms, which includes every fourth term of (a); while (c) gives a sequence of arbitrary length which either includes the terms derived before or interpolates between them. Of course only the magnitude of corresponding terms in (a), (b) and (c) are the same unless the particular conditions given by (1.17) are satisfied.

The harmonic wavelet algorithm shown in Fig. 1.2 has no redundancy in the sense that only the minimum number of wavelet coefficients are generated that are needed to reconstruct the starting signal from its wavelet decomposition. For vibration analysis this gives a coarse time-frequency map for which interpolation between adjacent ordinates on the map may be helpful. That interpolation is provided by making either calculation (a) or (c) above. Since the length of the input sequence may mean that N is very large, a computational strategy based on (c) is usually preferable. Typically a sequence length of 500 terms may be all that is needed to determine the time axis with sufficient resolution, whereas N may be many times larger. Since there may be at least 500 frequency levels to be explored, very large arrays of wavelet coefficients may have to be stored in a realistic practical analysis. Because the basic harmonic wavelet has the simple form of Fourier transform defined by (1.10), it is not well localised in the time domain. Instead it decays only in proportion to time^{-1}, as shown in Fig. 1.3(a). When a frequency weighting function is introduced, for example Gaussian weighting, the improved localisation shown in Fig. 1.3(b) is achieved. This is important to ensure that clarity is achieved in time-frequency mapping, although redundant non-orthogonal wavelets are now being used to deconstruct the input signal.

1.8 STFT and HWT Comparison

The STFT works by calculating all the Fourier coefficients of a reduced length of data. The complete input signal is broken down into short time segments, the data for each of which is frequency analysed separately. Each segment produces Fourier coefficients for the full frequency range (zero to the Nyquist frequency). There

	Short-time Fourier transform	**Harmonic wavelet transform**
No. of data points	N	N
Sampling interval (seconds)	Δ	Δ
Nyquist frequency (Hz)	$1/2\Delta$	$1/2\Delta$
Step 1	Choose a short time segment of n points, $n < N$, covering a time span $n\Delta$	Compute FFT of the whole signal
Step 2	Compute FFT of this short time segment, if desired after weighting with an appropriate time window function and padding with p zeros	Choose a narrow frequency band, n points, $n < N/2$, covering a frequency span $n/N\Delta$
Step 2A		Compute IFFT of this narrow band, if desired after weighting with an appropriate frequency window function and padding with p zeros
Step 3	Plot the magnitude squared of the Fourier coefficients for the chosen short time segment	Plot the magnitude squared of the wavelet coefficients for the chosen narrow frequency band
Time range (seconds)	Arbitrary, short time segment, n points, $n < N$, covering the time span $n\Delta$	Whole time span, $T = N\Delta$
Frequency range (Hz)	Whole frequency range, 0 to Nyquist, which is 0 to $1/(2\Delta)$	Arbitrary, narrow frequency band covering the frequency span $n/(N\Delta)$
Time interval for T-F map (seconds)	Arbitrary, minimum is the sampling interval Δ	Equally spaced at interval $n\Delta/(n+p)$
Frequency interval for T-F map (Hz)	Equally spaced at interval $Nyquist/((n+p)/2) = 1/((n+p)\Delta)$	Arbitrary, minimum is $1/N\Delta$

TABLE 1.1. Comparison between the STFT and the Harmonic Wavelet method for generating time-frequency maps (from Newland [13])

are half as many complex Fourier coefficients as there are data points in the time segment chosen. Therefore the time span is arbitrary but frequencies are always calculated at equally spaced frequency intervals. Each frequency coefficient has the same (constant) bandwidth. If the time segment has n data points, the bandwidth of the Fourier coefficients is (Nyquist)$\times 2/n$, which is constant. To improve the smoothness over frequency of the time-frequency map, zeros may be added to the data in the time segment chosen before computing its frequency content. Effectively, the short time signal is then correlated with a wider range of harmonics.

In contrast, the harmonic wavelet method computes frequency coefficients for an arbitrarily chosen center frequency and frequency band but for a range of equally-spaced times which cover the whole time span of the complete signal. By adding zeros to the data in the chosen frequency band, smoothness over time may be improved by multiplying by the Fourier transforms of a wider range of test functions, but data for equally-spaced times is always produced. Therefore there is an essential duality between the STFT and harmonic wavelet methods as shown in Tab. 1.1. The STFT produces results for local, short time segments, covering the whole frequency spectrum in constant bandwidth steps. The harmonic wavelet method produces results for local, narrow frequency bandwidths, covering the whole duration of the record in constant time steps.

The fundamental advantage of the harmonic wavelet transform is that it offers a computationally-efficient route for a variable bandwidth frequency transform so that the time-frequency map can have a constant-Q or a variable-Q basis as desired. In contrast, a time-frequency map constructed by the STFT always has a constant bandwidth basis, giving the same frequency resolution at high frequencies as it gives at low frequencies. The STFT may therefore lead to a requirement for (much) more computation than is required by the harmonic wavelet route. The procedure defined in Tab. 1.1, based on the algorithm in Fig. 1.1, is for harmonic wavelets whose Fourier transform has the simple form defined by (1.9) and (1.10). Windowing modifies this rectangular block (1.10) into a smoothed profile, but no phase angles are introduced. However, if other non-harmonic wavelets are to be used, this can be done by replacing $W(\omega)$ in (1.10) by the appropriate complex function. The algorithm remains FFT-based with the essential computational advantage that this brings.

1.9 Choice of Wavelet

A frequent question is: how do I choose the best wavelet? Even for the restricted class of harmonic wavelets with Gaussian windowing described here, the same question has to be answered. It takes this form: what should the bandwidth of the window be and how should this be related to the centre frequency of the wavelet? To the author's knowledge, there is (still) no clear answer to this question. Any signal can be decomposed by many different families of orthogonal wavelets and for signal processing in real time, computational efficiency may be the essential feature. However for the analysis of vibration records, that is usually not important compared with the need to achieve a time-frequency map with high resolution. The effective deconstruction of such a record requires a narrow bandwidth to achieve frequency resolution, and a wide bandwidth to achieve time resolution. This raises the possibility of making more than one map of the same signal, using different bandwidth strategies for each different map.

It is very important to realise that wavelet time-frequency maps look quite different when different wavelets are used. Similarly, a time-frequency map produced by the STFT depends on the segment lengths into which the input signal is divided and the data window used. Using long segments gives good frequency resolution,

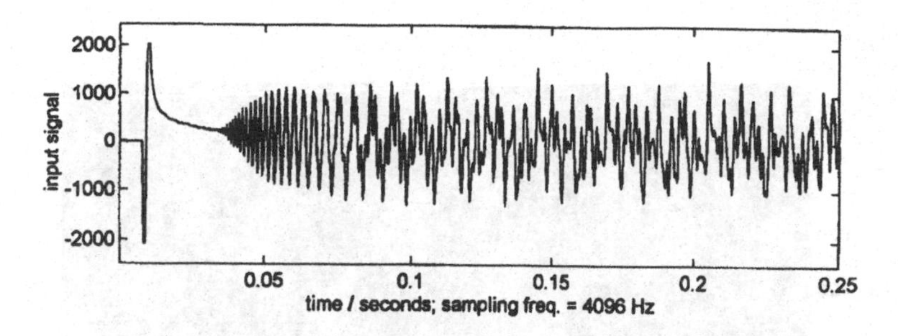

FIGURE 1.4. Sample acceleration time history for the impulse response at one end of an elastic beam

but poor time definition. A short segment length gives poor frequency resolution but now good time definition. As shown by Hodges et al. [2], the appearance of a time-frequency map produced by the STFT is very sensitive to segment length.

1.10 Example of Wavelet Selection

To demonstrate these ideas, an experimentally-obtained vibration signal is analysed. I am grateful to my colleague Dr. Jim Woodhouse for generating this by measuring the impulse response of a beam. A 7.2 m long mild-steel beam of rectangular cross-section 32.1 × 6.3 mm was suspended horizontally on light cords. One end was tapped lightly in a direction perpendicular to the flat side of the beam. The hammer had a soft tip designed so that only low frequency vibrations were generated (up to about 1 kHz). An accelerometer attached to the beam close to the point of impact measured beam response and this data was captured by a data logger. The sampling frequency was 4096 Hz and so the Nyquist frequency was 2048 Hz.

This signal, a short length of which is reproduced in Fig. 1.4, is interesting because it disguises the passage backwards and forwards along the beam of bending waves of different frequencies. When the impulse is applied first, bending waves travel along the beam until they are reflected at the free end. Then they return to the point of impact, before being reflected again at the first end of the beam. Because the group velocity of bending waves depends on frequency (velocity proportional to frequency$^{\frac{1}{2}}$), groups of high frequency waves travel faster than low frequency waves. Therefore a time-frequency map should show more frequent reflections for high frequencies than for low frequencies. This behaviour is, of course, not at all evident from the time-domain response record in Fig. 1.4.

The result of analysing this signal using Gaussian windowed harmonic wavelets is shown in Fig. 1.5. For this map, all wavelets occupy a frequency block width of 256 Hz and the Gaussian window is set to have a width corresponding to a 'standard deviation' of 1/8 of this band. Calculations have been made for wavelets whose centre frequencies advance in increments of 8 Hz. A surface view of the same map is shown in Fig. 1.6.

To illustrate the dependence of these results on the choice of wavelet used, figs. 1.7 and 1.8 are time-frequency maps for the same data but analysed by different wavelets. Fig. 1.7 uses a wavelet with a narrower bandwidth; Fig. 1.8 uses a

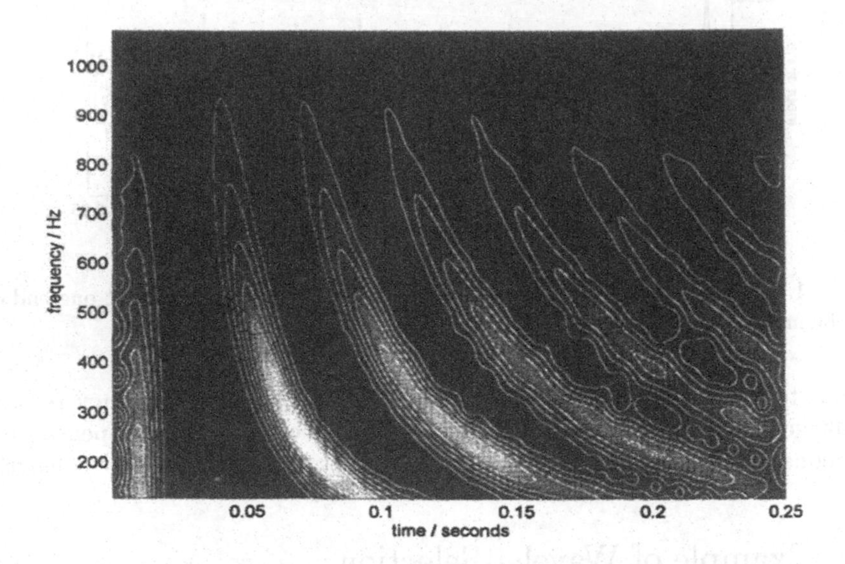

FIGURE 1.5. Time-frequency map for the sample time history in Fig. 1.4, computed using harmonic wavelets with a block width 256 Hz, frequency increment 8 Hz, and Gaussian window (half) width 32 Hz

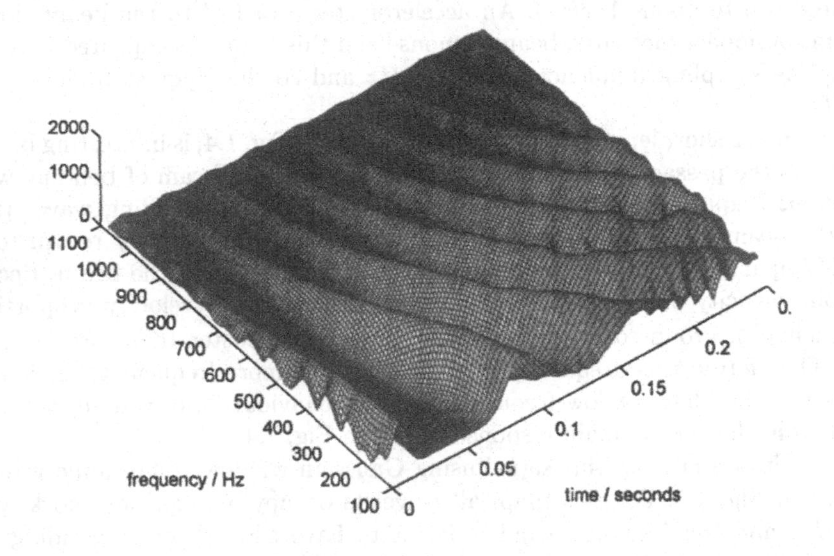

FIGURE 1.6. Surface diagram of the map in Fig. 1.5 (absolute values of the coefficients a_r, from Fig. 1.1, are plotted against the time integer r, scaled over the record length 0.25 seconds, for different centre frequencies)

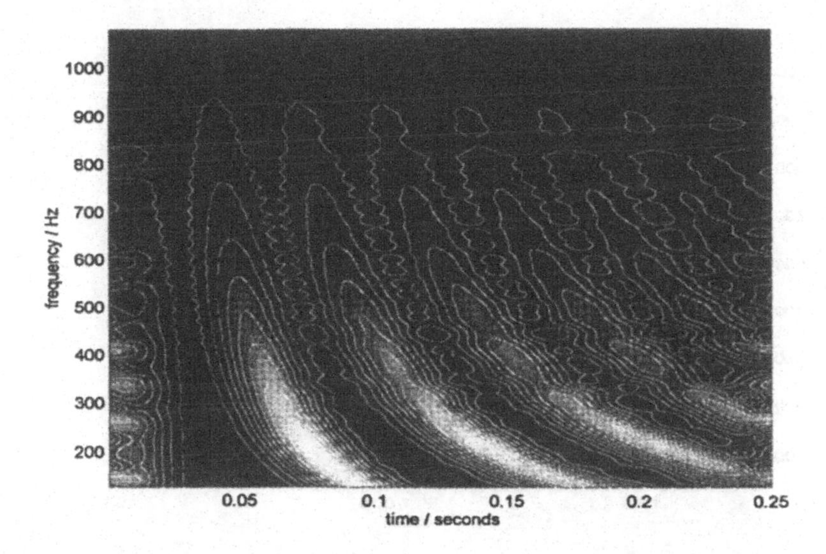

FIGURE 1.7. Time-frequency map for the same data as in Fig. 1.5 except that harmonic wavelets of reduced Gaussian (half) bandwidth 16 Hz have been used. The same frequency increment of 8 Hz and block width 256 Hz were used

wavelet with a wider bandwidth. For the purpose of these illustrations, the wavelets block widths are kept constant (independent of their centre frequency) at 256 Hz and the (half) width of the Gaussian windows set at 16 Hz for Fig. 1.7 and 128 Hz for Fig. 1.8.

To obtain the clarity shown in Figs. 1.5, 1.7 and 1.8, the input signal of 1024 terms is analysed to generate 30, 720 terms whose magnitudes are plotted as the 256 (horizontally) ×120 (vertically) data points in each of the figures. This very large redundancy is necessary in order to achieve the resolution shown in the figures.

1.11 Interpretation of Wavelet Phase

The algorithm in Fig. 1.1 computes N wavelet coefficients for each frequency band to generate the wavelet array A, each of whose rows is a_r, $r = 0$ to $N-1$. Each (complex) wavelet occupies a different position in the row (which is along the time axis), all wavelets being equally spaced. The phase of each of the a_r defines the ratio of the magnitude of the imaginary part (odd wavelet) to that of the real part (even wavelet) at the relevant position.

Suppose that the signal to be analysed is a pure harmonic of constant amplitude and phase. Its Fourier transform $F(\omega)$ may then be represented by the delta function $\delta(\omega - \omega_0)$ if its frequency is ω_0. Let $W(\omega)$ be the shape of the frequency window which has width $2b$. Also use the algorithm in Fig. 1.1 to compute (discrete values of) $a(t)$. However first modify the algorithm so that there are never any leading zeros in the series A_k. This amounts to shifting the frequencies by $\Omega - b$ where Ω is the center frequency of the window. The corresponding (continuous) wavelet

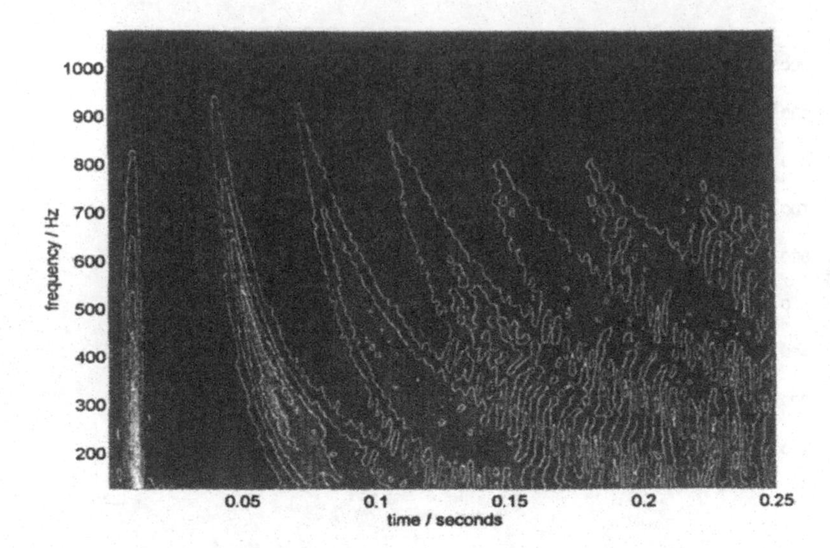

FIGURE 1.8. Time-frequency map for the same data as in Fig. 1.5 except that harmonic wavelets of increased Gaussian (half) bandwidth 128 Hz have been used. The same frequency increment of 8 Hz and block width of 256 Hz were used

coefficient is then $a(t)$ where

$$\begin{aligned} a(t) &= \int_{-\infty}^{\infty} \delta(\omega - b - \omega_0 + \Omega) W(\omega_0 - \Omega) exp(i\omega t)\, d\omega \\ &= W(\omega_0 - \Omega) exp\, i(b + \omega_0 - \Omega)t. \end{aligned} \quad (1.18)$$

From (1.18), the phase angle is seen to be

$$\phi = (b + \omega_0 - \Omega)t \quad (1.19)$$

provided that the window overlaps ω_0 which it does for

$$\omega_0 - b < \Omega < \omega_0 + b. \quad (1.20)$$

Instead of mapping the variation of wavelet phase ϕ with frequency and time, the dependence of ϕ on time t can be eliminated from (1.19) by plotting the gradient of phase along the time axis, which is

$$\frac{\partial \phi}{\partial t} = (b + \omega_0 - \Omega) \quad (1.21)$$

provided that (1.20) is satisfied (otherwise the wavelet coefficient is identically zero and there is no phase). This is achieved by plotting differential phase, computed in the Matlab® language by replacing each column of the array $angle(A(:,:))$ by the difference of adjacent columns to give $B(:,j) = angle(A(:,j)) - angle(A(:,j-1))$.

Fig. 1.9 shows an example. It is a mesh diagram of $B(1:64, 1:64)$ plotted for a constant amplitude harmonic. Time t is plotted along the x-axis, frequency Ω is plotted along the y-axis and differential phase, representing $\partial\phi/\partial t$, along the z-axis. Phase is everywhere zero except in the frequency band defined by (1.20). Within

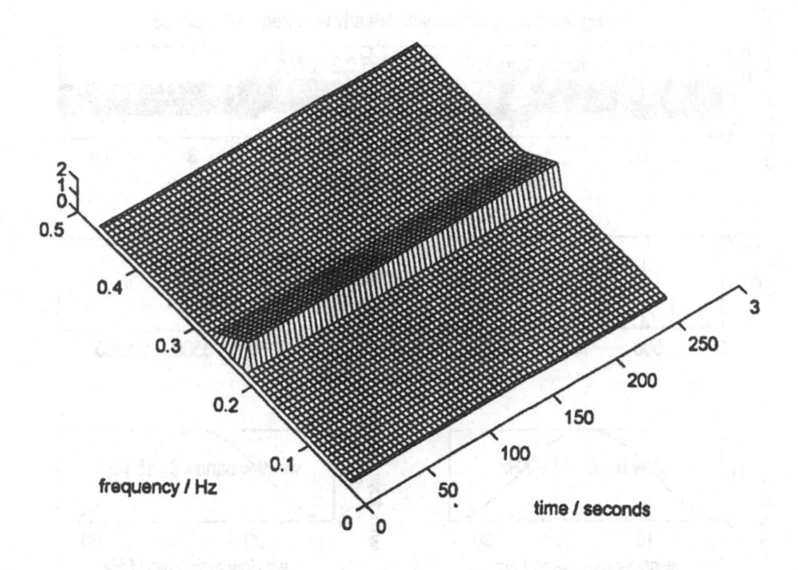

FIGURE 1.9. Mesh diagram of the differential phase $B(1:64,1:64)$ describing a constant amplitude harmonic of fixed frequency

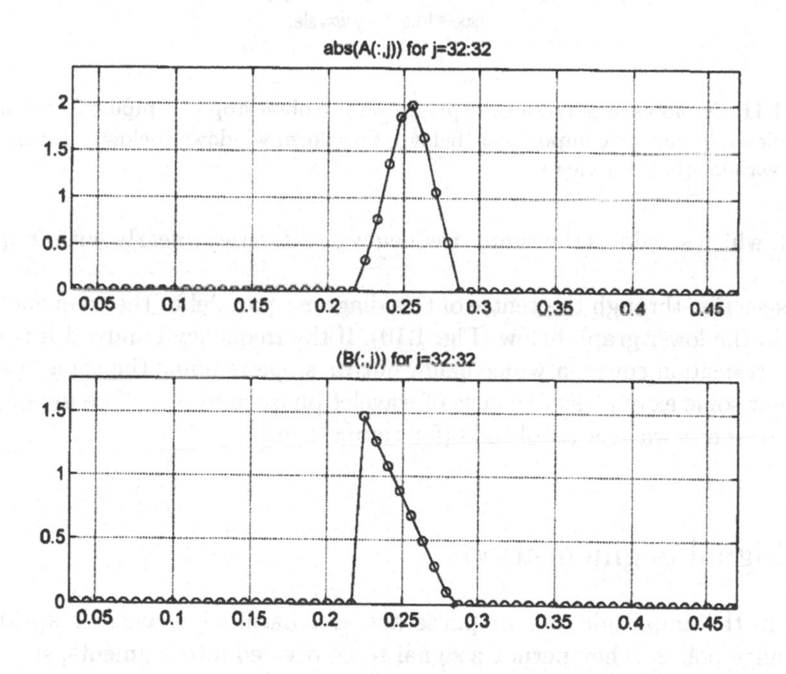

FIGURE 1.10. Cross-sections through the mesh diagram of wavelet amplitude (upper graph) and differential phase (lower graph) for a constant amplitude harmonic of fixed frequency

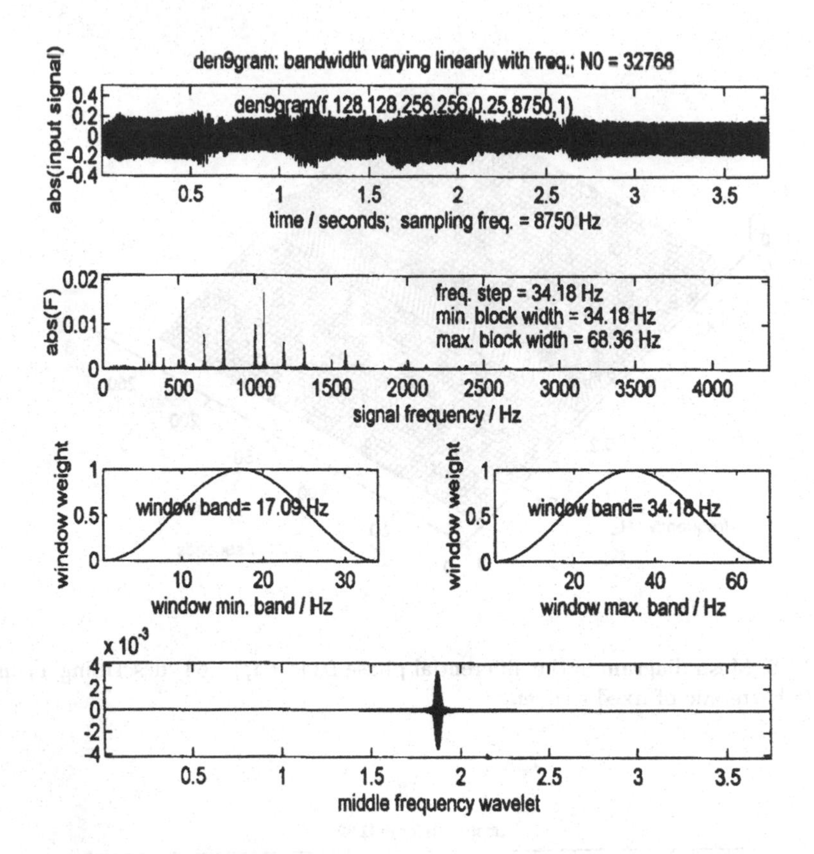

FIGURE 1.11. Six notes of a hymn tune played on the oboe stop of a pipe organ: time history (top view), frequency composition (below), transform windows (below), and specimen analysing wavelet (bottom view)

this band, which straddles the center frequency ω_0, it varies linearly with frequency Ω.

A cross-section through the centre of this diagram, parallel to the frequency axis, is shown in the lower graph below (Fig. 1.10). If the frequency bandwidth is wider, the phase transition covers a wider band, but its slope remains the same. Later in this chapter some examples of the use of wavelet phase are given. The upper graph in Fig. 1.10 shows wavelet amplitude (for comparison).

1.12 Signal Segmentation

Changes in the amplitude and/or phase across a harmonic wavelet map identify signal change points. They permit a signal to be divided into segments, suggesting where one regime may end and another begin. An example is a piece of music (see Tait and Findlay [15]). Fig. 1.11 is the time history of 6 notes of the melody line of the hymn tune Darwall's 148th played by the author on the oboe stop of a pipe organ. Because of the rich tonal content of the oboe stop, each note contains many harmonics so that the wavelet map appears as shown in Fig. 1.13. The white

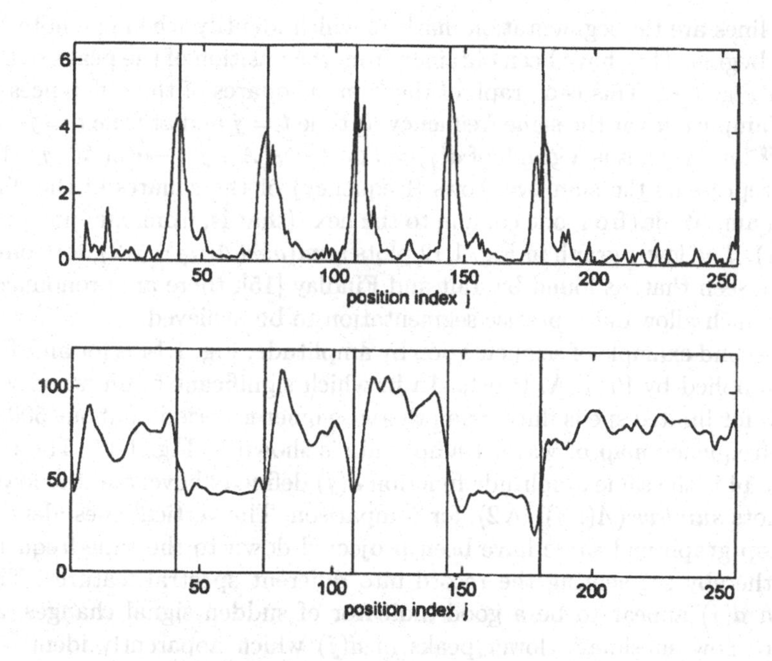

FIGURE 1.12. Wavelet amplitude discriminator $d(j)$ (top view) with segmentation markers plotted; underneath is the sum of squares of wavelet amplitudes for comparison

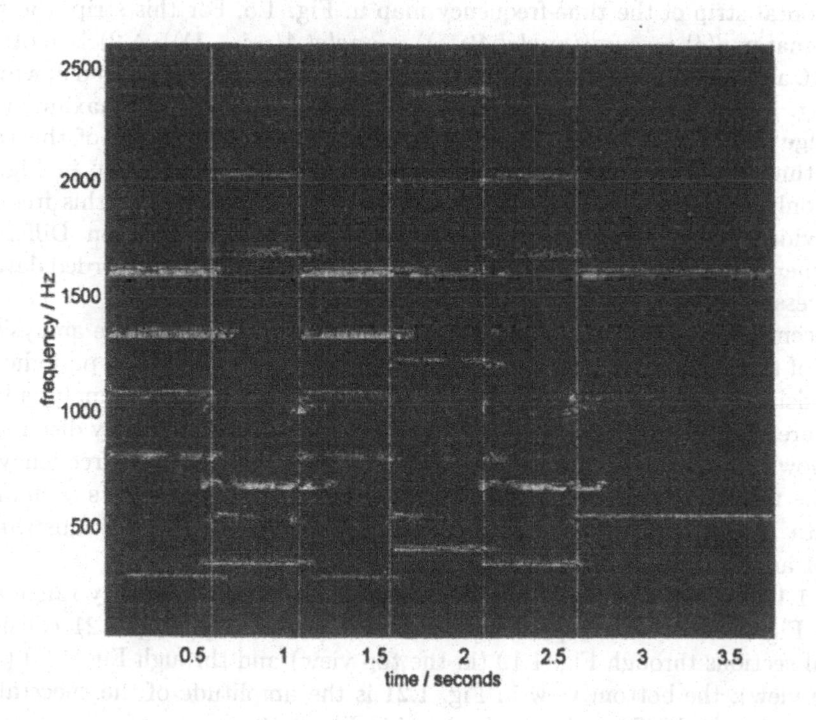

FIGURE 1.13. Segmentation markers from Fig. 1.12 overlayed on the harmonic wavelet map identify a beginning and end for each of the 6 separate notes

vertical lines are the segmentation markers which identify when one note ends and another begins. They have been obtained from the position of the peaks of the upper curve in Fig. 1.12. This is a graph of the sum of squares of the differences between wavelet amplitudes at the same frequency at time $t = j$ and at time $t = j-1$. In the Matlab® language, it is a graph of $d(j) = sum((abs(A(:,j)) - abs(A(:,j-1))).\wedge 2)$.

This represents the sum over rows (frequency) of the squares of the differences between amplitudes from one column to the next (that is, from one time position to the next). The lower graph in Fig. 1.12 plots $sum(abs(A(:,j)).\wedge 2)$, for comparison. It can be seen that, as found by Tait and Findlay [15], there are pronounced peaks in $d(j)$ which allow quite precise segmentation to be achieved.

As a second example of segmentation by amplitude, Fig. 1.14 is for an EEG signal kindly supplied by Prof. A. Procházka in which significant brain activity is being recorded for limited spells during an otherwise quieter period of about 500 seconds. A time-frequency map of wavelet amplitude is shown in Fig. 1.15. The top graph in Fig. 1.14 is the same amplitude function $d(j)$ defined above, and the lower graph again plots $sum(abs(A(:,j)).\wedge 2)$, for comparison. The vertical lines identify peaks in the top graph and these have been projected down to the time-frequency map below, thereby segmenting the record into different spectral features. The sharp peaks in $d(j)$ appear to be a good indicator of sudden signal changes, although there are now subsidiary, lower peaks in $d(j)$ which apparently identify smaller changes in brain activity.

The same method may be applied to the change of phase as position is varied at constant frequency. An interesting example is provided by the experimental beam impulse data for which results were shown in Figs. 1.5 and 1.6. Fig. 1.17 shows a horizontal strip of the time-frequency map in Fig. 1.5. For this strip, the phase discriminator $d(j) = sum((angle(A(:,j)) - angle(A(:,j-1))).\wedge 2)$ is plotted in Fig. 1.16 and vertical lines drawn through the peaks of $d(j)$ are projected downwards onto Fig. 1.17. The peaks (in the upper view) in Fig. 1.16 identify maximum rates of change of phase with time. These are seen to fall at the centre of the valleys of the time-frequency map of wavelet amplitude. Although the map in Fig. 1.17 covers only a narrow frequency band, reflected bursts of energy in this frequency band evidently have significant rates of change of phase with position. Differential phase therefore provides a good discriminator for segmenting the recorded data into its successive reflections of bending energy.

A recent paper by Loughlin and Bernard [3] gives results from the analysis of a record of the response of a second-order lightly-damped system to a periodic train of impulsive forces, Fig. 1.18. The results of wide and narrow band analysis by the STFT are shown and compared with the Cohen-Posch time-frequency distribution. It is shown that the Cohen-Posch TFD gives a more detailed time-frequency map than the results derived by the authors using the STFT. The results of analysing the data in Fig. 1.18 by the complex harmonic wavelet algorithm illustrated in Fig. 1.1 are shown in Figs. 1.19, 1.20, and 1.21 below.

Fig. 1.19 is a wavelet amplitude map covering the whole frequency range of the signal; Fig. 1.20 is a corresponding differential phase map; Fig. 1.21 consists of vertical sections through Fig. 1.19 (in the top view) and through Fig. 1.20 (in the middle view); the bottom view in Fig. 1.21 is the amplitude of the spectral lines computed by the FFT for the whole signal in Fig. 1.18.

Looking first at Fig. 1.21, the main peak of amplitude is at the frequency of the principal spectral line and the subsidiary peaks (to either side of the main

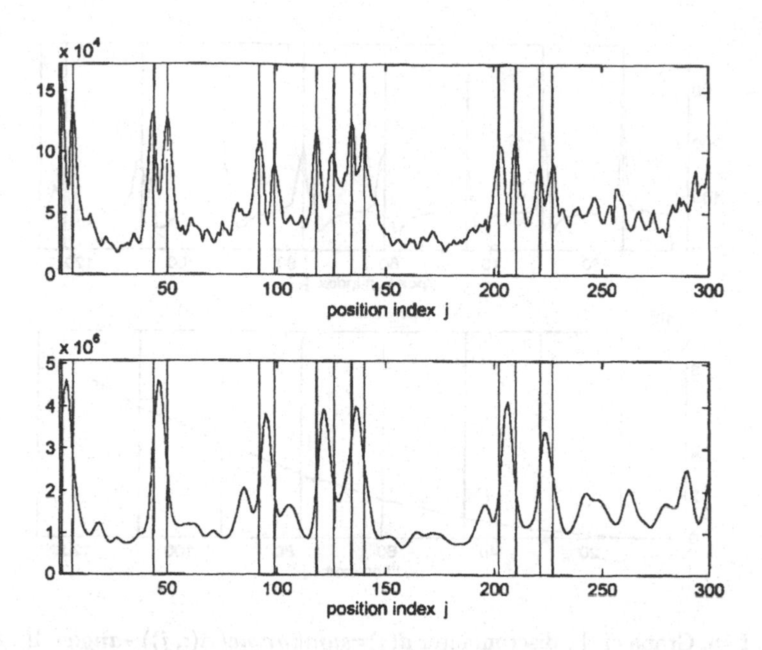

FIGURE 1.14. Amplitude discriminator $d(j)$ (upper view) for segmentation of the EEG signal whose harmonic wavelet map is shown in Fig. 1.15; the sum of squares of wavelet amplitudes is plotted below (from Newland [14])

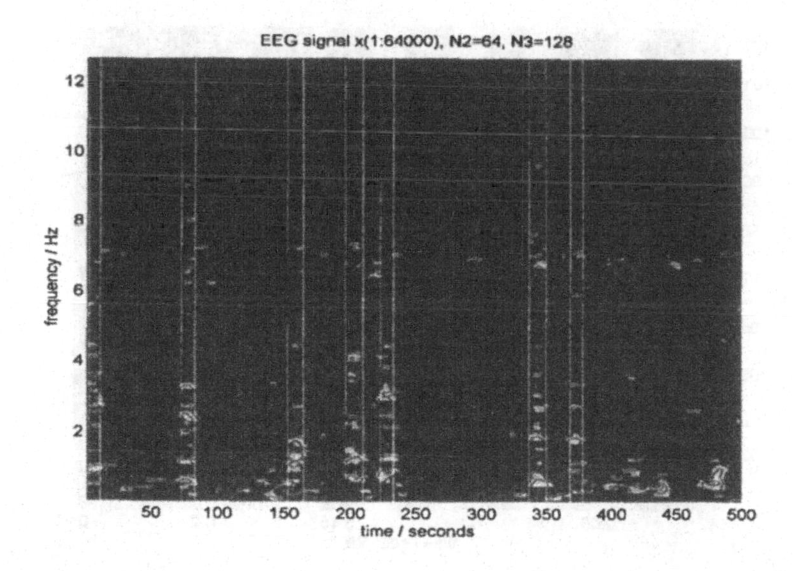

FIGURE 1.15. Segmentation by harmonic wavelet amplitude of an EEG signal based on peaks in the discriminator $sum((abs(A(:,j)) - abs(A(:,j-1))). \wedge 2)$

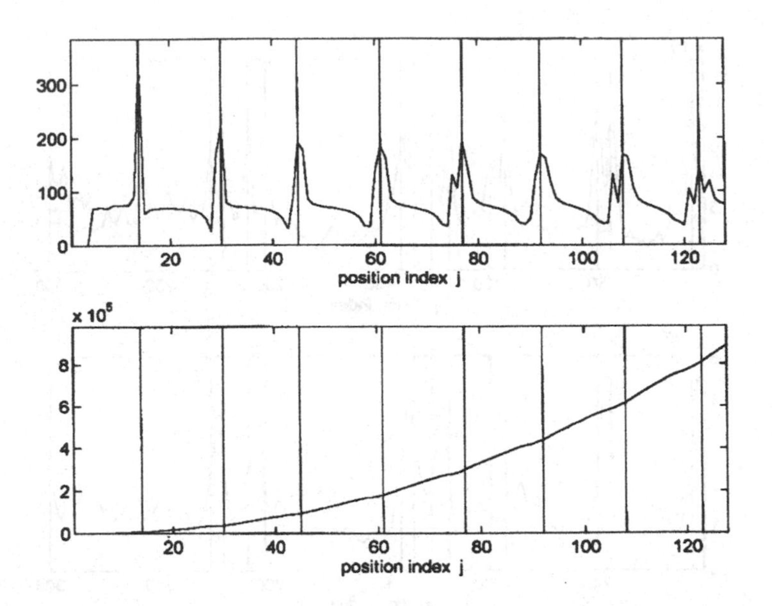

FIGURE 1.16. Graph of the discriminator $d(j)$=*sum*((*angle*(A(:, j))−*angle*(A(:, j−1))).∧2) for a portion of the time-frequency map in Fig.1.5 (upper view) and *sum*((*angle*(A(:, j))).∧2) for the same data (lower view) (from Newland [14])

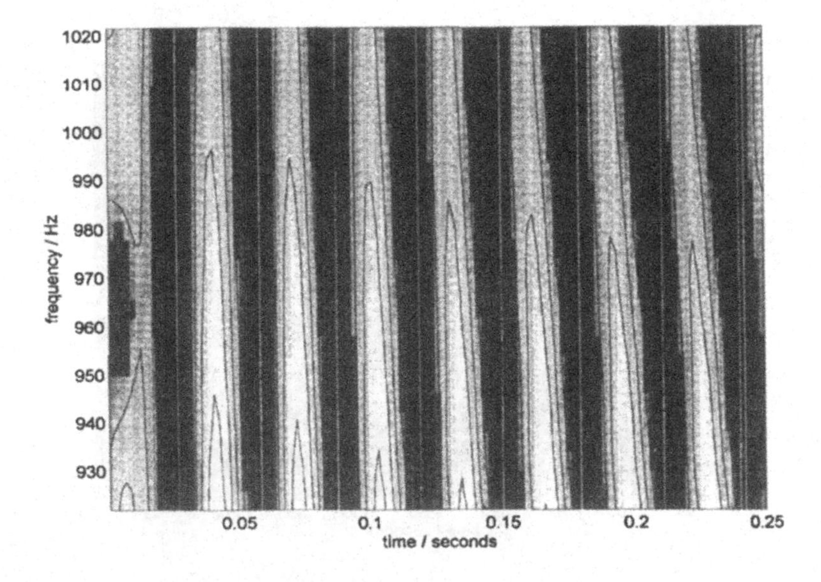

FIGURE 1.17. Segmentation of part of the recorded beam impulse data in Fig. 1.5 based on peaks in $d(j)$ in Fig. 1.16 (from Newland [14])

peak) are at the frequencies of the subsidiary spectral peaks. These show clearly in the contour map in Fig. 1.19. Changes of phase occur with the same spacing along the frequency axis (which is the x-axis in Fig. 1.21) as the spacing between spectral peaks; each spectral peak lies approximately midway between the peaks in the middle graph in Fig. 1.21. This is evident from Fig. 1.20 in which there are the same number of local peaks as there are spectral lines in Fig. 1.21. By comparing the peaks in the middle graph in Fig. 1.21 with the shape of the graph in Fig. 1.10, the saw-shaped changes in phase can be seen to correspond to the presence of the separate harmonics of the signal's Fourier spectrum. This appearance depends on the algorithm used to compute the wavelet coefficients that are plotted. Fig. 1.1 is used, but after first modifying it to remove leading zeros from the sequence A_k before computing its IFFT in the last step of the algorithm. As already described this step leads to equation (1.21), with the result shown in Fig. 1.10.

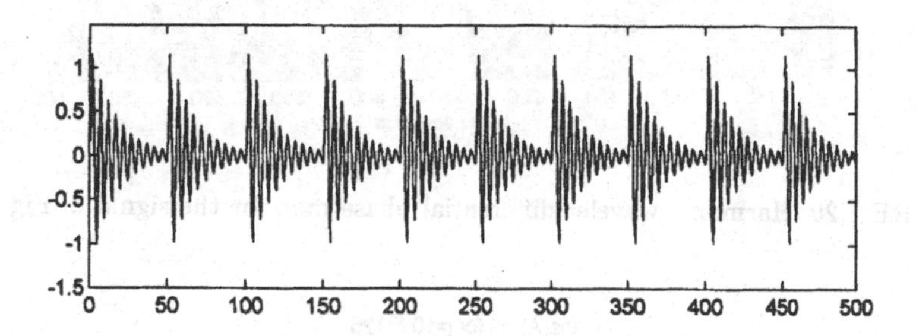

FIGURE 1.18. Response of a lightly-damped oscillator when subjected to periodic impulsive inputs

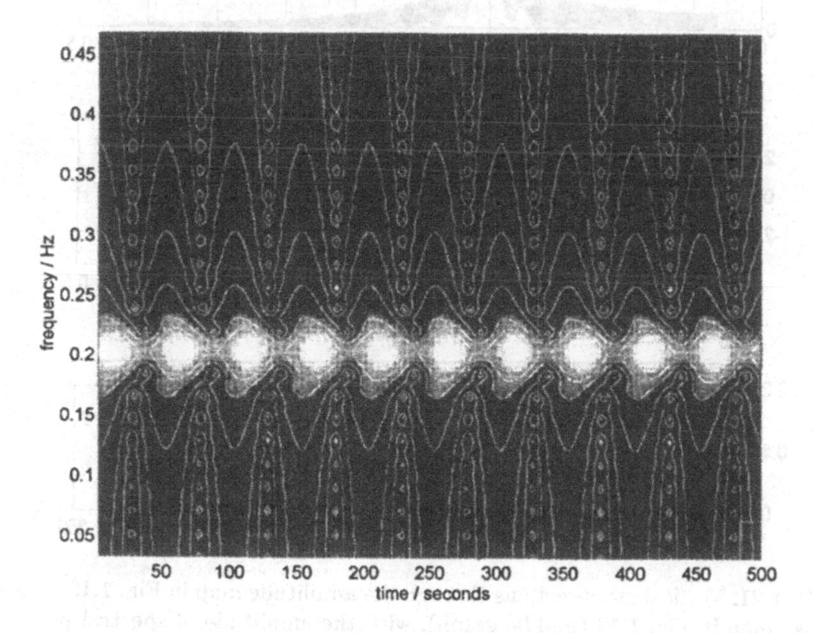

FIGURE 1.19. Harmonic wavelet amplitude map for the signal in Fig. 1.18

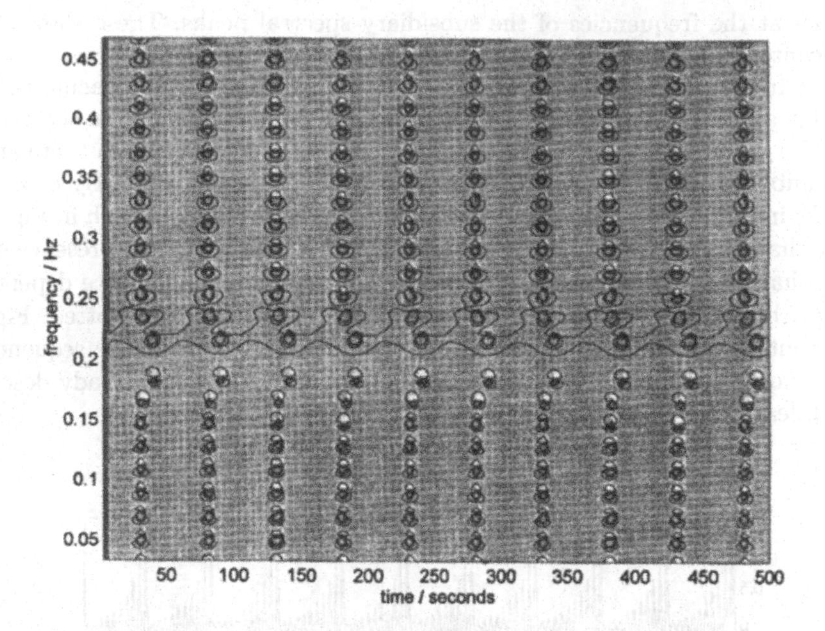

FIGURE 1.20. Harmonic wavelet differential phase map for the signal in Fig. 1.18

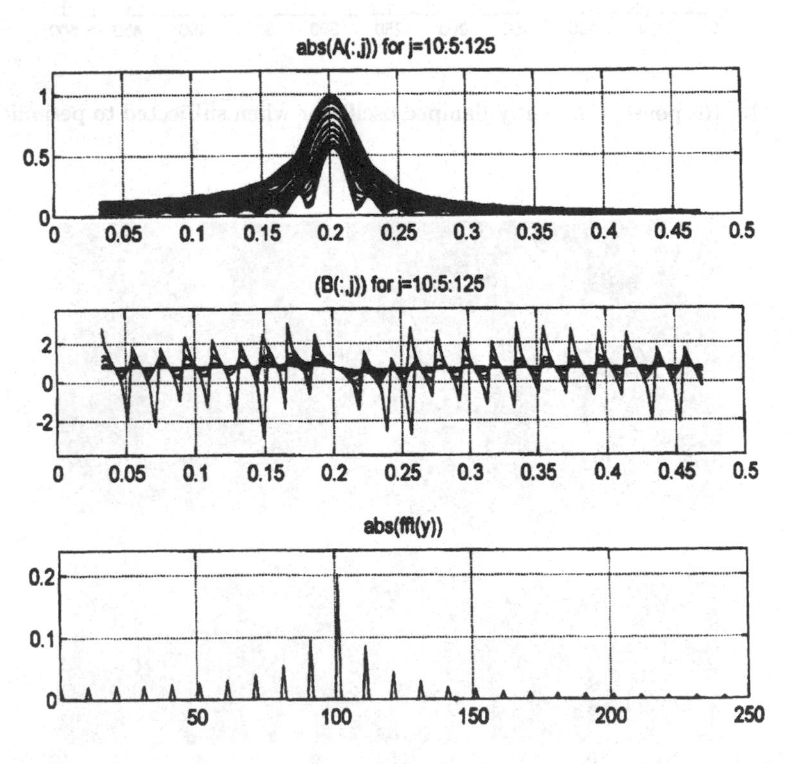

FIGURE 1.21. Vertical cross-sections through the amplitude map in Fig. 1.19 (top graph), the phase map in Fig. 1.20 (middle graph), with the amplitude of spectral peaks for the complete signal in Fig. 1.18 plotted below (bottom graph)

1.13 References

[1] L. Cohen. *Time-Frequency Analysis.* Prentice Hall, New Jersey, 1995.

[2] C.H. Hodges, J. Power, and J. Woodhouse. The Use of the Sonogram in Structural Acoustics and an Application to the Vibrations of Cylindrical Shells. *J.Sound Vib.*, 101:203–218, 1985.

[3] P.J. Loughlin and G.D. Bernard. Cohen-Posch (positive) time-frequency distributions and their application to machine vibration analysis. *Mech. Systems & Signal Processing*, 11:561–576, 1997.

[4] W.D. Mark. Spectral analysis of the convolution and filtering of non-stationary stochastic processes. *J. Sound Vib.*, 11:19–63, 1970.

[5] D.E. Newland. *Mechanical Vibration Analysis and Computation.* Addison Wesley Longman, 1989.

[6] D.E. Newland. Harmonic Wavelet Analysis. *Proc. R. Soc. Lond. A*, 443:203–225, 1993.

[7] D.E. Newland. *Random Vibrations, Spectral and Wavelet Analysis*, 3rd edition. Addison Wesley Longman, 1993.

[8] D.E. Newland. Harmonic and Musical Wavelets. *Proc. R. Soc. Lond. A*, 444:605–620, 1994.

[9] D.E. Newland. Wavelet Analysis of Vibration, Part 1: Theory. *J. Vibration & Acoustics, Trans. ASME*, 116:409–416, 1994.

[10] D.E. Newland. Wavelet Analysis of Vibration, Part 2: Wavelet Maps. *J. Vibration & Acoustics, Trans. ASME*, 116:417–425, 1994.

[11] D.E. Newland. Wavelet Theory and Applications. In *Proc. 3rd Int. Congress on Sound and Vibration*, Montreal, Canada (published by Int. Science Publications, AL, USA), pp. 695–713, 1994. Reprinted as Wavelet Analysis of Vibration Signals, Part 1. *Int. J. Acoustics and Vib.*, 1(1):11–16, 1997.

[12] D.E. Newland. Time-Frequency Analysis by Harmonic Wavelets and by the Short-Time Fourier Transform. In *Proc. 4th Int. Congress on Sound and Vibration*, St. Petersburg, Russia (published by Int. Science Publications, AL, USA), 3, pp. 1975–1982, 1996. Reprinted as Wavelet Analysis of Vibration Signals, Part 2. *Int. J. Acoustics and Vib.*, 2(1):21–27, 1997.

[13] D.E. Newland. Practical signal analysis: Do wavelets make any difference? In *Proc. 1997 ASME Design Engineering Technical Conferences, 16th Biennial Conf. on Vibration and Noise*, Paper DETC97/VIB–4135 (CD ROM ISBN 0 7918 1243 X), Sacramento, California, 1997.

[14] D.E. Newland. Application of Harmonic Wavelets to Time-Frequency Mapping. In *Proc. 5th Int. Congress on Vibration and Acoustics*, 4, pp. 2043–2054 (Paper No. 260, CD ROM ISBN 1 876346 06 X), University of Adelaide, Australia, 1997.

[15] C. Tait and W. Findlay. Wavelet Analysis for Onset Detection. In *Proc. Int. Computer Music Conf*, ICMA, Hong Kong, pp. 500–503, 1996.

[16] J. Ville. Théorie et Applications de la Notion de Signal Analytique. *Câbles et Transmission*, 2:61–74, 1948.

[17] E.P. Wigner. On the Quantum Correction for Thermodynamic Equilibrium. *Physical Review Letters*, 40:749–759, 1932.

Matlab® is a registered trademark of The MathWorks Inc.

2

Wavelet Transforms in Image Processing

Nick Kingsbury[1]
Julian Magarey[2]

ABSTRACT
This chapter is designed to be partly tutorial in nature and partly a summary of recent work by the authors in applying wavelets to various image processing problems. The tutorial part describes the filter-bank implementation of the discrete wavelet transform (DWT) and shows that most wavelets which permit perfect reconstruction are similar in shape and scale. We then discuss an important drawback of these wavelet transforms, which is that the distribution of energy between coefficients at different scales is very sensitive to shifts in the input data. We propose the Complex Wavelet Transform (CWT) as a solution to this problem and show how it may be applied in two dimensions. Finally we give brief details of applications of the CWT to motion estimation and image de-noising.
Keywords: Wavelets, complex wavelets, gabor filters, shift-invariance, motion estimation, image de-noising.

2.1 Introduction

Wavelets have become a popular tool for image compression research, although they have yet to make a big impact on image compression standards, most of which still use the discrete cosine transform (DCT) as their basic energy compaction (or decorrelation) process. A good review of wavelets and their application to compression may be found in Rioul and Vetterli [13] and in-depth coverage is given in the book by Vetterli and Kovacevic [16]. A recent issue of the Proceedings of the IEEE [6] has also been devoted to wavelets and includes many very readable articles by leading experts.

The conventional discrete wavelet transform (DWT) may be regarded as equivalent to filtering the input signal with a bank of bandpass filters, whose impulse responses are all approximately given by scaled versions of a *mother wavelet*. The scaling factor between adjacent filters is usually 2:1, leading to octave bandwidths and centre frequencies that are one octave apart. At the coarsest scale, a lowpass filter is also required to represent fully the lowest frequencies of the signal.

[1]University of Cambridge, Department of Engineering, Trumpington St., Cambridge, CB2 1PZ, United Kingdom, E-mail: ngk@eng.cam.ac.uk
[2]Centre for Sensor Signal and Information Processing, Technology Park Adelaide, The Levels, SA5095, Australia, E-mail: jfam@cssip.edu.au

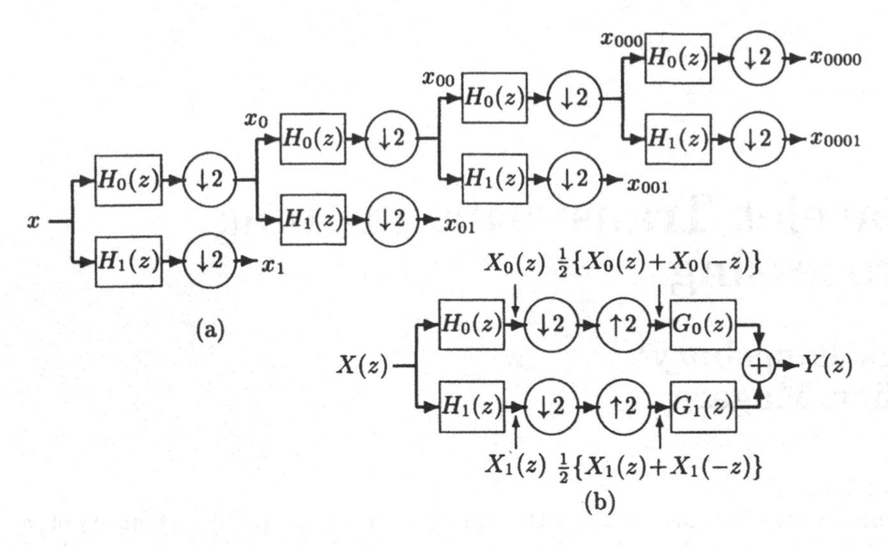

FIGURE 2.1. (a) Binary wavelet tree of lowpass (H_0) and highpass (H_1) filters; and (b) the 2-band reconstruction block

The outputs of the filters are usually maximally decimated so that the number of DWT output samples equals the number of input samples and the transform is invertable. The octave-band DWT is most efficiently implemented by Mallat's *dyadic wavelet decomposition tree* [12], a cascade of 2-band perfect-reconstruction filter banks, shown in Fig. 2.1a.

The main advantages of wavelets over the DCT are the absence of blocking artefacts and the multiscale nature of the DWT which allows near-optimal compression of features with a variety of different scales or sizes. Although wavelets are now a very popular alternative to the DCT for image compression, they have not yet found significant acceptance for image analysis tasks, in which information needs to be extracted from the image at a higher level than a simple frequency decomposition. We believe that a major reason for this is the sensitivity of the DWT to the precise position of image features with respect to the sampling grids used by the decimators at the filter outputs. For example, if an input step-function is slowly shifted past the wavelet filters, the energy distribution between different frequency bands fluctuates considerably. This is caused by aliasing in the decimators, which is only cancelled out when the correct reconstruction filter bank is used to invert the transform.

For sensible image analysis, we feel it is highly desirable that a transform should be *shift invariant*: i.e. the energy distribution between different bands should be independent of any shifts in the input signal (or at least approximately so). We have found that this may be achieved by generalising the usual wavelet filter bank so that it uses *complex-valued coefficients* rather than purely real ones, as is conventional. This leads to the Complex Wavelet Transform (CWT). Results are presented which show how the CWT may be applied in two areas of image processing: motion or displacement estimation, and the removal of unwanted noise (de-noising). It is likely that the CWT also has benefits in the area of texture segmentation, due to its strongly directional filters, but we have not yet investigated this.

This paper is arranged as follows. Section 2.2 discusses the conventional DWT as a filter-bank tree, and derives conditions for the filters to achieve perfect reconstruction. It then shows how this leads to a number of commonly used wavelets, and discusses their lack of shift invariance. Section 2.3 introduces our complex wavelet transform (CWT), designed to overcome this problem, and describes some important features when the CWT is extended to two dimensions, in particular the directional selectivity of the 2-D filters. Section 2.4 briefly describes how we have applied the CWT to motion estimation and de-noising and gives some results from this work.

2.2 The Discrete Wavelet Transform (DWT)

The discrete wavelet transform (DWT) is normally implemented by a binary tree of filters as shown for the one-dimensional (1-D) case in Fig. 2.1(a). The input signal x is split by filters H_0 and H_1 into a lowpass component x_0 and a highpass component x_1, both of which are decimated (down-sampled) by 2:1. The lowpass component is then split further into x_{00} and x_{01}, which are again decimated 2:1; and this process continues as far as required. The outputs of the DWT are the bandpass coefficients x_1, x_{01}, x_{001}, x_{0001}, ... , and the final lowpass coefficients $x_{00...00}$. Because of the decimation, the total output sample rate equals the input sample rate and there is thus no redundancy in the transform.

In order to reconstruct the signal, a pair of reconstruction filters G_0 and G_1 are used in the arrangement of Fig. 2.1(b), and usually the filters are designed such that the output signal $Y(z)$ is identical to the input signal $X(z)$. This is known as the condition for *perfect reconstruction*. Hence in Fig. 2.1(a), x_{000} may be reconstructed from x_{0000} and x_{0001}; and then x_{00} from x_{000} and x_{001}; and so on back to x, using an inverse tree of G filters.

The art of finding good wavelets lies in the design of the set of filters, $\{H_0\ H_1\ G_0\ G_1\}$, to achieve various tradeoffs between spatial and frequency domain characteristics while satisfying the perfect reconstruction (PR) condition.

2.2.1 Conditions for Perfect Reconstruction

In Fig. 2.1(b), the process of decimation and interpolation by 2:1 at the outputs of H_0 and H_1 effectively sets all odd samples of these signals to zero. For the lowpass branch, this is equivalent to multiplying $x_0(n)$ by $\frac{1}{2}(1 + (-1)^n)$. Hence $X_0(z)$ is converted to $\frac{1}{2}\{X_0(z)+X_0(-z)\}$. Similarly $X_1(z)$ is converted to $\frac{1}{2}\{X_1(z)+X_1(-z)\}$.

$$\begin{aligned}\therefore\ Y(z) &= \tfrac{1}{2}\{X_0(z) + X_0(-z)\}G_0(z) + \tfrac{1}{2}\{X_1(z) + X_1(-z)\}G_1(z) \\ &= \tfrac{1}{2}X(z)\{H_0(z)G_0(z) + H_1(z)G_1(z)\} + \\ &\quad \tfrac{1}{2}X(-z)\{H_0(-z)G_0(z) + H_1(-z)G_1(z)\} \qquad (2.1)\end{aligned}$$

The first PR condition requires aliasing cancellation and forces the above term in $X(-z)$ to be zero. Hence $H_0(-z)G_0(z) + H_1(-z)G_1(z) = 0$, which can be achieved if:

$$H_1(z) = z^{-k}G_0(-z) \text{ and } G_1(z) = z^k H_0(-z) \qquad (2.2)$$

where k must be odd (usually $k = \pm1$).

The second PR condition is that the transfer function from $X(z)$ to $Y(z)$ should be unity; i.e. $H_0(z)G_0(z) + H_1(z)G_1(z) = 2$. If we define a product filter $P(z) = H_0(z)G_0(z)$ and substitute the results from (2.2), then this condition becomes:

$$H_0(z)G_0(z) + H_1(z)G_1(z) = P(z) + P(-z) = 2 \tag{2.3}$$

This needs to be true for all z and, since the odd powers of z in $P(z)$ cancel with those in $P(-z)$, it requires that $p_0 = 1$ and that $p_n = 0$ for all n even and nonzero.

$P(z)$ is the transfer function of the lowpass branch in Fig. 2.1(b) (excluding the effects of the decimator and interpolator) and $P(-z)$ is that of the highpass branch. For compression applications $P(z)$ should be zero-phase to minimise distortions when the highpass branch is quantised to zero; so to obtain PR it must be of the form:

$$P(z) = \ldots + p_5 z^5 + p_3 z^3 + p_1 z + 1 + p_1 z^{-1} + p_3 z^{-3} + p_5 z^{-5} + \ldots \tag{2.4}$$

We may now define a design method for the PR filters to be as follows:

- Choose $p_1, p_3, p_5 \ldots$ in (2.4) to give a zero-phase lowpass product filter $P(z)$ with *good* characteristics (not trivial – see below).
- Factorize $P(z)$ into $H_0(z)$ and $G_0(z)$, preferably so that the two filters have similar lowpass frequency responses.
- Calculate $H_1(z)$ and $G_1(z)$ from Equations (2.2).

To simplify the tasks of choosing $P(z)$ and factorising it, based on the zero-phase symmetry we transform $P(z)$ into $P_t(Z)$ such that:

$$\begin{aligned} P(z) &= P_t(Z) = 1 + p_{t,1} Z + p_{t,3} Z^3 + p_{t,5} Z^5 + \ldots \\ \text{where } Z &= \tfrac{1}{2}(z + z^{-1}) \end{aligned} \tag{2.5}$$

To obtain the frequency response, let $z = e^{j\omega T_s}$:

$$\therefore\ Z = \tfrac{1}{2}(e^{j\omega T_s} + e^{-j\omega T_s}) = \cos(\omega T_s) \tag{2.6}$$

Hence Z is purely real, varying from 1 at $\omega T_s = 0$ to -1 at $\omega T_s = \pi$. Substituting this into $P_t(Z)$ gives the frequency response of P.

For smooth wavelets (after many levels of decomposition) Daubechies [1] has shown that $H_0(z)$ and $G_0(z)$ should have a number of zeros at $z = -1$ ($\omega T_s = \pi$). These will be zeros of $P(z)$ too, so $P_t(Z)$ needs zeros at $Z = -1$. In general more zeros at $z = -1$ produce smoother wavelets.

2.2.2 Some Wavelet Examples

We now illustrate this design method with examples of how it leads to some well-known wavelets as well as to some less common ones. In each case the analysis and reconstruction wavelets for 4 levels of decomposition are shown in Fig. 2.2. These two wavelets correspond to the impulse responses from x in Fig. 2.1(a) to x_{0001} via the H filters, and from x_{0001} back to x via the G filters, respectively.

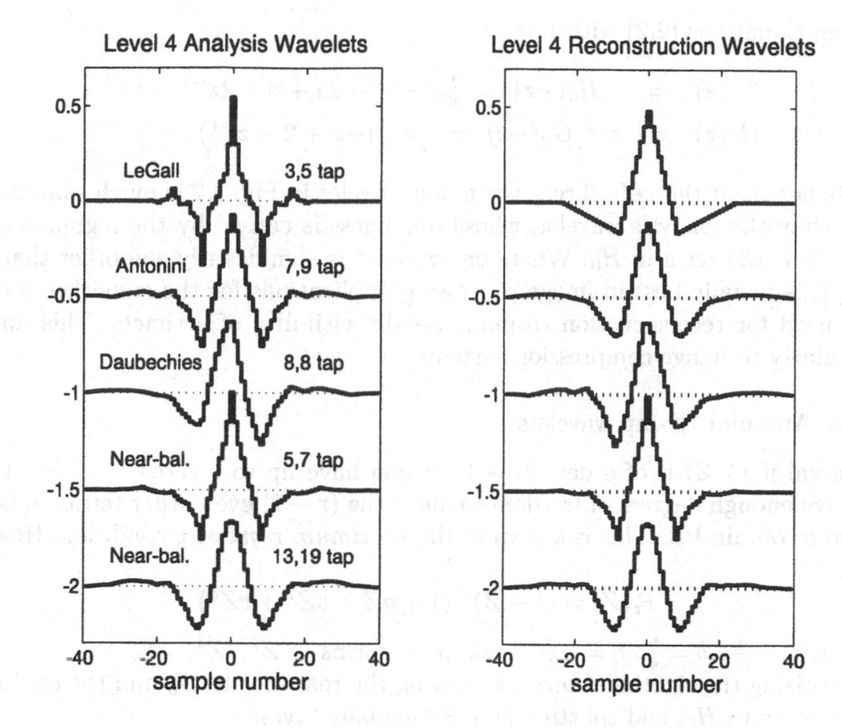

FIGURE 2.2. Comparison of a number of common wavelets

Case 1: Haar 2,2-tap wavelets

This uses the simplest possible $P_t(Z)$ with a single zero at $Z = -1$:

$$\begin{aligned} P_t(Z) &= 1 + Z \quad \text{and} \quad Z = \tfrac{1}{2}(z + z^{-1}) \\ \therefore\ P(z) &= \tfrac{1}{2}(z + 2 + z^{-1}) \\ &= \tfrac{1}{2}(z+1)(1+z^{-1}) = G_0(z)\ H_0(z) \end{aligned}$$

Using Equations (2.2) with $k = 1$:

$$\begin{aligned} G_1(z) &= z\ H_0(-z) = \tfrac{1}{2}z(1 - z^{-1}) = \tfrac{1}{2}(z - 1) \\ H_1(z) &= z^{-1}\ G_0(-z) = z^{-1}(-z+1) = z^{-1} - 1 \end{aligned}$$

Case 2: LeGall 3,5-tap wavelets

The next simplest $P_t(Z)$ is third order [7] (since even order terms must be zero for PR). The maximum number of zeros at $Z = -1$ is then two, leading to:

$$\begin{aligned} P_t(Z) &= (1+Z)^2\ (1+aZ) = 1 + (2+a)Z + (1+2a)Z^2 + aZ^3 \\ &= 1 + \tfrac{3}{2}Z - \tfrac{1}{2}Z^3 \quad \text{if } a = -\tfrac{1}{2} \text{ to suppress the term in } Z^2. \end{aligned}$$

Allocating $(1 + Z)$ to G_0 and $(1 + Z)(1 + aZ)$ to H_0, and again putting $Z = \frac{1}{2}(z + z^{-1})$ gives:

$$\begin{aligned} G_0(z) &= \tfrac{1}{2}(z + 2 + z^{-1}) \\ H_0(z) &= \tfrac{1}{8}(z + 2 + z^{-1})(-z + 4 - z^{-1}) \\ &= \tfrac{1}{8}(-z^2 + 2z + 6 + 2z^{-1} - z^{-2}) \end{aligned}$$

Using Equations (2.2) with $k = 1$:

$$\begin{aligned} G_1(z) &= z\, H_0(-z) = \tfrac{1}{8}z(-z^2 - 2z + 6 - 2z^{-1} - z^{-2}) \\ H_1(z) &= z^{-1}\, G_0(-z) = \tfrac{1}{2}z^{-1}(-z + 2 - z^{-1}) \end{aligned}$$

Note here that the LeGall reconstruction wavelet in Fig. 2.2 is much smoother in shape than the analysis wavelet, whose roughness is caused by the highpass effect of the $(1 + aZ)$ term in H_0. Where one wavelet is significantly smoother than the other, it is usually best in image processing applications for the smoother wavelet to be used for reconstruction to minimise the visibility of artifacts. This applies particularly to image compression systems.

Case 3: Antonini 7,9-tap wavelets

In general if $P_t(Z)$ is of order $(2r - 1)$, it can have up to r zeros at $Z = -1$ and still have enough degrees of freedom to allow the $(r - 1)$ even order terms to be set to zero to obtain PR. This is known as the *maximum regularity* condition. Hence if $r = 4$:

$$P_t(Z) = (1 + Z)^4\ (1 + aZ + bZ^2 + cZ^3)$$

where $a = -\frac{29}{16}$, $b = \frac{5}{4}$, $c = -\frac{5}{16}$ to suppress terms in Z^2, Z^4, Z^6.

Factorising the righthand term, allocating the real root to G_0 and the conjugate pair of roots to H_0, and splitting $(1 + Z)^4$ equally gives:

$$\begin{aligned} G_{0t}(Z) &= (1 + Z)^2\ (1 + \alpha Z) \\ H_{0t}(Z) &= (1 + Z)^2\ (1 + \beta Z + \gamma Z^2) \end{aligned}$$

where $\alpha = -0.5936$, $\beta = -1.2189$, $\gamma = 0.5265$.

Substituting $Z = \frac{1}{2}(z + z^{-1})$ and expanding factors gives $G_0(z)$ and $H_0(z)$, and then (2.2) with $k = 1$ gives $G_1(z)$ and $H_1(z)$, as before.

Note the improved smoothness of these wavelets in Fig. 2.2, compared to the simpler LeGall wavelets. However there is still some roughness in the analysis wavelet. These 7,9-tap wavelets posess the following attractive features:

- Good regularity – each filter has 2 factors $(1 + Z)$.
- All filters are linear (zero) phase.
- Not many taps – 7 for G_0 and H_1, 9 for H_0 and G_1.
- G_0 and H_0 are quite similar (also G_1 and H_1). The implications of this are discussed now.

Towards Orthogonality:

The LeGall and Antonini wavelets are *bi-orthogonal*, allowing perfect reconstruction, but the frequency responses of G_0 and H_0 are different (contrast their analysis and reconstruction impulse responses in Fig. 2.2). If we require an *orthogonal* DWT then the frequency response magnitudes of G_0, G_1 must be the same as H_0, H_1.

We cannot achieve *all* of:

- Equal (balanced) frequency responses

- Perfect reconstruction (PR)
- Linear phase for G_0, G_1 and H_0, H_1.

This is because, for PR, even powers of Z in $P_t(Z)$ must be zero, so $P_t(Z)$ must have an odd number of factors which cannot be distributed equally to G and H.

But there are three ways towards orthogonality:

Quadrature Mirror Wavelets remove the PR constraint and use a quadrature mirror filter (QMF) bank. Since these filters need to be of fairly high order to approximate PR well, we do not consider them further.

Daubechies Wavelets remove the linear phase constraint by splitting each factor in Z between $G_0(z)$ and $H_0(z)$.

Near-balanced Wavelets maintain linear phase for H and G, but reduce the number of roots at $Z = -1$ to allow a degree of freedom for making $H_0(e^{j\omega T_s})$ similar (but not identical) to $G_0(e^{j\omega T_s})$.

Case 4: Daubechies Wavelets

Each factor of $P_t(Z)$ may be factorised into a pair of factors in z, since:

$$(\alpha z + 1)(1 + \alpha z^{-1}) = \alpha z + (1 + \alpha^2) + \alpha z^{-1} \;=\; (1 + \alpha^2) + 2\alpha Z$$

For each factor of $P_t(Z)$, allocate one of its z subfactors to $H_0(z)$ and the other to $G_0(z)$. Hence H_0 and G_0 will no longer each be linear phase.

Since the subfactors occur in reciprocal pairs (with roots at $z = \alpha$ and α^{-1})

$$G_0(z) = H_0(z^{-1}) \quad \text{and} \quad G_0(e^{j\omega T_s}) = H_0(e^{-j\omega T_s})$$

so H_0 and G_0 are the time reverse of each other and the frequency response *magnitudes* are the same. Fig. 2.2 shows the $r = 4$ Daubechies wavelets, in which the z subfactors have been distributed between H_0 and G_0 so as to give the best approximation to linear phase. Note the time reverse property and the similarities to the two Antonini wavelets.

Case 5: Near-balanced Wavelets

Linear phase for H and G is desirable in image processing applications because the human eye is very sensitive to phase distortions at edges. It is preferable that any artifacts, due to severe quantisation or other distortions of wavelet coefficients, should be as symmetrical as possible about principal image features such as strong edges. We have developed [15] the following technique for obtaining linear (zero) phase H and G filters, perfect reconstruction and *approximate* symmetry (balance) between the analysis and reconstruction filters.

If $P_t(Z)$ is of order $2r - 1$, we may reduce the number of roots at $Z = -1$ from r to $r - 1$, to allow a degree of freedom for making $H_0(e^{j\omega T_s})$ similar to $G_0(e^{j\omega T_s})$.

If $r = 3$, we get:

$$P_t(Z) = (1 + Z)(1 + aZ + bZ^2) \,.\, (1 + Z)(1 + cZ)$$

If the terms in Z^2 and Z^4 are to be zero:

$$a = -\frac{(1+2c)^2}{2(1+c)^2} \quad \text{and} \quad b = \frac{c(1+2c)}{2(1+c)^2}$$

where c is a free parameter which may be adjusted to give maximum similarity between the left and right pairs of factors above for $Z = 1 \to -1$ ($\omega T_s = 0 \to \pi$).

We find that $c = -\frac{2}{7}$ produces good similarity and gives:

$$\begin{aligned} P_t(Z) &= \tfrac{1}{50}(50 + 41Z - 15Z^2 - 6Z^3) \,.\, \tfrac{1}{7}(7 + 5Z - 2Z^2) \\ &= H_{0t}(Z) \,.\, G_{0t}(Z) \end{aligned}$$

Using the transformation $Z = \frac{1}{2}(z + z^{-1})$ gives the simplest near-balanced wavelets with 5 and 7 tap filters. These wavelets are well balanced but have quite sharp cusps, as can be seen in Fig. 2.2.

Smoother near-balanced wavelets may be obtained by employing a higher order transformation from Z to z, such as:

$$Z = pz^3 + (\tfrac{1}{2} - p)(z + z^{-1}) + pz^{-3}$$

Four zeros on the unit circle near $z = -1$ are achieved for each $(Z + 1)$ factor when $p = -\frac{3}{32}$. When substituted into $P_t(Z)$, this gives relatively high order filters with 13 and 19 taps (although 2 taps of each filter are zero). Fig. 2.2 shows the improved smoothness of these filters.

One potentially important advantage of these near-balanced wavelets is that all the filter coefficients may be rational (see the $H_{0t}(Z)$ and $G_{0t}(Z)$ polynomials given above), which allows perfect reconstruction using fixed-point arithmetic. Furthermore they may be implemented efficiently using ladder filter structures. The Antonini and Daubechies coefficients are based on irrational quadratic roots and thus tend to need high precision arithmetic.

We have shown how a variety of wavelets may be designed and Fig. 2.2 illustrates the similarity of shape between them, which arises because they are all attempting to partition the frequency domain in similar ways. However there is one major problem with all of these wavelets, which we discuss next.

2.2.3 Shift Invariance

When we analyse the Fourier spectrum of a signal, we expect the energy in each frequency bin to be *invariant* to any shifts in time or space. It would be desirable if wavelet transforms behaved similarly with respect to scale instead of frequency, but unfortunately real wavelet transforms do not have this property.

Consider a step function input signal, analysed with the DWT using Antonini 7,9 tap filters. The step response at wavelet level 4 is shown in Fig. 2.3(a), assuming that wavelet coefficients are computed at the full input sampling rate. In practice, they are computed at 1/16 of this rate, yielding samples at points such as those of the crosses in Fig. 2.3(a). If the input step is shifted relative to the output sampling grid then this is equivalent to sampling the step response with a different horizontal offset; e.g. for an offset of 8 input samples, we obtain samples at the circles in Fig. 2.3(a).

Now, comparing the total energy of the samples at the crosses (which are all quite small in amplitude) with that of the samples at the circles (two of which are rather

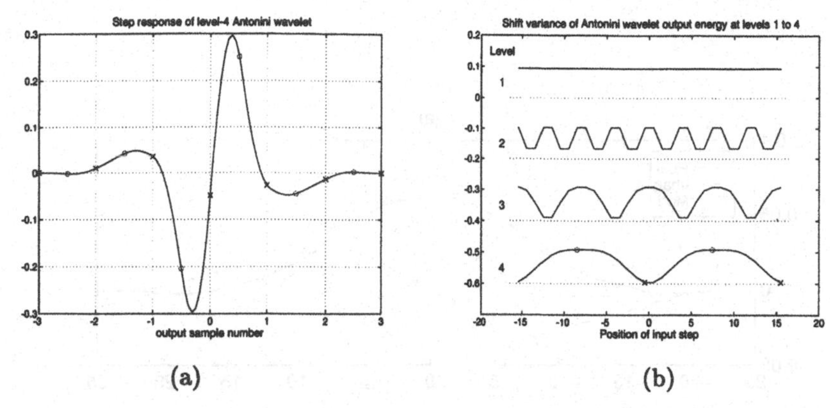

FIGURE 2.3. Step responses at level 4 of Antonini 7,9 tap wavelets (a) and their shift variance properties (b), showing the variation of energy in each level of wavelet coefficient for a unit step input as the position of the input step is shifted

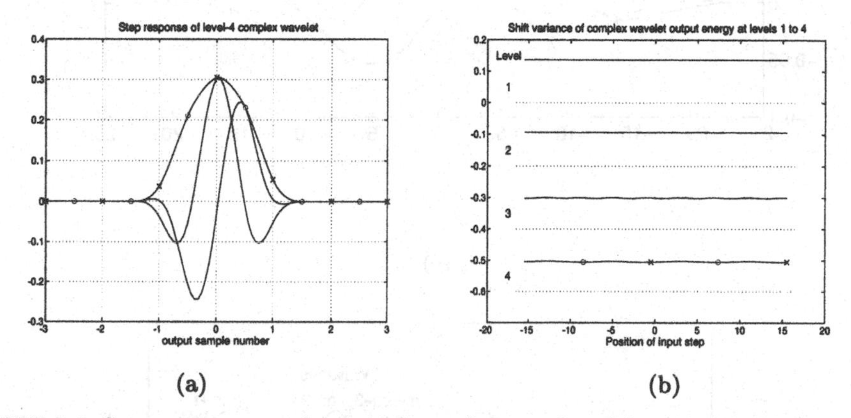

FIGURE 2.4. Step responses at level 4 of complex wavelets (a), and their shift variance properties (b), showing the variation of energy in each level of wavelet coefficient for a unit step input as the position of the input step is shifted

large), we find a big difference. This illustrates a significant drawback to using the standard DWT as a tool for analysing signals - the energy distribution between the various wavelet subbands depends critically on the position of key features of the signal relative to the wavelet subsampling grid, whereas we would like it to depend on just the shapes of the features.

This problem is illustrated more generally in Fig. 2.3(b) which shows how the total energy at each wavelet level varies as the input step is shifted. The period of each variation equals the subsampling period at that level, and the crosses and circles show the energies at level 4 corresponding to the shift positions shown by the same symbols in Fig. 2.3(a).

Hence we conclude that real DWTs are not good at extracting higher level features from images (e.g. edge detection or pattern matching). This problem can be avoided by not decimating the DWT outputs, but it produces considerable data redundancy. A better solution is the *Complex Wavelet Transform*.

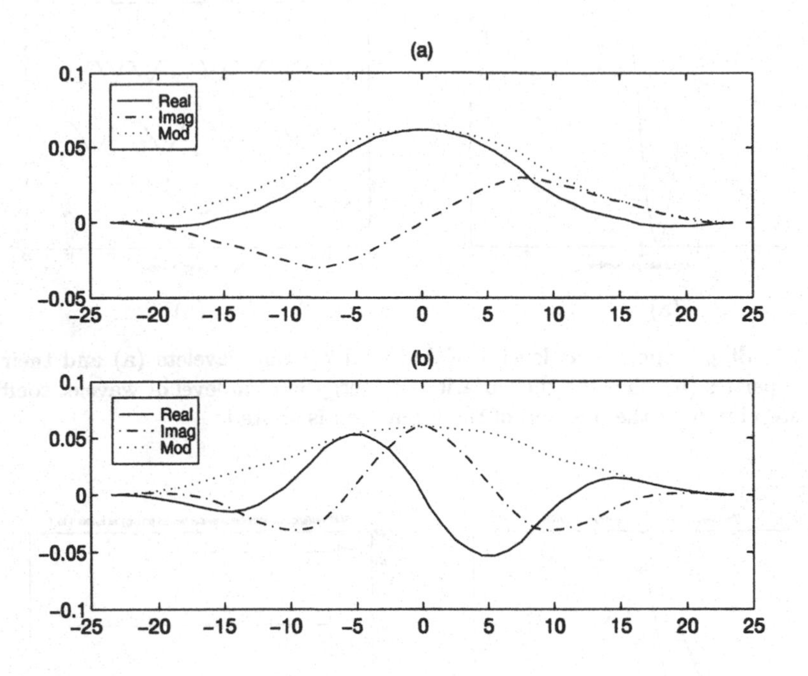

(c)

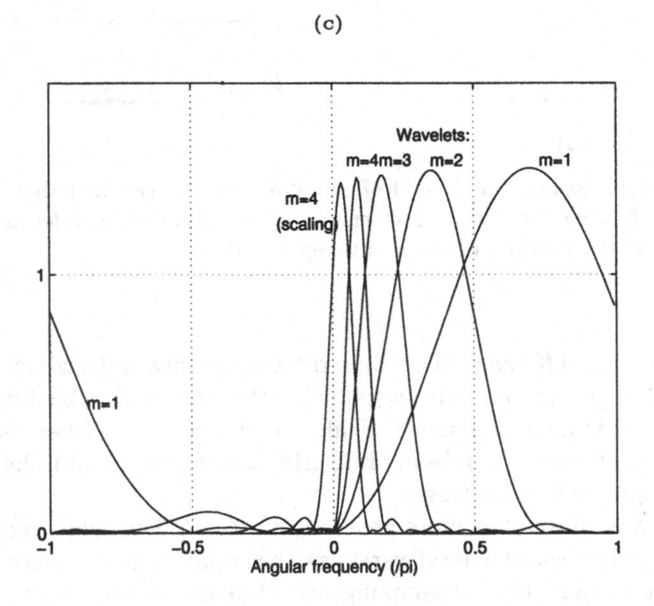

FIGURE 2.5. Complex wavelet responses: (a) level 4 lowpass impulse response (scaling function); (b) level 4 highpass impulse response (wavelet); (c) frequency responses of the wavelet filters at levels $m = 1, 2, 3, 4$ and of the scaling filter at $m = 4$

2.3 The Complex Wavelet Transform (CWT)

The structure of the CWT is the same as in Fig. 2.1(a), except that the CWT filters have complex coefficients and generate complex output samples. Since the output sampling rates are unchanged from the DWT, but each sample contains a real and imaginary part, a redundancy of 2:1 is introduced (we shall show later that this becomes 4:1 in two dimensions).

We may design the complex filters such that the *magnitudes* of their step responses vary slowly with input shift – only the *phases* vary rapidly. This is shown in Fig. 2.4(a), in which the real and imaginary parts of a typical complex wavelet step response are superimposed and the uppermost curve represents the magnitude of the response. Note that the real part is an odd function while the imaginary part is even. This wavelet was derived from the following simple 4,4 tap filters:

$$\begin{aligned} h_0 &= [\ 1-j,\ \ 4-j,\ \ 4+j,\ \ 1+j\ ]/10 \\ h_1 &= [\ -3-8j,\ \ 15+8j,\ \ -15+8j,\ \ 3-8j\ ]/48 \end{aligned} \tag{2.7}$$

Figure 2.4(b) plots the energy at each level vs. input step position for this CWT and, in contrast to Fig. 2.3(b), shows that it is approximately constant at all levels. Hence the energy of each CWT band may be made *shift invariant*. This arises because the power profile of the complex wavelet is smooth enough to satisfy Nyquist's sampling criterion at the output sampling rate for each level, which is not possible for real wavelets, whose power must dip to zero at every zero-crossing of the wavelet.

The other key property of the complex wavelets is that their phases vary approximately linearly with input shift (as with Fourier coefficients). Hence, based on phase shifts, efficient displacement estimation is possible and interpolation between consecutive complex samples can be relatively simple and accurate. In [14], Simoncelli et al. have shown that shift invariance of energy distribution is equivalent to the property of interpolability.

An important additional feature that we have introduced to the CWT is a prefilter $F(z)$ which precedes the filter tree of Fig. 2.1(a). The purpose of $F(z)$ is to simulate the effect of an infinite filter tree to the left of the input x so that the wavelets at all levels of the tree to the right of x are *perfectly scaled* versions of a single *mother wavelet*. The pre-filter coefficients $f(n)$ are found by solving for values such that:

$$h_{0f}(2n) = \lambda f(n)$$

where $h_{0f}(2n)$ are the even coefficients of the filter $F(z)\ H_0(z)$, and λ is a constant. When expressed as a matrix multiplication this becomes a simple eigenvector problem.

Using h_0 from equation (2.7) yields a rational valued even-length prefilter f. However to maintain even length for $F(z)\ H_0(z)$ and $F(z)\ H_1(z)$ requires an odd-length prefilter. The following odd-length filter may be shown [11] to give approximately the same performance:

$$f = [\ -j,\ \ 5,\ \ j\ ]/5 \tag{2.8}$$

Figure 2.5 shows the impulse responses of the wavelet and scaling function at level 4, and also the frequency responses of the wavelet filters for levels 1 to 4 and the level 4 scaling filter. The 3-tap prefilter f is incorporated for these results. Note

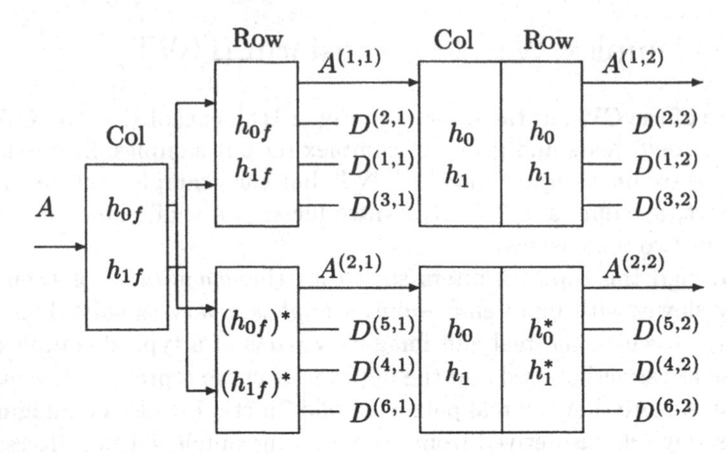

FIGURE 2.6. The 2-D CWT filter structure over levels 1 and 2, incorporating the pre-filters f at level 1

the approximate modulated-Gaussian form of the impulse responses, and the near-perfect scaling of the frequency responses. The sidelobes in the frequency domain are all low, and negative frequencies are almost completely suppressed (which is possible only with complex filters).

2.3.1 Two Dimensional CWT

The 2-D CWT is implemented separably (as is usual for the 2-D DWT), so that only 1-D convolutions are required. The result is three bandpass subimages at each level together with one lowpass image on which the process is iterated [12]. The 2-D equivalent wavelet filters are products of the 1-D lowpass and bandpass filters and they approximate 2-D Gabor filters (due to the modulated-Gaussian form of the 1-D filters). Because the 1-D filters pass only the positive half of the spectrum (Fig. 2.5(c)), the 2-D filters formed from them cover only the first quadrant of the 2-D frequency domain. However real images contain significant information in both the first and second quadrants of the frequency domain (the third and fourth quadrants are conjugate mirrors of these), so we must include a parallel set of filters to cover negative horizontal frequencies. Conjugating h_0 and h_1 reflects their frequency responses about $\omega = 0$, so the parallel filter set uses the same 2-D building block except that the row filters are h_0^* and h_1^* instead of h_0 and h_1. Figure 2.6 shows the complete 2-D CWT structure over 2 levels, incorporating the prefilters f to achieve perfect scaling.

Each level m of the tree produces 6 complex-valued bandpass subimages $\{D^{(n,m)}, n = 1, \dots, 6\}$ as well as two lowpass subimages $A^{(1,m)}$ and $A^{(2,m)}$ on which subsequent stages iterate. The overall redundancy of the 2-d CDWT is 4 to 1, i.e. 4 real or imaginary components are produced for every input pixel, regardless of how many levels are computed. This redundancy is important for interpolation.

Figure 2.7 shows the real part (a) and imaginary part (b) of the impulse responses of the six 2-D bandpass filters at level 4. These filters exhibit strong directionality and are oriented at approximate angles of 75, 45, 15, −15, −45, −75 degrees. Such directionality cannot be achieved with separable *real* filters; it requires complex filters for a separable implementation. Simple contours of the frequency responses of the 2-D filters at levels 3 and 4 are shown in Fig. 2.8. The relative positions of the

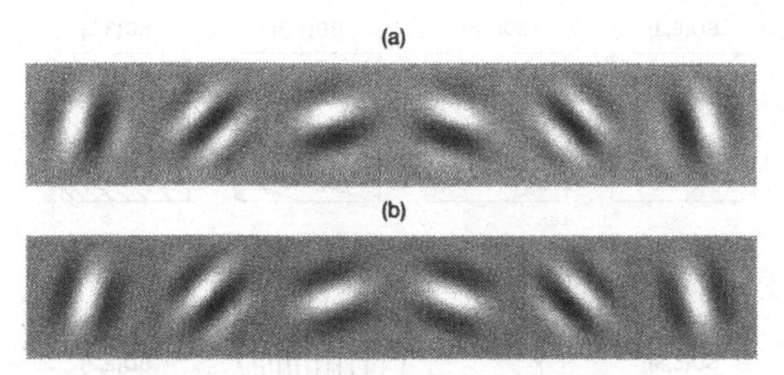

FIGURE 2.7. Greyscale plot (grey = 0) of the impulse responses of the six wavelets at level 4 of the 2-D CWT: (a) real part; (b) imaginary part

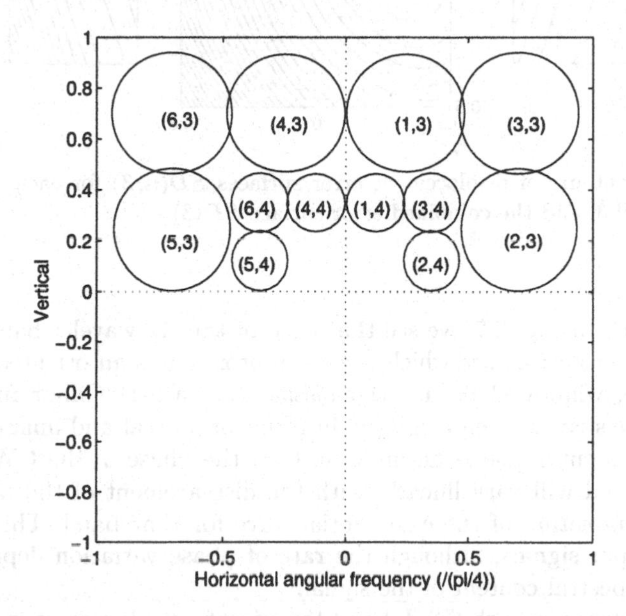

FIGURE 2.8. Contours of the frequency responses of the 2-D CWT wavelet filters at levels $m = 3$ and 4

responses are determined by the ratio of the centre frequencies of the 1-D highpass and lowpass filters at each level. With the chosen filters and the perfect scaling prefilter, this ratio is 3:1 at every level and leads to the near-uniform spacing of the bands shown in Fig. 2.8. Similar subbands for levels 1 and 2 appear on enlarged scales outside of the subbands shown; and subbands for lower levels can exist at smaller scales nearer to the origin.

2.4 Applications of the CWT in Image Processing

2.4.1 Motion Estimation

We have already observed that the 1-D CWT generates coefficients whose phases are approximately linearly related to the displacement of the input signal. In 2-D this still applies, but we have to specify the direction of the displacement.

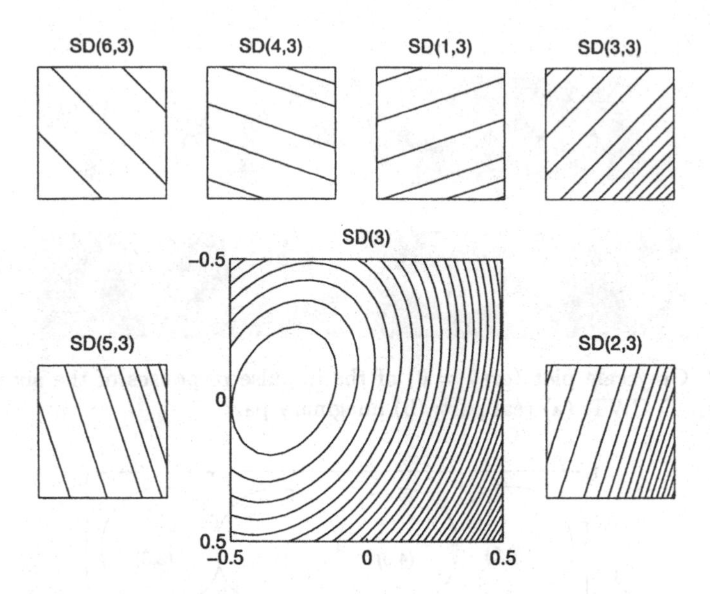

FIGURE 2.9. Contours of displacement error surfaces $SD(n,3)$ for each of the 6 CWT subbands at level 3, and the combined error surface $SD(3)$

Referring back to Fig. 2.7, we see that each of the six wavelet bands at a given level has an impulse response which closely approximates an oriented complex exponential wave, windowed by a 2-D gaussian (i.e. a 2-D Gabor function). The wave appears as stripes when displayed in terms of its real and imaginary parts in Fig. 2.7. If the input signal is an impulse, then the phase of the CWT coefficient from a given band will vary linearly with the displacement of the impulse *in the direction of propagation* of the exponential wave for that band. This also applies to arbitrary input signals, although the rate of phase variation depends to some extent on the spectral content of the signal.

We can therefore use each CWT band to estimate displacement in the direction of that band, by measuring the phase difference between equivalent coefficients from two consecutive frames of the image sequence. It is then necessary to combine the displacement estimates from all bands to arrive at an overall displacement vector.

To do this, we generate the parameters of a parabolic error surface for each band, such that the surface minimum represents the line of displacement vectors which all would give the phase difference that is measured, and the curvature of the surface shows approximately how the error between the two equivalent coefficients would increase for displacements that differ from these. Fig. 2.9 shows contours of six of these parabolic surfaces $SD(n,3)$ at CWT level 3, and we see how the contours are aligned with the stripes of each corresponding band in Fig. 2.7. To find the displacement vector which minimises the total error from all bands, we just add the six surfaces to obtain the quadratic surface $SD(3)$, shown in the centre of Fig. 2.9, and locate the minimum of the combined surface. If each surface is represented by a second-order polynomial in displacements u and v, which requires six coefficients, then the surfaces are added just by adding their corresponding coefficients. The overall surface will have only one minimum (since it is still quadratic) that may be calculated simply from the six coefficients for that surface, without a search being

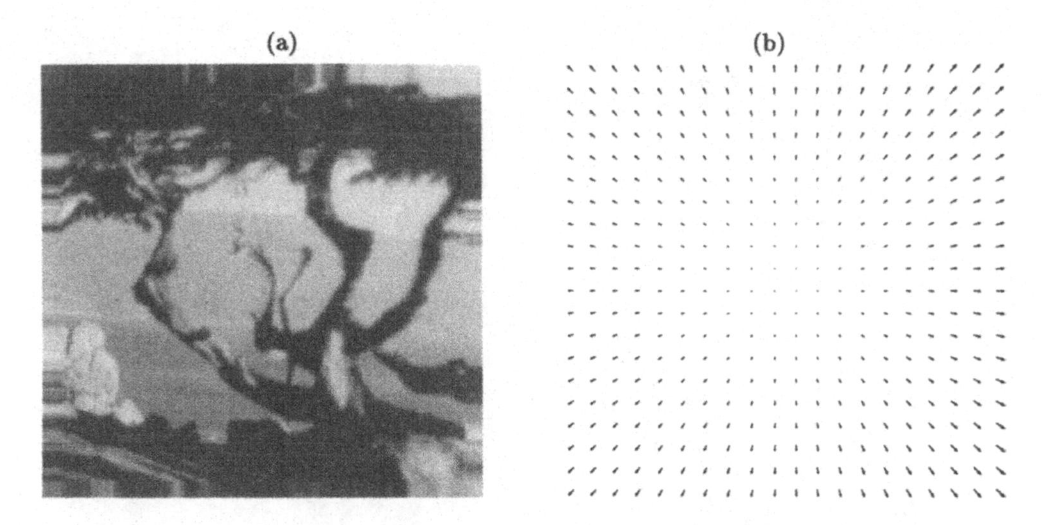

FIGURE 2.10. (a) Diverging tree test image; (b) its true motion field, subsampled to one vector per 8×8 block, and scaled up by 2

necessary. Hence the displacement vector at each sampling location within each wavelet scale may be found.

An additional by-product of the above method, is an elliptic confidence measure that accompanies each displacement vector. The ellipse corresponds to a contour of the final error surface, and its radius in any given direction indicates the level of uncertainty in the displacement vector in that direction. The major axis of the ellipse aligns with the direction of greatest uncertainty, which tends to be parallel to any nearby straight edge feature in the images. The confidence ellipse is defined by the three curvature coefficients of the surface polynomial. It is very useful for permitting the vector fields from consecutive CWT levels to be combined in an optimum way and for allowing optimal smoothing of the motion field to minimise aperture effects.

Further details of the algorithm are given in [8, 9, 10, 11], including a number of refinements to optimise the performance.

Fig. 2.10 shows frame 20 from the standard motion test sequence, Diverging Tree, together with its true (known) motion field. Fig. 2.11 shows the reconstructed motion fields, using three methods of motion estimation: (a) our CWT method; (b) a heirarchical gradient-based method [5]; and (c) a heirarchical block matching scheme with half-pel resolution. The three methods produce motion fields of differing resolutions, so all motion fields have been calculated to the finest resolution available and then interpolated up to full resolution. These fields were then compared to the full resolution known field and an error image was produced, based on the *error angle* defined by Fleet and Jepson [3], which is akin to a relative error measure without the exaggeration of errors in small true-motion vectors. These error images form the background of each image in Fig. 2.11, with darker regions indicating larger errors. The motion fields, subsampled 8:1 in each direction, are superimposed on the error images. From the general lack of dark regions in Fig. 2.11(a), we see the good performance of the CWT method.

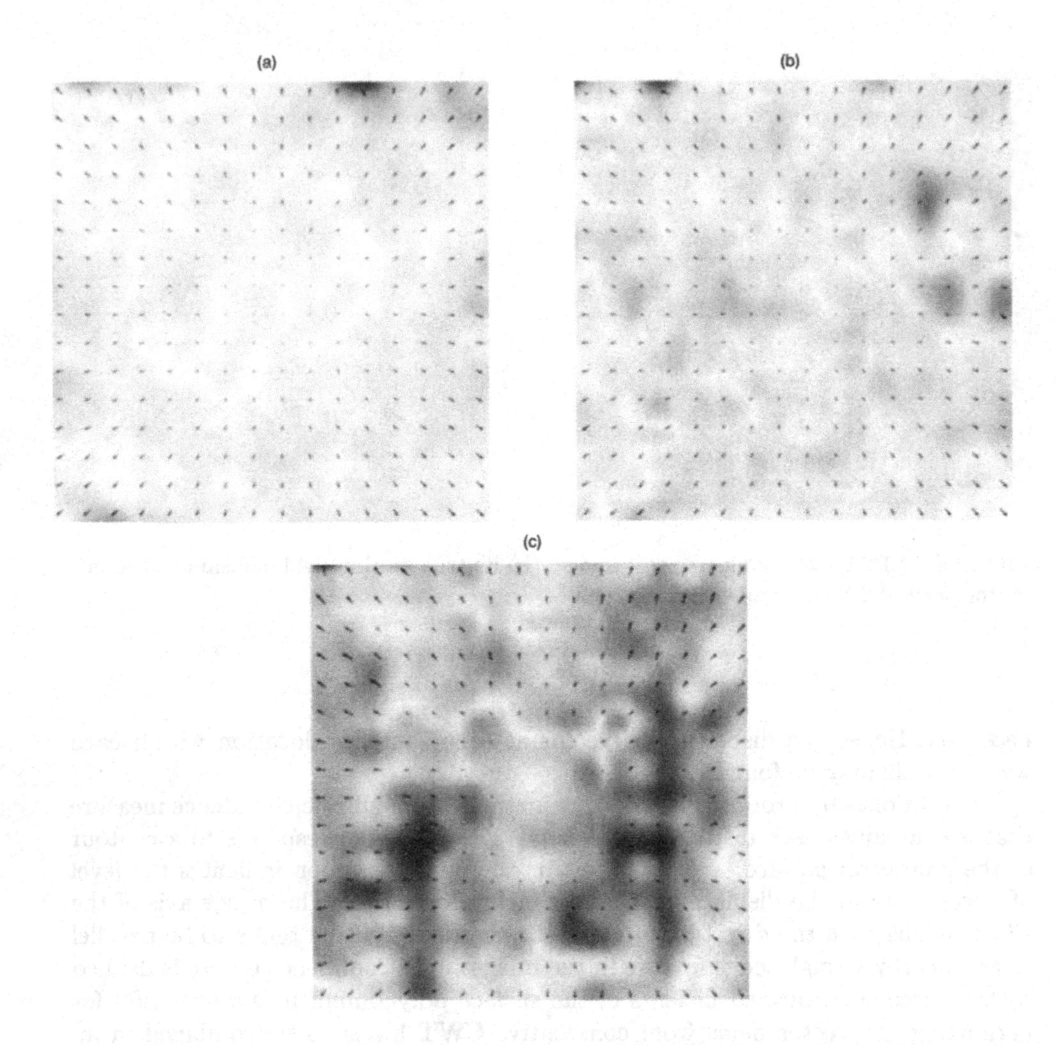

FIGURE 2.11. Motion fields (8-pel resolution, scaled up by 2) superimposed on motion error images for the diverging tree: (a) CWT method; (b) gradient-based method; (c) fractional block matching method

The CWT method is highly robust to intensity fluctuations between frames of a sequence since it derives its motion estimates from bandpass subimages only [10]. The depth of the coarsest level in the wavelet decomposition determines the low-frequency cut-off limit, below which all frequency components are largely ignored. The phase-based methods are insensitive to *scaling* fluctuations as well as additive offsets between frames. The method also shows good resilience to additive noise, since the multiscale directionally selective filters allow all low-amplitude CWT coefficients with poor SNR to be largely ignored while those with good SNR are retained.

The CWT displacement estimation algorithm has a number of application areas including video coding, stereo optics (3-D scene understanding), and image registration. Its key features are good *true-motion* accuracy, fine resolution, and high resilience to noise and intensity fluctuations.

2.4.2 De-Noising

The second main application area of the CWT that we give is that of noise reduction in images. Here we shall just consider still images, but the techniques are equally applicable to video sequences. De-noising is an area that has been studied quite extensively for the DWT and a good discussion is given by Donoho [2].

The basic technique is to transform the noisy input image into a domain in which the main signal energy is concentrated into as few coefficients as possible, while the noise energy is distributed more uniformly over all coefficients. By suppressing the amplitudes of the lower energy coefficients, the noise is attenuated more than the signal and hence the signal-to-noise ratio (SNR) is improved. The image may then be recovered by an inverse transformation.

Thus the key question is which transform is optimum for typical images. The discrete cosine transform (DCT) is well known to be the near-optimum block transform for typical images, but it is outperformed by the DWT because higher frequency image details are compressed better by short fine-scale wavelet basis functions than by the higher frequency DCT basis functions which are less spatially compact. The overlapped nature of wavelets also leads to better compression than the non-overlapped DCT.

However the DWT is not shift invariant (see Section 2.3), so when a given wavelet coefficient is quite low in amplitude, we cannot be sure whether this means that there is little signal energy at that scale and location, or whether the coefficient is small just because of the particular position of the input feature relative to the wavelet sampling grid (Fig. 2.3(a)). Hence the decision process for suppressing noise (but not signal) is impaired. One way to avoid this problem is to use the undecimated version of the DWT, but this generates substantial redundancy in the wavelet domain, $4M{:}1$ where M is the number of wavelet levels.

We feel that a better way is to use the CWT, whose redundancy is limited to 4:1, but is sufficient to provide shift invariance. The complex coefficients may now be selectively suppressed, based on their absolute magnitudes, which give a much more reliable estimate of input feature energy than the DWT coefficients. An additional advantage of the CWT lies in the directional selectivity of its filters: in particular, they can separate diagonal features near $+45°$ from those near $-45°$, which cannot be achieved by separable real wavelets.

The one major drawback to using the CWT lies in the difficulty of performing the inverse CWT so as to achieve perfect reconstruction. The difficulty arises when we try to design the 2-band reconstruction block of Fig. 2.1(b). We desire all four of the complex filters in this block to pass only positive frequency components and to reject negative frequency components. Hence this block *cannot* make $Y(z) \equiv X(z)$ (since it has frequency selectivity) and so PR for just this block is not feasible. Note that PR for the block is possible, but the filters no longer have desirable frequency domain properties. However PR for the whole CWT with good properties is potentially possible because the final operation in the inverse CWT is to take just the real part of the result, which is equivalent to reintroducing all the negative frequency components as if they came from another pair of trees containing complex conjugates of all the H and G filters.

We have an approximately PR solution for the CWT and its inverse, which is achieved by design in the frequency domain to ensure that all filters and their conjugate mirrors combine to produce a flat frequency response with zero phase.

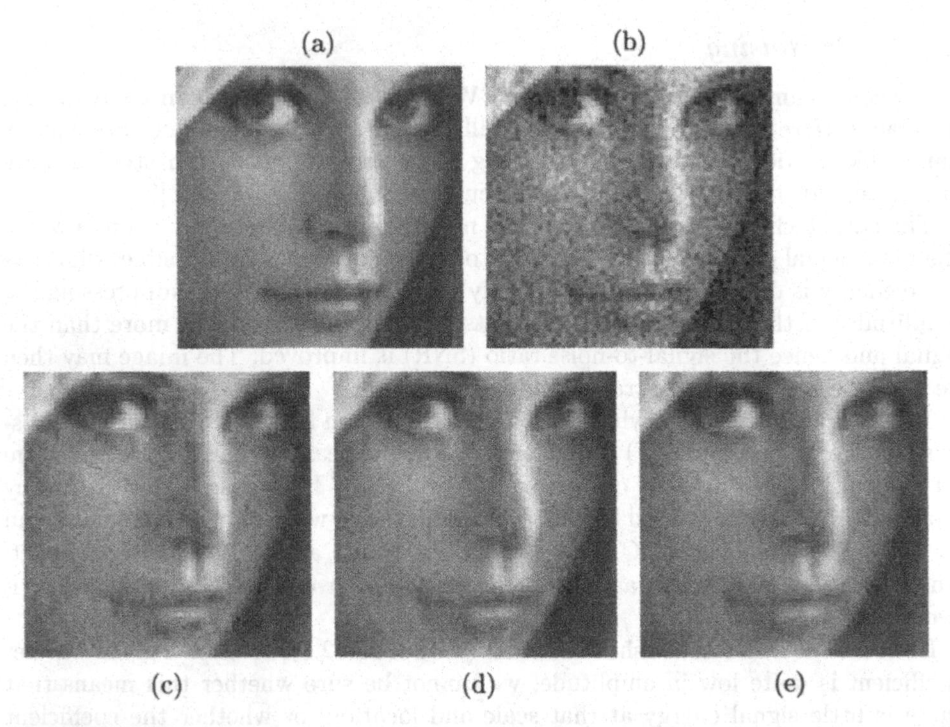

FIGURE 2.12. (a) Original 64 × 64 portion of Lena image; (b) with white gaussian noise added to give PSNR = 22.16 dB; (c) denoised with 13,19-tap DWT, PSNR = 26.35 dB; (d) denoised with CWT, PSNR = 27.72 dB; (e) denoised with undecimated WT, PSNR = 27.57 dB

However in the spatial domain these filters are of infinite length in theory, and in practice require 20 taps each to approximate PR well. This compares unfavourably with the 4-tap CWT filters, used for motion estimation.

Very recently [4], we have also developed a *Dual-Tree* implementation of the CWT which employs both odd-length and even-length PR biorthogonal filter sets alternately in two trees of real filters, such that the outputs of the two trees form the real and imaginary parts of the CWT coefficients. The whole transform is invertible with perfect reconstruction and the two trees provide approximate shift invariance.

We implement the filters using a ladder filter structure. For the even length filter set, we selected (6,10)-tap filters, $H_{0e}(z) = \frac{1}{2}(1 + z^{-1}) + \frac{1}{16}(-1 + z^{-1})(z^2 - z^{-2})$ and $H_{1e}(z) = \frac{1}{2}(-1 + z^{-1}) + \frac{1}{8}(z^2 - z^{-2})\, H_{0e}(z)$; and for odd length we chose the (13,19)-tap filters (Case 5 above) using the transformation function $Z = \frac{1}{32}(-3z^3 + 19z + 19z^{-1} - 3z^{-3})$ in the polynomials $H_{0o}(Z) = \frac{7}{10} + \frac{1}{10}Z(5 - 2Z)$ and $H_{1o}(Z) = \frac{1}{7}(5 - 2Z) - \frac{3}{7}Z\, H_{0o}(Z)$. (Many other filter choices exist, but these seem to offer a good tradeoff between performance and complexity.)

Fig. 2.12 shows results using the Dual-Tree CWT. Image (a) is the original (a central 64×64 pel portion of the 256×256 Lena image) and (b) is the original plus gaussian white noise of standard deviation 20 units (on a 0 to 255 image scale), yielding a peak SNR (PSNR) of 22.16 dB. Image (d) is the result of denoising image (b) using the Dual-Tree CWT and a soft thresholding method which suppresses all complex wavelet coefficients with low amplitude. For comparison we show images (c) and (e) which were obtained using the same soft thresholding with the real DWT in its decimated and undecimated forms respectively. (c) shows significantly worse artifacts than (d), while (e) is slightly more blurred than (d) and requires substantially more computation. Hence the CWT looks attractive for this application.

In the soft thresholding method, lower amplitude noisy coefficients were suppressed by scaling them with the gain function:

$$g(x) = \begin{cases} \frac{1}{2}[1 - \cos(\pi|x| \,/\, 2t)] & \text{for } |x| < 2t \\ 1 & \text{otherwise} \end{cases}$$

where $|x|$ is the amplitude of the coefficient. We have found this law gives better subjective results than either a hard gain decision of 0 or 1, or a Weiner gain of $1 - t^2/|x|^2$. In all cases the thresholds t at each level were selected so as to get minimum mean-squared error from the original (clean) image.

2.5 Conclusions

We have derived conditions for perfect reconstruction using filter banks with real filter coefficients, and shown that this leads to wavelets which all exhibit a lack of *shift invariance*. This means that for many image analysis tasks, these wavelets are unsuitable because the energy decomposition of the transformation process depends on the precise location of input features relative to the wavelet subsampling grids.

While the use of the undecimated wavelet transform is one solution to this problem, we have introduced an alternative, the *complex* wavelet transform, in which the redundancy is substantially less, as is the computation too. The CWT is shift invariant and also possesses other advantages over real transforms such as directionally selective diagonal filters and complex output coefficients whose phase shifts may be used to determine local displacements in the input signal.

To illustrate this we have given examples of application of complex wavelets to the problems of motion estimation and de-noising of images, showing useful performance advantages over more conventional approaches.

2.6 References

[1] I. Daubechies. The wavelet transform, time-frequency localisation and signal analysis. *IEEE Trans. on Information Theory*, 36(5):961–1005, 1990.

[2] D.L. Donoho. De-noising by soft thresholding, *IEEE Trans. Information Theory*, 41:613–627, 1995.

[3] D.J. Fleet and A.D. Jepson. Computation of component image velocity from local phase information. *Intern. J. Computer Vision*, 5:77–104, 1990.

[4] N.G. Kingsbury. A Complex Wavelet Transform with Perfect Reconstruction using Low-complexity Gabor-like Filters. Submitted to *IEEE Signal Processing Letters*, 1997.

[5] A.C. Kokaram and S.J. Godsill. A system for reconstruction of missing data in image sequences using sampled 3D AR models and MRF motion priors. *Computer Vision - ECCV '96, Springer Lecture Notes in Comp. Sc.*, II, pp. 613–624, 1996.

[6] J. Kovacevic and I. Daubechies (eds.). Special Issue on Wavelets. *Proc. IEEE*, April 1996.

[7] D. Le Gall and A. Tabatabai. Subband coding of digital images using symmetric short kernel filters and arithmetic coding techniques. In *Proc. IEEE Conf. on ASSP*, p. 761, 1988.

[8] J.F.A. Magarey and N.G. Kingsbury. Motion estimation using complex wavelets. In *Proc. IEEE Conf. on ASSP*, paper MDSP 13.10, Atlanta, 1996.

[9] J.F.A. Magarey and N.G. Kingsbury. An improved motion estimation algorithm using complex wavelets. In *Proc. IEEE Conference on Image Processing*, paper 17A3.08, 1, p. 969, Lausanne, 1996.

[10] J.F.A. Magarey and N.G. Kingsbury. Motion estimation using a complex-valued wavelet transform. Submitted to *IEEE Trans. on Signal Processing, special issue on wavelet applications*, 1997.

[11] J.F.A. Magarey. Motion estimation using complex wavelets. Ph.D. thesis, Cambridge University Department of Engineering, 1997.

[12] S.G. Mallat. A theory for multiresolution signal decomposition: The wavelet representation. *IEEE Trans. on Pattern Anal. and Machine Int.*, 11(7):674–693, 1989.

[13] O. Rioul and M. Vetterli. Wavelets and Signal Processing. *IEEE Signal Processing Mag.*, 8(4):14–38, 1991.

[14] E.P. Simoncelli, W.T. Freeman, E.H. Adelson, and D.J. Heeger. Shiftable multiscale transforms. *IEEE Trans. on Information Theory*, 38(2):587–607, 1992.

[15] D.B.H. Tay and N.G. Kingsbury. Flexible design of multidimensional perfect reconstruction FIR 2-band filters using transformations of variables. *IEEE Trans. on Image Processing*, 2(4):466–480, 1993.

[16] M. Vetterli and J. Kovacevic. *Wavelets and Subband Coding*. Prentice Hall, 1995.

3

General Sub–Band DCT: Fast Approximate Adaptive Algorithm

Abdulnasir Hossen[1]
Ulrich Heute[2]

ABSTRACT
The discrete cosine transform (DCT) has a variety of applications in image and speech coding [1, 2]. The idea of the approximate subband DFT (SB–DFT) [3, 4] is applied in [5] and [6] to the DCT. In this paper the basic idea of the subband DCT (SB–DCT) is reviewed. The calculation of the approximate SB–DCT is done with the aid of a fast cosine–transform method [7]. The subband DCT is generalized such that any band out of M subbands can be calculated. For this purpose the fast DCT method [7] is modified to calculate the fast DST. A general analysis of the errors due to the approximation is presented. The adaptive capability of the SB–DFT given in [8] is added to the SB–DCT to calculate adaptively the band of the dominant energy. The two–dimensional (2–D) SB–DCT is also investigated in this work. Applications of the general adaptive SB–DCT in speech cepstrum analysis and in echo detection are included.

3.1 Introduction

The discrete cosine transform (DCT) and the discrete sine transform (DST) of an N–point sequence $x(n)$, with $n \in \{0, 1, ..., N-1\}$ are defined respectively as:

$$C(k) = \sum_{n=0}^{N-1} 2x(n)\cos(\frac{\pi k(2n+1)}{2N}) \quad k \in \{0, 1, ..., N-1\} \tag{3.1}$$

$$S(k) = \sum_{n=0}^{N-1} 2x(n)\sin(\frac{\pi k(2n+1)}{2N}) \quad k \in \{0, 1, ..., N-1\} \tag{3.2}$$

A fast cosine transform given by Makhoul [7] can be computed for an N–point real signal according to the following procedure:

1. Compute $v(n)$ from $x(n)$ using:

$$v(n) \quad = \quad x(2n) \qquad\qquad 0 \le n \le [\frac{N-1}{2}]$$

---

[1]Applied Science University, Computer Science and Information Systems Department, 11931 Amman, Jordan
[2]University of Kiel, Institute for Network and System Theory, Kaiserstr. 2, 24143 Kiel, Germany, E-mail: uh@techfak.uni-kiel.d400.de

$$= x(2N-2n-1) \quad [\frac{N+1}{2}] \leq n \leq N-1 \tag{3.3}$$

where $[a]$ denotes the integer part of a.

2. Find the DFT $V(k)$ of $v(n)$.
3. Find the real part of $2V(k)\exp(\frac{-j\pi k}{2N})$.

In this work the use of the fast algorithm of Makhoul [7] is extended to find the DST of Eq. (3.2) according to the following procedure:

1. Compute $v(n)$ from $x(n)$ using:

$$\begin{aligned} v(n) &= x(2n) & 0 \leq n \leq [\frac{N-1}{2}] \\ &= -x(2N-2n-1) & [\frac{N+1}{2}] \leq n \leq N-1 \end{aligned} \tag{3.4}$$

2. Find the DFT $V(k)$ of $v(n)$.
3. Find the imaginary part of $2V(k)\exp(\frac{-j\pi k}{2N})$.

The SB–DFT idea is applied to the discrete cosine transform [5] for image–coding applications and then in [6] for speech analysis and echo detection applications. In [5] and [6] only the low–frequency band is calculated with a higher speed. In this paper the idea of the SB–DCT is generalized to calculate any band, not only the low–frequency band. The extension of the algorithm to the two–dimensional case is also presented. Furthermore, an adaptation of the algorithm is performed to find and then calculate the band of the dominant energy among the subbands of the decomposed signal. Application examples in speech cepstrum analysis and in echo detection are included.

3.2 Sub–Band DCT

The basic idea of the SB DCT is to decompose the length–N data sequence $x(n)$ into two subsequences each of length $N/2$ according to

$$\begin{aligned} g(n) &= 1/2[x(2n) + x(2n+1)] \\ h(n) &= 1/2[x(2n) - x(2n+1)] \end{aligned} \tag{3.5}$$

Eq. (3.1) can be written as

$$C(k) = \sum_{n=0}^{N/2-1} 2x(2n)\cos(\frac{\pi k(4n+1)}{2N}) + \sum_{n=0}^{N/2-1} 2x(2n+1)\cos(\frac{\pi k(4n+3)}{2N}) \tag{3.6}$$

Using a simple mathematical reformulation and with the aid of Eqs.(3.5) and (3.6) we get

$$\begin{aligned} C(k) &= 2\cos(\frac{\pi k}{2N}) \sum_{n=0}^{N/2-1} 2g(n)\cos(\frac{\pi k(2n+1)}{N}) \\ &+ 2\sin(\frac{\pi k}{2N}) \sum_{n=0}^{N/2-1} 2h(n)\sin(\frac{\pi k(2n+1)}{N}) \end{aligned} \tag{3.7}$$

or

$$C(k) = 2\cos(\frac{\pi k}{(2N)})C_g(k) + 2\sin(\frac{\pi k}{2N})S_h(k) \tag{3.8}$$

where $C_g(k)$ and $S_h(k)$ are the $N/2$–point DCT and DST of $g(n)$ and $h(n)$, respectively. Fig. 3.1 shows the two–band decomposition of the SB–DCT.

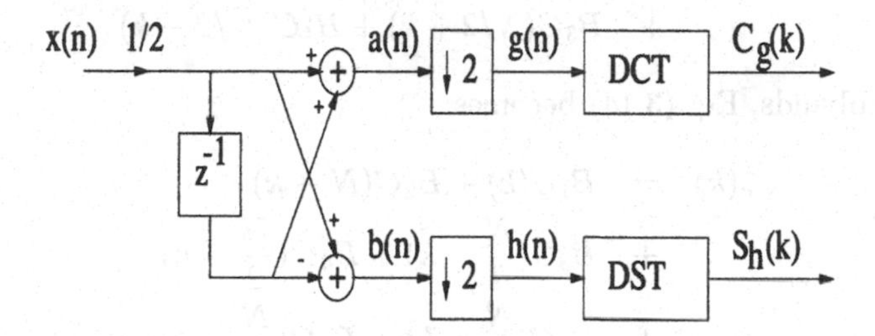

FIGURE 3.1. Two–band decomposition of the subband DCT

3.2.1 *Approximation and Generalization*

Eq. (3.8) can be approximated by calculating only the first term, corresponding to the low–frequency band as:

$$\hat{C}(k) = 2\cos(\frac{\pi k}{2N})C_g(k) \quad k \in \{0, 1, ..., \frac{N}{2} - 1\} \tag{3.9}$$

If the high–frequency band is to be calculated only, Eq. (3.8) becomes:

$$\hat{C}(k) = 2\sin(\frac{\pi k}{2N})S_h(k) \quad k \in \{\frac{N}{2}, \frac{N}{2} + 1, ..., N - 1\} \tag{3.10}$$

Indeed two main types of errors appear during the approximation: linear distortions and aliasing. Combining the two Eqs. (3.8), (3.9) and relating $S_h(k)$ with $C(N-k)$, we obtain:

$$\frac{2\hat{C}(k)}{\cos(\frac{\pi k}{2N})} = C(k) - \frac{\cos(\frac{\pi(N-k)}{2N})}{\cos(\frac{\pi k}{2N})}C(N-k) \tag{3.11}$$

The left–hand side of this equation is the approximated transform after a compensation of the linear distortion . The second term in the right–hand side is due to the aliasing error created by non–zero transform points $C(N-k)$. The decomposition process can be also repeated to the two sequences $g(n)$ and $h(n)$ in Eq. (3.5) and the same procedure can be followed to get

$$\begin{aligned} \hat{C}(k) &= 4\cos(\frac{\pi k}{2N})\cos(\frac{\pi k}{N})C_{gg}(k) + 4\cos(\frac{\pi k}{2N})\sin(\frac{\pi k}{N})S_{gh}(k) \\ &+ 4\sin(\frac{\pi k}{2N})\cos(\frac{\pi k}{N})S_{hg}(k) - 4\sin(\frac{\pi k}{2N})\sin(\frac{\pi k}{N})C_{hh}(k) \end{aligned} \tag{3.12}$$

$C_{gg}(k)$ and $C_{hh}(k)$ represent two $(N/4)$–point DCT's, while $S_{gh}(k)$ and $S_{hg}(k)$ represent two $(N/4)$–point DST's. Calculating only the first term of Eq. (3.12) yields:

$$\hat{C}(k) = 4\cos(\frac{\pi k}{2N})\cos(\frac{\pi k}{N})C_{gg}(k) \tag{3.13}$$

where $C_{gg}(k)$ is the discrete cosine transform of the low–low frequency band. For $m = 2$ decomposition stages, we obtain:

$$\begin{aligned} \hat{C}_c(k) &= B_1C(k) + B_2C(N-k) \\ &+ B_3C(N/2+k) + B_4C(N/2-k) \end{aligned} \tag{3.14}$$

For $M = 8$ subbands, Eq. (3.14) becomes:

$$\begin{aligned} \hat{C}_c(k) &= B_1C(k) + B_2C(N-k) \\ &+ B_3C(\frac{N}{2}-k) + B_4C(\frac{N}{2}+k) \\ &+ B_5C(\frac{N}{4}-k) + B_6C(\frac{N}{4}+k) \\ &+ B_7C(\frac{3N}{4}-k) + B_8C(\frac{3N}{4}+k) \end{aligned} \tag{3.15}$$

Generally, in addition to the first two coefficients multiplied by $C(k)$ and $C(N-k)$ there are $M-2$ aliasing coefficients, which are multiplied by the transform points $C(l)$, with $l = (\frac{2rN}{M} \pm k)$, where $r \in \{1, 2, ..., M/2-1\}$. Those coefficients B_j can be shown to be:

$$B_j = s(l)\prod_{i=1}^{i=m}\frac{f_i(\pi il/(2N))}{f_i(\pi ik/(2N))}(-1)^{t(i)(j+1)} \tag{3.16}$$

where

$$\begin{aligned} s(l) &= 1 \qquad \text{for even r} \\ &= -1 \qquad \text{for odd r} \end{aligned} \tag{3.17}$$

with $f_i(u) = \cos(u)$ and $t(i) = 0$ for low–pass decomposition in stage i and $f_i(u) = \sin(u)$ and $t(i) = 1$ for high–pass decomposition in stage i. For $N = 16$ and $m = 2$ and if only one band is calculated, Table 3.1 shows the aliasing effects of points $C(N-k)$ and $C(\frac{N}{2}+k)$ and $C(\frac{N}{2}-k)$ on points $C(k)$, assuming that the ratio of the transform points causing aliasing to the true components in the calculated band is fixed to 0.1.

Calculated Points	Effect of $C(N-k)$	Effect of $C(\frac{N}{2}+k)$	Effect of $C(\frac{N}{2}-k)$
$C(0)$	0	0	0
$C(1)$	−0.0098	+0.0127	−0.0155
$C(2)$	−0.0199	+0.0235	−0.0351
$C(3)$	−0.0303	+0.0329	−0.0616
$C(4)$	−0.0414	0	0
$C(5)$	−0.0535	−0.022	+0.0725
$C(6)$	−0.0668	−0.0097	+0.0489
$C(7)$	−0.0821	−0.0025	+0.0256
$C(8)$	0	0	0
$C(9)$	+0.0821	−0.0025	−0.0256
$C(10)$	+0.0668	−0.0097	−0.0489
$C(11)$	+0.0535	−0.022	−0.0725
$C(12)$	+0.0414	0	0
$C(13)$	+0.0303	+0.0329	+0.0616
$C(14)$	+0.0199	+0.0235	+0.0351
$C(15)$	+0.0098	+0.0127	+0.0155

TABLE 3.1. Aliasing effects of other bands on the calculated band for $N = 16$ and $M = 4$

3.2.2 Complexity Results

The approximated SB–DCT is compared with the exact DCT [7] in terms of program running times. Fig. 3.2 shows this comparison (done using a 486 processor with 66 MHz) for different numbers of input points N and for all possible m. The values of the execution–time corresponding to $m = 0$ are the times required by the full–band DCT of length N.

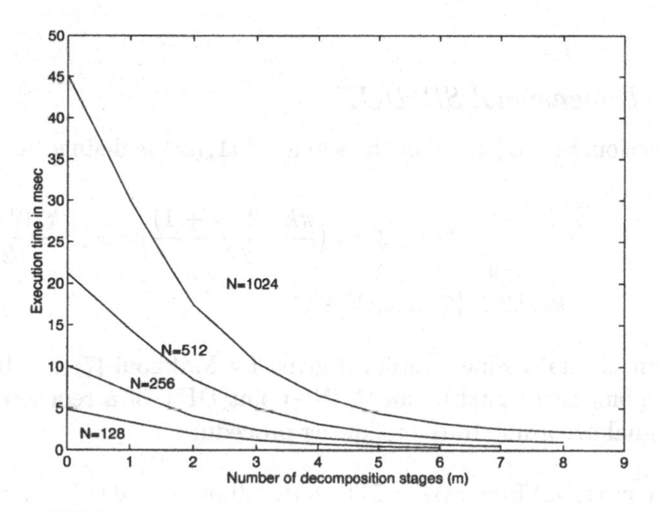

FIGURE 3.2. Running time comparison of SB–DCT

3.2.3 Adaptive SB–DCT

If there is no a-priori information about the concentration of the signal energy in the different frequency bands, a simple algorithm used for SB–DFT computations [8] can be included also in the SB–DCT. A comparison between the energy of the low- and high-frequency subsequences $g(n)$ and $h(n)$ given by Eq. (3.5) is performed by finding:

$$sgn(B) = sgn \sum_{n=0}^{N/2-1} |g(n)| - |h(n)| \tag{3.18}$$

According to $sgn(B)$, the decision will be taken: If B is positive, the low-frequency band will be calculated, and if B is negative, the high-frequency band will be calculated. Fig. 3.3 shows the effect of the adaptive algorithm on the execution time of the SB–DCT method, for $N = 128$ and for varying m.

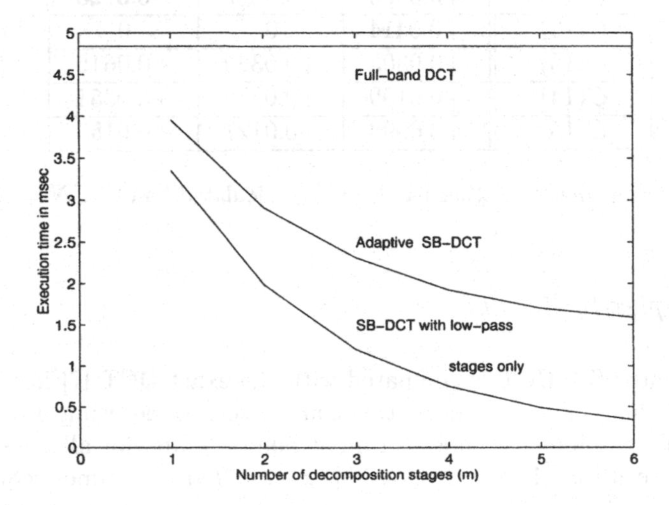

FIGURE 3.3. Running time comparison of 1–D DCT methods

3.2.4 Two–Dimensional SB–DCT

The two-dimensional (2–D) DCT of the signal $x(n1, n2)$ is defined as [7]:

$$C(k1,k2) = \sum_{n1=0}^{N-1}\sum_{n2=0}^{N-1} 4x(n1,n2)\cos(\frac{\pi k1(2\,n1+1)}{2N})\ \cos(\frac{\pi k2(2\,n2+1)}{2N})$$
$$k1, k2 \in \{0, 1, ..., N-1\} \tag{3.19}$$

A fast two–dimensional cosine transform given by Makhoul [7] can be computed for an (N*N)–point real signal by an (N*N)–point DFT of a reordered version of the original signal according to the following procedure:

1. Compute $v(n1, n2)$ from $x(n1, n2)$ by a two dimensional extension of Eq. (3.3)

$$v(n) = x(2n1, 2n2) \qquad 0 \le n1 \le [\frac{N-1}{2}],\quad 0 \le n2 \le [\frac{N-1}{2}]$$

$$
\begin{aligned}
&= x(2N-2n1-1, 2n2) \quad [\frac{N+1}{2}] \le n1 \le N-1, \quad 0 \le n2 \le [\frac{N-1}{2}] \\
&= x(2n1, 2N-2n2-1) \quad 0 \le n1 \le [\frac{N-1}{2}], \quad [\frac{N+1}{2}] \le n2 \le N-1 \\
&= x(2N-2n1-1, 2N-2n2-1) \quad [\frac{N+1}{2}] \le n1 \le N-1, \\
&\qquad [\frac{N+1}{2}] \le n2 \le N-1
\end{aligned}
$$

2. Find the 2–D DFT $V(k1, k2)$ of $v(n1, n2)$
3. Find $C(k1, k2)$ from:

$$
\begin{aligned}
C(k1,k2) &= 2Real\{W_{4N}^{k1}[W_{4N}^{k2}V(k1,k2) \\
&+ W_{4N}^{-k2}V(k1,N-k2)]\}
\end{aligned} \tag{3.20}
$$

The computation of the 2–D DCT using a separable subband decomposition in both spatial variables needs to calculate the following four subsequences:

$$
\begin{aligned}
x(2n1,2n2) &= g_{ll}(n1,n2) + g_{lh}(n1,n2) + g_{hl}(n1,n2) + g_{hh}(n1,n2) \\
x(2n1,2n2+1) &= g_{ll}(n1,n2) - g_{lh}(n1,n2) + g_{hl}(n1,n2) - g_{hh}(n1,n2) \\
x(2n1+1,2n2) &= g_{ll}(n1,n2) + g_{lh}(n1,n2) - g_{hl}(n1,n2) - g_{hh}(n1,n2) \\
x(2n1+1,2n2+1) &= g_{ll}(n1,n2) - g_{lh}(n1,n2) - g_{hl}(n1,n2) + g_{hh}(n1,n2)
\end{aligned} \tag{3.21}
$$

Following the same procedure used for the one–dimensional signal, an approximate SB–DCT of the low–low (low–pass filtered in both directions) band after compensating the linear distortion is found to be:

$$
\begin{aligned}
\frac{4\hat{C}(k1,k2)}{\cos(\frac{\pi k1}{2N})\cos(\frac{\pi k2}{2N})} &= C(k) - \frac{\cos(\frac{\pi(N-k1)}{2N})\cos(\frac{\pi(N-k2)}{2N})}{\cos(\frac{\pi k1}{2N})\cos(\frac{\pi k2}{2N})} C(N-k1,N-k2) \\
&+ \frac{\cos(\frac{\pi(N-k1)}{2N})\cos(\frac{\pi(k2)}{2N})}{\cos(\frac{\pi k1}{2N})\cos(\frac{\pi k2}{2N})} C(N-k1,k2) + \frac{\cos(\frac{\pi(k1)}{2N})\cos(\frac{\pi(N-k2)}{2N})}{\cos(\frac{\pi k1}{2N})\cos(\frac{\pi k2}{2N})} C(k1,N-k2)
\end{aligned} \tag{3.22}
$$

The first term on the right side of Eq. (3.22) is the exact transform while the other three terms are due to aliasing. Figure 3.4 shows the running time comparison of the 2–D full–band DCT for an input signal of size (128∗128) with the 2–D SB–DCT (with its adaptive version) for different values of m.

3.3 Applications

The DFT–based real and complex cepstrum (*RFC* and *CFC*) of a signal x are defined as:

$$
\begin{aligned}
RFC &= Real(IFFT(ln(Abs(FFT(x))))) \\
CFC &= Real(IFFT(ln(FFT(x))))
\end{aligned} \tag{3.23}
$$

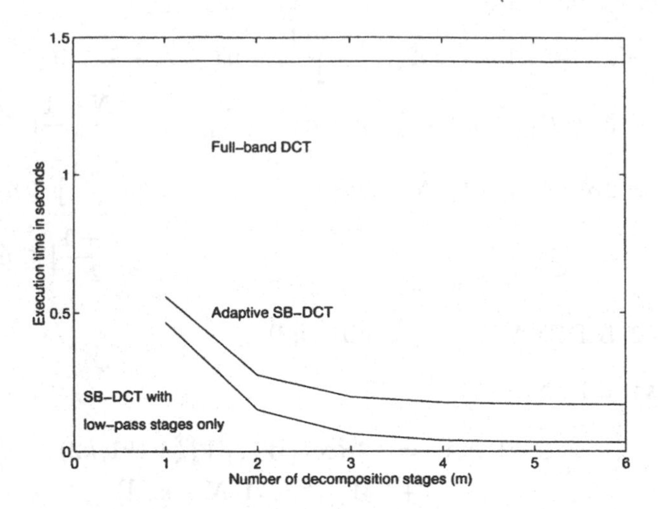

FIGURE 3.4. Running time comparison of 2–D DCT methods

In [10], it is shown that using the DCT instead of the FFT does not degrade the information contained in the cepstrum while substantially reducing the computational complexity, so the DCT–based cepstrum (real or complex, meaning the same in case of the DCT) (termed *RCC* and *CCC*, respectively) according to [10] are

$$RCC = CCC = Real(IDCT(ln(Abs(DCT(x))))) \tag{3.24}$$

In this work the SB–DCT is used instead of the full–band DCT in computation of both the real DCT–based cepstrum and the complex DCT–based cepstrum. So Eq. (3.24) can be changed to

$$RSCC = CSCC = Real(IDCT(ln(Abs(SB\text{–}DCT(x))))) \tag{3.25}$$

Comparing this equation with the corresponding DCT–based cepstrum Eq. (3.24), reduction of computational complexity is caused by the following facts:

1. A SB–DCT is calculated instead of a full–band DCT;
2. a smaller–size IDCT is needed instead of the full–band IDCT;
3. the *ln* and *Abs* functions are computed for smaller–size sequences.

3.3.1 Approximate Speech Cepstrum

The real cepstrum of a voiced speech segment contains impulses at the multiples of the pitch period, while an unvoiced speech cepstrum contains no such impulses [9]. Figure 3.5 shows the *RCC* and the *RSCC* for two different speech segments (voiced and unvoiced) of a signal sampled at 16 kHz. In all cases the speech signal is windowed by a Hamming window of the wanted segment size before applying the DCT or the SB–DCT. We can conclude from this figure that the SB–DCT determines correctly the mode of excitation. The pitch period is seen to be the same for the voiced segment by applying either *RCC* or *RSCC*.

3.3.2 Approximate Echo Detection

The SB–DCT is applied in [6] instead of the full–band DCT in computation of the complex cepstrum and applied in the problem of echo detection. In this work the

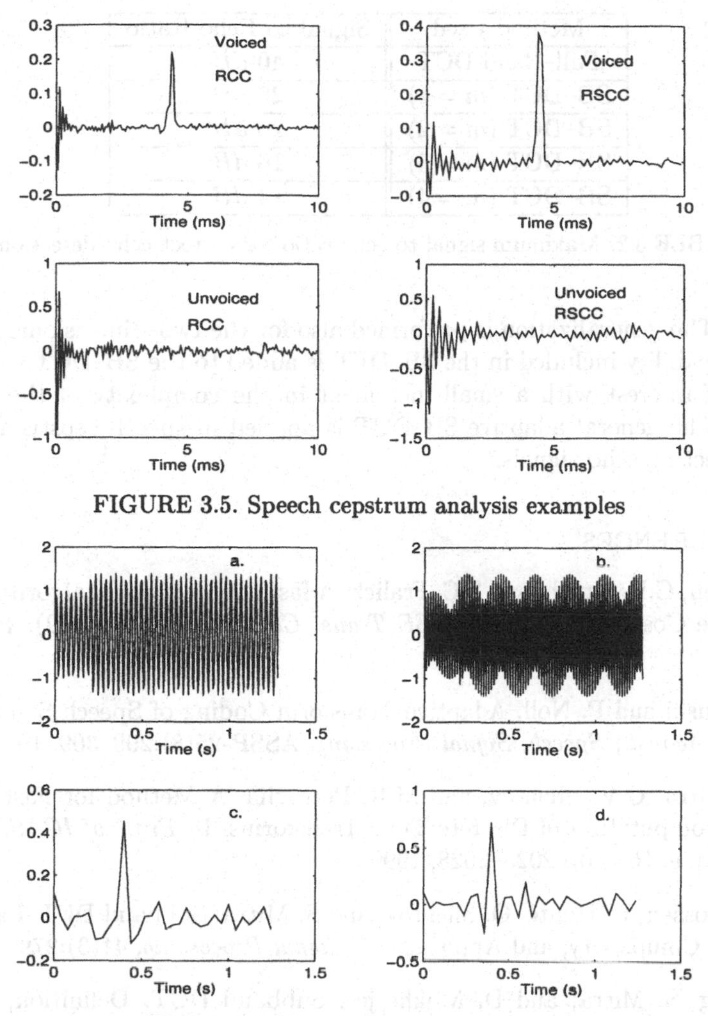

FIGURE 3.5. Speech cepstrum analysis examples

FIGURE 3.6. Echo detection examples

same simulation test given by [11] and used in [6] is repeated using the general adaptive SB–DCT algorithm. Results of such an application of the SB–DCT are shown in Fig. 3.6 for different frequency bands. Fig. 3.6.a and Fig. 3.6.b show an echo signal with amplitude 0.5 and at a position of 0.4 seconds after the beginning of the signal. The signal frequencies in those figures are 35 Hz and 65 Hz respectively with a sampling frequency of 200 Hz. Fig. 3.6.c and Fig. 3.6.d are the results of echo detection of Fig. 3.6.a and Fig. 3.6.b respectively.

Table 3.2 shows the efficiency of the SB–DCT cepstra in detecting echo signals in terms of the maximum signal to echo ratio or in other words in terms of the minimum detectable echo for different numbers of decomposition stages.

3.4 Conclusion

A fast and approximate subband–DCT method is investigated in this paper. The method is analyzed for the general case in which any band out of M subbands is

Method Used	Signal to Echo Ratio
Full–Band DCT	40 dB
SB–DCT ($m = 1$)	26 dB
SB–DCT ($m = 2$)	20 dB
SB–DCT ($m = 3$)	16 dB
SB–DCT ($m = 4$)	14 dB

TABLE 3.2. Maximum signal to echo ratio for correct echo detection

computed. The generalization is performed also for the two–dimensional case. The adaptive capability included in the SB–DFT is added to the SB–DCT to calculate the band of interest with a small increment in the complexity of the SB–DCT algorithm. The general adaptive SB–DCT is applied in speech cepstrum analysis and for detecting echo signals.

3.5 References

[1] W. Chen, C.H. Smith, and S.C. Fralick. A fast computational algorithm for the Discrete Cosine Transform. *IEEE Trans. Commun.*, COM-25(9): 1004–1009, 1977.

[2] R. Zelinski and P. Noll. Adaptive Transform Coding of Speech Signals. *IEEE Trans. Acoust., Speech, Signal Processing*, ASSP-25(8):299–309, 1977.

[3] S.K. Mitra, O.V. Shentov, and M.R. Petraglia. A Method for Fast Approximate Computation of Discrete Time Transforms. In *Proc. of ICASSP'90*, Albuquerque, NM, pp.2025–2028, 1990.

[4] A.N. Hossen, U. Heute, O. Shentov, and S. Mitra. Subband DFT–Part II: Accuracy, Complexity, and Applications. *Signal Processing*, 41(3):279–294, 1995.

[5] S. Jung, S. Mitra, and D. Mukherjee. Subband DCT: Definition, Analysis, and Applications. *IEEE Trans. on Circuits and Systems for Video Technology*, 6(3), 1996.

[6] A.N. Hossen and U. Heute. Fast Approximate DCT: Basic–Idea, Error Analysis, Applications. In *Proceedings of ICASSP-97*, Munich, Germany, 1997.

[7] J. Makhoul. A Fast Cosine Transform in One and Two Dimensions. *IEEE Transaction Vol.*, ASSP-28(1):27–34, 1980.

[8] A.N. Hossen and U. Heute. Fully Adaptive Evaluation of the Sub–Band DFT. In *Proc. of ISCAS'93*, Chicago, pp.655–658, 1993.

[9] A.V. Oppenheim and R.W. Schafer. *Discrete–Time Signal Processing*, Prentice Hall, 1989.

[10] H. Hassanein and M. Rudko. On the Use of Discrete Cosine Transform in Cepstral Analysis. *IEEE Trans.*, ASSP–32(4), 1984.

[11] Matlab Signal Processing Toolbox, The MathWorks, 1994.

4

Higher-Order Statistics in Signal Processing

Asoke K. Nandi[1]

ABSTRACT
First some definitions and properties with respect to higher-order statistics are highlighted to throw some light on the reasons for strong interests of the signal processing community over the last ten years in this research field. Later blind source separation, blind system identification, time-delay estimation, and blind equalization are introduced as examples of some of the current signal processing applications.

4.1 Introduction

Until the mid-1980's, signal processing - signal analysis, system identification, signal estimation problems, etc. - was primarily based on second-order statistical information. Autocorrelations and cross-correlations are examples of second-order statistics (SOS). The power spectrum which is widely used and contains useful information is again based on the second-order statistics in that the power spectrum is the one-dimensional Fourier transform of the autocorrelation function. As Gaussian processes exist and a Gaussian probability density function (pdf) is completely characterized by its first two moments, the analysis of linear systems and signals has so far been quite effective in many circumstances. It has nevertheless been limited by the assumptions of Gaussianity, minimum phase systems, linear systems, etc.

When signals are non-Gaussian the first two moments do not define their pdf and consequently higher- order statistics (HOS), namely of order greater than two, can reveal other information about them than SOS alone can. Ideally the entire pdf is needed to characterize a non-Gaussian signal. In practice this is not available but the pdf may be characterized by its moments. It should however be noted that some distributions do not possess finite moments of all orders. As an example, Cauchy distribution, defined as

$$p(x) = \frac{1}{\pi\beta} \quad \frac{1}{1+(\frac{x-\alpha}{\beta})^2} \qquad -\infty < x < \infty \tag{4.1}$$

has all its moments, including the mean, undefined. Also some distributions give rise to finite moments but these moments do not uniquely define the distributions.

[1]Signal Processing Division, Department of Electronic and Electrical Engineering, University of Strathclyde, 204 George Street, Glasgow, G1 1XW, United Kingdom, E-mail: asoke@eee.strath.ac.uk

For example, the log-normal distribution is not determined by its moments [16]. As an example of the fact that different distributions can have the same set of moments, consider

$$dF(x) = \gamma \exp(-\alpha x^{\lambda}) \left\{1 + \epsilon \sin(\beta x^{\lambda})\right\} \; dx \tag{4.2}$$

for $0 \leq x \leq \infty$, $\alpha > 0$, $0 < \lambda < 1/2$, and $|\epsilon| < 1$. The interesting thing about this set of distributions (obtained for different values of ϵ) is that they all have the same set of moments for all allowed values of ϵ in the range $|\epsilon| < 1$ [32] because

$$\int_0^{\infty} x^n \exp(-\alpha x^{\lambda}) \sin(\beta x^{\lambda}) \; dx = 0 \tag{4.3}$$

Thus it is clear that the moments, even when they exist for all orders, do not necessarily determine the pdf completely. Only under certain conditions will a set of moments determine a pdf uniquely. It is rather fortunate that these conditions are satisfied by most of the distributions arising commonly. For practical purposes, the knowledge of moments may be considered equivalent to the knowledge of the pdf. Thus distributions that have a finite number of the lower moments in common will, in a sense, be close approximations to each other. In practice, approximations of this kind often turn out to be remarkably good, even when only the first three or four moments are equated [30].

Since the mid-1980's there has been a tremendous growth in HOS research activities and publications in the signal processing community. Almost all of these investigations in HOS require moments or cumulants of orders up to four or their spectra. Advantages of the analysis of systems or signals using HOS over SOS have been highlighted in many places (for example, see extensive references in [34]). Among its many benefits is its ability to recognize Gaussian and non-Gaussian signals, linear and non-linear systems, minimum and non-minimum phase systems, (quadratic, cubic, etc.) phase coupling, higher-order time-frequency distributions, etc. As cumulants of order three or higher of Gaussian noise vanish completely, the bispectrum of a non-Gaussian signal and an additive, uncorrelated Gaussian noise will filter out the Gaussian noise part and consequently will represent only third-order cumulant sequence of the signal. For excellent overviews and tutorials on higher-order statistics in the context of signal processing as well as in their relation to second-order statistics, see ref. [34]. Recently special issues have appeared on the Applications of Higher Order Statistics in the Journal of the Franklin Institute [24] and on Higher-Order Statistics in the EURASIP journal of Signal Processing [33] and a book on the subject has also been published [28].

4.2 Moments and Cumulants

4.2.1 Definitions

Let the cumulative distribution function (cdf) of x be denoted by $F(x)$. The central moment (about the mean) of order ν of x is defined by

$$\mu_{\nu} = \int_{-\infty}^{\infty} (x - m)^{\nu} \; dF \tag{4.4}$$

for $\nu = 1, 2, 3, 4, ...$, where m, the mean of x, is given by $\int_{-\infty}^{\infty} x dF$, $\mu_0 = 1$ and $\mu_1 = 0$. As noted earlier, not all distributions have finite moments of all orders; for example, the Cauchy distribution belongs to this class. In the following it is assumed that distributions are zero-mean. One can also introduce the characteristic function, for real values of t,

$$\phi(t) = \int_{-\infty}^{\infty} \exp(jtx) \; dF = \sum_{\nu=0}^{\infty} \mu_\nu (jt)^\nu / \nu! \tag{4.5}$$

where $j = \sqrt{-1}$ and μ_ν is the moment of order ν about the origin. Hence coefficients of $(jt)^\nu/\nu!$ in the power series expansion of the $\phi(t)$ represent moments. Moments are thus one set of descriptive constants of a distribution. In general, moments may not completely determine the distribution even when moments of all orders exist. For example, the log-normal distribution is not uniquely determined by its moments.

Cumulants make up another set of descriptive constants. If one were to express $\phi(t)$ as follows,

$$\phi(t) = \int_{-\infty}^{\infty} \exp(jtx) \; dF = \exp\left\{ \sum_{\nu=1}^{\infty} C_\nu (jt)^\nu / \nu! \right\} \tag{4.6}$$

then the C_ν's are the cumulants of x and these are the coefficients of $(jt)^\nu/\nu!$ in the power series expansion of the natural logarithm of $\phi(t)$, $\ln \phi(t)$. The cumulants, except for the C_1, are invariant under the shift of the origin, a property that is not shared by the moments.

Cumulants and moments are different though clearly related (as seen through the characteristic function). Cumulants are not directly estimable by summatory or integrative processes, and to find them it is necessary either to derive them from the characteristic function or to find the moments first. For zero-mean distributions, the first three central moments and the corresponding cumulants are identical but they begin to differ from order four - i.e. $C_1 = \mu_1 = 0$, $C_2 = \mu_2$, $C_3 = \mu_3$, and $C_4 = \mu_4 - 3\mu_2^4$. For zero- mean Gaussian distributions, $C_1 = 0$ (zero-mean), $C_2 = \sigma^2$ (variance), and $C_\nu = 0$ for $\nu > 2$. On the other hand for Poisson distributions, $C_\nu = \lambda$ (mean) for all values of ν.

For a zero-mean, real, stationary time-series $\{x(k)\}$ the second-order moment sequence (autocorrelations) is defined as

$$M_2(k) = M_{xx}(k) = \mathcal{E}\{x(i)x(i+k)\} \tag{4.7}$$

where $\mathcal{E}\{.\}$ is the expectation operator and i is the time index. In this case the second- order cumulants, $C_2(k)$, are the same as $M_2(k)$, i.e. $C_2(k) = C_{xx}(k) = M_2(k) \quad \forall \quad k$. The third-order moment sequence is defined by

$$M_3(k,m) = M_{xxx}(k,m) = \mathcal{E}\{x(i)x(i+k)x(i+m)\} \tag{4.8}$$

and again $C_3(k,m) = C_{xxx}(k,m) = M_3(k,m) \quad \forall \quad k,m$ where $C_3(.,.)$ is the third-order cumulant sequence. The fourth-order moment sequence is defined as

$$M_4(k,m,n) = M_{xxxx}(k,m,n) = \mathcal{E}\{x(i)x(i+k)x(i+m)x(i+n)\} \tag{4.9}$$

and the fourth-order cumulants are

$$C_4(k,m,n) = C_{xxxx}(k,m,n)$$
$$= M_4(k,m,n) - C_2(k)C_2(m-n) - C_2(m)C_2(k-n) - C_2(n)C_2(m-k)$$

As can be seen the fourth-order moments are different from the fourth-order cumulants.

4.2.2 *Moment and Cumulant Estimation*

In practice, a finite number of data samples are available - $\{x(i), i = 1, 2, \ldots, N\}$. These are assumed to be samples from a real, zero-mean, stationary process. The sample estimates at second-order are given by

$$\hat{M}_2(k) = \frac{1}{N_3} \sum_{i=N_1}^{N_2} x(i)x(i+k) \tag{4.10}$$

and

$$\hat{C}_2(k) = \hat{M}_2(k) \tag{4.11}$$

where $|k| < N$, $N_1 = \begin{cases} 1 & , \text{if } \ k \geq 0 \\ -k+1 & , \text{if } \ k < 0 \end{cases}$, $N_2 = \begin{cases} N-k & , \text{if } \ k \geq 0 \\ N & , \text{if } \ k < 0 \end{cases}$
If N_3 is set to the actual number of terms in the summation, namely $(N_2 - N_1 + 1)$, unbiased estimates are obtained. Usually N_3 is set to N, the number of data samples, to obtain asymptotically unbiased estimates. Similarly sample estimates of third-order moments and cumulants are given by

$$\hat{M}_3(k,m) = \frac{1}{N_3} \sum_{i=N_1}^{N_2} x(i)x(i+k)x(i+m) \tag{4.12}$$

and

$$\hat{C}_3(k,m) = \hat{M}_3(k,m) \tag{4.13}$$

where N_1 and N_2 take up different values from those in the second-order case. Such estimates are known to be consistent under some weak conditions. For large sample numbers N, the variance of the third-order cumulants can be expected as follows

$$variance\{C_3(k,m)\} \propto \frac{1}{N} \tag{4.14}$$

Finally the fourth-order moments are estimated by

$$\hat{M}_4(k,m,n) = \frac{1}{N_3} \sum_{i=N_1}^{N_2} x(i)x(i+k)x(i+m)x(i+n) \tag{4.15}$$

where N_1 and N_2 take up different values from those in the second-order as well as third-order cases and the fourth-order cumulants can be written as

$$\hat{C}_4(k,m,n) = \hat{M}_4(k,m,n) - \hat{M}_2(k)\hat{M}_2(m-n) - \hat{M}_2(m)\hat{M}_2(k-n) - \hat{M}_2(n)\hat{M}_2(m-k) \tag{4.16}$$

As these assume that the processes are zero-mean, in practice the sample mean is removed before calculating moments and cumulants. Mean square convergence and asymptotic normality of the sample cumulant estimates under some mixing conditions are given in [4].

Thus standard estimation method evaluates third-order moments as

$$\begin{aligned}\hat{M}_3(k,m) &= \frac{1}{N_3}\sum_{i=N_1}^{N_2} x(i)x(i+k)x(i+m) \\ &= \frac{1}{N_3}\sum_{i=N_1}^{N_2} z_{k,m}(i) \end{aligned} \qquad (4.17)$$

where $z_{k,m}(i) \equiv x(i)x(i+k)x(i+m)$. This last formulation demonstrates that the standard evaluation employs the mean estimator (of $z_{k,m}(i)$). Sometimes the time-series data are segmented and the set of required cumulants in each of these segments is estimated separately using the mean estimator, and then for the final estimate of a cumulant the mean of the same is calculated over all the segments. Accuracy of methods based on higher-order cumulants depends on, among others, the accuracy of estimates of the cumulants. By their very nature, estimates of third-order cumulants of a given set of data samples tend to be more variable than the autocorrelations (second-order cumulants) of the data. Any error in the values of cumulants estimated from finite segments of a time-series will be reflected as larger variance in other higher-order estimates.

Numerous algorithms employing HOS have been proposed for applications in areas such as array processing, blind system identification, time-delay estimation, blind deconvolution and equalization, interference cancellation, etc. Generally these use higher-order moments or cumulants of a given set of data samples. One of the difficulties with HOS is the increased computational complexity. One reason is that, for a given number of data samples, HOS computation requires more multiplications than the corresponding SOS calculation.

Another important reason lies in the fact that, for a given number of data samples, variances of the higher-order cumulant estimates are generally larger than that of the second-order cumulant estimates. Consequently, to obtain estimates of comparable variance, one needs to employ a greater number of samples for HOS calculations in comparison to SOS calculations. Using a moderate number of samples, standard estimates of such cumulants are of comparatively high variance and to make these algorithms practical one needs to obtain some lower variance sub-asymptotic estimates.

Recently the problem of robust estimation of second and higher-order cumulants has been addressed [1, 2, 17, 18, 23]. The mean estimator is the one utilized in applications to date. A number of estimators including the mean, median, biweight, and wave were compared using random data. It has been argued that, for not too large number of samples, the mean estimator is not optimal and this has been supported by extensive simulations. Also were developed some generalized trimmed mean estimators for moments and these appear to perform better than the standard estimator in small number of samples in simulations as well as in the estimates of the bispectrum using real data [18, 23]. Another important issue relating to the effects of finite register length (quantization noise) on the cumulant estimates are being considered (see, for example, [17]).

4.2.3 Spectral Estimation

Let a zero-mean, real, stationary time-series $\{x(k)\}$ represent the observed signal. It is well known that the power spectrum of this signal can be defined as the one-dimensional Fourier transform of the autocorrelations (second-order cumulants) of the signal. Therefore,

$$S_2(\omega_1) = \sum_m C_2(m) \exp\{-j\omega_1 m\} \tag{4.18}$$

where

$$C_2(m) = \mathcal{E}\{x(k)x(k+m)\} \tag{4.19}$$

and ω_1 is the frequency.

Similarly, the bispectrum (based on the third-order statistics) of the $x(k)$ can be defined as the two-dimensional Fourier transform of the third-order cumulants, i.e.

$$S_3(\omega_1, \omega_2) = \sum_m \sum_n C_3(m,n) \exp[-j(m\omega_1 + n\omega_2)] \tag{4.20}$$

where the $C_3(m,n) = \mathcal{E}\{x(k)x(k+m)x(k+n)\}$ is the third-order cumulant sequence. Correspondingly, the trispectrum (based on the fourth-order statistics) of the $\{x(k)\}$ can be defined as the three-dimensional Fourier transform of the fourth-order cumulants, i.e.

$$S_3(\omega_1, \omega_2, \omega_3) = \sum_m \sum_n \sum_l C_4(m,n,l) \exp[-j(m\omega_1 + n\omega_2 + l\omega_3)] \tag{4.21}$$

where the $C_4(m,n,l)$ is the fourth-order cumulant sequence.

However, just as the power spectrum can be estimated from the Fourier transform of the signal (rather than its autocorrelations), one can estimate the bispectrum and trispectrum from the same Fourier transform. The difference will be in the variance of the resulting estimates. It should be noted that consistent estimators of higher-order cumulant spectra via spectral windows and the limiting behaviour of certain functionals of higher-order spectra have been studied.

4.3 Pictorial Motivation for HOS

Three time-series corresponding to independent and identically distributed (i.i.d.) exponential, Gaussian and uniform random variables (r.v.) are simulated in MATLAB [21] and each of these has 4096 samples of zero mean and unit variance. Figure 4.1 shows the estimated power spectrum (the top one corresponds to exponential, the middle one to Gaussian and the bottom one to uniform) versus the frequency index, while Figure 4.2 (o corresponds to exponential, x corresponds to Gaussian and + corresponds to uniform) shows the estimated second-order cumulant versus the lag.

It is clear that in both of these two figures, which represent second-order statistical information, exponential, Gaussian and uniform r.v. are not differentiated. Figure 4.3 shows the histograms of these three sets of r.v. from which the differences are visually obvious.

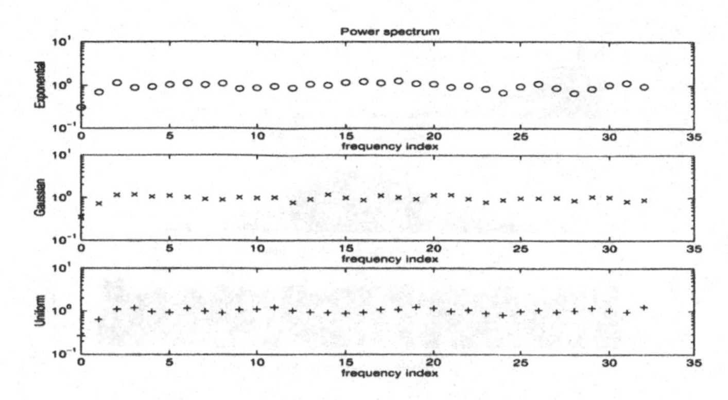

FIGURE 4.1. Power spectrum of random signals

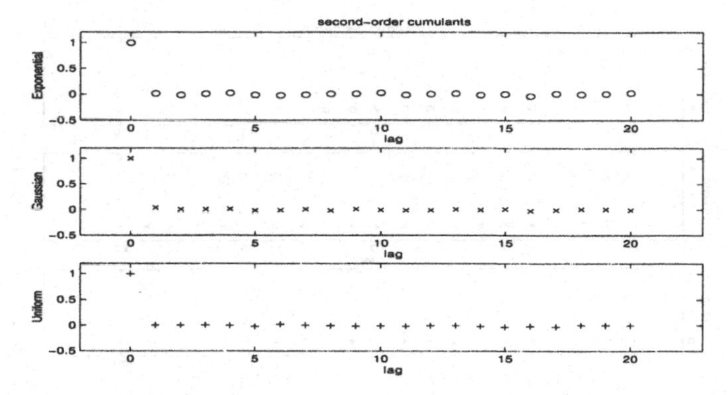

FIGURE 4.2. Autocorrelations of random signals

Figure 4.4 shows the estimated third-order cumulants, $\hat{C}_3(k,k)$, versus the lag, k. In this figure, exponential r.v. are clearly distinguished from Gaussian and uniform r.v. Figure 4.5 shows the estimated fourth-order cumulants, $\hat{C}_4(k,k,k)$, versus the lag, k. Unlike in the last figure, now it is clear that all three sets of r.v. are differentiated in this figure. The reasons for the above are obvious from Table 4.1, which records theoretical values of cumulants of up to order four for these three types of r.v. of zero mean and unit variance, and from Table 4.2, which presents estimated values of cumulants up to order four at zero lag. In particular all cumulants of i.i.d., Gaussian r.v. beyond order two are theoretically zero.

	Exponential	Gaussian	Uniform
C_1 or mean	0	0	0
$C_2(k)$	$\begin{cases} 1, & for\ k=0 \\ 0, & otherwise \end{cases}$	$\begin{cases} 1, & for\ k=0 \\ 0, & otherwise \end{cases}$	$\begin{cases} 1, & for\ k=0 \\ 0, & otherwise \end{cases}$
$C_3(k,k)$	$\begin{cases} 2, & for\ k=0 \\ 0, & otherwise \end{cases}$	0	0
$C_4(k,k,k)$	$\begin{cases} 6, & for\ k=0 \\ 0, & otherwise \end{cases}$	0	$\begin{cases} -1.2, & for\ k=0 \\ 0, & otherwise \end{cases}$

TABLE 4.1. Theoretical values of cumulants of random signals

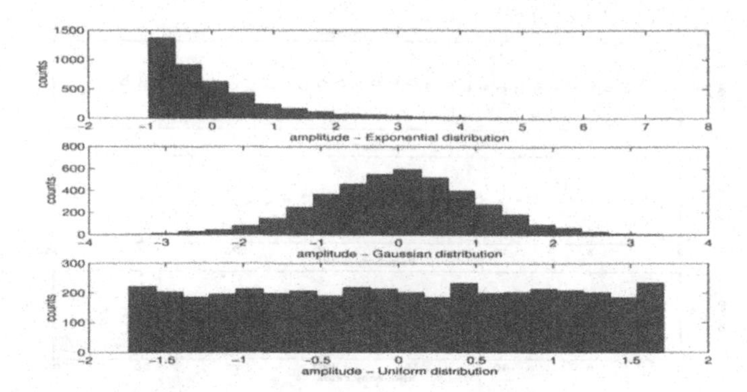

FIGURE 4.3. Histograms of random signals

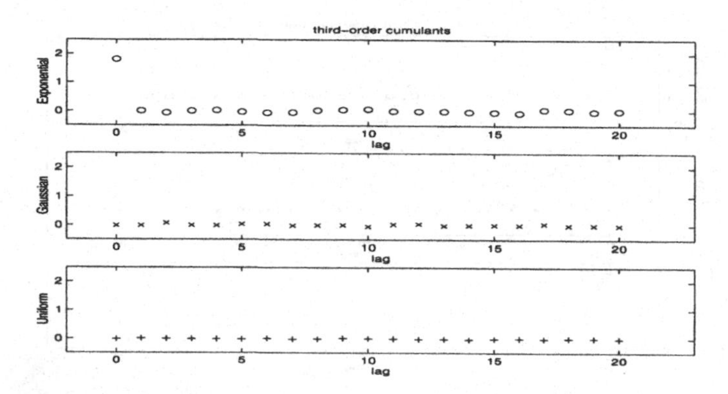

FIGURE 4.4. Third-order diagonal cumulants of random signals

4.4 Blind Source Separation

The goal of the blind source separation (BSS) is the reconstruction of a set of unknown source signals from the observation of another set of linear mixture of the sources when the mixture matrix is not known. As neither the source signals nor the linear transformation matrix from the sources to observed signals are known, this problem is referred to as BSS. In reality, the observed signals are corrupted by noise and this must be considered in the solution of the problem. This particular problem of multichannel processing arises in many signal processing areas. A classical example is the problem of separating competing speakers using multiple microphone measurements. Similar problems can be found in many applications including passive sonar array processing, multichannel data communications, radar, audio, and biomedical.

Estimated cumulant	Exponential	Gaussian	Uniform
$\hat{C}_1$	0.0	0.0	0.0
$\hat{C}_2(0)$	1.0	1.0	1.0
$\hat{C}_3(0,0)$	1.8	−0.0	−0.0
$\hat{C}_4(0,0,0)$	4.2	0.1	−1.2

TABLE 4.2. Estimated values of cumulants at zero-lag of random signals

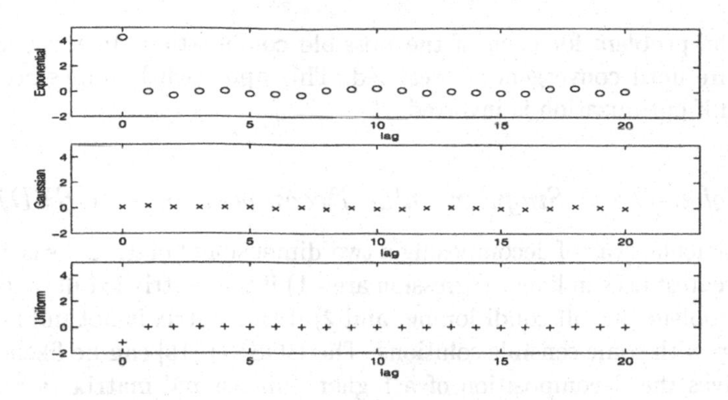

FIGURE 4.5. Fourth-order diagonal cumulants of random signals

The BSS problem, in its simplest form, can be formulated in matrix form as $Y = MX + N$, where Y represents the noisy observed signals, M is the mixing matrix, X represents the source signals, and N denotes the additive noise. The aim of BSS is to retrieve X from the knowledge of the observed signals Y only. Methods for BSS turn the assumption of the statistical independence of signals into the criterion for signal separation [7]. It can be shown that statistical independence is a sufficient condition for signal separation provided that source signals are not jointly Gaussian. Some of the problems or uncertainties with current methods are discussed below.

Several approaches - either using second-order or HOS - have been proposed to solve such problems. While some methods are based on the second-order statistical independence (i.e. just the second-order correlation being zero), other more recent methods [15, 19, 25, 35, 36, 37] are based on HOS independence and these HOS based methods require more constraints than their lower order counterparts. In general HOS based methods appear to offer better performance than SOS based methods, but most of these HOS based methods solve the problem when there are two sources and two channels (sensors or receivers) (see for example, [25, 35]). Some of the most recent methods can be found in [6, 11, 25].

4.4.1 Constrained Optimization

A number of methods depend on (constrained) optimization. As is well known, multidimensional optimization has many difficulties; this leads to less robust algorithms. Additionally, quite often these are of high computational complexity although attempts are made to overcome problems like local optima. To avoid such problems along the line of what has been achieved for the algorithm of [19], singular value decomposition (SVD) and some other previously published works have been used in [25] to provide a closed form solution which has resulted in an algorithm which is more robust as well as of less computational complexity.

4.4.2 Two Source-Sensor Scenario

Most of the algorithms solve the BSS problem when there are two sources and two sensors. These need to be extended for multiple (≥ 2) sources and sensors. In such cases one can consider all possible combinations of two source-sensor scenarios and solve the problem for each of the possible combinations in turn, and repeat the procedure until convergence is reached. This approach becomes horrendously complicated if optimization is involved.

4.4.3 Higher-Order Singular Value Decomposition (HOSVD)

SVD is a particular way of decomposing a two-dimensional matrix. As is well known two of its greatest uses in linear regression are - 1) if the matrix is full rank, it avoids many the problems like ill-conditioning, and 2) if the matrix is not full rank, it will still come up with some sensible solutions. The HOSVD [15] can be likened to SVD but it involves the decomposition of a higher dimensional matrix (a tensor, say) which naturally arises in higher-order cumulants. As BSS methods use higher-order cumulants, their connection with the HOSVD is clear. Much work on their uniqueness and consistency is yet to be done. For example, despite the multidimensionality of the fourth-order cumulant tensor, the decomposition is carried out in many steps and in each step no more than two dimensional matrices are involved. This is excellent news for systolic array implementation, but is the decomposition unique and does the order matter?

4.4.4 General Theoretical Framework

All BSS methods turn the assumption that signals are statistically independent into the criterion for signal separation and consider the cases when the linear transformation matrix M is full rank. Very recently a general theoretical framework for BSS has been considered [5] whether the matrix M is full rank or not but no particular method has been offered. This paper discusses two major issues - separability and separation principles. Separability is an intrinsic property of the linear mixture matrix M and it indicates under what conditions it is possible to separate sources. Given a set of separable sources, separation principles refer to the general principles that can be applied to perform the task. These results can be considered as "existence theorems" and based on these ideas, specific methods for the BSS will have to be developed; these methods would extend the class of soluble problems.

4.5 Blind Identification of Linear Time Invariant Systems

Parametric modelling methods using SOS are theoretically well founded and frequently practised. If the driving noise sequence is Gaussian, a non-minimum phase system will be identified incorrectly as a minimum phase system. The main motivation behind the use of HOS is to recover both the magnitude and the correct phase response of a system; therefore the system must be driven by a non-Gaussian i.i.d. noise process.

There are many blind estimation methods - adaptive as well as non-adaptive - which use only the system output data with no knowledge of the input except for it being non-Gaussian and i.i.d. Most of these can be found in ref. [34]. For Moving Average (MA) systems, there are closed form solutions, optimization methods - nonlinear least squares approach as well as linear programming approach, the Giannakis and Mendel (GM) method, and even methods based on ARMA and

AR approximations. For ARMA systems, there are methods for estimation of parameters from the system impulse response, residual time-series, the double $c(q,k)$ algorithm, the Q-slice algorithm, and those via second and fourth-order cumulant matching. Along with system parameter estimation, system order estimation is important and a few of the new methods can be found in [8, 9]. Some examples are given for MA models for which the system output, $x(n)$, can be written as

$$x(n) = \sum_{k=0}^{q} b(k)\omega(n-k) \tag{4.22}$$

where $\omega(.)$ is the system input, and $b(.)$ is the system impulse response, and q is the order of the system.

4.5.1 MA Models - Second and Third-Order Cumulants Method

In this case,

$$\begin{aligned} C_{xxx}(\tau,\tau) &= \mathcal{E}\{x(n)x(n+\tau)x(n+\tau)\} \\ &= \gamma_{3\omega}\sum_{n=0}^{q} b(n)b(n+\tau)b(n+\tau) \\ &= \gamma_{3\omega}\sum_{n} b(n)h(n+\tau) \end{aligned} \tag{4.23}$$

where $\gamma_{3\omega} = C_{\omega\omega\omega}(0,0)$ and $h(n+\tau) \equiv b(n+\tau)b(n+\tau)$. By taking z-transform, one obtains

$$\begin{aligned} C_{xxx}(z) &= \gamma_{3\omega}B(z^{-1})H(z) \\ &= \gamma_{3\omega}B(z^{-1})(B(z)*B(z)) \end{aligned} \tag{4.24}$$

Now,

$$\begin{aligned} C_{xx}(\tau) &= \mathcal{E}\{x(n)x(n+\tau)\} \\ &= \gamma_{2\omega}\sum_{n} b(n)b(n+\tau) \end{aligned} \tag{4.25}$$

where $\gamma_{2\omega} = C_{\omega\omega}(0)$ and its z-transform is given by

$$C_{xx}(z) = \gamma_{2\omega}B(z^{-1})B(z) \tag{4.26}$$

Now,

$$\begin{aligned} H(z)C_{xx}(z) &= \gamma_{2\omega}B(z^{-1})B(z)H(z) \\ &= \frac{\gamma_{2\omega}}{\gamma_{3\omega}}B(z)C_{xxx}(z) \end{aligned} \tag{4.27}$$

or,

$$H(z)C_{xx}(z) = \epsilon_{3\omega}B(z)C_{xxx}(z) \tag{4.28}$$

where $\epsilon_{3\omega} = \gamma_{2\omega}/\gamma_{3\omega}$.

In time-domain this becomes

$$\sum_{k=0}^{q} b^2(k)C_{xx}(n-k) = \sum_{k=0}^{q} \epsilon_{3\omega} b(k) C_{xxx}(n-k, n-k) \tag{4.29}$$

for $n = -q, -q+1, \ldots, 2q$. This is known as the GM equation which combines second and third-order statistics. This represents a set of $(3q+1)$ equations in $(2q+1)$ unknowns (e.g. $b^2(k)$ for $k = 1, 2, \ldots, q$ and $\epsilon_{3\omega} b(k)$ for $k = 0, 1, \ldots, q$).

In the presence of noise the observed sequence $y(n) = x(n) + g(n)$, where $g(n)$ is the additive, i.i.d. noise. For noisy observations the corresponding GM equations [28] are

$$\sum_{k=0}^{q} b^2(k)C_{yy}(n-k) = \sum_{k=0}^{q} \epsilon_{3\omega} b(k) C_{yyy}(n-k, n-k) \tag{4.30}$$

but $(q+1)$ equations involving $C_{yy}(0)$, which includes additive noise, must be eliminated. This results in a set of $2q$ simultaneous linear equations in $(2q+1)$ unknowns and they cannot be solved. Having observed this, Tugnait [28] appends

$$b(q)C_{yy}(n) = \sum_{k=0}^{q} \epsilon_{3\omega} b(k) C_{yyy}(k-n, q) \tag{4.31}$$

for $n = -q, \ldots, -1, 1, \ldots, q$. Now there are altogether $4q$ equations and still $(2q+1)$ unknowns resulting in an over-determined set of equations.

4.5.2 MA Models - Second and Fourth-Order Cumulants Method

Let us define $\epsilon_{4\omega} = \frac{\gamma_{2\omega}}{\gamma_{4\omega}}$, where $\gamma_{4\omega} \equiv C_{\omega\omega\omega\omega}(0,0,0)$. Along similar lines to the last subsection there are two sets of equations [28]

$$\sum_{k=0}^{q} b^3(k)C_{yy}(n-k) = \sum_{k=0}^{q} \epsilon_{4\omega} b(k) C_{yyyy}(n-k, n-k, n-k) \tag{4.32}$$

for $n = -q, \ldots, 2q$, while $(n-k) \neq 0$, and

$$b^2(q)C_{yy}(n) = \sum_{k=0}^{q} \epsilon_{4\omega} b(k) C_{yyyy}(k-n, q, 0) \tag{4.33}$$

for $n = -q, \ldots, q$ while $n \neq 0$. As before, there are $4q$ equations in $(2q+1)$ unknowns. There are other ideas which combine second, third and fourth-order cumulants (for example, see [20]).

4.5.3 Simulations

Consider the MA system with $b(0) = 1$, $b(1) = 0.9$, $b(2) = 0.385$ and $b(3) = -0.771$. This is a non-minimum phase system with one zero outside the unit circle on the z-plane at $z = 1.6667$. This system was excited with i.i.d., zero-mean, one-sided, negative exponential noise and the output was contaminated with i.i.d., zero-mean, Gaussian noise. Monte Carlo simulations were made to obtain one hundred

Estimated coefficient	N = 256	N = 1024
based on second and third-order cumulants		
$\hat{b}(1)$	1	1
$\hat{b}(2)$	0.878 ± 0.190	0.882 ± 0.196
$\hat{b}(3)$	0.409 ± 0.171	0.415 ± 0.180
$\hat{b}(4)$	-0.744 ± 0.134	-0.748 ± 0.138
based on second and fourth-order cumulants		
$\hat{b}(1)$	1	1
$\hat{b}(2)$	-0.240 ± 9.568	1.087 ± 1.256
$\hat{b}(3)$	1.109 ± 7.795	0.627 ± 0.984
$\hat{b}(4)$	-0.828 ± 4.255	-0.829 ± 0.564

TABLE 4.3. Results of blind identification of a MA system.

sets of such output. System parameters were estimated using second and third-order cumulants as well as using second and fourth-order cumulants. In each case, results corresponding to different lengths of the output sequence are presented in Table 4.3. In the case of second and third-order cumulants both 256-sample long output and 1024-sample long output are adequate to obtain sensible parameter estimates. However in the case of second and fourth-order cumulants, the variances of the parameter estimates are too large. Although the variance from 1024-sample is significantly smaller than the same from 256-sample, the length of the output has to be significantly increased, due to the comparatively large estimation variance of fourth-order cumulants, to make the method more effective.

4.6 Time-Delay Estimation

Applications of time-delay estimation appear in radar, sonar, biomedical engineering, flow measurement, etc. In its simplest form, there are two observed signals, $x(n)$ and $y(n)$,

$$\begin{aligned} x(n) &= s(n) + g_x(n) \\ y(n) &= A\,s(n-D) + g_y(n) \end{aligned} \tag{4.34}$$

where $s(n)$ is the zero-mean source signal, D is the delay to be estimated, A is the relative amplitude, and $g_x(n)$ and $g_y(n)$ are zero-mean additive noise in $x(n)$ and $y(n)$ respectively.

4.6.1 Cross-Correlation Based Method

The cross-correlation between $x(n)$ and $y(n)$ is given by

$$C_{xy}(\tau) = A\,C_{ss}(\tau - D) + C_{g_x g_y}(\tau) \tag{4.35}$$

where $C_{ss}(\cdot)$ is the auto-correlation of the signal $s(n)$ and $C_{g_x g_y}(\cdot)$ is the cross-correlation between the two noise signals. If the noises are uncorrelated, $C_{xy}(\tau)$ will have a peak at $\tau = D$. Hence, D can be estimated from the position of the maximum peak in $C_{xy}(\tau)$. To sharpen the peak, the estimated cross-correlation is multiplied by a window function; one popular window function is the 'Maximum-likelihood' window. This is implemented in Higher-Order Spectral Analysis Toolbox for use with MATLAB [22].

4.6.2 Cross-Cumulant Based Method

As described below, the cross-correlation based method shows a strong, undesired peak if the noise processes are correlated. Depending on the relative strength of the noise process and the underlying signal, the peak arising from the desired signal time-delay may not be detectable. As long as the noise processes are Gaussian, even when they are correlated, and the signals are non-Gaussian third-order cumulants can be used to suppress noise effects.

Let d be the maximum expected integer delay. One can write [28]

$$y(n) = \sum_{i=-d}^{+d} a(i)\, x(n-i) + g(n) \tag{4.36}$$

where

$$a(i) = \begin{cases} 0, & \text{if} \quad i \neq D \\ 1, & \text{if} \quad i = D \end{cases} \tag{4.37}$$

The third-order cumulants can be written as

$$C_{xxx}(m,n) = \mathcal{E}\{x(k)x(k+n)x(k+m)\} \tag{4.38}$$

and

$$C_{yxx}(m,n) = \mathcal{E}\{y(k)x(k+n)x(k+m)\} = \sum_{i=-d}^{+d} a(i)\, \mathcal{E}\{x(k-i)x(k+m)x(k+n)\} \tag{4.39}$$

or

$$C_{yxx}(m,n) = \sum_{i=-d}^{+d} a(i)\, C_{xxx}(m+i, n+i) \tag{4.40}$$

This set of linear equations can be solved for $a(i)$'s and the delay is the index i for which $|a(i)|$ is the largest.

4.6.3 Simulations

To demonstrate the efficiency of the above two methods, two noisy sensor signals have been generated. Before noise corruption, the signal at sensor 1 is i.i.d., zero-mean, one-sided negative exponentially distributed with variance of one and skewness of two. Similarly, before noise corruption, the signal at sensor 2 is a delayed (by 16 units of sampling time) version of the signal at sensor 1. Both sensor signals contain 1024 samples each. Figure 4.6 (the top part) shows the windowed cross-correlation as a function of the lag. There are two clear, large peaks. One peak corresponds to a delay of 10 which is the signal delay to be estimated and this estimate is correct. However, there is also another strong peak at a delay of 5 which is due to the presence of spatially correlated noise. The problem with this method lies in deciding which is the correct time-delay and in the resulting confusion. The same two sensor signals are analysed through the cross-cumulant method and the resulting coefficients are plotted in Figure 4.6 (the bottom part). There is only one clear peak in this figure and it finds the correct delay with no confusion. The essential benefit of the cross-cumulant (third-order) method over the cross-correlation (second-order) method is the 'removal' of spatially correlated additive Gaussian noise to produce an unambiguous estimate of the delay.

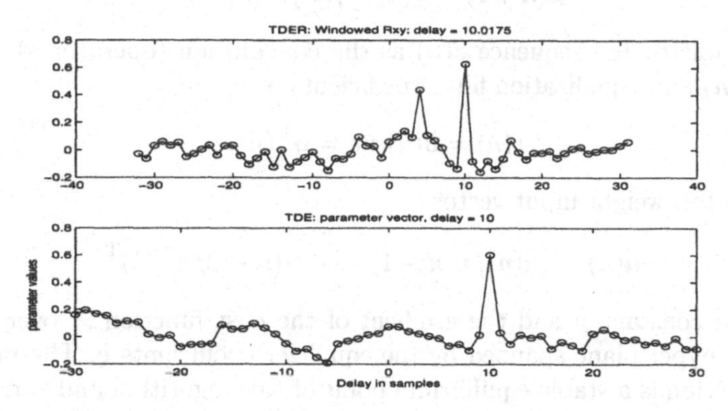

FIGURE 4.6. Time delay of random signals

4.7 Blind Equalization

Often in applications like communications, sonar, radar, etc. an unknown signal propagates through a multi-path environment of unknown transfer function. Blind equalization is the reconstruction of the original signal from a multi-path version of it. Clearly some distinguishable characteristics of either the original signal or the transmission channel are necessary to carry out the equalization uniquely. Most of the algorithms proposed in the past tend to deal with specific applications where some properties of the two signals are known. Algorithms, which do not require an initial or pre-arranged training sequence in the data stream, are called ***blind equalization*** algorithms. There are other applications of blind deconvolution like recovering sparse, reflectivity sequences from seismic or ultrasonic non-destructive testing data (see [26]).

Now there are many blind deconvolution algorithms based on HOS. In essence, blind deconvolution algorithms are adaptive filtering algorithms designed in such a way that they do not need the external supply of a desired response to generate the error signal in the output of the the adaptive equalization filter. The algorithm itself generates an estimate of the desired response by applying a nonlinear transformation on sequences involved in the adaptation. Depending on where the nonlinear transformation is applied, three important families of blind equalization algorithms are generated: 1) polycepstra algorithms - the nonlinearity is in the input of the adaptive filter, 2) other algorithms where the nonlinearity is inside the adaptive filter, i.e. a nonlinear filter - Volterra filter, neural network, etc., and 3) Bussgang algorithms - the nonlinearity is in the output of the adaptive filter. A great deal of work has been done in all of these areas and still more research is being carried out.

4.7.1 Bussgang Type Algorithms

The Bussgang type algorithms use, similar to the Least Mean Square, an iterative update of the equalizer coefficients $\mathbf{e}(n)$ (see Fig. 4.7) [3]:

$$\mathbf{e}(n+1) = \mathbf{e}(n) - \mu\nabla_f\mathcal{F}(z(n)) \tag{4.41}$$

with the reconstructed sequence $z(n)$ as the convolution (operator $*$) of received sequence $u(n)$ and equalization filter coefficients $\mathbf{e}$

$$z(n) = u(n) * e = \mathbf{u}^{\mathrm{T}}(n)\mathbf{e} \tag{4.42}$$

the $(M \times 1)$ tap-weight input vector

$$\mathbf{u}(n) = (u(n), u(n-1), \cdots, u(n-M+1))^{\mathrm{T}} \tag{4.43}$$

the adaption constant μ and the gradient of the cost function $\mathcal{F}$ over the multi-dimensional hyper-plane spanned by the equalizer coefficients $\mathbf{e}$. The minimum of the cost-function is a stable equilibrium point of this algorithm and corresponds to finding of the zero point of the error function ψ which is obtained after the partial differentiation of the cost function

$$\mathbf{e}(n+1) = \mathbf{e}(n) - \mu\psi\Big(\mathbf{u}(n)^{\mathrm{T}}\mathbf{e}(n)\Big)\mathbf{u}^*(n) \tag{4.44}$$

The equilibrium point for a zero updating of the filter coefficients averaged over the time (statistical expectation operator $\mathcal{E}\{\bullet\}$) is therefore described by:

$$\mathcal{E}\{\psi\Big(\mathbf{u}^{\mathrm{T}}(n)\mathbf{e}(n)\Big)\mathbf{u}^*(n)\} = 0 \tag{4.45}$$

Based on this principle exist a variety of cost functions $\mathcal{F}$ (for example, see [12, 13, 29, 31]), which are able to converge to the equilibrium. These are generalized by Godard to the Godard class with the common expression:

$$\mathcal{F}_p(z(n)) = \frac{1}{2p}\left(|z(n)|^p - R_p\right)^2 \tag{4.46}$$

The additional knowledge needed about the input signal is the modulation dependent constant R_p:

$$R_p = \frac{\mathcal{E}\{|x(n)|^{2p}\}}{\mathcal{E}\{|x(n)|^p\}}, \qquad p = 1, 2, \ldots \tag{4.47}$$

This cost function is based on the so called ***Bussgang Equilibrium*** [13]. From the last equation one obtains

- the Sato algorithm for $p = 1$, and
- the Constant Modulus Algorithm (CMA) for $p = 2$.

Further variations, like the Stop-and-Go [12] or the Benveniste-Goursat algorithm [31], based on this principle exist as well. An important fact is that the convergence to the global minimum of the cost function cannot be guaranteed [10].

4.7.2 Maximum Kurtosis Criterion

A new era in blind equalization started in 1990 with the super exponential algorithms based on the *maximum kurtosis criterion*, which was published by SHALVI and WEINSTEIN [30]. The technique uses the instantaneous second and fourth-order cumulants at zero lags of the equalizer output $z(n)$ to minimize the Inter Symbol Interference (ISI) of the System Impulse Response (SIR) (see Fig. 4.7). The fourth-order cumulant at zero lags is given by

$$C_{zzzz}(0,0,0) = \mathcal{E}\{|z(n)|^4\} - 2\left(\mathcal{E}\{|z(n)|^2\}\right)^2 - \left|\mathcal{E}\{z^2(n)\}\right|^2 \quad (4.48)$$

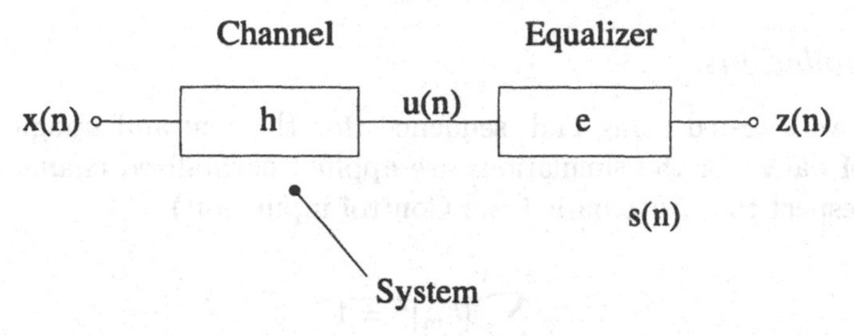

FIGURE 4.7. System concerns channel and equalizer

The constrained maximization of the fourth-order cumulant of the equalizer output leads to the unique global minimum of the cost function [30] which corresponds to the best equalization quality. JELONNEK, BOSS and KAMMEYER in [14] introduced an extra filter which has the same length as the equalizer. This reference filter is known and can be updated. As the input, $u(n)$, to this filter is known, its output $y(n)$ can be computed. Given that

$$z(n) = u(n) * e(n) = \mathbf{u}^{\mathrm{T}}(n)\mathbf{e}\,, \quad z^*(n) = \mathbf{e}^{\mathrm{H}}\mathbf{u}^*(n) \quad (4.49)$$

where $\mathbf{e}$ is the $(M \times 1)$ equalizer coefficient vector, the fourth-order cross-cumulant of $z(n)$ and $y(n)$ can be written in the form:

$$\begin{aligned} C_{zyzy}(0,0,0) \quad &= \mathbf{e}^{\mathrm{H}}\Big(\mathcal{E}\{\mathbf{u}^*(n)y^*(n)y(n)\mathbf{u}^{\mathrm{T}}(n)\} - \\ &\mathcal{E}\{\mathbf{u}^*(n)y(n)\}\mathcal{E}\{y^*(n)\mathbf{u}^{\mathrm{T}}(n)\} - \\ &-\mathcal{E}\{y^*(n)y(n)\}\mathcal{E}\{\mathbf{u}^*(n)\mathbf{u}^{\mathrm{T}}(n)\} - \\ &-\mathcal{E}\{\mathbf{u}^*(n)y^*(n)\}\mathcal{E}\{y(n)\mathbf{u}^{\mathrm{T}}(n)\}\Big)\mathbf{e} \\ &= \mathbf{e}^{\mathrm{H}}\mathbf{C}_{uyuy}\mathbf{e} \end{aligned} \quad (4.50)$$

The matrix inside the brackets, which has got the structure of a cross-cumulant matrix, may be denoted as $\mathbf{C}_{uyuy}$. Thus the fourth-order cumulant of the equalizer output depends on the equalization filter coefficients as well as the known fourth-order cross-cumulant of the observed output and the reference filter output. The application of the maximum cross-cumulant criterion can be written as:

$$\max\left|\mathbf{e}^{\mathrm{H}}\mathbf{C}_{uyuy}\mathbf{e}\right| \quad (4.51)$$

with the constraint that $\mathcal{E}\{|x(n)|^2\} = \mathcal{E}\{|z(n)|^2\}$. This is a classical eigenvalue problem (maximization of the Rayleigh coefficient). The constraint is satisfied, if

$$\mathcal{E}\{|z(n)|^2\} = \left|\mathbf{e}^{\mathrm{H}}\mathcal{E}\{\mathbf{u}(n)\mathbf{u}^{\mathrm{H}}(n)\}\mathbf{e}\right| \implies \mathcal{E}\{|x(n)|^2\} \quad (4.52)$$

The maximization, satisfying all conditions, is the generalized eigenvalue problem

$$\mathbf{C}_{uyuy}\mathbf{e} = \lambda_{\max}\mathcal{E}\{\mathbf{u}(n)\mathbf{u}^{\mathrm{H}}(n)\}\mathbf{e} \tag{4.53}$$

Simulations show, that the convergence of the iteration[2] process is excellent (typically, 3 iterations are enough). The ISI of the SIR is removed with a super-exponential behaviour. The algorithm, based on the eigenvector decomposition of the fourth-order cross-cumulant $\mathbf{C}_{uyuy}$, is called *the EigenVector Algorithm (EVA)*. It realizes a convex cost function and therefore converges to the global minimum independent of the initialization set [14, 30].

4.7.3 Simulations

Algorithms were tested using i.i.d. sequences for the real and imaginary part of 16-ary-QAM data. For the simulations are applied normalized channels (which is done with respect to a Automatic Gain Control input unit)

$$\sum_{m=0}^{M-1} |h_m|^2 = 1 \tag{4.54}$$

with M complex coefficients. Channel input and output sequences therefore had the same variance. After the channel, the output signal was contaminated by an additive, zero-mean, i.i.d., Gaussian noise of specified Signal-to-Noise-Ratio. A simple channel model [12] was used. The following criteria are used to evaluate the performance

- the smoothed Mean Square Error (MSE) after shift-, phase- and amplitude synchronization

$$\mathrm{MSE}(n) = \frac{1}{k}\sum_{n-k+1}^{n} |x(n) - z(n)|^2\,, \quad n \geq k \tag{4.55}$$

- the residual ISI of the SIR

$$\mathrm{ISI}(n) = \frac{1}{|s(k_0,n)|^2}\sum_{\substack{k \\ k\neq k_0}} |s(k,n)|^2 \tag{4.56}$$

  with k being the iteration index and the maximum magnitude of the SIR given by

$$|s(k_0,n)| = \max_k |s(k,n)| \tag{4.57}$$

Results from simulations of the channel are presented in Table 4.4 from which it is clear that EVA has a much better performance than CMA both in terms of minimum MSE (MMSE) and its convergence in sample numbers. However EVA is far more expensive in computation than CMA.

---

[2] Iterations are necessary, because $z(n) = f(\mathbf{e})$ and to comply with Eq. (4.52).

[3] Conditions are: SNR=40dB, $\mu = .001$, 5000 samples, equalization filter length $L = 20$, EVA on its own with a growing window up to 5000 samples

Algorithm	MMSE	MISI	MSE_{2500}
CMA	8.6e-3	2.9e-3	2.0e-1
EVA	3.8e-3	1.2e-2	1.0e-2

TABLE 4.4. Minimum Mean Square Error (MMSE) at final stage of the algorithm, Minimum Inter Symbol Interference (MISI) and the MSE after 2500 sample as criteria for convergence speed of the CMA and EVA algorithms [3]

4.8 Closing Remarks

There are significant interests in theoretical developments as well as applications of HOS. A great deal of work in a number of traditional signal processing areas is under way. Investigations into the effects of finite word-length, variability in cumulant estimates, filtering, and others need to be considered in detail for successful application of HOS. Research results on higher-order cyclostationary processes have been omitted from this discussion. It should be remembered that already HOS have been successfully applied in many areas. The HOS research field is a developing area and consequently one expects to see new theoretical ideas as well as new applications in the future.

Acknowledgments: Many authors have contributed and are still continuing to contribute to these exciting developments. The author is grateful to many colleagues for discussions and would like to thank the organizers of the First European Conference on Signal Analysis and Prediction, held in Prague during June, 1997, for the invitation to present a tutorial paper [27]. This work was partly supported by the Engineering and Physical Sciences Research Council of U.K. under the research grant number GR/J82744.

4.9 References

[1] P.-O. Amblard and J.-M. Brossier. Adaptive estimation of the fourth-order cumulant of a white stochastic process. *Signal Processing*, 42:37-43, 1995.

[2] S.N. Batalama and D. Kazakos. On the robust estimation of the autocorrelation coefficients of stationary sequences. *IEEE Trans. Signal Processing*, 44:2508-2520, 1996.

[3] A. Benveniste, M. Goursat and G. Ruget. Robust Identification of a Nonminimum Phase System: Blind Adjustment of Linear Equalizer in Data Communications, *IEEE Trans. on Automatic Control*, 25:385-399, 1980.

[4] D.R. Brillinger. *Time series: Data analysis and theory*, Holden-Day Inc., San Francisco, USA, 1981.

[5] X.-R. Cao and R.-W. Liu. General approach to blind source separation. *IEEE Trans. Signal Processing*, 44:562-571, 1996.

[6] J.-F. Cardoso and B.H. Laheld. Equivariant adaptive source separation. *IEEE Trans. Signal Processing*, 44:3017-3030, 1996.

[7] P. Comon. Independent component analysis: A new concept? *Signal Processing*, 36:287-314, 1994.

[8] J.R. Dickie and A.K. Nandi. A comparative study of AR order selection methods. *Signal Processing*, 40:239-255, 1994.

[9] J.R. Dickie and A.K. Nandi. AR modelling of skewed signals using third-order cumulants. *IEE Proceedings - Part VIS*, 142:78-86, 1995.

[10] Z. Ding and C.R. Johnson, Jr. Existing Gap Between Theory and Application of Blind Equalization, *SPIE Adaptive Signal Processing*, 1565:154-165, 1991.

[11] F. Harroy and J.-L. Lacoume. Maximum likelihood estimators and Cramer-Rao bounds in source separation. *Signal Processing*, 55:167-177, 1996.

[12] D. Hatzinakos. Stop-and-Go Sign Algorithms for Blind Equalization, *SPIE Adaptive Signal Processing*, 1565:118-129, 1991.

[13] S. Haykin, *Adaptive Filter Theory*, Prentice-Hall International Editions, 1996.

[14] B. Jelonnek, D. Boss and K.-D. Kammeyer. Generalized Eigenvector Algorithm for Blind Equalization, *Signal Processing*, 61:237-264, 1997.

[15] L.D. Lathauwer, B.D. Moor, and J. Vandewalle. Blind source separation by higher-order singular value decomposition. *Proceedings EUSIPCO-94*, 1:175-178, 1994.

[16] R. Leipnik. The lognormal distribution and strong non-uniqueness of the moment problem. *Theory Prob. Appl.*, 26:850-852, 1981.

[17] G.C.W. Leung and D. Hatzinakos. Implementation aspects of various higher-order statistics estimators. *J. Franklin Inst.*, 333B:349-367, 1996.

[18] D. Mämpel, A.K. Nandi, and K. Schelhorn. Unified approach to trimmed mean estimation and its application to bispectrum of EEG signals. *J. Franklin Inst.*, 333B:369-383, 1996.

[19] A. Mansour and C. Jutten. Fourth-order criteria for blind source separation. *IEEE Trans. Signal Processing*, 43:2022-2025, 1995.

[20] J.K. Martin and A.K. Nandi. Blind system identification using second, third and fourth order cumulants. *J. Franklin Inst.*, 333B:1-14, 1996.

[21] The MathWorks Inc. *Matlab Reference Guide*, 1995.

[22] The MathWorks Inc. *Higher-Order Spectral Analysis Toolbox*, 1995.

[23] A.K. Nandi and D. Mämpel. Improved estimation of third order cumulants. *Frequenz*, 49:156-160, 1995.

[24] A.K. Nandi. Guest editor of the Special Issue on Applications of Higher Order Statistics. *J. Franklin Inst.*, 333B:311-452, 1996.

[25] A.K. Nandi and V. Zarzoso. Fourth-order cumulant based blind source separation. *IEEE Signal Processing Letters*, 3:312-314, 1996.

[26] A.K. Nandi, D. Mämpel, and B. Roscher. Blind deconvolution of ultrasonic signals in non-destructive testing applications. *IEEE Trans. Signal Processing*, 45:1382-1390, 1997.

[27] A.K. Nandi. Higher order statistics in signal processing. *Proceedings of ECSAP-97*, 1-8, 1997.

[28] C.L. Nikias and A.P. Petropulu. *Higher-order spectra analysis.* Prentice Hall, Englewood Cliffs, New Jersey, USA, First edition, 1993.

[29] J.G. Proakis and C.L. Nikias. Blind Equalization, *SPIE Adaptive Signal Processing*, 1565:76-87, 1991.

[30] O. Shalvi and E. Weinstein. New Criteria for Blind Deconvolution of Nonminimum Phase Systems (Channels), *IEEE Trans. on Information Theory*, 36:312-321, 1990.

[31] J.J. Shynk, R.P. Gooch, G. Krishnamurthy and C.K. Chan. A Comparative Performance Study of Several Blind Equalization Algorithms, *SPIE Adaptive Signal Processing*, 1565:102-117, 1991.

[32] A. Stuart and J. K. Ord. *Kendall's Advanced Theory of Statistics.* Charles Griffin and Company, London, Volume 1, Fifth edition, 1987.

[33] A. Swami and G.B. Giannakis. Guest editors of the Special Issue on Higher-Order Statistics. *Signal Processing*, 53:89-255, 1996.

[34] A. Swami, G.B. Giannakis, and G. Zhou. Bibliography on higher-order statistics. *Signal Processing*, 60:65-126, 1997.

[35] D. Yellin and E. Weinstein. Multichannel signal separation: Methods and analysis. *IEEE Trans. Signal Processing*, 44:106-118, 1996.

[36] V. Zarzoso and A.K. Nandi. The potential of decorrelation in blind separation of sources based on cumulants. *Proceedings of ECSAP-97*, 293:296, 1997.

[37] V. Zarzoso, A.K. Nandi, and E. Bachakarakis. Maternal and foetal ECG separation using blind source separation methods. *IMA Journal of Mathematics Applied in Medicine & Biology*, 14:207-225, 1997.

5

Constrained Optimization Using Geometric Algebra and its Application to Signal Analysis

Joan Lasenby[1]
Anthony N. Lasenby[2]

ABSTRACT
In this paper we discuss a mathematical system based on the algebras of Grassmann and Clifford [4, 1], called geometric algebra [6]. It is shown how geometric algebra can be used to carry out, in a simple manner, various complex manipulations relevant to matrix-based problems, including that of optimization. In particular we look at how differentiation of certain matrix functions with respect to the matrix, can easily be achieved. The encoding of structure into such problems will be discussed and applied to a multi-source signal separation problem. Other applications are also discussed.

5.1 Introduction

Let $\mathcal{G}_n$ denote the geometric algebra of n-dimensions – this is a graded linear space. As well as vector addition and scalar multiplication we have a non-commutative product which is associative and distributive over addition – this is the **geometric** or **Clifford** product. A further distinguishing feature of the algebra is that any vector squares to give a scalar. The geometric product of two vectors $\boldsymbol{a}$ and $\boldsymbol{b}$ is written $\boldsymbol{ab}$ and can be expressed as a sum of its symmetric and antisymmetric parts

$$\boldsymbol{ab} = \boldsymbol{a}\cdot\boldsymbol{b} + \boldsymbol{a}\wedge\boldsymbol{b} \tag{5.1}$$

where the inner product $\boldsymbol{a}\cdot\boldsymbol{b}$ and the outer product $\boldsymbol{a}\wedge\boldsymbol{b}$ are defined by

$$\boldsymbol{a}\cdot\boldsymbol{b} = \frac{1}{2}(\boldsymbol{ab}+\boldsymbol{ba}) \qquad \boldsymbol{a}\wedge\boldsymbol{b} = \frac{1}{2}(\boldsymbol{ab}-\boldsymbol{ba}) \tag{5.2}$$

The inner product of two vectors is the standard *scalar* or *dot* product and produces a scalar. The outer or wedge product of two vectors is a new quantity we call a **bivector**. We think of a bivector as a directed area in the plane containing $\boldsymbol{a}$ and $\boldsymbol{b}$, formed by sweeping $\boldsymbol{a}$ along $\boldsymbol{b}$ – see Figure 5.1. Thus, $\boldsymbol{b}\wedge\boldsymbol{a}$ will have the opposite orientation making the wedge product anticommutative. The outer product is immediately generalizable to higher dimensions – for example, $(\boldsymbol{a}\wedge\boldsymbol{b})\wedge\boldsymbol{c}$, a **trivector**, is interpreted as the oriented volume formed by sweeping the area

[1]University of Cambridge, Department of Engineering, Trumpington St., Cambridge, CB2 1PZ, United Kingdom, E-mail: jl@eng.cam.ac.uk

[2]MRAO, Cavendish Laboratory, Madingley Road, Cambridge, CB3 OHE, United Kingdom, E-mail: anthony@mrao.cam.ac.uk

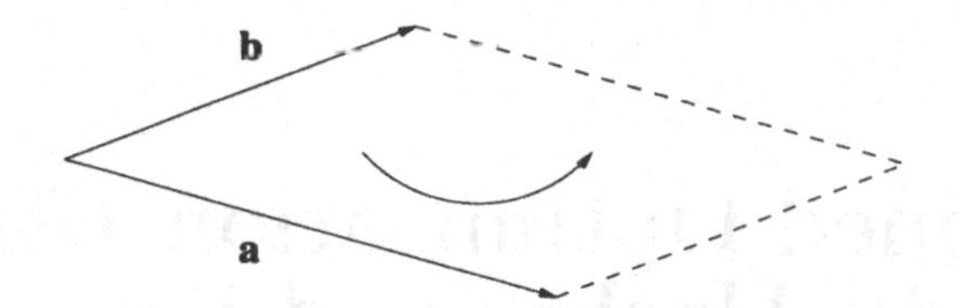

FIGURE 5.1. The directed area, or bivector, $a \wedge b$.

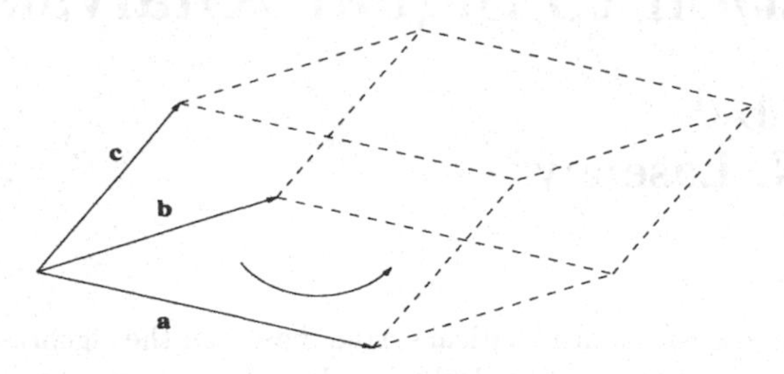

FIGURE 5.2. The oriented volume, or trivector, $a \wedge b \wedge c$.

$a \wedge b$ along vector c, see Figure 5.2. The outer product of k vectors is a ***k*-blade** – the term ***k*-blade** is used for quantities that can be written as the wedge product of k vectors, while ***k*-vector** indicates that the quantity may be the sum of several k-blades (a k-vector can not always be written as a k-blade). In what follows we will sometimes refer to the dimensionality of a k-vector as its **grade**: for example, a scalar has grade 0, a vector grade 1, a bivector grade 2, etc. A **multivector** will mean a sum of quantities of any grade and a multivector is said to be *homogeneous* if it contains terms of only a single grade. The geometric algebra provides a means of manipulating multivectors which allows us to keep track of different grade objects simultaneously – much as one does with complex number operations. In a space of 3 dimensions we can construct a trivector $a \wedge b \wedge c$, but no 4-vectors exist since there is no possibility of sweeping the volume element $a \wedge b \wedge c$ over a 4th dimension. The highest grade element in a space is called the **pseudoscalar**. The unit pseudoscalar is denoted by I. In 2 and 3 dimensions (with a Euclidean metric) the pseudoscalar is denoted by i and it is easy to show that $i^2 = -1$. We can generalize the definitions of inner and outer products given in Equation (5.2). For two homogeneous multivectors A_r and B_s (i.e. multivectors of grades r and s respectively), we define the inner and outer products as

$$A_r \cdot B_s = \langle A_r B_s \rangle_{|r-s|} \tag{5.3}$$

$$A_r \wedge B_s = \langle A_r B_s \rangle_{r+s} \tag{5.4}$$

Where $\langle M \rangle_t$ denotes the t-grade part of the multivector M. Thus, the inner product produces an $|r-s|$-vector – which means it effectively reduces the grade of B_s by r; and the outer product gives an $(r+s)$-vector, therefore increasing the grade of B_s by r. This is an extension of the general principle that dotting with a vector *lowers* the grade of a multivector by 1 and wedging with a vector *raises* the grade of a multivector by 1. In the following sections we will frequently evaluate a vector dotted with a bivector, according to the above this produces a vector and is explicitly given by

$$a \cdot (b \wedge c) = (a \cdot b)c - (a \cdot c)b \tag{5.5}$$

The operation of **reversion** reverses the order of vectors in any multivector. The reverse of A is written as $\tilde{A}$, and for a product we have

$$(AB)^{\tilde{}} = \tilde{B}\tilde{A}. \tag{5.6}$$

Another concept which will be used elsewhere in this chapter is that of the *reciprocal frame*. Given a set of linearly independent vectors $\{e_k\}$ (where no assumption of orthonormality is made), we can form a **reciprocal frame**, $\{e^k\}$, which is such that

$$e^k \cdot e_j = \delta_j^k \tag{5.7}$$

where $\delta_j^k = 1$ if $k = j$ and 0 otherwise. For details of the explicit construction of such a reciprocal frame in n-dimensions see [6]. In three dimensions this is a very simple operation and the reciprocal frame vectors for a linearly independent set of vectors $\{e_k\}$, $k = 1, .., 3$, are as follows

$$\begin{aligned} e^1 &= \frac{1}{\alpha} i e_2 \wedge e_3 \\ e^2 &= \frac{1}{\alpha} i e_3 \wedge e_1 \\ e^3 &= \frac{1}{\alpha} i e_1 \wedge e_2 \end{aligned} \tag{5.8}$$

where $i\alpha = e_3 \wedge e_2 \wedge e_1$.

In the following sections the summation convention will be used unless otherwise stated, i.e. repeated indices are summed over.

5.2 Linear Algebra and Multivector Calculus

Geometric algebra (GA) has very powerful associated linear algebra and calculus frameworks and is therefore a natural language for the study of linear functions and non-orthonormal frames. Consider a linear function $f(a)$ which maps vectors to vectors in the same space. It is then possible to extend f to act linearly on multivectors. This extension of f (the outermorphism) is written as $\underline{f}$ and is given by

$$\underline{f}(a_1 \wedge a_2 \wedge \ldots \wedge a_r) \equiv f(a_1) \wedge f(a_2) \wedge \ldots \wedge f(a_r) \tag{5.9}$$

f is thus grade-preserving since it maps an r-vector to an r-vector. The **adjoint** to $\underline{f}$ is written as $\overline{f}$ and can be defined implicitly via the relation $\underline{f}(a) \cdot b = \overline{f}(b) \cdot a$ for any vectors a and b – it is also possible to define the adjoint in an explicit manner using derivatives, but we will not discuss this here. Note that the linear function f is frame-independent, however, if we have a reference frame with basis vectors $\{\sigma_j\}$ ($\{\sigma_j\}$ are used for the canonical basis vectors in GA and it is the convention not to write them in bold) then we can define scalar elements F_{ij}, where

$$F_{ij} = \sigma_i \cdot \underline{f}(\sigma_j) \tag{5.10}$$

We then see that the F_{ij} can be interpreted as the elements of a matrix F – in this way we have turned a frame-independent quantity into a frame-dependent quantity. Since $\sigma_i \cdot \underline{f}(\sigma_j) = \overline{f}(\sigma_i) \cdot \sigma_j$ we see that the adjoint corresponds to the transpose of the matrix.

The outermorphism preserves grade, and therefore the pseudoscalar of the space must be mapped onto some multiple of itself. The scale factor in this mapping is the **determinant** of $\underline{f}$;

$$\underline{f}(I) = \det(\underline{f})I \tag{5.11}$$

This is much simpler than many definitions of the determinant. Using this definition, most properties of determinants can be established with little effort. A simple example of this is seen by considering the determinant of the product of two matrices P and Q. Suppose P and Q are represented by the linear functions $\underline{f}$ and $\underline{g}$;

$$\begin{aligned} \det(\underline{fg})I &= (\underline{fg})(I) \\ &= \underline{f}[\underline{g}(I)] \\ &= \det(\underline{g})\underline{f}(I) \\ &= \det(\underline{g})\det(\underline{f})I \end{aligned} \tag{5.12}$$

We therefore see that the result $\det(PQ) = \det(P)\det(Q)$ drops out trivially from our definition of the determinant. In addition, it is straightforward to obtain expressions for the inverse of a function (if it exists) and the inverse of an adjoint;

$$\underline{f}^{-1}(A) = \det(\underline{f})^{-1} I\overline{f}(I^{-1}A) \tag{5.13}$$

$$\overline{f}^{-1}(A) = \det(\underline{f})^{-1} I\underline{f}(I^{-1}A) \tag{5.14}$$

where A is any arbitrary multivector.

Within geometric algebra it is possible to differentiate with respect to any multivector quantity [6] – in practice this turns out to be very useful in many optimization processes in signal analysis. Here we will give a brief description of multivector differentiation and a few of the standard results.

If X is a mixed-grade multivector, $X = \sum_r X_r$, and $F(X)$ is a general multivector-valued function of X, then the derivative of F in the A 'direction' (where A has the same grades as X) is written as $A * \partial_X F(X)$ (here we use $*$ to denote the scalar part of the product of two multivectors, i.e. $A * B \equiv \langle AB \rangle$ – where $\langle\ \rangle$ is shorthand for the scalar part, $\langle\ \rangle_0$), and is defined as

$$A * \partial_X F(X) \equiv \lim_{\tau \to 0} \frac{F(X + \tau A) - F(X)}{\tau} \tag{5.15}$$

We impose the constraint that A must have the same grades as X so that the differentiation in terms of the limit makes some physical sense. If X contains no terms of grade-r and A_r is a homogeneous multivector, then we define $A_r * \partial_X = 0$. This definition of the derivative also ensures that the operator $A * \partial_X$ is a scalar operator and satisfies all of the usual partial derivative properties. We can now use the above definition of the directional derivative to formulate a general expression for the multivector derivative ∂_X without reference to one particular direction. This is accomplished by introducing an arbitrary frame $\{e_j\}$ and extending this to a basis (vectors, bivectors, etc..) for the entire algebra, $\{e_J\}$. Then ∂_X is defined as

$$\partial_X \equiv \sum_J e^J e_J * \partial_X \tag{5.16}$$

where $\{e^J\}$ is an extended basis built out of the reciprocal frame. The directional derivative, $e_J * \partial_X$, is only non-zero when e_J is one of the grades contained in X (as previously discussed) so that ∂_X inherits the multivector properties of its argument X. Although we have here defined the multivector derivative using an extended basis, it should be noted that the sum over all the basis ensures that ∂_X is independent of the choice of $\{e_j\}$ and so all of the properties of ∂_X can be formulated in a frame-free manner. One of the most useful results concerning multivector derivatives is

$$\partial_X \langle XB \rangle = \mathrm{P}_X(B) \tag{5.17}$$

where $\mathrm{P}_X(B)$ is the projection of the multivector B onto the grades contained in X. We can see this as follows. Since

$$\partial_X \langle XB \rangle = \sum_J e^J e_J * \partial_X \langle XB \rangle \tag{5.18}$$

and $e_J * \partial_X = 0$ if e_J is not a grade of X, we see that

$$\begin{aligned} \partial_X \langle XB \rangle &= \sum_{J'} e^{J'} \lim_{\tau \to 0} \frac{\langle (X + \tau e'_J)B \rangle - \langle XB \rangle}{\tau} \\ &= \sum_{J'} e^{J'} \langle e_{J'} B \rangle \\ &= P_X(B) \end{aligned} \tag{5.19}$$

where the sum over J' is over those grades contained in X. Using this result the following relations can be shown to hold

$$\partial_X \langle \tilde{X} B \rangle = \mathrm{P}_X(\tilde{B}) \tag{5.20}$$

$$\partial_{\tilde{X}} \langle \tilde{X} B \rangle = \mathrm{P}_{\tilde{X}}(B) = \mathrm{P}_X(B) \tag{5.21}$$

$$\partial_\psi \langle M \psi^{-1} \rangle = -\psi^{-1} \mathrm{P}_\psi(M) \psi^{-1} \quad \text{for general multivectors } \psi \text{ and } M \tag{5.22}$$

It is often convenient to indicate, via an overdot, the quantity on which ∂_X operates. For example, $\dot{\partial}_X A \dot{B}$ means that the derivative part of ∂_X acts on B. A complete discussion of the geometric calculus is given in [6], where the results in Equations (5.20) and (5.21) are proved. The result in Equation (5.22) is discussed in [2].

One often wants to take the derivative, ∂_a, with respect to a vector quantity $\boldsymbol{a}$. If we replace A by $\boldsymbol{a}$, e^J by e^j and e_J by e_j in Equation (5.16) and use the definition in Equation (5.15) we see that the differential operator ∂_a can be written as

$$\partial_a = e^i \frac{\partial}{\partial a^i} \qquad \text{where} \qquad \boldsymbol{a} = a^i e_i \tag{5.23}$$

This will be used several times in the following sections. Note that we will not write vectors in bold when they appear as subscripts in the vector derivative. At this point it is appropriate to give some explanation of the expansion of $\boldsymbol{a}$ used above, namely $\boldsymbol{a} = a^i e_i$. Recall that our basis $\{e_i\}$ was not necessarily an orthonormal basis and that we were able to define a reciprocal basis $\{e^i\}$, such that $e^i \cdot e_j = \delta_{ij}$. Consider the component of the vector $\boldsymbol{a}$ in the e^j direction – this is given by $\boldsymbol{a} \cdot e^j$ and we will call this component a^j, where the upstairs index tells us that it is the component of a in the direction of the jth basis vector of the *reciprocal* frame. Similarly, the component of a in the e_j direction is given by $a \cdot e_j$ which we call a_j. We note the useful identities

$$a = (a \cdot e_j)e^j \qquad \text{and} \qquad a = (a \cdot e^j)e_j \tag{5.24}$$

Of course, if the basis is an orthonormal basis, say $\{\sigma_i\}$, then $\sigma^i = \sigma_i$ and we would, without ambiguity, write $a = a_i \sigma_i$.

This ability to differentiate with respect to any multivector is often very useful – for example, one can easily differentiate with respect to rotations (represented by scalar plus bivector) in order to find the optimal motion in a given problem. To complete our calculus it is now necessary to look at the formulation of functional differentiation in GA. In [3] it is shown that we can differentiate with respect to a linear function according to the following rule:

$$\partial_{\underline{f}(a)}\{\underline{f}(b) \cdot c\} = (a \cdot b)c \tag{5.25}$$

for any vectors a, b, c. This is the basic result and can be extended to the case where we have general multivectors as follows:

$$\partial_{\underline{f}(a)}\langle \underline{f}(A)B\rangle = \sum_r \langle \underline{f}(a \cdot A_r)B_r\rangle_1 \tag{5.26}$$

where the right hand side represents a sum of vector parts.

Using this functional differentiation we can differentiate expressions which would be considerably harder to deal with in the usual matrix formulation – an example of this is the differentiation of $\det P$ with respect to a matrix P. This is rather hard to do conventionally but reasonably straightforward in GA. If $\underline{f}$ represents P, then from the definition of the determinant in Equation (5.11) we see that

$$\partial_{\underline{f}(a)} \det(\underline{f}) = \partial_{\underline{f}(a)} \underline{f}(I) I^{-1} \tag{5.27}$$

From the formula for functional differentiation in Equation (5.26) and from the definition of the inverse given in Equation (5.13), we can evaluate the above to give

$$\begin{aligned} \partial_{\underline{f}(a)} \det(\underline{f}) &= \underline{f}(a \cdot I) I^{-1} \\ &= \det(\underline{f}) \overline{f}^{-1}(a) \end{aligned} \tag{5.28}$$

which agrees with the standard result.

5.3 Optimization and Preserving Matrix Structure

Equipped with the basic results of functional differentiation and some results which tell us how to implement the chain rule, it is possible to apply these techniques to more complicated problems. The aim in this chapter will be to use this framework in order to optimize expressions containing structured matrices. Firstly we consider a simple example: find the matrix R, such that R is cyclic Toeplitz, and which maximizes

$$p(x) = \frac{1}{(2\pi)^{\frac{n}{2}}} |R|^{-\frac{1}{2}} \exp\{-\frac{1}{2} x^T R^{-1} x\} \tag{5.29}$$

For data $\boldsymbol{x}$ coming from an underlying zero-mean Gaussian process we expect R to be the estimate of the autocorrelation matrix. Here we want to optimize only over the space of *allowed* R's, i.e. those that are cyclic Toeplitz. An $n \times n$ cyclic Toeplitz matrix R can be written in terms of a generating vector $\boldsymbol{a} = a_1 e_1 + a_2 e_2 + \ldots + a_n e_n$, where $\{e_i\}$ is a basis for the n-dimensional space – for example:

$$R = \begin{bmatrix} a_1 & a_2 & a_3 & \cdots & a_n \\ a_n & a_1 & a_2 & \cdots & a_{n-1} \\ \vdots & \vdots & \vdots & & \vdots \\ a_2 & a_3 & a_4 & \cdots & a_1 \end{bmatrix} \tag{5.30}$$

Consider another set of basis vectors $\{\sigma_i\}$ such that $\sigma_i \cdot \sigma_j = \delta_{ij}$ and $\sigma_i \cdot e_j = 0$ for all i, j – we then think of our linear function, f, representing R^{-1}, as mapping from σ-space to e-space. The corresponding matrix is then $R^{-1}{}_{ij} = e_i \cdot \underline{f}(\sigma_j) = \overline{f}(e_i) \cdot \sigma_j$. In the geometric algebra formulation of Hamiltonian dynamics one uses quantities called *doubling bivectors*, and it turns out that similar quantities can be used here to encode the Toeplitz nature of the function – i.e. the fact that each row of the matrix is just a shifted version of the first. We define $J_k = e_k \wedge \sigma_k$ and $\boldsymbol{J} = \sum_k J_k$, so that the effect of $\boldsymbol{J}$ is to take e_k to σ_k and vice versa;

$$e_k \cdot \boldsymbol{J} = \sigma_k, \quad \sigma_k \cdot \boldsymbol{J} = -e_k \tag{5.31}$$

which can be verified using the result in Equation (5.5). By modifying $\boldsymbol{J}$ we can produce a shifted vector for use in describing R^{-1}. If we define

$$S_k = \sum_{j=1}^{n} e_j \wedge \sigma_{[j+k-2]_n + 1} \tag{5.32}$$

(where $[..]_n$ indicates modulo-n) then each row of the matrix can be written as $\boldsymbol{a} \cdot S_k$ – in terms of the adjoint of f this gives

$$\overline{f}(e_k) = \boldsymbol{a} \cdot S_k \tag{5.33}$$

With this characterization of R^{-1} we can differentiate Equation (5.29) straightforwardly with respect to $\boldsymbol{a}$ and set this to zero to give

$$\begin{aligned} \partial_a p(\boldsymbol{x}) &= \partial_{\underline{f}(\sigma_k)} p(\boldsymbol{x}) \cdot \partial_a \underline{f}(\sigma_k) \\ &= \frac{1}{(2\pi)^{\frac{n}{2}}} (\det f)^{\frac{1}{2}} \exp\{-\frac{1}{2} \tilde{\boldsymbol{x}} \cdot \underline{f}(\boldsymbol{x})\} \times \\ & \quad \{\underline{f}^{-1}(e_k) \cdot S_k - (e_k \cdot \tilde{\boldsymbol{x}}) \boldsymbol{x} \cdot S_k\} \\ &= 0 \end{aligned} \tag{5.34}$$

In the above expression $\tilde{\boldsymbol{x}} = x_j e_j$ and $\boldsymbol{x} = x_j \sigma_j$, and we have used the *chain rule* in the differentiation. Here, the chain rule is applied in a straightforward manner – i.e. we know that the derivative with respect to $\boldsymbol{a}$ produces a vector; since the derivative of $p(\boldsymbol{x})$ wrt $\underline{f}$ gives a vector and the derivative of $\underline{f}(\sigma_k)$ wrt $\boldsymbol{a}$ will give a bivector, it follows that we must dot these together to produce a vector as required. Thus, we see that the solution to this is given by

$$\underline{f}^{-1}(e_k) \cdot S_k = (e_k \cdot \tilde{\boldsymbol{x}}) \boldsymbol{x} \cdot S_k \tag{5.35}$$

Since $\underline{f}$ was constructed to represent R^{-1}, it follows that $\underline{f}^{-1}$ (mapping from e-space to σ-space) will represent R. The above then reduces to

$$a_j = \frac{1}{n}(e_k \cdot \tilde{x}) x \cdot (e_j \cdot S_k) = \frac{1}{n} x_k x_{[j+k-2]_n+1} \tag{5.36}$$

This indeed gives the required estimate of the autocorrelation matrix. Carrying out a similar operation with matrices is possible but harder. A general $m \times n$ Toeplitz matrix can be similarly characterized by its two generating vectors and two doubling bivectors and it is therefore possible to use the above techniques to form derivatives of the more complicated expressions involving such general Toeplitz matrices.

Once a given characteristic has been encoded in terms of linear functions – as in Equation (5.33) for the cyclic Toeplitz case – the procedures of differentiation and functional differentiation are straightforward and offer an attractive new technique for dealing with structure in matrices. However, we note here that it is the encoding of the structure into the linear function which is the hardest stage of the problem and that there is, as yet, no set recipe for doing this.

5.4 Application to Signal Separation

In order to illustrate the above techniques we will look at one particular formulation of a signal separation problem with scalar mixing. Suppose we have n sources, represented by time series $s_i(t)$, $i = 1,..,n$, and N measured estimates, r_i, $i = 1,...,N$. The process of obtaining the estimates can be modelled as passing the source signals through a *mixer* followed by an *estimator*. The mixer is modelled by an $n \times n$ matrix so that

$$\begin{bmatrix} x_1(t) \\ x_2(t) \\ \vdots \\ x_n(t) \end{bmatrix} = \begin{bmatrix} a_{11} & a_{12} & \cdots & a_{1n} \\ a_{21} & a_{22} & \cdots & a_{2n} \\ \vdots & \vdots & \vdots & \vdots \\ a_{n1} & a_{n2} & \cdots & a_{nn} \end{bmatrix} \begin{bmatrix} s_1(t) \\ s_2(t) \\ \vdots \\ s_n(t) \end{bmatrix} \tag{5.37}$$

which can also be written as $\boldsymbol{x}(t) = A\boldsymbol{s}(t)$. If we then take expectation values of $\boldsymbol{x}(t)\boldsymbol{x}(t+q)^T$ and make the assumption that the signals are uncorrelated, it is possible to write the measured correlations as

$$\boldsymbol{r}_q = A'\boldsymbol{s}_q \tag{5.38}$$

In this $\boldsymbol{r}_q = [r_{11}(q), r_{12}(q), \ldots, r_{nn}(q)]^T$ (an $n^2 \times 1$ vector), and $r_{ij}(q) = E[x_i(t)x_j(t+q)]$. The vector $\boldsymbol{s}_q$ is given by $[s_{11}(q), s_{22}(q), \ldots, s_{nn}(q)]^T$ (an $n \times 1$ vector) with $s_{ii}(q) = E[s_i(t)s_i(t+q)]$ (no summation convention). The matrix A' is given by

$$A' = [\boldsymbol{a}_1 \otimes \boldsymbol{a}_1, \boldsymbol{a}_2 \otimes \boldsymbol{a}_2, ...] \tag{5.39}$$

where $\boldsymbol{a}_i$ is the vector formed from the ith column of matrix A and $\otimes$ is the Kronecker product, such that

$$\begin{aligned} \boldsymbol{a}_1 \otimes \boldsymbol{a}_1 &= [a_{11}a_{11}, a_{11}a_{21}, ..., a_{11}a_{n1}, a_{21}a_{11}, a_{21}a_{21}, \\ &\quad ..., a_{21}a_{n1}, ..., a_{n1}a_{11}, a_{n1}a_{21}, ..., a_{n1}a_{n1}]^T \end{aligned} \tag{5.40}$$

We are now able to formulate this in the form of a least squares problem: taking, say, m lags, we want to minimize

$$\mathcal{E} = \sum_{q=1}^{m} (\boldsymbol{r}_q - A'\boldsymbol{s}_q)^2 \tag{5.41}$$

i.e. find the A' and $\boldsymbol{s}_q$ which minimize $\mathcal{E}$ such that A' has the structure given in Equation (5.39). To do this our approach is to differentiate with respect to the $\{\boldsymbol{s}_q\}$ and the $\{\boldsymbol{a}_i\}$ and for this we need to express the matrix A' as a linear function. Suppose we write the vector $\boldsymbol{a}_i$ as $\boldsymbol{a}_i = a_{1i}\sigma_1 + a_{2i}\sigma_2 + \ldots + a_{ni}\sigma_n$, we then want to look for a linear function, $\underline{f}$, representing A', such that $\underline{f}$ maps σ-space onto e-space, as before. It is reasonably straightforward to see that we can write the ith column of the matrix as

$$\begin{aligned} \underline{f}(\sigma_i) &= (\sigma_{\tilde{j}}\cdot\boldsymbol{a}_i)(\sigma_{\{j\}_n}\cdot\boldsymbol{a}_i)e_j \\ &= a_{\tilde{j}i}a_{\{j\}_n i}e_j \end{aligned} \tag{5.42}$$

where $\tilde{j} = int(\frac{j-1}{n}) + 1$ (int refers to *integer part*), $\{j\}_n = [j-1]_n + 1$ and j runs from 1 to n^2. Equation (5.41) can then be written as

$$\mathcal{E} = \sum_{q=1}^{m} (\boldsymbol{r}_q - \underline{f}(\boldsymbol{s}_q))^2 \tag{5.43}$$

where we now think of $\boldsymbol{s}_q$ as the vector $s_{11}(q)\sigma_1 + s_{22}(q)\sigma_2 + \ldots + s_{nn}(q)\sigma_n$, and $\boldsymbol{r}_q$ as the vector $r_{11}(q)e_1 + r_{12}(q)e_2 + \ldots + r_{nn}(q)e_{n^2}$, so that $\underline{f}$ takes σ-space onto e-space. Differentiating with respect to $\boldsymbol{s}_q$ is straightforward and gives the standard solution

$$\boldsymbol{s}_q = (\overline{f}\underline{f})^{-1}\overline{f}(\boldsymbol{r}_q) \tag{5.44}$$

Differentiating with respect to the vectors $\boldsymbol{a}_i$ is achieved by first evaluating the expression obtained from the chain rule

$$\partial_{a_i}\mathcal{E} = (\partial_{\underline{f}(\sigma_j)}\mathcal{E})\cdot(\partial_{a_i}\underline{f}(\sigma_j)) = 0 \tag{5.45}$$

Evaluating the derivatives gives us the following equations (one for each $i = 1, ..., n$)

$$\partial_{a_i}\mathcal{E} = 2\left\{\sum_{q=1}^{m}(e_i\cdot\boldsymbol{s}_q)(\underline{f}(\boldsymbol{s}_q) - \boldsymbol{r}_q)\right\}\cdot B_i = 0 \tag{5.46}$$

where B_i is a bivector given by

$$a_{\{k\}_n i}\sigma_{\tilde{k}}\wedge e_k + a_{\tilde{k}i}\sigma_{\{k\}_n}\wedge e_k \tag{5.47}$$

We can now try to solve Equations (5.44) and (5.47) or use the derivatives in a gradient-based optimization scheme. Having analytic derivatives is useful in many schemes for several reasons: firstly we may be able to solve the resulting equations, if not directly then often iteratively, and secondly, they provide more robustness in search methods than do differencing techniques. The example given here of scalar mixing is simply illustrative of the technique, there are many matrix-based methods that solve this problem from an eigen-decomposition approach.

5.5 Conclusions

The aim of this paper has been to outline the general principles of optimization in signal processing problems using linear functions and the geometric algebra framework instead of matrix calculus. In general, once we have successfully encoded a desired structure into a linear function the subsequent differentiation is straightforward – it is this encoding of structure that varies from problem to problem and for which there is no easy recipe. We have given two examples of problems which can be addressed by these techniques, firstly a simple example of 1D zero-mean Gaussian data, for which the answer was known. Secondly, the question of multi-source signal separation was put into a form to which these methods could be applied. There are many other instances in signal processing where the encoding of structure may be of great use. Further areas we intend to investigate are power spectrum estimation in the case where we have missing data and array processing problems – both problems will involve the manipulation of complicated block-Toeplitz matrices, so that the starting point will be the encoding of such block-Toeplitz nature into a linear function.

Acknowledgments: JL is supported by a Royal Society University Research Fellowship.

5.6 References

[1] W.K. Clifford. Applications of Grassmann's extensive algebra. *Am. J. Math.* 1: 350–358, 1878.

[2] C.J.L. Doran. Geometric Algebra and its Applications to Mathematical Physics. *Ph.D. Thesis, University of Cambridge*, 1994.

[3] C.J.L. Doran, A.N. Lasenby and S.F. Gull. Geometric Algebra: Applications in Engineering. In W.E. Baylis, editor, *Geometric (Clifford) Algebras in Physics.* Birkhauser Boston, 65–79, 1996.

[4] H. Grassmann. Der Ort der Hamilton'schen Quaternionen in der Ausdehnungslehre. *Math. Ann.*, 12: 375, 1877.

[5] D. Hestenes. New Foundations for Classical Mechanics. *D. Reidel*, Dordrecht, 1986.

[6] D. Hestenes and G. Sobczyk. Clifford Algebra to Geometric Calculus: A unified language for mathematics and physics. *D. Reidel*, Dordrecht, 1984.

6

Processing of Non-Stationary Vibrations Using the Affine Wigner Distribution

Marcelo Iribarren[1]
Cesar San Martin[1]
Pedro Saavedra[2]

ABSTRACT
This paper deals with the processing of non-stationary signals applied to rotating-machines condition monitoring. The advantages of applying the affine Wigner distribution is explored. First a brief synthesis of this technique and other related time-frequency distributions is given to exhibit their advantages, in particular when faults to diagnose are characterized by complex changes in spectrum or by weak non-stationarities in the vibration signal. Testing on synthesized and real signals shows it as a promising tool when compared to the Wavelet transform or smoothed pseudo Wigner-Ville distribution.

6.1 Introduction

In the industrial environment, the digital processing of signals related to vibrations and noise produced by rotating machinery (pumps, motors, generators, etc.) is becoming an everyday tool. The main purpose is to predict their mechanical condition (condition monitoring) and to repair them only when strictly necessary (predictive maintenance). Signals could be provided by either in-situ, pre-programmed measurements or by a continuous monitoring and alarm generation system related to a distributed control system. The main objective of improving the vibration analysis techniques is cost reduction of the maintenance and operation of industrial systems.

There are several failures involving machines with rotating elements, where a failure could be detected in an early stage by looking for non stationary signals of different kinds: evolving harmonics, wide-band evolving signals or transients.

Most of the commercial systems nowadays devoted to vibration monitoring are Fourier-based; that is, those using spectral analysis based on the Fast Fourier Transform (FFT). However, Fourier analysis has an intrinsic limitation, it does not allow high time and frequency resolution simultaneously. This is a serious drawback when analyzing evolving or transient signals.

---

[1]University of Concepción, Dept. of Electrical Engineering, P. O. Box 53-C, Concepción, Chile, E-mails: {miribarr, csanmart}@manet.die.udec.cl

[2]U. of Concepción, Dept. of Mechanical Engineering, P. O. Box 53-C, Concepción, Chile

A better representation to detect a non-stationary signal is obtained by describing it as a joint time-frequency distribution (TFD). It is well known, however, that there is no single TFR which is the best for all problems. It is still an unresolved problem how to determine the optimum TFR for a given signal class and analysis task. There have been some answers in the past (see e.g. [4, 22, 15]), but each of them is restricted to some special signal classes or problems.

The better known TFD is the Short-Time Fourier Transform. Some recently developed TFD's have improved its limitations. Between them are the Wavelet transform [16], the S-transform [19, 10], the Wigner-Ville distribution [20, 17], or the affine Wigner Distribution [5].

Since it has been impossible to find a TFR which is optimum for all signals and problems, some authors have proposed the use of data adaptive TFRs. In [13] a locally optimum TFR is proposed where for each point in the time-frequency plane another TFR kernel is used by maximizing the generalized ambiguity function. To reduce the computation expenses, in [2] a globally optimum TFR based on an energy concentration measure is described. In the general context of classification and detection, in [6] a method to find a unique optimum TFR based on a different criterion is proposed.

Although in theory data-adaptive TFR seem to be promising tools, the present work takes a conventional approach. It explores the advantages of applying to machine monitoring one of the newest and lesser known non-adaptive TFRs, the smoothed pseudo-affine Wigner Distribution (PAWD). First a brief synthesis of this technique and other related time-frequency distributions is given to exhibit their advantages, in particular when the faults to be diagnosed are characterized by complex changes in spectrum or by weak non-stationarities in the vibration signal. Then the smoothed PAWD, implemented in a Matlab environment, is applied to synthesized and real signals representing these features. Its application to real signals acquired from a test bench is under way. The results are compared against TFR´s implemented and analyzed in previous works, that is the STFT, the smoothed pseudo Wigner-Ville distribution and the Wavelet transform.

6.2 Joint Time-Frequency Distributions

6.2.1 Linear Distributions

The most known linear TFD is obtained by applying the FFT to overlapped time windows. Thus is generated the Short Time Fourier Transform (STFT), given by:

$$STFT\,(t,f) \;=\; \int s\,(\tau)\,w\,(t-\tau)\exp^{-2\pi\tau}\,d\tau \tag{6.1}$$

where w is the weighting function. Eq. (6.1) can be rewritten as:

$$STFT\,(t,f) = FFT\,(s\,(\tau)\,w\,(\tau-t)) \tag{6.2}$$

which is the common algorithm to calculate the STFT, i.e. the FFT is calculated for each time t having τ as the independent variable. It is assumed that the signal is stationary during the time interval determined by the time window. The square of STFT's magnitude is known as the *spectrogram*. Its main limitation is the inherent trade-off between time and frequency resolution.

A linear distribution better adapted to the analysis of non-stationary or transient signals is the *Wavelet transform* (WT) [11, 12]. The WT has a variable time-frequency resolution, i.e. a greater time resolution at high frequencies and a better frequency resolution at low frequencies. This WT property is very useful to detect transients, which are usually signals constrained in time but wide band in frequency. However its very linearity precludes desirable theoretical properties such as correct marginal distributions or perfect localization.

The continuous Wavelet transform is given by :

$$WT(t,a) = a^{-1/2} \int s(\tau)\, \psi\left(\frac{\tau - t}{a}\right) d\tau, \qquad a = \frac{f_0}{f} \tag{6.3}$$

where a is the scale parameter, t the time shift and ψ is the wavelet function. By expressing Eq. (6.3) in the frequency plane, a fast algorithm to calculate the WT by means of the FFT is obtained

$$\begin{aligned} WT(t,a) &= a^{1/2} \int S(\nu)\, \Psi(a\nu)\, e^{2\pi j\nu t} d\nu \\ WT(t,a) &= a^{1/2} IFFT\left(S(\nu)\, \Psi(a\nu)\right) \\ WT(t,a) &= a^{1/2} IFFT\left(FFT(s(t))\, \Psi(a\nu)\right) \end{aligned} \tag{6.4}$$

By taking the square of WT's magnitude the *scalogram* is obtained. It provides a distribution of signal's energy but in a time-scale plane, the scale, a, being the inverse of frequency.

The WT could be discretized by using a dyadic grid and the resulting basis functions may then be orthonormal. The so obtained discrete WT (DWT) is equivalent to an octave filter bank and, therefore, is known as the *multiresolution analysis*. This analysis over a dyadic grid, although useful for some applications, has a poor simultaneous time-frequency resolution. If the goal is higher resolution spectral analysis, it is necessary to subdivide the octave filters by using non-integer powers of two (or voices) during the scale parameter discretization. The trade-off is that a non-orthogonal basis function set is obtained.

To implement this higher resolution time-frequency analysis, normally the *à trous* algorithm is used[18]. But a compromise between resolution and computational burden has to be made. In most of the cases 12 voices per octave provides a reasonable trade-off [11].

6.2.2 *Quadratic Distributions*

A large class of time-frequency distributions of a signal *s(t)* is given by Cohen's class of TFRs:

$$C(t,f) = \frac{1}{2\pi} \int\int\int e^{-j\theta t - 2\pi j\tau f - j\theta u} \phi(\theta,\tau)\; s\left(u + \frac{\tau}{2}\right) s^*\left(u - \frac{\tau}{2}\right) d\tau\, du\, d\theta \tag{6.5}$$

where the function ϕ is the kernel defining the distribution properties.

If in the general expression of a TFD of the Cohen class given by Eq. (6.5), the kernel takes an unitary value, a *Wigner-Ville distribution* (WVD) is obtained, which is given by

$$WVD(t,f) = \frac{1}{2\pi} \int s\left(t + \frac{\tau}{2}\right) s^*\left(t - \frac{\tau}{2}\right) e^{-j2\pi f t} d\tau \tag{6.6}$$

The WVD can be regarded as theoretically optimal in that it satisfies a maximum number of desirable mathematical properties and features optimal time-frequency concentration. However, the WVD is never positive for the total range of time and frequency. Moreover, the WVD, the same as the STFT, is ***covariant to time and frequency shifting.*** This covariance means that time shifting and frequency modulation of a signal *s(t)* is converted to time and frequency shifting of the TFD. That is,

$$s\,(t-t_0)\,e^{-j2\pi f_0 t} \quad \rightarrow \quad WVD\,(t-t_0,\, f-f_0) \tag{6.7}$$

Distributions with this property, in general, provide good performance in analyzing narrow band signals or local harmonic signals but have difficulties in revealing both low frequency components and high frequency transients.

Eq. (6.6) indicates that the WVD is a bilinear (or quadratic) representation in the time-frequency plane. This quadratic characteristic translates to the following distribution properties [7]:

- The WVD of the sum of two signals is equal to the sum of their respective WVD's plus crossed terms resulting from the fact that the cross-correlation of both signals is different from zero. These cross-terms are always real.
- The WVD could have negative values generated by the cross-terms.
- Taking the WVD of the analytical signal related to the original real signal, i.e. adding the Hilbert transform of the signal as its imaginary part, the negative part of the spectrum is eliminated and thus the cross-terms between positive and negative frequencies.

The WVD given by Eq. (6.6) can be also implemented using the FFT as follows:

$$WVD\,(t,f) = \frac{1}{2\pi} FFT\,[s\,(t+\tau/2)\;\; s^*\,(t-\tau/2)] \tag{6.8}$$

In the case of multicomponent signals, the cross or interference-terms introduced by the WVD generate spectrograms that are difficult to interpret. This problem has been partially solved by the introduction of a sliding window in the WVD calculation; so generating the pseudo WVD (PWVD) and the smoothed PWVD, depending on whether the selected window is applied in the time or frequency domain, respectively. Thus:

$$SPWVD\,(t,f) = \int g\,(u-t)\,PWVD\,(u,f)\,du \tag{6.9}$$

where

$$PWVD\,(t,f) = \frac{1}{2\pi}\int h\,(\tau)\,s\,(t+\tau/2)\;\; s^*\,(t-\tau/2)\,e^{-j2\pi f\tau}\,d\tau \tag{6.10}$$

The sliding window usually used is a Gaussian one because it has an optimal time-frequency concentration and yet it avoids negative terms [3]. Quite generally the smoothing tends to produce the following effects:

1. a (desired) partial attenuation of crossed-terms;
2. an (undesired) broadening of signal terms, i.e., a loss of time-frequency concentration;
3. a (sometimes undesired) loss of some of the nice mathematical properties of the WVD.

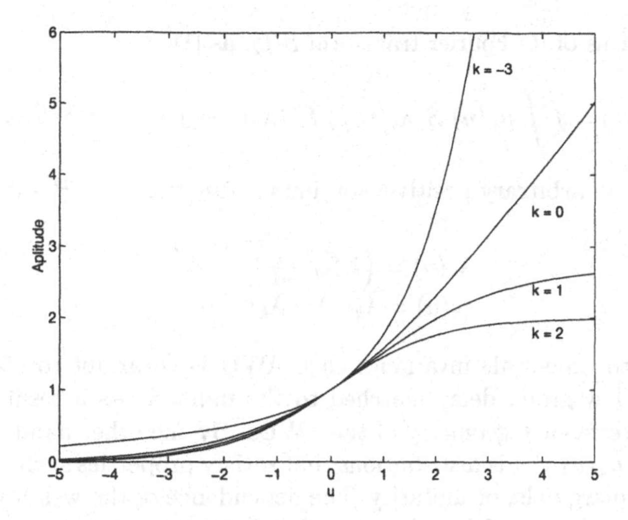

FIGURE 6.1. Plotting of the weigthing function $\lambda(u)$ for different values of parameter k

Another quadratic TFR which retains many desirable mathematical properties and yet attains a partial attenuation of crossed terms is the *Choi-Williams distribution* (CWD) [4]. In particular, the CWD satisfies the marginal properties and is thus an "energy distribution". On the other hand, it has been shown that the validity of the marginal properties places a limit on the crossed-terms attenuation of any TFR [8] and in case of the CWD also decreases its ability to detect low intensity transients [9]. A generalization of the CWD, the family of *reduced interference distributions* [21], satisfies also the finite support properties. Other extensions of the CWD (e.g., the generalized exponential distribution and the Butterworth distribution) are introduced in [15].

6.3 Affine Wigner Distribution

Most TFDs of current interest belong to either (or both) Cohen's class or the affine class. A TFD is *affine covariant* or time-scale covariant if translations and dilations in time lead to translations and scaling in frequency. To differentiate distributions belonging to the affine class from the ones belonging to the Cohen class they are named *time-scale distributions* (TSD). The property can be schematized by:

$$a^{-1/2}s\left(\frac{t-t_0}{a}\right) \quad \rightarrow \quad E_s\left(\frac{t-t_0}{a}, af\right) \tag{6.11}$$

where E is the squared magnitude of the TSD, e.g. the continuous Wavelet transform. This property makes affine distributions natural for a host of applications, including wideband radar and sonar, and self-similar signal analysis.

The *affine Wigner distributions* (AWD) are a numerous group of time-scale invariant distributions. The k-th affine Wigner distribution of an analytic signal, *s(t)*, is defined in terms of its Fourier transform *S(f)*, as [1]:

$$P_s^{(k)}(t,f) = f \int \mu_k(u)\ S(\lambda_k(u)f)\ S^*(\lambda_k(-u)f) e^{-i2\pi t f \xi_k(u)} du \tag{6.12}$$

where $\mu_k(u)$ is an arbitrary positive, continuous function, $k \in R$ and:

$$\begin{aligned} \lambda_k(u) &= \left(k \tfrac{e^{-u}-1}{e^{-ku}-1}\right)^{1/(k-1)} \\ \xi_k(u) &= \lambda_k(u) - \lambda_k(-u) \end{aligned} \tag{6.13}$$

In addition to time-scale invariance each AWD is covariant to transformations along a power-law group delay matched to the index k. As a result, the index k controls the interference geometry of the AWDs. By the other hand, proper choice of the function $\mu_k(u)$ yields distributions that satisfy properties such as correct time and frequency marginals, or unitarity. The dependence of the weighting function λ from the parameters u and k is shown in the Fig. 6.1.

The AWD being a quadratic distribution, it also generates troublesome crossed-components. Thus, in an analogous way the Wigner-Ville TFD, it has been defined a ***pseudo affine Wigner distribution*** (PAWD) by including a time window in Eq. (6.12). In contrast to the pseudo Wigner case, however, this windowing must be frequency dependent, to ensure that the resulting TFD remains affine covariant. As a result, the smoothing in frequency direction is proportional to bandwidth, rather than being constant-bandwidth as in the pseudo Wigner distribution. Thus, by using as the weighting function a wavelet ψe^{-2jx} the resulting PAWD can be formulated in terms of the Wavelet transform of the signal, E_w, [5] :

$$E_p^{(k)}(t,f) = \int \frac{\mu_k(u)}{\sqrt{\lambda_k(u)\lambda_k(-u)}} E_w(t,a_+)E_w^*(t,a_-)du \tag{6.14}$$

where :

$$a_+ = \lambda_k(u)f\ , \qquad a_- = \lambda_k(-u)f$$

The distribution given by Eq. (6.14) is one of the transformations that follows the properties given by Bertrand and Bertrand (1992) for the affine time-frequency distributions. The amount of interference suppression in the frequency direction will depend on the wavelet used.

In Eq. (6.14) the Wavelet transform is evaluated in time and scale. Since the scale parameter λ is a function of k, then k pseudo-affine distributions are generated. By choosing k=2 the ***Pseudo-Affine Wigner Distribution*** is obtained.

It is interesting to note that because the Wavelet transform can be also calculated using a linear frequency scale, not a logarithmic one, a narrower grid at low frequencies and wider at high frequencies can be used. In the Wavelet evaluation, the function λ will indicate the way by which the frequency axis will be scanned and, therefore, could be used to adapt the distribution performance to the signal.

By introducing in Eq. (6.14) a smoothing in the frequency direction, the *smoothed PAWD* is obtained. In this case the smoothing can be introduced by a frequency-independent window such as the Gaussian window. This smoothed distribution can still have a resolution exceeding that of the scalogram. In addition, changing the range of the μ parameter, the smoothed PAWD allows a continuous transition between affine Wigner distributions and scalograms.

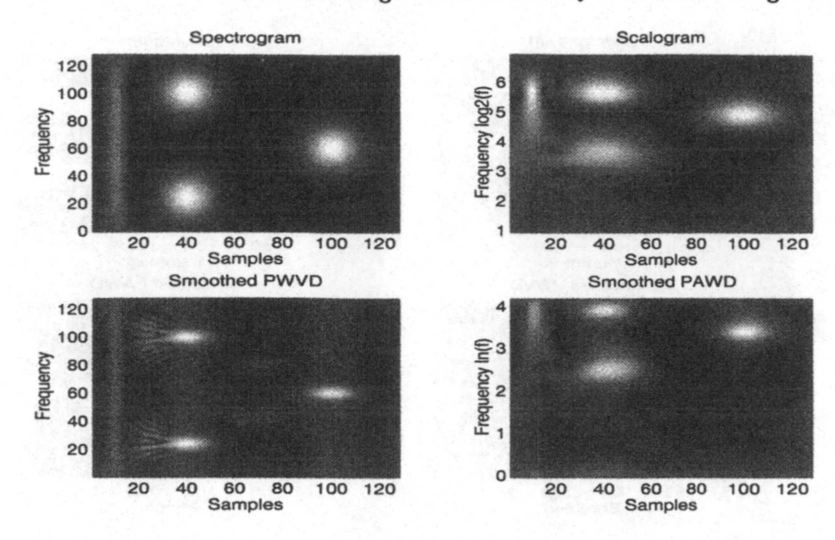

FIGURE 6.2. Time-frequency and time-scale distributions of a test signal consisting of three Gaussian-weigthed tones centered at different frequencies plus an impulsive transient at sample 10

6.3.1 Implementation of the Affine Wigner Distribution

Looking at Eq. (6.14) we can state the following steps to calculate the PAWD :

- Calculate the Wavelet transform
- Rescale the transform WT(t,f) to WT(t,$\lambda(\pm u)f$)
- Calculate the inner product or convolution of Eq. (6.14).

The change of scale implies that the WT be calculated as a function of frequency, not scale, and then be rescaled to the proper λ values. The implementation is greatly simplified by using the *discrete Mellin transform* [14], which converts dilation operations into multiplications by a phase factor. The Mellin transform of a positive function *p(f)* is defined by

$$\begin{aligned} M_p(\beta) &= \textstyle\int_0^\infty p(x)\, f^{\beta-1} dx \\ p(x) &= \tfrac{1}{2\pi j} \textstyle\int_0^\infty M_p(\beta)\, x^{-\beta} d\beta \end{aligned} \tag{6.15}$$

where β is the independent variable. The change of variable $x = e^u$ shows that the Mellin transform is closely related to the Fourier or Laplace transform. The Mellin transform has a scaling property given by

$$M_{p(ax)}(\beta) = a^{-\beta}\, M_{p(x)}(\beta) \tag{6.16}$$

where a is a positive constant different from zero. Therefore, since λ is a positive constant different from zero, it is possible to scale the wavelet transform of the function under analysis by using the Mellin transform. Furthermore, assuming $x{=}e^y$ and β as purely imaginary, a fast algorithm for the Mellin transform (FMT), using the FFT, can be implemented

$$\begin{aligned} &M\left[f(x=e^y); \beta = j\omega\right] = \textstyle\int_{-\infty}^{+\infty} f(e^y) e^{j\omega y} dy \\ &M\left[f(e^y); j\omega\right] = IFFT\left[f(e^y)\right] \\ &f(e^y) = FFT\left[M\left[f(e^y); j\omega\right]\right] \end{aligned} \tag{6.17}$$

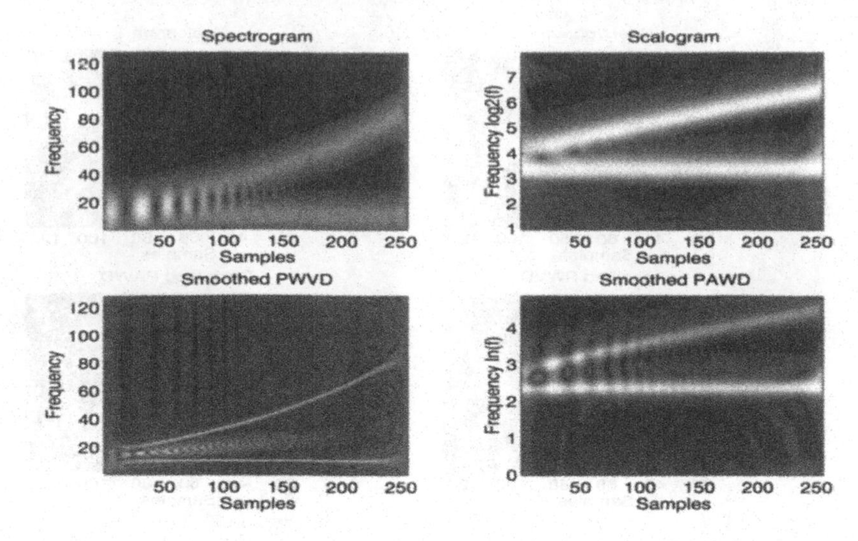

FIGURE 6.3. Time-frequency and time-scale distributions of a test signal consisting of a synthesized signal composed of a low frequency tone and an exponential chirp

Thus to calculate the FMT of a function, it is only required that the signal values $s(x)$ be calculated for exponentially spaced x values. Therefore, to calculate the smoothed PAWD through Eq. (6.14), the WT has to be calculated for a linear time-scale and an exponential frequency-scale. The FMT is applied to each temporal sample, arriving to the following expression :

$$\begin{aligned} E_p(t,f) = Re\{\textstyle\sum_{u=-U}^{U} & \frac{\mu_k(u)}{\sqrt{\lambda_k(u)\lambda_k(-u)}} \\ & FFT\left[(\lambda_k(u))^{-j\omega}\, IFFT\,(WT_s(t,f))\right] \\ & FFT^*\left[(\lambda_k(-u))^{-j\omega}\, IFFT\,(WT_s(t,f))\right]\} \end{aligned} \tag{6.18}$$

As showed by Eq. (6.18) the algorithm used to calculate the smoothed PAWD introduces an higher computational cost compared to the WT or the SPWVD. The calculation involves an array of N by N, where N is the amount of samples. By running Matlab in a Pentium PC the calculation of smoothed PAWD is restricted to samples of 256 points.

6.4 Application to Synthesized and Real Signals

In this section, the time-frequency distributions described above are applied to two synthesized signals and two practical examples. The four time-frequency distributions used are visually represented in a similar manner to make direct comparation easier. The maximum energy level in each distribution has been normalized to 1.0, as the intention is to examine the pattern of the energy distribution rather than the actual energy levels.

6.4.1 Results with Synthesized Signals

In order to show the main features of the selected tools, they are first applied to synthetic signals. The Figures 6.2 and 6.3 show the spectrogram, SPWVD, scalogram and SPAWD for two test signals, respectively. The spectrogram is computed with an overlap time window using a Hanning weighting function and the window length is selected in order to provide the best trade-off between time and frequency resolution. The SPWVD is computed with a Gaussian window in time and frequency. The SPAWD, as well as the scalogram, are computed by using a Morlet wavelet. A Gaussian window smooths the time scale. It was found that a μ parameter in the range [-4, 4] provides the best SPAWD performance.

The first test signal is a synthetic signal composed by three Gaussian-weighted tones plus an impulse. Two of the weighted-tones are centered at the same time position to verify the effectiveness of crossed-terms elimination in frequency. The SPAWD provides a performance similar to the WT and has a better cross-term elimination than the SPWVD, as shown in Fig. 6.2. It is interesting to note the fact that the WT and the SPAWD are not able to show the flat spectrum of an impulse, which should cover the whole frequency range under study.

The WT has not been plotted by using a linear scale because the change of resolution as a function of frequency distorts the scalogram. In case of the SPAWD, if using an exponential frequency scale, the low frequency components appear with a higher intensity.

A second synthesized signal, composed of a tone at low frequency and an exponential chirp, is included to emphasize the differences in time-frequency resolution of the different tools and the change of resolution both in time and frequency. In Fig. 6.3 can be clearly appreciated the good time-frequency resolution of the WVD counterweighted by only a partial elimination of the cross-terms. This interference is more severe while the components are closer.

Tests with other synthetic signals like harmonic tones, weighted tones, crossed chirps, impulses, or combinations of some of them, consistently provides a good performance from the SPAWD.

6.4.2 Results with Real Signals

6.4.2.1 Analysis of an Electrical Motor

The third vibration signal was acquired from an electrical motor in steady state rotating at 1800 rpm. Figure 6.4 shows time-frequency representations of this signal. These representations, specially SPAWD show a noisy signal with an amplitude modulation of the rotating speed frequency by a subharmonic component of this frequency. Furthermore we observe an impulse at time 0.2 s. As shown in 6.4, the SPAWD shows a better behavior than the WT to detect transient signals and it discriminates with the amplitude modulation with higher resolution than the STFT. As expected the SPWVD, owing to the crossed terms, fails to represent in a good way both features simultaneously.

6.4.2.2 Analysis of a Crown-Pinion System

The fourth signal, shown in Fig. 6.5, was generated by the coupling system between a ball mill and its motor in the Chuquicamata copper mine. It was acquired with

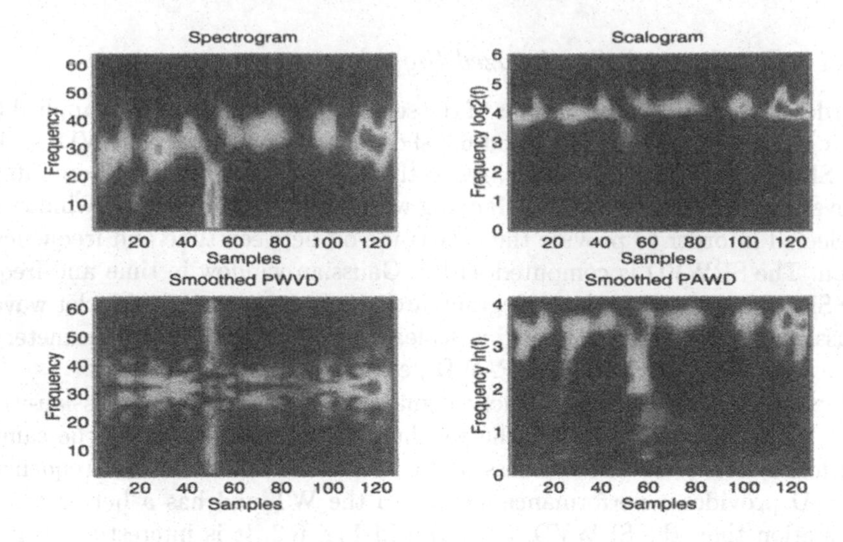

FIGURE 6.4. Time-frequency and time-scale distributions of a signal consisting of a simulated bearing defect in a rotating machine. The defect is visualized as the impulsive transient at sample 50

a data logger using a sampling frequency of 512 Hz and an observation time of 0.5 sec.. The crown of the coupling system has $N = 22$ teeth and it is rotating at $f_r = 200$ rpm. Thus, the gear frequency given by $f_g = N * f_r$, equals 73.33 Hz. A previous analysis by using the Fourier spectrum showed a small component at the rotation speed f_r, and an amplitude modulation of the gear frequency produced by an unknown component centered at 8.6 Hz. The main goal of using time-frequency representations was to observe the temporal characteristics of the unknown signal in order to be able to identify its source.

The normal vibrations at the gears in ball mills are components at the gear tooth meshing frequency and its harmonics, with some random modulation of the amplitude and/or phase of f_g due to moving loading inside the ball mill. The SPAWD of the signal showen in Fig. 6.5 allowed us to infer that the unknown component at 8.6 Hz is a torsional natural frequency of the mill excited by a high spot in the crown gear.

6.5 Conclusions

The vibration analysis techniques currently in use in machine fault detection have limitations which restrict their application in the diagnosis and prognosis of faults. Time-frequency representations can help mainly in the detection of non-stationary events. However it is well known that there is no single time-frequency representation which is the best for all problems. The same statement applies to the problem of analysing vibration signals produced by rotating machinery. The different tools tested to date are complementary and no one scores as the best.

For example, the Wavelet transform adapts better to non-stationary signal analy-

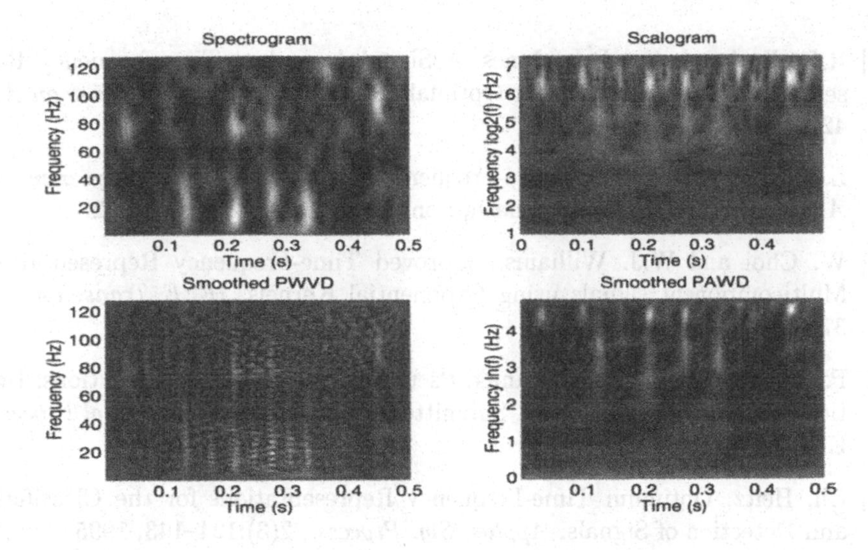

FIGURE 6.5. Time-frequency and time-scale distributions of a signal generated by the reduction gear system coupling a motor to a ball mill in Chuquicamata copper mine

sis than the STFT. The smoothed pseudo-Wigner-Ville distribution has the highest time-frequency resolution which is paid for by accepting interference produced by crossed terms.

The present work has explored the potential of the smoothed-pseudo-affine Wigner distribution. The WT and the SPAWD time-scale distributions, being of the affine class, both show some analogous properties and similar performance in many cases. However, the SPAWD consistently shows a higher time-frequency resolution.

Some cases have been explored where the SPAWD improves the time-scale representation of the WT at the cost of admitting of a small amount of interference owing to cross-terms. Anyway, the SPAWD presents less interference owing to the cross-terms than the SPWVD.

In brief, the SPAWD is a tool with good characteristics for the processing of non-stationary signals, like vibrations produced by rotating-machines applied to condition monitoring. However, any exhaustive analysis of faults has to use a complementary set of tools. The four member set tried in this work (i.e., STFT, WT, SPWVD and SPAWD) has a good potential.

Acknowledgments: This work has been funded by the Research Council of the University of Concepción, under the project Nr.95.92.31-1.1. A special acknowledgement is due to Dr. P. Gonçalvès in providing preprint material of his papers.

6.6 References

[1] J. Bertrand and P. Bertrand. Affine Time-Frequency Distributions. In *Time-Frequency Signal Analysis*. Edited by B. Boashash, Longman Cheschire Pub., 1992.

[2] R.G. Barianuk and D.L. Joncs. A Signal-dependent Time-frequency Representation: Fast Algorithm for Optimal Kernel Design. *IEEE Trans. on ASSP*, 42(1):134–146, 1994.

[3] L. Cohen. A Primer on Time-Frequency Analysis. In *Time-Frequency Signal Analysis*. Edited by B. Boashash, Longman Cheschire Pub., 1992.

[4] W. Choi and W.J. Williams. Improved Time-Frequency Representation of Multicomponent signals using Exponential Kernels. *IEEE Trans. on ASSP*, 37(6):862–871, 1989.

[5] P. Gonçalvès and R.G. Barianuk. Pseudo Affine Wigner Distributions: Definition and Kernel Formulation. Submitted to *IEEE Trans. on Signal Processing*, EDICS No. SP-2.3.1., 1996.

[6] Ch. Heitz. Optimum Time-Frequency Representations for the Classification and Detection of Signals. *Applied Sig. Process.*, 2(3):124–143, 1995.

[7] F. Hlawatsch and G.F. Boudreaux-Bartels. Linear and Quadratic Time-Frequency Signal Representations. *IEEE Signal Processing Magazine*, 21–67, 1053–5888/92, 1992.

[8] F. Hlawatsch and P. Flandrin. The Interference Structure of the Wigner Distribution and Related Time-Frequency Signal Representations. In *The Wigner Distribution - Theory and Applications in Signal Processing*. Edited by W. Mecklenbräuker, Elsevier, Amsterdam, 1995.

[9] M. Iribarren et al. Perspectives of Time-Frequency Distributions to the Analysis of Rotating Machinery Vibrations, In *Proc. IX Conference on Numerical Methods and its Applications*, Bariloche, Argentina (in Spanish), 1995.

[10] M. Iribarren, I. Foppiano, and P. Saavedra. Contributions of new Time-Frequency Distributions to Non Stationary Signal Detection, In *Proc. XII Chilean Conference on Automatic Control*, pp. 279–286, Santiago de Chile, 1996.

[11] M. Iribarren, E. Zabala, and P. Saavedra. Wavelet Signal Processing of Nonstationary Vibrations for Machinery Condition Monitoring (in Spanish), In *Proc. 7th. Latin-American Conference on Automatic Control*, pp. 803–809, Buenos Aires, Argentina, 1996.

[12] Zh. Liu. Detection of Transient Signal by Optimal Choice of a Wavelet, In *Proc. IEEE Int. Conf. on ASSP*, pp. 1553–1556, Dallas, Texas, 1994.

[13] D.L. Jones and T.W. Parks. A High Resolution Data-Adaptive Time-Frequency Representation, *IEEE trans. on ASSP*, 38(12):2127–2135, 1990.

[14] J.P. Ovarlez, J. Bertrand, and P. Bertrand. Computations of Affine Time-Frequency Distributions using the Fast Mellin Transform, In *Proc. IEEE Int. Conf. on ASSP*, pp. V117–V120, San Francisco, California, 1992.

[15] A. Papandreou and G.F. Boudreaux-Bartels. Generalization of the Choi-Williams Distribution and the Butterworth Distribution for Detection. *IEEE Trans on ASSP*, 41(1):463–472, 1993.

[16] O. Rioul and M. Vetterli, Wavelets and Signal Processing. *IEEE Signal Processing Magazine*, 8(4):14–38, 1991.

[17] P. Saavedra and M. Iribarren. Machinery Dynamic Analysis using the Wigner-Ville Distribution (in Spanish). In *Proc. V Argentinian Congress on Computational Mechanics*, Tucuman, Argentina, 1996.

[18] M.J. Shensa. The Discrete Wavelet Transform: Wedding the A-trous and Mallat Algorithms. *IEEE Trans. on Signal Processing*, 40(10):2464–2482, 1992.

[19] R.G. Stockwell, L. Mansinha, and R. P. Lowe. Localization of the Complex Spectrum: The S transform. *IEEE Trans. on Signal Processing*, 44(4):998–1001, 1996.

[20] E.F. Velez. Spectral Estimation based on the Wigner-Ville Representation. *Signal Processing*, 20(4):325-347, 1990.

[21] W.J. Williams. Reduced Interference Distributions: Biological Applications and Interpretations. In *Proceedings of the IEEE*, 84(9):1264–1280, 1996.

[22] Y. Zhao, L.E. Altes, and R.J. Marks. The Use of Cone-Shaped Kernels for Generalized Time-Frequency Representations of Nonstationary Signals. *IEEE Trans. on ASSP*, 38(7):1084-1091, 1990.

7

Coarsely Quantized Spectral Estimation of Radio Astronomic Sources in Highly Corruptive Environments

Rodolphe Weber[1]
Christian Faye[2]

ABSTRACT
This chapter is devoted to an application of non-Gaussian signal detection. In radio astronomy, given the huge flow of processed data, quantized correlators are widely used. Unfortunately, the occurrence of non-Gaussian interference can strongly alter the shape of the estimated spectra. In this chapter, the effect of a sine wave interference on a 1-bit correlator is first analysed. Then, a new interference detection criterion is proposed. It uses the real-time capabilities of quantized correlators and makes a statistical comparison between contaminated and non-contaminated quantized correlation functions. Thus, the final spectral estimation can be preserved by blanking the correlator in real-time. Simulations, using synthetic and actual data, are presented. This technique of real time detection can significantly improve the quality of spectral line observations.

7.1 Introduction

The negative impact of radio frequency interference (RFI) on the quality of spectral line observations is a matter of increasing concern for the radio astronomy community [3]. To preserve radio astronomical observation capabilities, one solution is to directly process the incoming signal. This eliminates, or at least minimizes, the effect of the harmful radio interference. In spite of the needs, very few solutions (e.g. [2, 7]) have been proposed in practice.

Moreover, the problem is compounded by the non-linearities caused by the coarsely quantized correlators (1 to 3 bits) used to estimate spectra. When the interference power increases, the induced distortions become progressively too important and make spectral estimations completely unusable (see Fig. 7.1). In this case, it becomes necessary to detect the presence of such an RFI before the spectral estimation process. Then, the data acquisition can be suspended momentarily and the previous data kept undistorted.

In this chapter, the impact of a sine wave RFI on a 1-bit spectral estimation

---

[1]Laboratoire d'Electronique, Signaux et Images, Université d'Orléans, 45067 Orléans, France, E-mail: weber@lesi.univ-orleans.fr

[2]Equipe Traitement des Images et Signaux, 6 Avenue du Ponceau, 95014 Cergy-Pontoise, France

process is first analysed. The results of this study yield the minimum detection performances necessary for a useful RFI excision. Then, the concept of a statistical detector criterion is introduced and its real time implementation is described. Simulation studies, based on actual data, show significant improvement of cosmic spectral observations in highly corruptive environments.

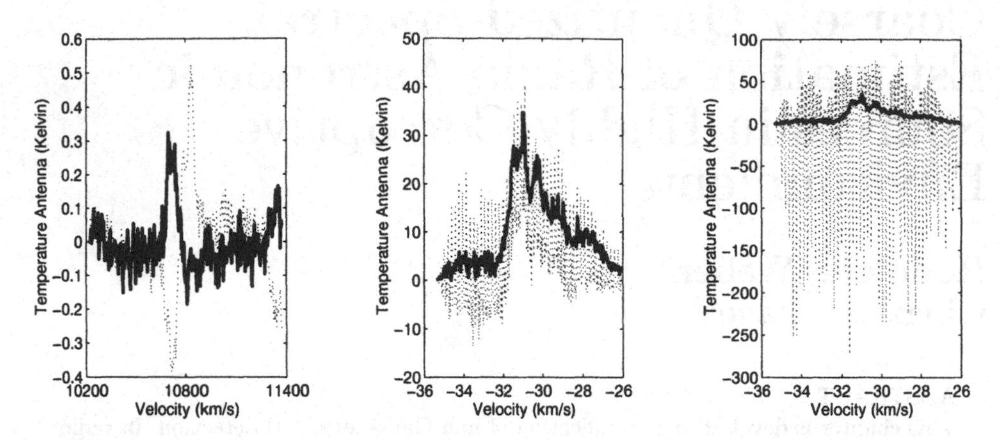

FIGURE 7.1. Spectra of the OH radical measured with the Nançay decimetric radio telescope and polluted by spread spectrum RFI. The continuous line represents the expected profile and the dotted line represents the contaminated one. The negative values of the Antenna Temperature are due to an offset profile removal

7.2 1-bit Correlator vs Sine Wave RFI

7.2.1 *Principle of a Quantized Correlator*

In our case, the spectral estimation process through a quantized correlator can be divided in four steps:

- The quantization of the measured signal $s(t)$: denote l_i as the quantization levels (see examples in Tab. 7.1).
- The computation of the quantized autocorrelation function, $R(\tau)$: denote $g(a,b)$ as the quantized product between level a and level b. When $l_i < a \leq l_{i+1}$ and $l_j < b \leq l_{j+1}$, then $g(a,b) = m_{i,j}$. The set of $m_{i,j}$ defines the quantized product table (see examples in Tab. 7.1).
- The removal of the correlation distortion induced by the coarse quantization: without RFI, the electric signal measured by a radio telescope is the sum of several Gaussian contributions, from which the very weak Gaussian part due to the cosmic source must be extracted. Under this Gaussian hypothesis, a correction function can be derived from the features of the quantized correlator [5]:

$$R(g(.),\overline{\rho},\sigma^2) = \begin{cases} \sum_i \sum_j m_{i,j} P_{i,j}(\overline{\rho},\sigma^2) & ;\overline{\rho} \neq 1 \\ \sum_i m_{i,i} P_i(\sigma^2) & ;\overline{\rho} = 1 \end{cases} \tag{7.1}$$

  where $P_{i,j}(\overline{\rho},\sigma^2)$ is the joint probability that two Gaussian variables with zero mean, equal variance σ^2 and normalized correlation coefficient $\overline{\rho} = \frac{\rho}{\sigma^2}$ belong to the $[l_i, l_{i+1}] \times [l_j, l_{j+1}]$ area. $P_i(\sigma^2)$ is the probability that a Gaussian variable zero mean and with variance σ^2, belongs to the $[l_i, l_{i+1}]$ interval.

In the 1-bit case, the relation between the true correlation, $\rho(\tau)$, and the computed one, $R(\tau)$, is given by:

$$\rho(\tau) = \rho(0)\sin\left(\frac{\pi}{2}R(\tau)\right) \tag{7.2}$$

- The Fourier transform of the corrected autocorrelation function, $\rho(\tau)$.

When $s(t)$ is free of RFI (termed the $\mathcal{H}_o$ hypothesis throughout this paper), this process gives a correct spectral estimation. Only its sensitivity (in terms of estimation variance)is slightly diminished compared to an unquantized correlator (see Tab. 7.1).

$g(a,b)$ **Quantization levels l_i ($\sigma^2 = 1$)**	1-bit (0)	Nançay (0 0.436 1.017 1.67)
Product table $m_{i,j}$	-1 1 1 -1	1 3 6 8 1 2 4 6 0 1 2 3 0 0 1 1
Sensitivity compared to an unquantized correlator	63%	95%

TABLE 7.1. Examples of technical features of quantized correlators. For the Nançay Radio Telescope, only 1/2 of the quantization levels and 1/4 of the product table are given

7.2.2 Impact of a Sine Wave RFI on the Spectral Estimation

Now, let us consider $s(t)$ as the sum of a sine wave interference, $i(t)$, and a white Gaussian signal, $s_o(t)$, representing the measured signal under the $\mathcal{H}_o$ hypothesis. Let us define the ISR (Interference to Signal Ratio) as the interference power divided by the $s_o(t)$ power (ie. the total system noise power in the receiver bandwidth).

We wish to estimate the power spectral density PSD of $s(t)$ by using the 1-bit correlator (Table 7.1). In this case, we compute the quantization of $s(t)$ first. Then, we apply a classical spectral estimation by correlation and Fourier transform of the correlation function. The complete proof is given in [6].

The resulting power spectral density estimation through a 1-bit correlator is given by:

$$PSD(f) \propto S_{ISR} + H_{1,ISR}\delta(f \pm f_c) - \sum_{n\geq 1} |H_{2n+1,ISR}|\,\delta(f \pm (2n+1)f_c) \tag{7.3}$$

where $\delta(f)$ is the Dirac delta. S_{ISR} represents the continuous level due to $s_o(t)$ (see Eq. 7.4), $H_{1,ISR}$ and $H_{2n+1,ISR}$ represent respectively the fundamental and the odd harmonics levels of the sine wave RFI with frequency f_c (see Eq. 7.5):

$$S_{ISR} = \left(1 - \sin\left(\frac{\pi}{2}\sum_{m=1}^{+\infty} C_{2m+1}^{ISR}\right)\right) \tag{7.4}$$

$$H_{2n+1,ISR} = \int_0^1 \sin\left(\frac{\pi}{2}\sum_{m=1}^{+\infty} C_{2m+1}^{ISR}\cdot\cos(2\pi m\tau)\right)\cos(2\pi(2n+1)\tau)d\tau \tag{7.5}$$

$$C_m^{ISR} = \frac{2}{\pi^2}\left(\int_0^{\pi} \operatorname{erf}\left(\sqrt{ISR}\sin(\psi)\right)\sin(m\psi)d\psi\right)^2 \tag{7.6}$$

The behaviour of these values as a function of ISR is plotted in Fig. 7.2. It appears that, for an ISR below -10 dB the correlator behavior remains linear. Notably, harmonics can be neglected, as their level is typically 50 dB below the s_o power[3]. This allows suppression or reduction of the interfering signal by off-line processing (an example is given in [1]).

On the contrary, for an ISR exceeding -10 dB, the effects of interference are exacerbated by the non-linearity of coarse quantization. In particular, the estimation of the continuous level is wrong and the harmonics generated have harmful ***negative*** power levels. Simulations with other coarse quantization schemes have shown similar results [6].

Thus, it becomes necessary to detect the presence of RFI before conducting the spectral estimation. The acquisition process must be momentarily suspended to prevent contamination of previously stored data. Consequently the final spectrum is preserved, free of distorsions, and too large RFI are suppressed. In this framework, the principle of a real time RFI detector is described in the following section.

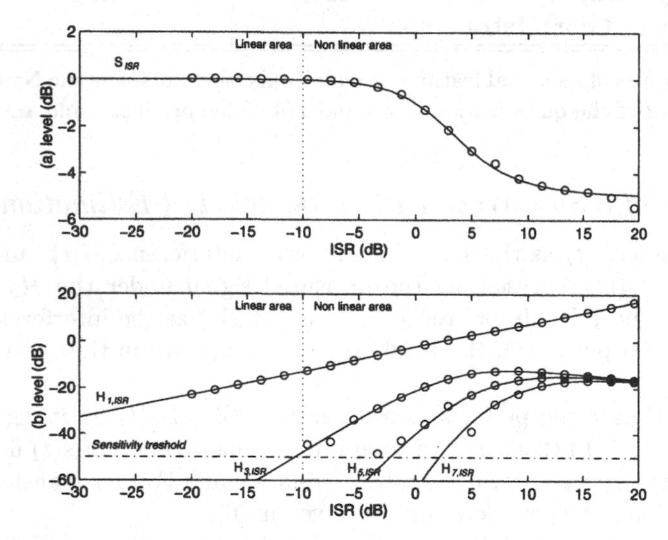

FIGURE 7.2. The spectral behaviour of a 1-bit correlator as a function of ISR. (a) the continuous level, S_{ISR}. (b) the fundamental and the odd harmonics levels of the sine wave RFI, $H_{2n+1,ISR}$

7.3 RFI Detection

7.3.1 Introduction

In this section, the concept of a spectral detector is introduced. It is based on the comparison between the spectral and statistical profiles of the measured —and pos-

[3]The interference level detrimental to radioastronomy adopted by the ITU (see, e.g., Recommendation ITU-R RA 769) gives a response equal to 0.1 of the rms noise fluctuations after 2000 seconds of integration. For a spectral line observation in L-band, a typical bandwidth is 20 kHz. The ISR is then -48 dB.

sibly disturbed— signal $s(t)$, and that of an undisturbed Gaussian signal delivered by a noise generator placed at the input of the receiver. One of the main advantages of the proposed detector is its tolerance of short and medium term non stationarity. This usually affects the measured noise due to changing observation conditions, namely antenna or earth rotation. It uses the concept of a generalized chi-square test [4] applied to second order moments of $s(t)$. The proposed implementation takes advantage of the real time capabilities of existing correlators.

7.3.2 *Application of* χ^2

Let s_i be the discrete version of the measured signal $s(t)$. From the non-linear function $g(a,b)$ and from an appropriate choice of Q time-lags τ_k, a set of Q *sample mean* generalized moments, w_k, are computed on N successive samples:

$$w_k = \frac{1}{N}\sum_{i=0}^{N-1} g(s_i, s_{i-\tau_k}) \tag{7.7}$$

where $k = 1, \ldots, Q$.

Let $\mathbf{R}$ denote the Q-dimensional vector formed by the juxtaposition of the *sample mean* w_k:

$$\mathbf{R} = [w_1, w_2, \cdots, w_Q]^t \tag{7.8}$$

where $[\ ]^t$ symbolizes the transpose operator.

In fact, the proposed vector $\mathbf{R}$ corresponds to Q time-lags of the quantized correlator and it is instantaneously obtained through the correlator buffers after N clock cycles.

When the number of samples, N, is large, the vector $\mathbf{R}$ converges to a multivariate Gaussian variable with *ensemble average* mean vector, $\mathbf{R_o}$, and *ensemble average* covariance matrix $\mathbf{\Gamma_o}$. The test function $C(s)$ consists in computing the quadratic error between the measured $\mathbf{R}$ and the expected $\mathbf{R_o}$. To normalize and take the statistical links between the w_k into account, this quadratic error is weighted by the inverse of the covariance matrix $\mathbf{\Gamma_o}$. Namely,

$$C(s) = (\mathbf{R} - \mathbf{R_o})^t \mathbf{\Gamma_o}^{-1} (\mathbf{R} - \mathbf{R_o}) \tag{7.9}$$

Under the $\mathcal{H}_o$ hypothesis, The test function $C(s)$ is distributed asymptotically as a central chi-sqared variate whose degree of freedom is related to vector size Q.

In the application presented here, only the second order statistics of $s(t)$ are tested. Nevertheless, it is also possible to exploit higher order statistics of $s(t)$ through a correlator. For example tests on 3rd or 4th order statistics can be performed by feeding the correlator with $s(t)$ and its square version, $s^2(t)$:

$$\begin{aligned} w_k^3 &= \tfrac{1}{N}\textstyle\sum_{j=0}^{N-1} g(s_i, \alpha s_{i-k}^2 + \beta) \\ w_k^4 &= \tfrac{1}{N}\textstyle\sum_{j=0}^{N-1} g(\alpha s_i^2 + \beta, \alpha s_{i-k}^2 + \beta) \end{aligned} \tag{7.10}$$

where α and β are parameters used to center and to normalize $s^2(t)$ in relation to the quantization levels. Unfortunately, simulations for low ISR (<-10 dB) have shown that performance is not improved compared with the second order case given by Eq. 7.7. In fact, such modifications of $\mathbf{R}$ increase its variance without increasing the difference between itself and the reference vector $\mathbf{R_o}$.

7.3.3 Evaluation of the Referencc Vector $\mathbf{R_o}$ and the Covariance Matrix $\mathbf{\Gamma_o}$

Under the $\mathcal{H}_o$ hypothesis, the measured signal is assumed to be Gaussian. $\mathbf{R_o}$ and $\mathbf{\Gamma_o}$ are characterized by the total power σ^2 and the normalized autocorrelation function, $\overline{\rho}(\tau)$.

From a practical point of view, $\overline{\rho}(\tau)$ depends mainly on the receiver (e.g. the successive filter shapes). Other contributions to the final spectral shape can be considered either locally white (e.g. the ground noise) or negligible (e.g. the cosmic source). Moreover, $\overline{\rho}(\tau)$ being stationary over large durations (>1 hour), it will be estimated once at the beginning of the observation by using a noise generator as a virtual non-contaminated source (see Section 7.3.4).

With regard to σ^2, due to antenna or earth rotation, stationarity can not be guaranteed beyond a few minutes. It is therefore necessary to take into account the current σ^2 value when computing $\mathcal{C}(s)$.

Because of the quantization process, the dependence of $\mathbf{R_o}$ and $\mathbf{\Gamma_o}$ on σ^2 is not simple. Moreover, the number of operations required to estimate the test function $\mathcal{C}(s)$ is proportional to Q^2. This can restrain the real time capabilities for large Q. To circumvent these drawbacks, it is necessary, firstly, to find a simple way to take into account the dependence on σ^2 and, secondly, to reduce the impact of the $Q \times Q$ matrix $\mathbf{\Gamma}_o^{-1}$ on the computational time.

Knowing the normalized autocorrelation function $\overline{\rho}(\tau)$ for each involved time lag, the reference vector $\mathbf{R}_o$ depends only on σ^2:

$$\mathbf{R_o}(\sigma^2) = \left[\sum_i \sum_j m_{i,j} P_{i,j}(\overline{\rho}(\tau_1), \sigma^2), \ldots, \sum_i \sum_j m_{i,j} P_{i,j}(\overline{\rho}(\tau_Q), \sigma^2) \right]^t \tag{7.11}$$

So, the $\mathbf{R}_o$ variation with σ^2 will be given by Q (one for each component of the vector), second order polynomial approximations of Eq. 7.11. These approximations must be valid within the expected fluctuation domain of σ^2 under the $\mathcal{H}_o$ hypothesis.

A general formula for $\mathbf{\Gamma}_o$ is complicated by the correlation between the samples s_i. A solution would be to neglect the dependence between samples and to simply consider the noise under the $\mathcal{H}_o$ hypothesis as a white noise despite its band limitation. This simplification induces a change in the weighting factors of the errors between the $\mathbf{R}$ and $\mathbf{R}_o$ components and makes the criterion less than optimal.

Under either of these hypotheses, the matrix $\mathbf{\Gamma}_o$ becomes diagonal:

$$\mathbf{\Gamma_o}(\sigma^2) = \frac{1}{N} \sum_i \sum_j m_{i,j}^2 P_{i,j}(0, \sigma^2) \mathbf{I} \tag{7.12}$$

where $\mathbf{I}$ is the unit matrix.

Therefore, no matrix inversion is needed and the number of operations necessary to compute the test function $\mathcal{C}(s)$ becomes proportional to Q. Furthermore, its variation with σ^2 can be computed as a second order polynomial approximation. $\mathcal{C}(s)$ becomes a classical quadratic mean error test.

7.3.4 Implementation

In this section, a method to implement the previous detection criterion is proposed (see Fig. 7.3). The criterion uses the first Q time lags ($\tau_k = k$) of the quantized correlation function $R(\tau)$. To initialize the detector, knowledge of the receiver spectral shape is necessary. Thus, at the beginning of the observation, the receiver is directly connected to a noise generator so that the $\mathcal{H}_o$ hypothesis is forced. The quantized autocorrelation $R(\tau)$ is estimated and the true normalized autorrelation, $\overline{\rho}(\tau)$, is deduced from $R(\tau)$ by applying the correlation correction function. Then, the coefficients of the second order polynomial approximations, used to evaluate the $\mathbf{R_o}$ and $\mathbf{\Gamma_o}$ dependance on σ^2, are computed and stored.

The receiving system is then connected to the antenna, and observation starts. At each clock cycle, the Q values of the quantized product for the Q first time lags are computed by the correlator and stored into Q shift registers of size N. For each time-lag, these shift registers represent a moving window on the N last quantized products used to compute the final vector $\mathbf{R}$.

The component w_o, which is the result of the null time-lag and is an estimate of the input power σ^2 (see note 4), is used to update $\mathbf{R_o}$ and $\mathbf{\Gamma_o}$. Then, the test function $C(s)$ is computed and its value is compared with the predefined detection level λ. If the test function value is less than λ, no RFI is detected, and the "first in" quantized products are sent to the final integration. If the criterion value is greater, an RFI is detected and the final sum is suspended until the test function value comes down below the detection level again.

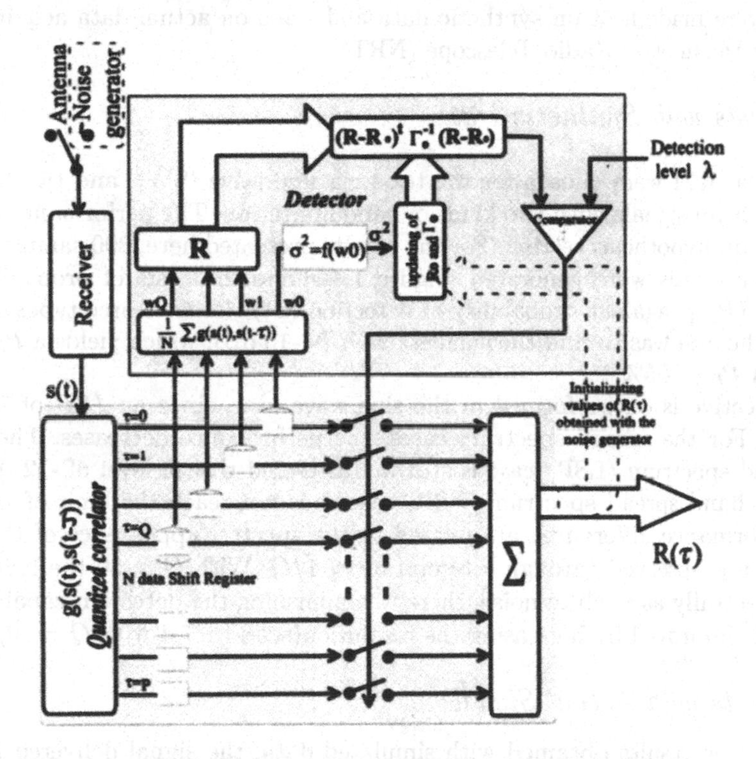

FIGURE 7.3. Implementation of the detector

For a given RFI, the choice of the size Q strongly determines the detector performance. From a spectral point of view, the detector carries out a comparison

[4] w_o and σ^2 are linked together by $w_o = \sum_i m_{i,i} P_i(\sigma^2)$.

between an estimated spectrum and a reference spectrum with a spectral resolution inversely proportional to Q. By using values of Q which are too small, the risk is the smoothing of relevant spectral features of the RFI and therefore reduction in the quadratic error between $\mathbf{R}$ and the reference $\mathbf{R}_o$. In contrast, large values of Q may reduce the detector performance because of a large induced variance[5]. In fact, the optimal value of Q must be chosen as a function of the RFI profile and the observational context.

Nevertheless, for multiple RFI detection or blind detection (no a priori information on RFI), the proposed detector can be modified to perform multiresolution criteria: the detector is sized for the largest value of Q (highest resolution) and criteria with intermediate resolution are obtained recursively.

7.4 Performance

7.4.1 Introduction

The purpose of this section is to evaluate the performance of the proposed detector. The number of samples, N, has been fixed at 10000. All the numerical processes, including the quantized autocorrelation, have been simulated numerically. The quantization parameters are those of the Nançay autocorrelator (see Tab. 7.1). The tests were made first on synthetic data and, then on actual data acquired by the Nançay Decimetric Radio Telescope (NRT).

7.4.2 Tests with Synthetic Data

Three typical RFI were chosen for the tests : a sine wave (SW), and two filtered spread spectrum signals with two kinds of band limitation. The performance analysis is based on hypothesis testing. For the results presented here, 800 sample paths for each hypothesis were generated. Figure 7.4 shows the plots of probability of false alarm (P_{FA}) against probability of detection (P_D) for the three types of RFI retained. The aim was to find the smallest ISR (<-10 dB), which yields a P_D near 95% with a P_{FA} of 5%.

This objective is outperformed in the sine wave case, since an ISR of -17 dB is reached. For the spread spectrum cases, the performance decreases. The large band spread spectrum (LSP) case is still well detected with a level of -12 dB but the narrow band spread spectrum (NSP) case does not reach the limit of -10 dB. These performance differences are related to the spectral appearance of the RFI when they are observed through a resolution of $1/Q$. With $Q = 30$ the NSP case appears spectrally as a white noise, thereby diminishing the detection capabilities. Detection is improved by increasing the resolution(see Fig. 7.4 with $Q = 70$).

7.4.3 Tests with Actual Signals

To validate the results obtained with simulated data, the signal delivered by the receiver was sampled at a rate of 200 kHz for 10 s. A dedicated RFI generator was turned on at random times and for random durations. The power of RFI was adjusted to deliver an ISR of -10 dB. The same algorithm (defined in Section 7.3.4)

[5]The variance of chi-square law is proportional to the degree of freedom, which is approximatively Q in the present case.

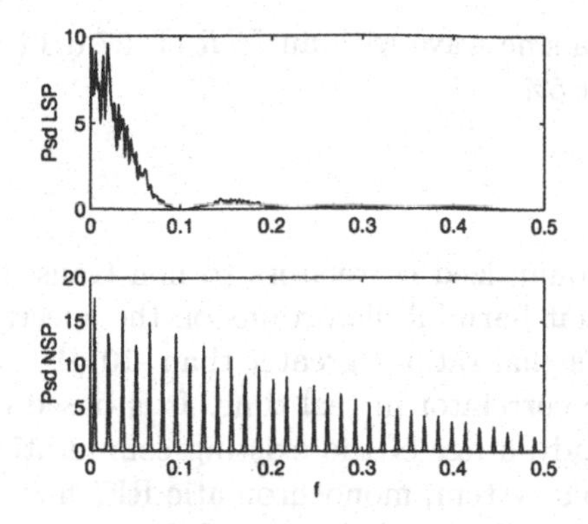

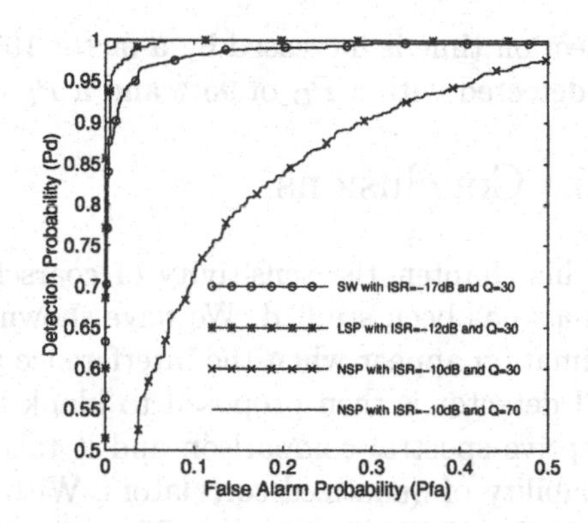

FIGURE 7.4. P_{FA} **against** P_D **for the three types of RFI : a sine wave (SW), a narrow band spread spectrum RFI (NSP), a large band spread spectrum RFI (LSP).** $N = 10000$ **samples are used for each one of the 100 trials and** Q **is the size of the test vector** $\mathbf{R}$

was applied to the stored data, the detection threshold being chosen to guarantee a P_{FA} of 5%. The detection window was fixed at $N = 10000$ samples (50 ms). When an RFI was detected, the corresponding data were discarded from the final integration. The resulting spectra are shown in Fig. 7.5.

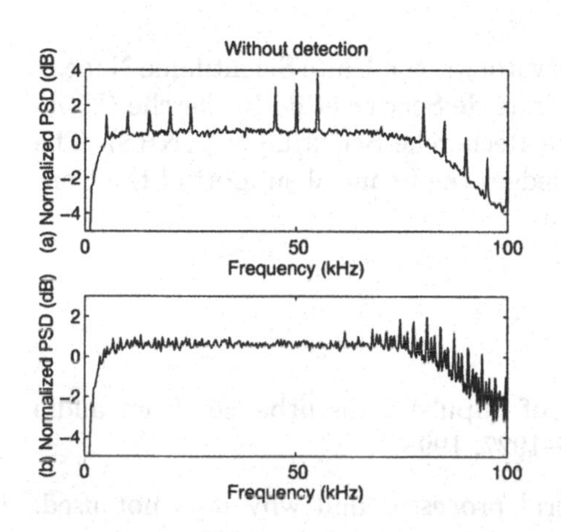

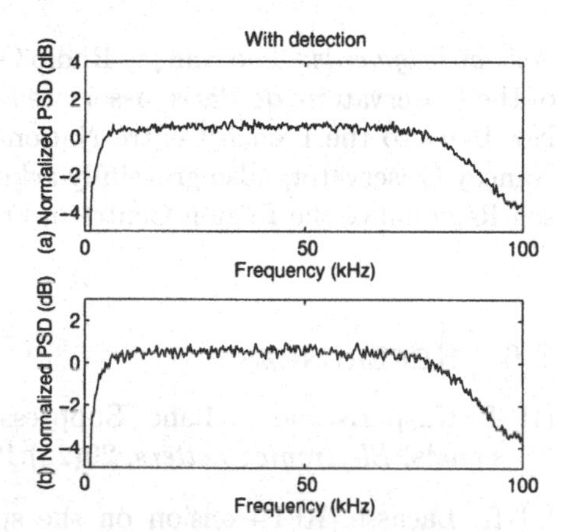

FIGURE 7.5. Experimental spectra from the decimetric Nançay radiotelescope receiver. They are measured for 10 s on a 100 kHz band without (left) and with (right) time-blanking processing. Detection time = 50 ms (equivalent to $N = 10000$ **samples),** $P_{FA} = 5\%$**,** $Q = 20$**,** $ISR = -10$ **dB: (a) 12 sine waves RFI (SW), (b) 2 narrow band spread spectrum RFI (NSP)**

7.4.4 Asymptotic Performances

In this section, the asymptotic performance is evaluated as a function of ISR and N. As shown in Section 7.4.2, the dependance on Q is strongly linked with the RFI spectral appearance and is not included in this analysis. Equation 7.12 shows that $\mathbf{\Gamma}_o^{-1}$ is proportional to N. When ISR is low (<-10 dB) the quantized correlator can be considered as linear. Consequently, the term $\mathbf{R} - \mathbf{R_o}$ of Eq. 7.9 is proportional to ISR. Thus, the criterion $\mathcal{C}(s)$ is proportional to $N(ISR)^2$. For example, if the detection time is increased by a factor 100, a sinc wave with an ISR of -27 dB can be detected with a P_D of 95% and a P_{FA} of 5%.

7.5 Conclusions

In this chapter, the sensitivity of coarsely quantized correlators to non-Gaussian signals has been studied. We have shown that harmful distortions on the spectral estimation appear when the interference to signal ratio is greater than -10 dB. An RFI detector is then proposed to blank the correlator in real time. It is based on adaptive spectral comparison and it takes advantage of the existing computation capability of quantized correlators. With this system, monochromatic RFI having an ISR=-17dB are detected. This occurs with $Q = 30$ time-lags of the quantized correlation function estimated over only $N = 10000$ samples. The method described in the paper is thus useful for rapid identification of rather strong (compared to radio astronomic thresholds) interference. To limit the impact of the RFI spectral appearance, a multi-frequency strategy is proposed. A later version of this detector could adopt a multi-scale strategy with the aim of minimizing the receiver blanking as a function of the RFI duration.

Acknowledgments: The Nançay Radio Observatory is the Unité Scientifique Nançay of the Observatoire de Paris, associated as Unité de Service et de Recherche (USR) No. B704 to the French Centre National de Recherche Scientifique (CNRS). The Nançay Observatory also gratefully acknowledges the financial support of the Conseil Régional of the Région Centre in France.

7.6 References

[1] T. Kasparis and J. Lane. Suppression of impulsive disturbances from audio signals. *Electronics Letters*, 29(22):1926–1927, 1993.

[2] R. Lacasse. RFI excision on the spectral processor and why it is not used. *Workshop on New Generation Digital Correlators*, N.R.A.O., 1993.

[3] D. McNally. *The vanishing universe - adverse environmental impacts on astronomy.* Cambridge University Press, Part III, pp. 71–113, 1994.

[4] E. Moulines, J.W. Dalle Molle, K. Choukri, and M. Charbit. Testing that a stationary time-series is Gaussian: time-domain vs. frequency-domain approaches. *IEEE Proc. on H.O.S.*, 336–340, 1993.

[5] A.R. Thomson, J.M. Moran, and G.W. Swenson. *Interferometry and synthesis in radio astronomy.* John Wiley & Son, 1986.

[6] R. Weber. Ph.D Thesis, Université de Paris XI Orsay, France, 1996.

[7] R. Weber and C. Faye. Détecteur temps réel de signaux cyclostationnaires. Principe et implémentation, In *GRETSI'97 Proc.*, pp.1021-1024, France, 1997.

8

Wavelet Transform Implementation for Star-Shaped Modelling

Andrei Doncescu[1]
Jean-Paul Gourret[1]

ABSTRACT
We present in this paper a new mesh generation method based on wavelets. We have constructed filters associated with wavelet basis which was applied to $3D$ objects to obtain several levels of subdivision.

8.1 Introduction

Today, numerical simulation of object deformations is of interest in the field of image analysis and in the field of image synthesis. Mathematical models are often common to these fields. We can address the finite element method primarily developed to analyze the behavior of metallic structures and then used to synthesize object deformation in accordance with physical laws [6, 13]. We can address the wavelet method primarily developed to analyze the behavior of geophysical signals [14], to analyze images for image understanding [1] and compression [11], then used to synthesize shapes [9, 4] or illumination [8].

To deform a $3D$ objects the finite elements method is used. In general the objects are meshed by triangles, which introduce extraordinary points and the deformation calculation is not homogeneous. Also, a systematic recursive subdivision of facets is difficult because there is a loss of connectivity properties. The mesh developed in this paper is made up of 3-connected (a vertex is connected at the end of three bars) 1D-elements of approximately equal length which define "pentagons" and "hexagons" not necessarily planar. The one-dimensional elements are elastic bars. They can be translated without deformation, compressed or stretched. This may seem regressive compared with sophisticated two-dimensional patches developed today, but it gives many advantages for mesh generation and deformation calculations. The 1D-elements cover the external envelope of the faces.

As an example, we have implemented our algorithm to generate the mesh on digitized human faces. The faces are 3D solids, supposed star-shaped. We project a given level of the sphere on a digitized face and our algorithm deduces other levels

[1]University of La Rochelle, 17000 La Rochelle, France,
E-mails: {andrei, jean-paul.gourret}@mail.univ-lr.fr

FIGURE 8.1. The projection of the sphere on a digitized face

of meshing (see Fig. 8.1).

The new mesh generation is based on wavelets. We construct filters whose coefficients are associated with a wavelet basis. With these filters, and admitting an error on reconstruction, we are able to generate objects at several levels of sampling. At each level an object is represented by an approximation and detail information.

The methods of subdivision, developed in this paper and based on filters associated with the wavelet transform, hold the centers of gravity and each new level of subdivision adds new centers of gravity between the old centers. We could consider our algorithm as a method of interpolation.

8.2 Modelling

Our process needs to construct a dodecahedron which is a regular polyhedron and then construct a truncated icosahedron, which can also be termed as a semiregular polyhedron made up of 12 pentagons and 20 hexagons. This construction is very interesting because we know that a truncated icosahedron can be obtained from an icosahedron when dividing its edges by 3. This division is very simple to carry out, but we must start from an icosahedron for this process and the process cannot be repeated indefinitely. Indeed, our model needs an equal length for all edges on the sphere, a connectivity 3 and a homogeneous mesh. To transform level 1 to level 2 (and so on) we cannot truncate the truncated icosahedron because it is not possible

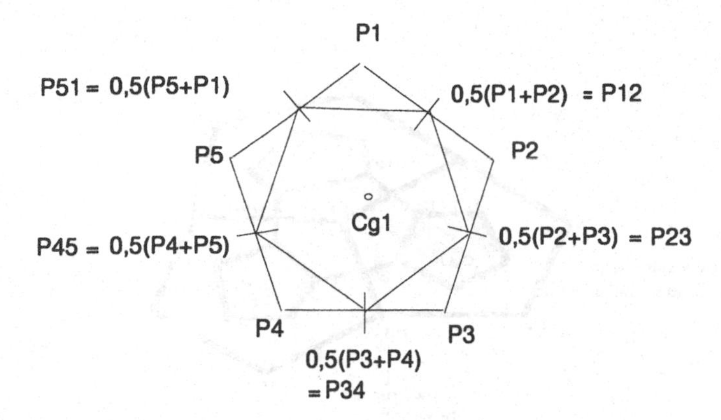

FIGURE 8.2. First step of the subdivision process

to keep the homogeneous condition.

So we choose another way to obtain a truncated icosahedron and the following levels. We start from a dodecahedron and we rotate and contract each pentagonal facet around the axis passing through its center of gravity and the center of the sphere. In this way we can maintain edge with connectivity 3, approximately equal edge length, and a homogeneous mesh. Polyhedra obtained with our process are "semi-regular polyedra". By definition we will put the word "polyhedra", "polygon", "hexagon", "pentagon" in inverted commas when the facets are not planar.

A possible way to transform level 0 to level 1 is described below. In this process the rotations are decomposed on the three axis coordinates. This method gives good results when the facets contain small deformations but the centers of gravity are always held whatever the amplitude of deformations.

We obtain rotation-contractions in two steps, by means of signal processing methods. First, by joining the middle of every edge of polygonal facets. Fig. 8.2 shows the subdivision of a polygonal facet $\underline{P} = (P_1\ P_2\ P_3\ P_4\ P_5)$ and the corresponding matrix form as given by:

$$\underline{M} = \underline{P}\underline{\underline{H_1}} \tag{8.1}$$

$$\underline{M} = \begin{pmatrix} P_1 & P_2 & P_3 & P_4 & P_5 \end{pmatrix} \begin{pmatrix} 0.5 & 0 & 0 & 0 & 0.5 \\ 0.5 & 0.5 & 0 & 0 & 0 \\ 0 & 0.5 & 0.5 & 0 & 0 \\ 0 & 0 & 0.5 & 0.5 & 0 \\ 0 & 0 & 0 & 0.5 & 0.5 \end{pmatrix} \tag{8.2}$$

$$\underline{M} = (0.5(P_1 + P_2) \quad 0.5(P_2 + P_3) \quad 0.5(P_3 + P_4) \quad 0.5(P_4 + P_5) \quad 0.5(P_5 + P_1)) \tag{8.3}$$

This step can be found to be Haar Wavelet Transform with a $\underline{\underline{H_1}}$ filter. Indeed the division of a segment edge is on one hand $P_{12} = tP_1 + (1-t)P_2$ and the middle is obtained for $t = 0.5$. On the other hand the filter associated with a Haar basis has two coefficient $h(0) = 0.5$ and $h(1) = 0.5$ which also give $P_{12} = h(0)P_1 + h(1)P_2$.

The axis of rotation is perpendicular to the polygonal facet at its center of gravity C_g. The first contraction is made around this axis. The second is made by contracting again each new polygonal facet (see Fig. 8.3). Doing this, each new vertex such as P_{12} is separated into two vertices P'_{12} and P''_{12}.

Every vertex has 3 neighbors and of course 3 edges. The subdivision from level 0

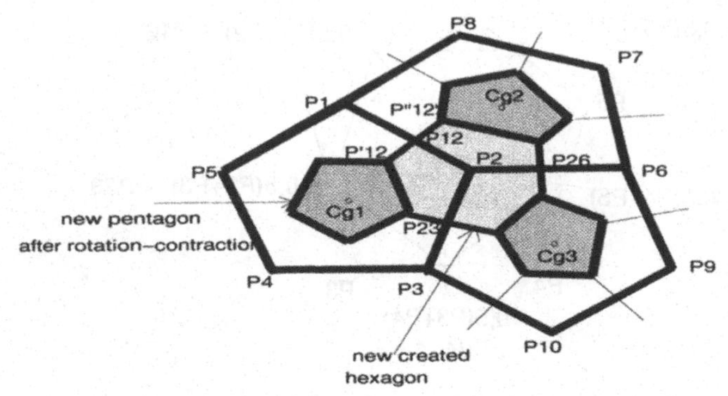

FIGURE 8.3. Second step of the subdivision process

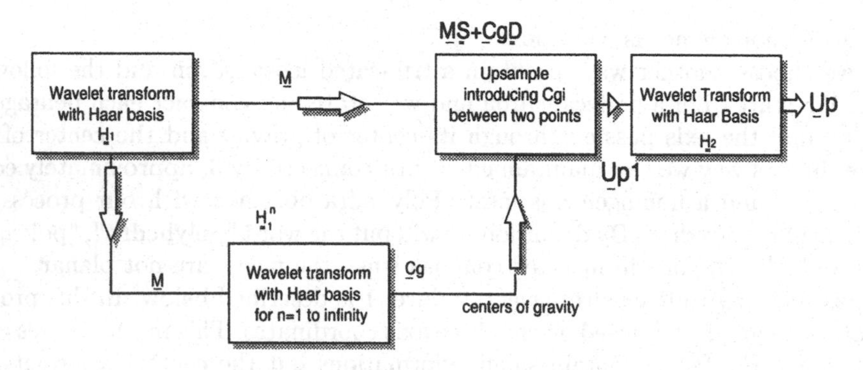

FIGURE 8.4. Subdivision with Haar basis

to 1 and more generally from level i to level $i+1$ constructs a new hexagon around each vertex.

This second contraction process can be found to be another Wavelet Transform. It is a Wavelet Transform associated with a Haar basis when $P_{12}^{'}$ is the center of $P_{12} - C_{g_1}$, or associated with another basis when $P_{12}^{'}$ is not at the center. The next relationships describe the splitting of the vertex P_{12} into $P_{12}^{'}$ and $P_{12}^{"}$.When rotation-contractions are carried out to infinity each polygonal face reaches the center of gravity (C_{g_1}). The centers of gravity can be calculated by applying $\underline{\underline{H_1}}^n$ to $\underline{P}$ as shown in Fig. 8.4.

The next relationship gives the matrix form to introduce these centers of gravity which are the point of convergence of each polygon when level reaches the infinity.

$$\underline{U_{p_1}} = \underline{M}\,\underline{\underline{S}} + \underline{C_g}\,\underline{\underline{D}} \tag{8.4}$$

where:

$$\underline{C_g} = \quad (C_{g_1} \quad C_{g_1} \quad C_{g_1} \quad C_{g_1} \quad C_{g_1}) = \underline{P} * \underline{\underline{C}}$$

$$\underline{\underline{C}} = \frac{1}{5}\begin{pmatrix} 1 & 1 & 1 & 1 & 1 \\ 1 & 1 & 1 & 1 & 1 \\ 1 & 1 & 1 & 1 & 1 \\ 1 & 1 & 1 & 1 & 1 \\ 1 & 1 & 1 & 1 & 1 \end{pmatrix}$$

$$\underline{\underline{D}} = \begin{pmatrix} 0 & 1 & 0 & 0 & 0 & 0 & 0 & 0 & 0 & 0 \\ 0 & 0 & 0 & 1 & 0 & 0 & 0 & 0 & 0 & 0 \\ 0 & 0 & 0 & 0 & 0 & 1 & 0 & 0 & 0 & 0 \\ 0 & 0 & 0 & 0 & 0 & 0 & 0 & 1 & 0 & 0 \\ 0 & 0 & 0 & 0 & 0 & 0 & 0 & 0 & 0 & 1 \end{pmatrix}$$

$$\underline{\underline{S}} = \begin{pmatrix} 1 & 0 & 0 & 0 & 0 & 0 & 0 & 0 & 0 & 0 \\ 0 & 0 & 1 & 0 & 0 & 0 & 0 & 0 & 0 & 0 \\ 0 & 0 & 0 & 0 & 1 & 0 & 0 & 0 & 0 & 0 \\ 0 & 0 & 0 & 0 & 0 & 0 & 1 & 0 & 0 & 0 \\ 0 & 0 & 0 & 0 & 0 & 0 & 0 & 0 & 1 & 0 \end{pmatrix}$$

The matrix form of the second contraction process around the center of gravity is :

$$\underline{U_p} = \underline{U_{p_1}}\ \underline{\underline{H_2}} \tag{8.5}$$

where the matrix $\underline{\underline{H_2}}$ is $[10 \times 5]$ because we upsample by C_{g_i} and we downsample one of two.

$$\underline{U_p} = \begin{pmatrix} 0.5(P_1+P_2) \\ C_{g_1} \\ 0.5(P_3+P_2) \\ C_{g_1} \\ 0.5(P_3+P_4) \\ C_{g_1} \\ 0.5(P_4+P_5) \\ C_{g_1} \\ 0.5(P_1+P_5) \\ C_{g_1} \end{pmatrix}^T \begin{pmatrix} 0.5 & 0 & 0 & 0 & 0 \\ 0.5 & 0 & 0 & 0 & 0 \\ 0 & 0.5 & 0 & 0 & 0 \\ 0 & 0.5 & 0 & 0 & 0 \\ 0 & 0 & 0.5 & 0 & 0 \\ 0 & 0 & 0.5 & 0 & 0 \\ 0 & 0 & 0 & 0.5 & 0 \\ 0 & 0 & 0 & 0.5 & 0 \\ 0 & 0 & 0 & 0 & 0.5 \\ 0 & 0 & 0 & 0 & 0.5 \end{pmatrix} \tag{8.6}$$

$$\underline{U_p} = \begin{pmatrix} 0.25(P_1+P_2)+0.5C_{g_1} \\ 0.25(P_3+P_2)+0.5C_{g_1} \\ 0.25(P_3+P_4)+0.5C_{g_1} \\ 0.25(P_4+P_5)+0.5C_{g_1} \\ 0.25(P_1+P_5)+0.5C_{g_1} \end{pmatrix}^T \tag{8.7}$$

$$\left.\begin{matrix} 0.25(P_1+P_2)+0.5C_{g_1} \\ 0.25(P_1+P_2)+0.5C_{g_2} \end{matrix}\right\} P_2 \textit{ split in two points}$$

After this process we obtain a truncated polyhedron composed of 12 pentagonal facets and 20 hexagonal facets obtained around every vertex of level 0. That is the basis of our approximation of a sphere named level 1.

For obtaining the successive levels (2,3,...) we use the same transformation on pentagons and hexagons. For the hexagons the matrices $\underline{\underline{D}}$ and $\underline{\underline{S}}$ have a size $[6 \times 12]$ instead of $[5 \times 10]$. the vectors $\underline{M}$ and $\underline{U_p}$ are of size $[1 \times 6]$ instead of $[1 \times 5]$, the vector $\underline{U_{p_1}}$ is of size $[1 \times 12]$ instead of $[1 \times 10]$, the Haar matrix $\underline{\underline{H_1}}$ is of size $[6 \times 6]$ instead of $[5 \times 5]$, the Haar matrix $\underline{\underline{H_2}}$ is of size $[12 \times 6]$ at the place $[10 \times 5]$, and the $\underline{\underline{C}}$ matrix is of size $[6 \times 6]$ multiplied by a coefficient 1/6.

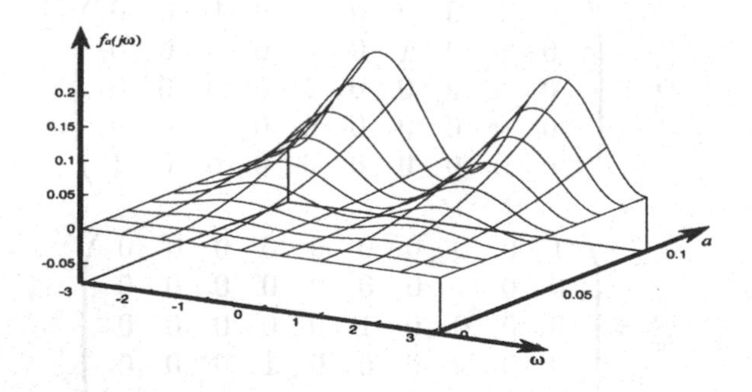

FIGURE 8.5. Error of reconstruction

8.3 Application of Wavelet Filter to Splitting

During the subdivision process we want to obtain the same length for each edge. However, the contraction in the second step with Haar basis $\underline{\underline{H}}_2$ gives an error in the length of edges.

The division of a segment edge is:

$$P_3 = tP_1 + (1-t)P_2$$

where P_1 corresponds to a C_{g_i}, P_2 corresponds to one middle point obtained from the first step and P_3 is obtained from this second step. With Haar basis $t = 0.5$, P_3 is the middle of the segment and we can write :

$$P_3 = h(0)P_1 + h(1)P_2 \tag{8.8}$$

where $(h(0) = 0.5, h(1) = 0.5)$ are the coefficients of the filter.

Because there is an error in the length of edges, we must define a new filter associated with another wavelet basis.

A filter is associated with a wavelet basis when the condition of reconstruction is satisfied and the filter must be high-pass, so the conditions:

$$|G(e^{j\omega})|^2 + |G(e^{j(\omega+\pi)})|^2 = 1 \tag{8.9}$$

$$|G(e^{j\pi})| = 1 \text{ and } |G(e^{j0})| = 0 \tag{8.10}$$

G is the Fourier Transform of the mirror filter of H. There are a lot of filters which satisfy the conditions described previously, but we also want to satisfy $\sum_k h(k) = 1$ and $\sum_k g(k) = 0$ and with a minimum number of coefficients.

If we use a filter with 2 coefficients, we obtain the Haar basis (t=0.5). If we use 3 coefficients it is impossible to satisfy the conditions. So we must use a filter with a minimum of 4 coefficients.

A filter with 4 coefficients, which we wish to examine in this work, is of the form:

$$g(0) = 0.5 - a \quad g(1) = -0.5 - a \quad g(2) = a \quad g(3) = a$$

The particular case of Haar basis is included when $a = 0$.

$$G(e^{j0}) = g(0) + g(1) + g(2) + g(3) = 0 \tag{8.11}$$

$$G(e^{j\pi}) = g(0) - g(1) + g(2) - g(3) = 1 \tag{8.12}$$

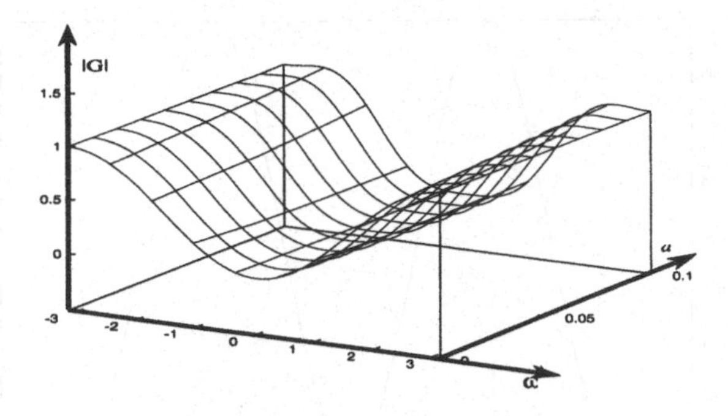

FIGURE 8.6. Modulus of G

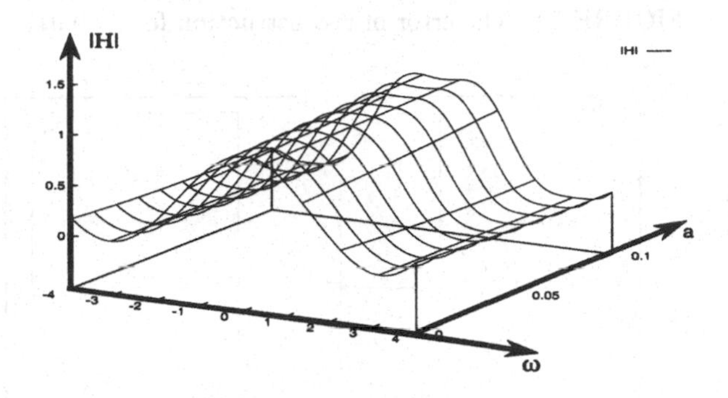

FIGURE 8.7. Modulus of H

With our coefficients, the relations $G(e^{j\pi}) = 1$ et $G(e^{j0}) = 0$ are satisfied. Fig. 8.6 provides evidence that our filter is high-pass.

The reconstruction formula is approximated by the function :

$$f_a(e^{j\omega}) = |G(e^{j\omega})|^2 + |G(e^{j(\omega+\pi)})|^2 - 1 \tag{8.13}$$

We have shown (see Figs. 8.5, 8.6) and 8.7) the influence of a on $f_a(e^{j\omega})$ and on the modulus of G and H.

The reconstruction seems valid for values $a < 0.05$. For $a > 0.05$ the error of reconstruction is too large because we cannot say that the basis is orthogonal. The error of reconstruction for $a = 0.017$ is shown in Fig. 8.8. Because the non-orthogonality the relationship:

$$\sum_k h(k)\bar{h}(k+2m) = \delta_{0,m} \tag{8.14}$$

is not perfectly verified. To transform level 0 to level 1 we have found $a = 0.09$ (to obtain equal length for the edges).

To transform level 1 to level 2 we have found $a = 0.017$. After level 2 we have used the same coefficient a = 0.017 . This choice is valid because we can continue to apply planar geometry to the 12 pentagons and 20 hexagons, which stay planar and we have the same length of edges for these polygons, at every level of subdivision. Other " hexagons" created after level 1 are not planar. So we put the word hexagon in inverted commas. Then it is trivial to find coefficients $h(i)$.

In the following we discuss level i ($i \geq 2$). For these the coefficients are:

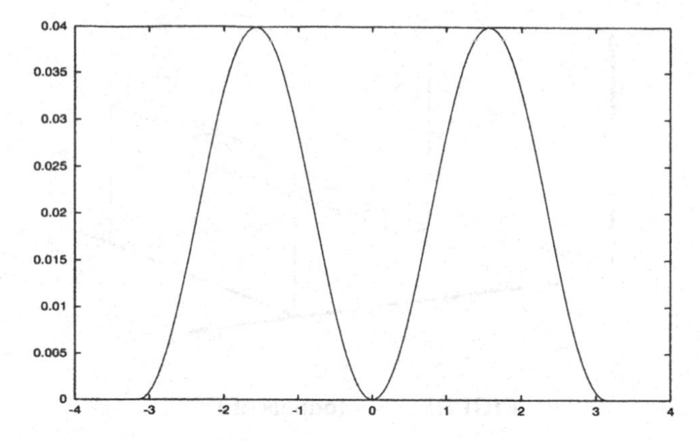

FIGURE 8.8. The error of reconstruction for a=0.017

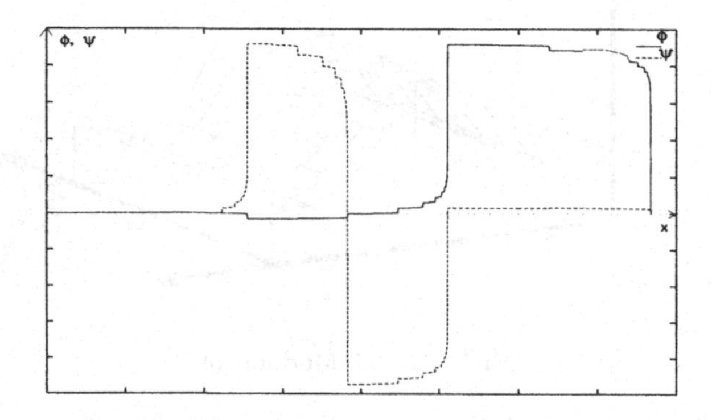

FIGURE 8.9. Functions ϕ and ψ

$$g(0) = 0.483 \quad g(1) = -0.517 \quad g(2) = 0.017 \quad g(3) = 0.017$$

We can deduce:

$$h(0) = -0.017 \quad h(1) = 0.017 \quad h(2) = 0.517 \quad h(3) = 0.483$$

The functions ϕ and ψ associated with these filters coefficients are shown in Fig. 8.9. They are obtained by a cascade algorithm.

8.4 Database Construction

We have previously described the subdivision process. This process consists of adding vertices, edges and "polygonal facets" in accordance with the following recursive relations:

Vertices:

$V_0 = 20$ number of vertices for the dodecahedron (level 0)
$V_1 = 60$ number of vertices for the truncated icosahedron (level 1)
$V_i = 6V_{i-2} + V_{i-1} = 3V_{i-1}$ (*level* $i \geq 2$)
$V_i = 20 \times 3^i$

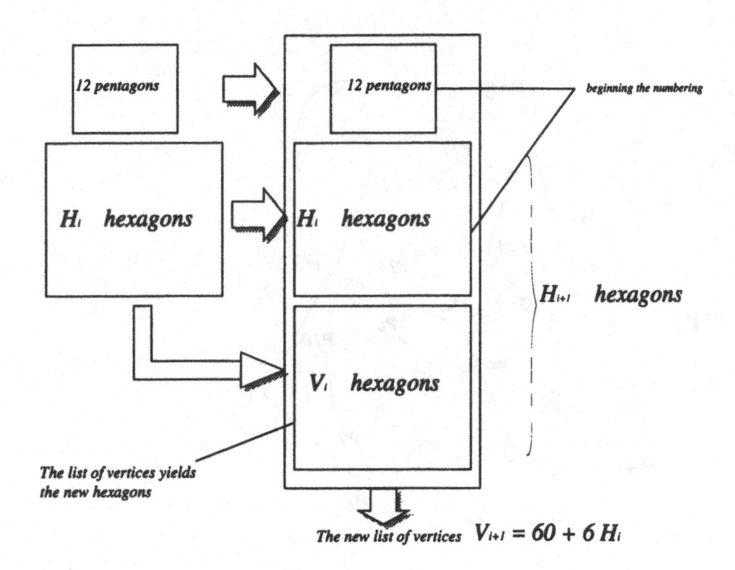

FIGURE 8.10. Pentagons and hexagons builder

Facets:
Pentagons
$P_i = 12$ level i, $i = 0, 1, 2, \ldots$
Hexagons
$H_0 = 0$ (level 0)
$H_1 = 20$ (level 1)
$H_i = 60 + 6H_{i-2} + H_{i-1}$ (*level* $i \geq 2$)
$H_i = V_{i-1} + H_{i-1} = 20\ 3^{i-1} + H_{i-1}$
$H_i = 20(3^{i-1} + 3^{i-2} + \ldots + 1) = 20\frac{3^i-1}{3-1} = 10(3^i - 1)$
Edges: generated by Euler relation: $V_i - E_i + F_i = 2$
which gives $E_i = V_i + F_i - 2$ with $F_i = P_i + H_i$
$E_i = 20\ 3^i + 12 + 10(3^i - 1) - 2 = 10 \times 3^{i+1}$

It appears that the transformation of level i to level $i+1$ keeps the number of pentagons ($P_i = 12$) because each pentagon is reproduced, keeps the number of hexagons H_i because each hexagon is reproduced and adds V_i hexagons because each vertex of level i creates a new hexagon in level $i+1$ (see Fig. 8.10). It is important to remark that at each level of subdivision the number of vertices and the number of hexagons is known.

The recursive process for constructing the database is shown in Fig. 8.11 and Fig. 8.12. We use in this paper a database adapted to our processes. We define two lists. A list with 3 fields:

- field 1 - number of the vertex
- field 2 - list of the 3 facets containing the vertex
- field 3 - position in each facet in accordance with a typical definition (counter clockwise, with a normal directed towards the exterior of the sphere); a list of facets

When we transform level i to level $i+1$:

1. we copy the list of facets with another numbering.

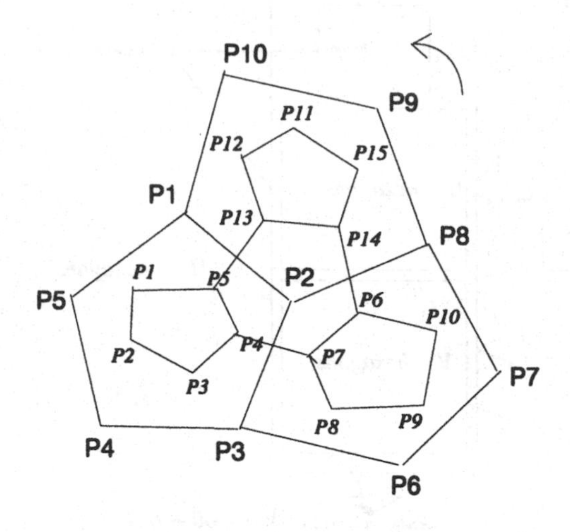

FIGURE 8.11. Vertices numbering

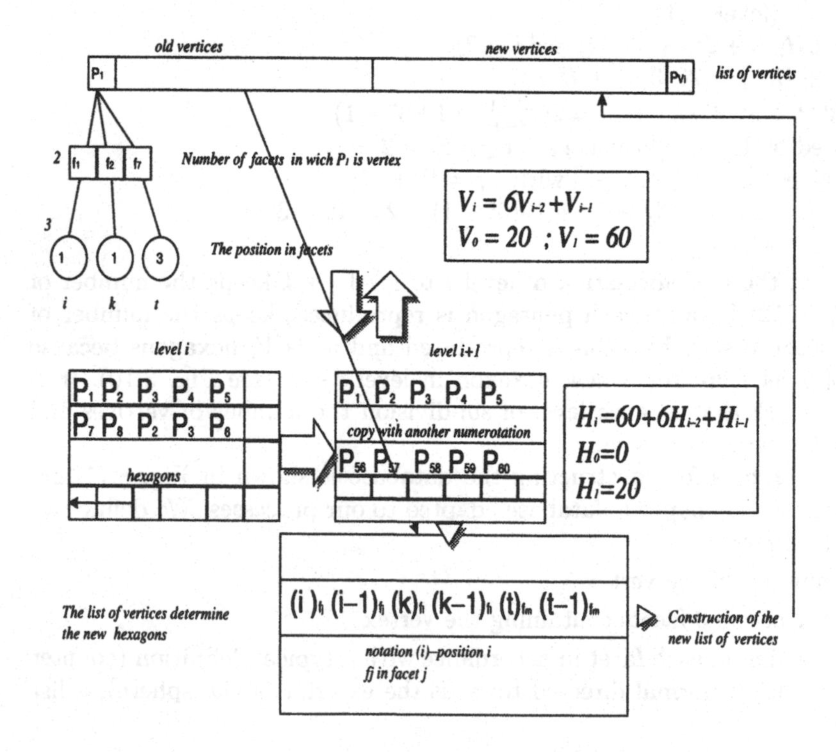

FIGURE 8.12. Recursive data structure

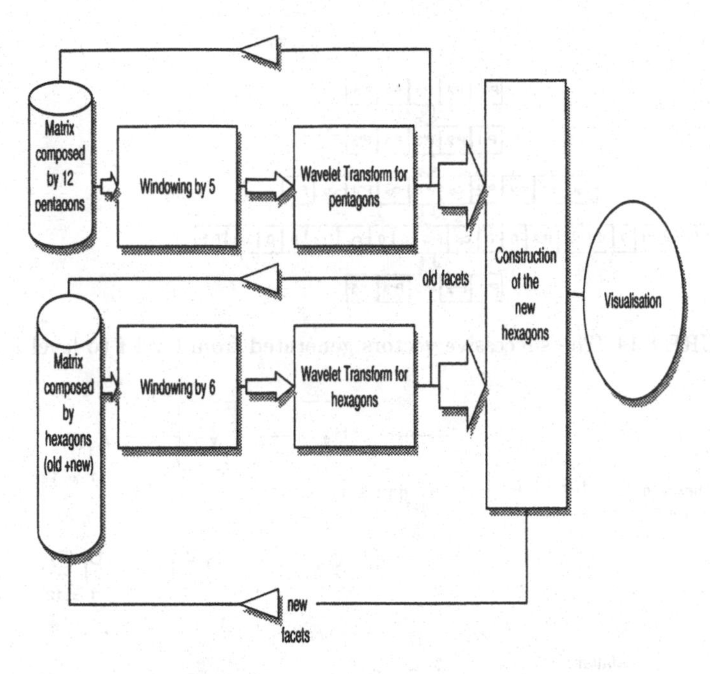

FIGURE 8.13. Data base model

2. starting from the known information (field 2 and 3) of the old vertices we construct the new list of hexagons.
3. starting from the complete new list of facets, we then construct the new list of vertices (fields 2 and 3).

Figure 8.13 summarizes our model including recursive database construction and recursive wavelet transform. It appears that the model works as a digital signal processor made up of two local polygonal transforms.

8.5 Implementation

When we transform level i to level $i+1$ of the sphere we change the resolution from $2^{-(j+1)}$ to 2^{-j} (j arbitrary). That is we obtain an approximation $A_j(\underline{P})$ in the space of approximation $V_{2^{-j}}$ from an approximation $A_{j+1}(\underline{P})$ in the space of approximation $V_{2^{-(j+1)}}$. Our model can be found as a Discrete Wavelet Transform (DWT). Indeed:

$$\underline{M} = \underline{P}\,\underline{\underline{H_1}} \tag{8.15}$$

$$\underline{A_{j+1}}(\underline{P}) = \underline{M}\,\underline{\underline{S}} + \underline{C_g}\,\underline{\underline{D}} \tag{8.16}$$

$$\underline{C_g} = \underline{P}\,\underline{\underline{C}} \tag{8.17}$$

$$\underline{A_j}\,(\underline{P}) = \underline{A_{j+1}}(\underline{P})\,\underline{\underline{\tilde{H}_3}}\,\underline{\underline{Down}} \tag{8.18}$$

starting from $\underline{P}$, the approximation at level i, of size $[1 \times 5]$ for a pentagon or $[1 \times 6]$ for an hexagon, we obtain $\underline{A_j}(\underline{P})$, the approximation at level $j+1$, of size $[1 \times 10]$ for a pentagon or $[1 \times 12]$ for hexagon from which we extract the new pentagon or

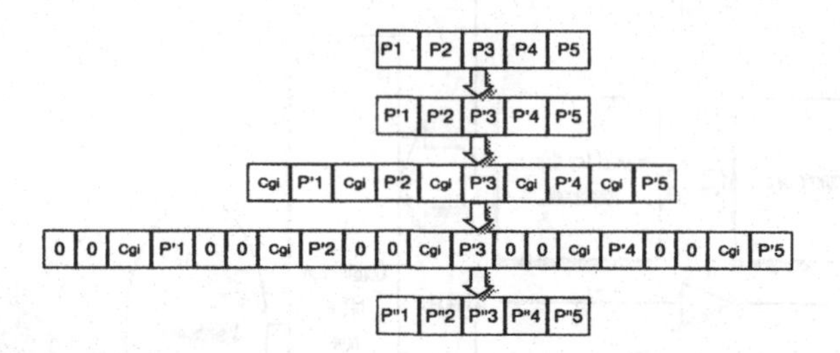

FIGURE 8.14. The successive vectors generated from level i to level $i + 1$

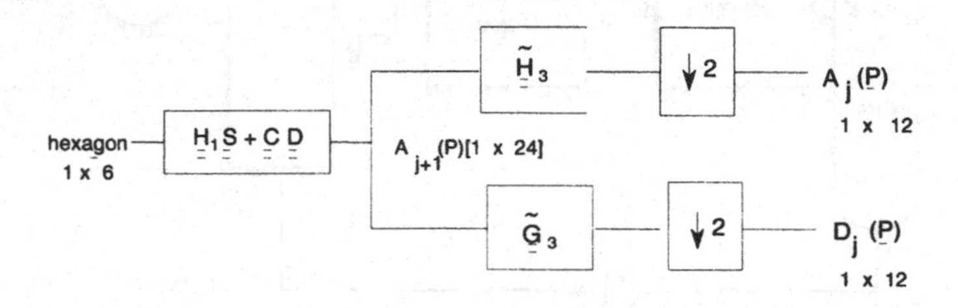

Mallat :

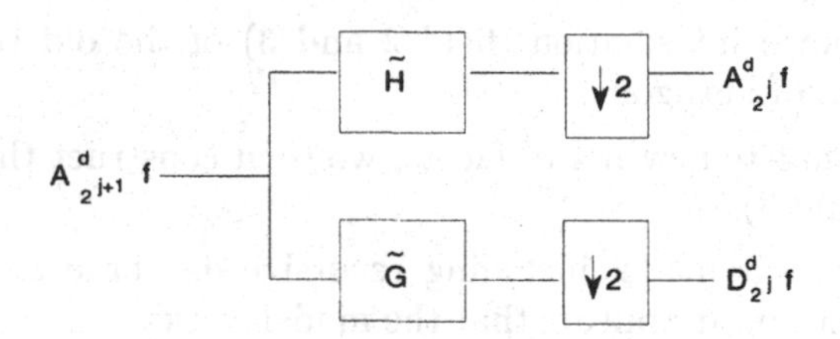

FIGURE 8.15. The discrete wavelet transform

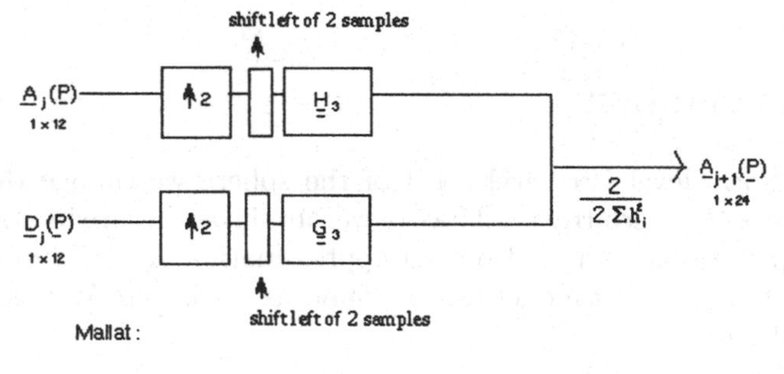

Mallat :

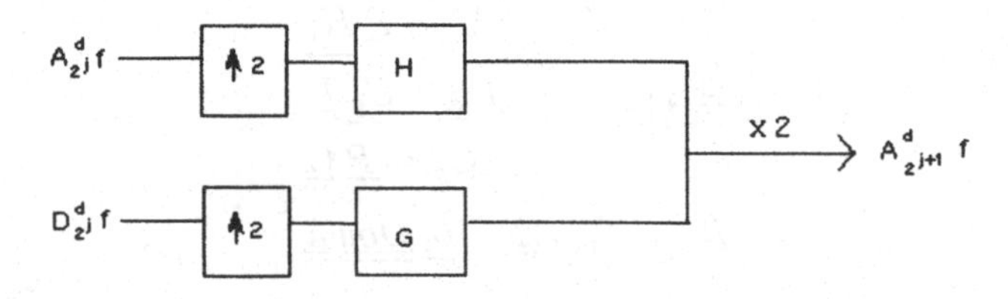

FIGURE 8.16. The inverse wavelet transform

hexagon. The matrix $\underline{\underline{Down}}$ keep one coefficient in two.

$$\underline{A_j}(\underline{P}) = \underline{P}\,(\underline{\underline{H_1}}\,\underline{\underline{S}} + \underline{\underline{C}}\,\underline{\underline{D}})\underline{\underline{\tilde{H}_3}}\,\underline{\underline{Down}} \tag{8.19}$$

In these relations for a pentagon, matrices are of size [5 × 5] for $\underline{\underline{H_1}}$, [5 × 20] for $\underline{\underline{S}}$, [5 × 5] for $\underline{\underline{C}}$, [5 × 20] for $\underline{\underline{D}}$, [20 × 20] for $\underline{\underline{\tilde{H}_3}}$. For a hexagon, matrices are of size [6 × 6] for $\underline{\underline{H_1}}$, [6 × 24] for $\underline{\underline{S}}$, [6 × 6] for $\underline{\underline{C}}$, [6 × 24] for $\underline{\underline{D}}$, [24 × 24] for $\underline{\underline{\tilde{H}_3}}$. $\underline{\underline{\tilde{H}_3}}$ is a block circulating matrix with a shift of one and whose coefficients are $\overline{h(0)}$, $\overline{h(1)}$, $h(2)$, $h(3)$.

For a hexagon we must invert the C_{g_i} and the $P_i^{'}$ in Fig. 8.14. In same way we

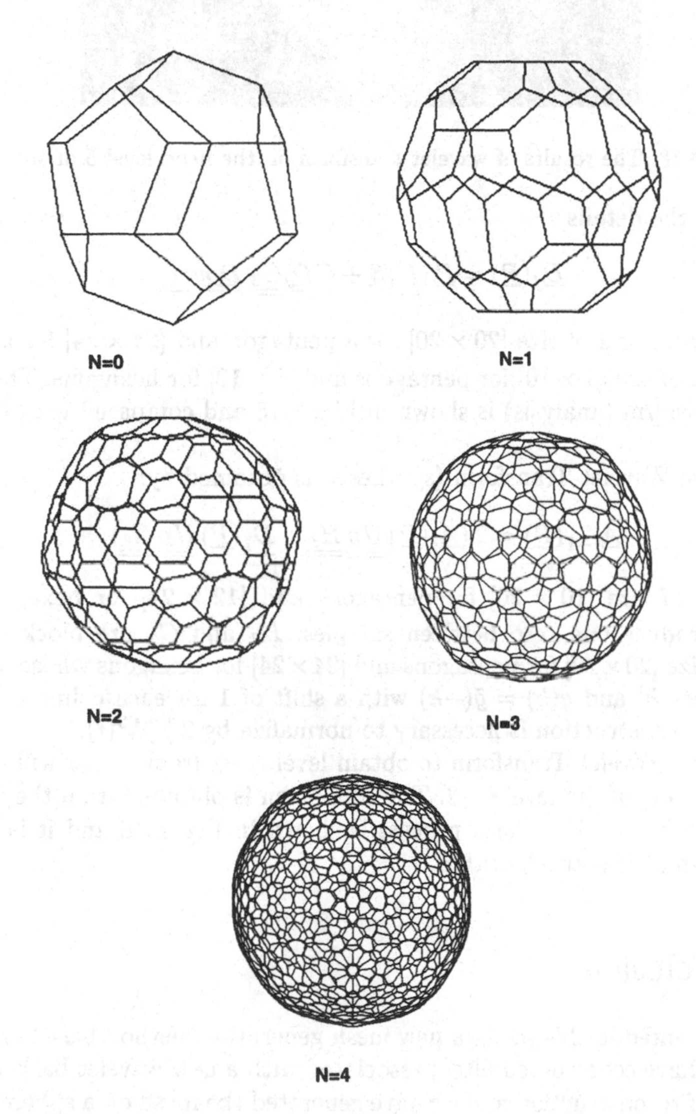

FIGURE 8.17. The results of the wavelet transform on the sphere (only processed polygons have an equal length)

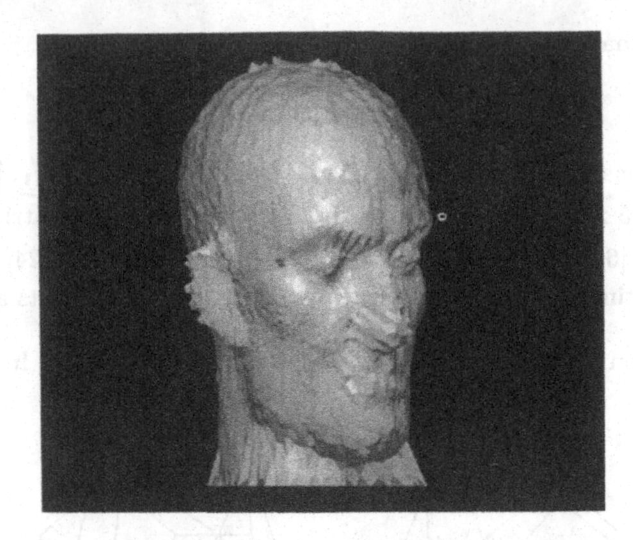

FIGURE 8.18. The results of wavelet transform on the head level 5 of subdivision

can write for the details

$$\underline{D}_j(\underline{P}) = \underline{P}\,(\underline{\underline{H_1}}\,\underline{S} + \underline{\underline{CD}})\underline{\underline{\tilde{G}_3}}\,\underline{\underline{Down}} \tag{8.20}$$

where $\underline{\underline{\tilde{G}_3}}$ is a matrix of size [20 × 20] for a pentagon and [24 × 24] for a hexagon, and $\underline{D}_j(\underline{P})$ is of size [1 × 10] for pentagons and [1 × 12] for hexagons. The Discrete Wavelet Transform (analysis) is shown in Fig. 8.15 and compared with the Mallat Algorithm.

The Inverse Wavelet Transform (synthesis) is obtained by:

$$\underline{A}_{j+1}(\underline{P}) = 2\,(\underline{A}_j(\underline{P})\,\underline{\underline{Up}}\,\underline{\underline{H_3}} + \underline{D}_j(\underline{P})\,\underline{\underline{Up}}\,\underline{\underline{G_3}}) \tag{8.21}$$

where $\underline{\underline{Up}}$ is of size [10 × 20] for pentagons and [12 × 24] for hexagons. These matrices introduce one zero between samples. H_3 and G_3 are block circulating matrices of size [20 × 20] for pentagons and [24 × 24] for hexagons whose coefficients are $h(k) = \tilde{h}(-k)$ and $g(k) = \tilde{g}(-k)$ with a shift of 1 for each column. To reduce the error of reconstruction is necessary to normalize by $2\sum h^2(i)$.

The Inverse Wavelet Transform to obtain level $i-1$ from level i will be applied only to the facets of the level $i-2$. The connexion is obtained from the level $i-1$. The Inverse Wavelet Transform process is shown in Fig. 8.16 and it is compared with the original Mallat Algorithm.

8.6 Conclusion

We have presented in this paper a new mesh generation method based on wavelets. For this, we have constructed filters associated with a new wavelet basis for a minimum error of reconstruction and we have generated the mesh on a sphere at several levels of sampling. The mesh uses 3-connected one-dimensional elements of approximately equal length which define "pentagons" and "hexagons" not necessarily planar. We show in Fig. 8.17 five levels of sphere subdivision.

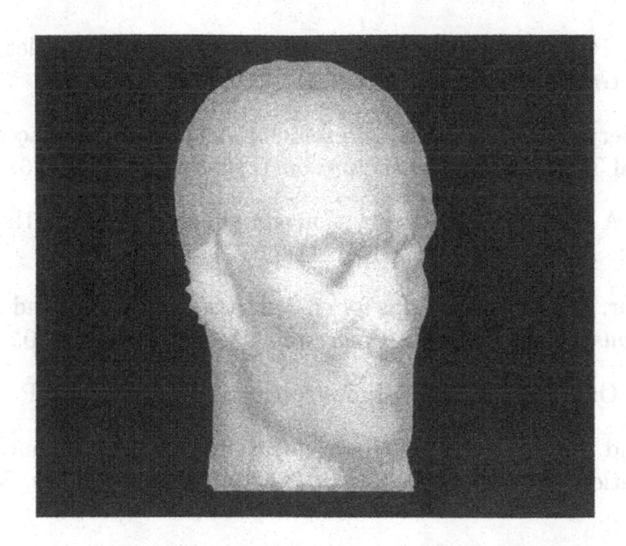

FIGURE 8.19. The results of wavelet transform on the head level 6 of subdivision

Although our algorithm can be used to represent the envelope of several kinds of solid objects, we have used this topology to model human faces [7]. In the case to apply our algorithm to the digitized face, the sphere of level 4 is projected on a digitized face and we present the levels 5 and 6 of subdivision (see Fig. 8.18, 8.19). The centres of gravity of the sphere and of the digitized face data base coincide, and we suppose the digitized face is star-shaped.

The modification of details to obtain new faces is the main interest of the Wavelet Transform in this application.

8.7 References

[1] A. Arneodo and al. Ondelettes, multifractales et turbulence de l'ADN aux croissances cristallines. *DIDEROT EDITEUR*, Paris, 1995.

[2] P.J. Burt and E. Adelson. The Laplacian pyramid as a compact image code. In *Proc. Int. Conf Acoust., Speech, Signal Processing*, pp. 191–195, 1977.

[3] I. Daubechies. The Wavelet Transform. *IEEE Trans. Inform. Theory*, 36:961–1005, 1990.

[4] M. Eck, T. DeRose, T. Duchamp, H. Hoppe, M. Lounsbery, and W. Stuetzer. Multiresolution analysis of arbitrary meshes In *Proc. SIGGRAPH'95, Computer Graphics*, pp. 174–182, 1995.

[5] D. Esteban and C. Galand. Application of quadrature mirror filters to split band voice coding schemes. *IEEE Trans. Comun.*, 31:532–540, 1983.

[6] J.P. Gourret, N. Magnenat-Thalmann, and D. Thalmann. Simulation of object and human skin deformations in a grasping task. in *Proc. SIGGRAPH'89, Computer Graphics*, 23(3):21– 31, 1989.

[7] J.P. Gourret and J. Khamlichi. A model for compression and classification of data structures. *Computers and Graphics*, 20(6):863–879, 1996.

[8] S. Gortler, P. Schroder, M. Cohen, and P. Hanrahan. Wavelet radiosity. In *Computer Graphics Proceedings*, pp. 213-220, 1993.

[9] M. Lounsbery et T. Derose. Multiresolution Analysis for Surfaces of Arbitrary Topological Type. *ACM Transaction on Graphics*, 16(1):34–73, 1997.

[10] S. Mallat. A theory for multiresolution signal decomposition: the wavelet representation. *IEEE Trans. on PAMI*, 11(7):674–693, 1989.

[11] H.S. Malvar. Lapped transforms for efficient transform/subband coding. *IEEE Trans. Acoust., Speech, Signal Processing*, 38(6):969–978, 1990.

[12] Y. Meyer. Ondelettes et algorithmes concurrents. *Hermann*, Paris, 1992.

[13] A. Pentland and J. Williams. Good vibrations: modal dynamics for graphics and animation. In *Proc.SIGGRAPH'89, Computer Graphics*, Vol.23(3):215–222, 1989.

[14] B. Torresani. Analyse continue par ondelettes. *InterEditions/CNRS*, Paris, 1995.

Part II

SIGNAL IDENTIFICATION AND MODELLING

9

The Bayesian Approach to Signal Modelling

Peter Rayner[1]
Bill Fitzgerald[1]

ABSTRACT
In this paper, an introduction to Bayesian methods in signal processing will be given. The paper starts by considering the important issues of model selection and parameter estimation. The important class of signal models, known as the General Linear Model, is introduced and the concept of marginal estimation of certain model parameter is developed. The techniques are illustrated for the problem of estimating sinusoidal frequency components in white Gaussian noise and for the general changepoint problem. Numerical integration methods are introduced based on Markov chain Monte Carlo techniques and the Gibbs sampler in particular and applications to audio restoration are presented.

9.1 Introduction

All scientific investigations of the so-called real world essentially involve three related parts, after deciding what it is that one wishes to investigate;

- The design of the experimental measuring apparatus.
- The measurement process itself.
- The analysis of the gathered data.

As such, we desire to make propositions about the real world, which we believe to be either true or false.

It is very easy to develop the perception that the data gathered are the measurements of the underlying physical phenomena that one wishes to infer. This causes endless confusion in the field of data analysis and is a manifestation of the confusion of the sample space with the hypothesis space. As a consequence, it is sometimes the case that data are analysed without thought ever being given to the underlying hypothesis that one is actually trying to infer!

Any form of experimental science requires the gathered data to be analysed in the best possible ways. This is even more obvious when one considers that some modern experimental facilities are extremely expensive to run and hence the gathered data

[1] University of Cambridge, Department of Engineering, Trumpington St., Cambridge, CB2 1PZ, United Kingdom, E-mails: {pjwr, wjf}@eng.cam.ac.uk

is very valuable. It is therefore senseless to use primitive analysis techniques on such expensive data. Even if the data analysis methods require lots of computer time this can, in many situations, be very worthwhile.

One of the major problems with experimental science is that very rarely can a series of logical deductions be made from data to hypothesis. This is because of experimental uncertainties, usually in the form of added noise.

In these situations we have to reason as best we can in situation of incomplete information - this is called scientific inference.

R.T. Cox, in 1946, [1], showed that any method of inference which satisfies simple rules for logical and consistent reasoning must be equivalent to the use of ordinary probability theory, as originally formulated by Bernoulli, Bayes and Laplace.

Accordingly, the conditional probability density function (p.d.f.), $p(H|D,I)$ summarises our inference about the hypothesis H given the data D and our prior knowledge I about H and the experimental setup. Since the numerical value of the probability assigned to any particular H is a measure of how much we believe that it is the true hypothesis, our best estimate is given by that H which maximises $p(H|D,I)$. The width, or spread, of this p.d.f. about the maximum tells us the reliability of the estimate: if the p.d.f. is sharply-peaked then we are confident of our prediction, but if it is broad then we are fairly uncertain about the true hypothesis.

Bayesian probability theory provides a unifying framework for data modelling and in this framework the overall aims are to find models that are well-matched to the data, and to use these models to make optimal predictions. The language of probability theory can be used to express inference of parameter values, optimisation of model complexity, model comparison and prediction.

If a model space includes a collection of models of different complexity or having different noise models, different preprocessing, etc., the evaluation of the appropriate probability makes it possible to select a model matched to the data.

Within a probabilistic framework, it is also possible to quantify information gains, for example, the information that some data convey about the model. The Bayesian viewpoint makes it possible to compare different models in the light of the data and to see which are most probable and best matched to the data.

In data modelling, once we have constructed a probability model and once we have computed the posterior distribution of the estimated parameters (or the point estimates if this is relevant) we must assess the fit of the model to the data. This is a crucial step in any analysis. If the model structure is 'wrong' we will get invalid inferences.

It is often the case that there exists more than one plausible model to account for the data and a sensitivity analysis should be performed. To ask the question 'Is the model true or false' is not very helpful since probability models in most data analysis problems are never 'true'.

9.2 Bayesian Inference

A fundamental task in signal processing and science in general is to develop models for signals which are observed and to determine whether the model function that one is using to describe the data, is actually appropriate for the particular problem under investigation, and if so, to extract values for the free parameters of the model.

We need a way to choose between several possible models. To attempt to solve this problem, one must enumerate the possible models and realise that in terms of real data the correct model may not be within the set chosen. All that we can do is compare various models within a set that we have defined to see which models are more plausible. This, and the problem of extracting values for model parameters, is a problem of scientific inference, and to carry out consistent reasoning and inference, one may use the Bayesian paradigm.

The framework can be broken down into essentially three levels of inference. At the first level it is assumed that one of the models within a chosen set, is the correct model with which to interpret the data and the problem of inference at this level consists of extracting values for the free parameters of the model, given the data. Bayes' theorem may be used to express the posterior probability of the parameters as:

$$p(\mathbf{w}|\mathbf{D}) = \frac{p(\mathbf{D}|\mathbf{w})\,p(\mathbf{w})}{p(\mathbf{D})} \tag{9.1}$$

where:

- $\mathbf{D}$ is a vector of observed data
- $\mathbf{w}$ is a vector of model parameters
- $p(\mathbf{w}|\mathbf{D})$ is the *Posterior probability*
- $p(\mathbf{D}|\mathbf{w})$ is the *Likelihood*
- $p(\mathbf{w})$ is the *Prior probability*
- $p(\mathbf{D})$ is the *Evidence*

The values of the free model parameters may be estimated from the posterior probability in a variety of ways but the two methods most commonly used are:

1.) The Maximum A-Posteriori (MAP) estimate in which the estimate is taken to be those values of the model parameters which maximise the posterior probability.

$$\hat{\mathbf{w}} = \max_{\mathbf{w}} \; p(\mathbf{w}|\mathbf{D})$$

2.) The Minimum Variance estimate in which the model parameters are chosen to minimise the variance of the estimate calculated from the posterior probability.

$$\hat{\mathbf{w}} = \min_{\mathbf{w}} \; E\{(\hat{\mathbf{w}} - \mathbf{w}|\mathbf{D})^2\}$$

$$\therefore\; E\{-2(\hat{\mathbf{w}} - \mathbf{w}|\mathbf{D})\} = 0$$

$$\hat{\mathbf{w}} = E\{\mathbf{w}|\mathbf{D}\}$$

i.e.

$$\hat{\mathbf{w}} = \int \mathbf{w}\, p(\mathbf{w}|\mathbf{D})\, d\mathbf{w}$$

In many problems it is not required to estimate all elements of the model parameter vector $\mathbf{w}$ and parameters which are of no interest are known as *Nuisance parameters.* A powerful feature of the Bayesian framework is that the nuisance parameters may be removed from consideration by *marginalisation.*

The parameter vector may be partitioned into 2 sets of parameters:
$\mathbf{w}_1$ - parameters of interest
$\mathbf{w}_2$ - nuisance parameters
The posterior probability can now be rewritten as:

$$p(\mathbf{w}|\mathbf{D}) \equiv p(\mathbf{w}_1, \mathbf{w}_2|\mathbf{D})$$

$$\therefore p(\mathbf{w}_1|\mathbf{D}) = \int_{\mathbf{w}_2} p(\mathbf{w}_1, \mathbf{w}_2|\mathbf{D})\, d\mathbf{w}_2$$

The second level of inference is that of choosing between several different models to determine the model which is most plausible, or the one that has the greatest evidence. Equation (9.1) for the posterior probability of the model parameters is implicitly conditioned on a particular model structure and in order to develop the model selection framework it is necessary to make this explicit.

$$p(\mathbf{w}|\mathbf{D}, M_i) = \frac{p(\mathbf{D}|\mathbf{w}, M_i)\, p(\mathbf{w}|M_i)}{p(\mathbf{D}|M_i)} \tag{9.2}$$

Selecting the most appropriate model for the observed data may be formulated in terms of maximising $P(M_i|\mathbf{D})$, the posterior probability density of the model, which may be obtained by applying Bayes' theorem to the evidence term in Eq. (9.2) to give:

$$P(M_i|\mathbf{D}) = \frac{p(\mathbf{D}|M_i)P(M_i)}{p(\mathbf{D})}$$

The denominator is a constant for all models, and if at this stage we have no reason to assign differing priors $P(M_i)$, then the choice of model structure reduces to that model structure which has the maximum evidence $p(\mathbf{D}|M_i)$. The evidence may be written as:

$$p(\mathbf{D}|M_i) = \int p(\mathbf{D}|\mathbf{w}, M_i)\, p(\mathbf{w}|M_i) d\mathbf{w}$$

Although this integral can be carried out analytically in some cases it is usually necessary to use numerical methods.

9.3 Parameter Estimation

In an estimation problem one assumes that the model is true for some unknown values of the model parameters, and one explores the constraints imposed on the parameters by the data, using Bayes' theorem. The hypothesis space for an estimation problem is therefore the set of possible values of the parameter vector $\mathbf{w}$, and it is this vector that will form the *hypothesis* that will be used in Bayes' theorem. The data form the sample space, and both the hypothesis space and the sample space may be either discrete or continuous.
In many real-world problems the observed data may be represented as:

$$d(n) = f(n) + e(n)$$

where $d(n)$ is a particular observed data value and $f(n)$ is a parameterised model for the data. The term $e(n)$ is a random term which may be considered as being due

to measurement error when observing the data or as error arising from incorrect modelling of the data. In the absence of any explicit knowledge to the contrary, the error term is usually assumed to be a zero-mean white Gaussian process. In general the data model $f(n)$ will be a nonlinear function of the model parameters.
We will now discuss an extremely important signal model which may be used in a very large number of applications.

9.3.1 *The General Linear Model*

Any data which may be described in terms of a linear combination of basis functions with an additive Gaussian noise component satisfies the ***general linear model.***
Suppose the observed data may be described by a model of the form:

$$d(n) = \sum_{q=1}^{Q} b_q g_q(n) + e(n) \quad \text{for} \quad 1 \leq n \leq N$$

where $g_q(n)$ is the value of a time dependent model function $g_q(t)$ evaluated at time t_n.
This can be written in the form of a matrix equation:

$$\mathbf{d} = \mathbf{G}\,\mathbf{b} + \mathbf{e} \tag{9.3}$$

where:

- $\mathbf{d}$ is an $N \times 1$ matrix of data points
- $\mathbf{e}$ is an $N \times 1$ vector of noise samples
- $\mathbf{G}$ is an $N \times Q$ matrix whose columns are the basis functions evaluated at each point in the time series
- $\mathbf{b}$ is a $Q \times 1$ linear coefficient vector.

Many of the ***standard*** signal processing structures are representatives of the general linear model structure and some examples are:
Sinusoidal Model

$$d(n) = \sum_{q=1}^{Q} \{a_q \sin(\omega_q t_n) + b_q \cos(\omega_q t_n)\} + e(n)$$

Autoregressive (AR) Model

$$d(n) = \sum_{q=1}^{Q} a_q\, d(n-q) + e(n)$$

Autoregressive with External Input (ARX) Model

$$d(n) = \sum_{q=1}^{Q} a_q\, d(n-q) + \sum_{q=1}^{Q} b_q\, u(n-q) + e(n)$$

where $u(n)$ is a known system input.

Nonlinear Autoregressive (NAR) Model

$$\begin{aligned} d(n) &= \sum_{q=1}^{Q} a_q\, d(n-q) \\ &+ \sum_{q_1=1}^{Q_1} \sum_{q_2=1}^{Q_2} a_{q_1 q_2}\, d(n-q_2) d(n-q_2) \\ &+ \ldots + e(n) \end{aligned}$$

Volterra Model

$$\begin{aligned} d(n) &= \sum_{q=0}^{Q} a_q\, u(n-q) \\ &+ \sum_{q_1=0}^{Q_1} \sum_{q_2=0}^{Q_2} a_{q_1 q_2}\, u(n-q_2) u(n-q_2) \\ &+ \ldots + e(n) \end{aligned}$$

Returning to the general linear model, the error term or innovations process is a zero-mean white Gaussian process $N(0, \sigma^2)$ defined by the probability density:

$$p(\mathbf{e}) = \left(2\,\pi\,\sigma^2\right)^{-\frac{N}{2}} \exp\left[-\frac{\mathbf{e}^{\mathrm{T}}\mathbf{e}}{2\,\sigma^2}\right]$$

and the data likelihood may be written as:

$$p\,(\mathbf{d} \mid \mathbf{w}_m,\, \sigma,\, \mathbf{b},\, M) = \left(2\,\pi\,\sigma^2\right)^{-\frac{N}{2}} \exp\left[-\frac{\mathbf{e}^{\mathrm{T}}\mathbf{e}}{2\,\sigma^2}\right]$$

where $\mathbf{w}_m$ denotes the parameters of the matrix of basis functions $\mathbf{G}$. Substituting from Eq. (9.3) for the general linear model gives:

$$\therefore p\,(\mathbf{d}|\mathbf{w}_m,\, \mathbf{b},\, \sigma,\, M) = \left(2\,\pi\,\sigma^2\right)^{-\frac{N}{2}} \exp\left[-\frac{(\mathbf{d}-\mathbf{G}\mathbf{b})^{\mathrm{T}}(\mathbf{d}-\mathbf{G}\mathbf{b})}{2\,\sigma^2}\right] \tag{9.4}$$

Application of Bayes' theorem gives the joint posterior probability density function for the various parameters as:

$$p\,(\mathbf{w}_m,\, \mathbf{b},\, \sigma|\mathbf{d},\, M) = \frac{p\,(\mathbf{d}|\mathbf{w}_m,\, \mathbf{b},\, \sigma,\, M)\; p(\mathbf{w}_m,\, \mathbf{b},\, \sigma|M)}{p(\mathbf{d}|M)}$$

and assuming that $\mathbf{w}_m$, $\mathbf{b}$ and σ are statistically independent:

$$p(\mathbf{w}_m,\, \mathbf{b},\, \sigma|M) = p(\mathbf{w}_m|M)\, p(\mathbf{b}|M)\, p(\sigma|M)$$

and the joint posterior density becomes:

$$p\,(\mathbf{w}_m,\, \mathbf{b},\, \sigma|\mathbf{d},\, M) = \frac{p\,(\mathbf{d}|\mathbf{w}_m,\, \mathbf{b},\, \sigma,\, M)\; p(\mathbf{w}_m|M)\, p(\mathbf{b}|M)\, p(\sigma|M)}{p(\mathbf{d}|M)}$$

Often there will be little prior knowledge concerning the noise variance σ^2 and it is then appropriate to use the Jeffreys' prior [11]:

$$p(\sigma|M) = \frac{1}{\sigma}$$

Similarly, in the absence of any prior knowledge concerning the model parameters uniform priors may be used.

$$p(\mathbf{w}_m|M) = k_w \quad \text{and} \quad p(\mathbf{b}|M) = k_b$$

Thus the posterior probability becomes:

$$p(\mathbf{w}_m, \mathbf{b}, \sigma|\mathbf{d}, M) = \frac{k_w\, k_b}{\sigma} \frac{p(\mathbf{d}|\mathbf{w}_m, \mathbf{b}, \sigma, M)}{p(\mathbf{d}|M)}$$

It should be noted that both the Jeffreys' prior and the uniform prior are *improper* in the sense that they are not normalised and this can raise both philosophical and practical problems [12].
If it is assumed that the solution required to the parameter estimation problem is the maximum of the posterior density function (i.e. MAP estimate) then the constant scaling terms and the evidence term $p(\mathbf{d}|M)$ are independent of the parameter values and the posterior may conveniently be written as:

$$p(\mathbf{w}_m, \mathbf{b}, \sigma|\mathbf{d}, M) = k\, p(\mathbf{d}|\mathbf{w}_m, \mathbf{b}, \sigma, M)$$

where:

$$k = \frac{k_w\, k_b}{\sigma\, p(\mathbf{d}|M)}$$

and its actual value is not required for the MAP solution.
Substituting for the data likelihood from Eq. (9.4) gives:

$$p(\mathbf{w}_m, \mathbf{b}, \sigma|\mathbf{d}, M) = k\, (2\,\pi\,\sigma^2)^{-\frac{N}{2}} \exp\left[-\frac{(\mathbf{d}-\mathbf{G}\mathbf{b})^{\mathrm{T}}(\mathbf{d}-\mathbf{G}\mathbf{b})}{2\,\sigma^2}\right] \tag{9.5}$$

The procedure now depends on the particular problem under consideration. In some cases the matrix $\mathbf{G}$ is completely specified so the parameter values $\mathbf{w}_m$ are known. Since $\mathbf{w}_m$ are no longer considered as random parameters they do not appear in the argument for the posterior density.

$$p(\mathbf{w}_m, \mathbf{b}, \sigma|\mathbf{d}, M) \equiv p(\mathbf{b}, \sigma|\mathbf{d}, M)$$

The objective is to estimate the coefficient vector $\mathbf{b}$. The noise standard deviation σ is a nuisance parameter which may integrated out

$$p(\mathbf{b}|\mathbf{d}, M) = \int_0^\infty p(\mathbf{b}, \sigma|\mathbf{d}, M)\, d\sigma$$

using the Gamma integral definition:

$$\int_0^\infty x^{\alpha-1} \exp(-Q\,x)\, dx = \frac{\Gamma(\alpha)}{Q^\alpha}$$

in conjunction with Eq. (9.5) to give:

$$p(\mathbf{b}|\mathbf{d}, M) = \frac{1}{2}\Gamma(\frac{N-1}{2}) \left[\frac{1}{2}(\mathbf{d} - \mathbf{Gb})^{\mathrm{T}}(\mathbf{d} - \mathbf{Gb})\right]^{-\frac{N-1}{2}}$$

The MAP estimate for the parameter vector **b** may be obtained by maximising the above to give:

$$\mathbf{b} = (\mathbf{G}^{\mathrm{T}}\,\mathbf{G})^{-1}\,\mathbf{G}^{\mathrm{T}}\,\mathbf{d}$$

This is the familiar least squares (minimum variance) estimate that is obtained from linear algebra.
The second application of the posterior probability, Eq. (9.5), is in determining the MAP estimate of the elements $\mathbf{w}_m$ of the model matrix **G** without inferring values for the coefficient vector **b** and the noise standard deviation σ. The coefficient vector **b** may be marginalised by integrating the posterior probability density, Eq. (9.5), with respect to **b**.

$$p(\sigma, \mathbf{w}_m \mid \mathbf{d}, M) = \int_{\Re^Q} p(\{\mathbf{b}, \omega, \sigma\} \mid \mathbf{d}, \mathbf{I})\, d\mathbf{b}$$

where $\Re^Q$ is an Q dimensional space of real numbers.
This may be achieved by use of the following standard integral:

$$\int_{\Re^Q} \exp\left[-\frac{1}{2\sigma^2}\left(\mathbf{x}^{\mathrm{T}}\mathbf{A}\mathbf{x} + \mathbf{x}^{\mathrm{T}}\mathbf{b} + c\right)\right] d\mathbf{x}$$

$$= \frac{(2\pi\sigma^2)^{Q/2}}{\sqrt{\det(\mathbf{A})}} \exp\left[-\frac{1}{2\sigma^2}\left(c - \frac{\mathbf{b}^{\mathrm{T}}\mathbf{A}^{-1}\mathbf{b}}{4}\right)\right]$$

$$\therefore p(\sigma, \mathbf{w}_m \mid \mathbf{d}, \mathbf{M}) \propto \frac{(2\pi\sigma^2)^{-(\frac{N-Q}{2})}}{\sqrt{\det(\mathbf{G}^{\mathrm{T}}\mathbf{G})}} \exp\left[-\left(\frac{\mathbf{d}^{\mathrm{T}}\mathbf{d} - \mathbf{d}^{\mathrm{T}}\mathbf{G}(\mathbf{G}^{\mathrm{T}}\mathbf{G})^{-1}\mathbf{G}^{\mathrm{T}}\mathbf{d}}{2\sigma^2}\right)\right]$$

The standard deviation may again be integrated out as a gamma integral to give:

$$p(\mathbf{w}_m \mid \mathbf{d}, M) \propto \frac{\left[\mathbf{d}^{\mathrm{T}}\mathbf{d} - \mathbf{d}^{\mathrm{T}}\mathbf{G}(\mathbf{G}^{\mathrm{T}}\mathbf{G})^{-1}\mathbf{G}^{\mathrm{T}}\mathbf{d}\right]^{\frac{-(N-Q)}{2}}}{\sqrt{\det(\mathbf{G}^{\mathrm{T}}\mathbf{G})}} \tag{9.6}$$

which is in the form of a student-t distribution.
Note that this is a function of $\mathbf{w}_m$ only. This means that there is no need to know about the standard deviation nor the values of the linear parameters in order to infer the values of $\mathbf{w}_m$. Here the integrals have been done analytically so the dimensionality of the parameter space was reduced for each parameter integrated out. This reduction of the dimensionality is a property of Bayesian marginal estimates and is a major advantage in many applications.

9.3.1.1 Frequency Estimation

As an example of the application of the general linear model, consider the detection of a single frequency.

$$f(t) = A\cos(\omega t) + B\sin(\omega t)$$

The data are assumed to consist of samples from the signal $f(t)$ corrupted with independent white zero mean Gaussian noise samples with standard deviation σ. The signal model just described agrees with the general linear model. The structure of the $\mathbf{G}$ matrix is:

$$\mathbf{G} = \begin{bmatrix} \cos(\omega t_1) & \sin(\omega t_1) \\ \cos(\omega t_2) & \sin(\omega t_2) \\ \cos(\omega t_3) & \sin(\omega t_3) \\ \vdots & \vdots \\ \cos(\omega t_N) & \sin(\omega t_N) \end{bmatrix}$$

and the linear coefficient vector is:

$$\mathbf{b} = \begin{bmatrix} A \\ B \end{bmatrix}$$

The general result, Eq. (9.6), for the marginalised posterior probability density for the parameters $\mathbf{w}_m$ of the model matrix $\mathbf{G}$ may be used to express the probability density for the angular frequency ω as:

$$p(\omega \mid \mathbf{d}, \mathbf{I}) \propto \frac{\left[\mathbf{d}^{\mathrm{T}}\mathbf{d} - \mathbf{d}^{\mathrm{T}}\mathbf{G}\left(\mathbf{G}^{\mathrm{T}}\mathbf{G}\right)^{-1}\mathbf{G}^{\mathrm{T}}\mathbf{d}\right]^{\frac{2-N}{2}}}{\sqrt{\det\left(\mathbf{G}^{\mathrm{T}}\mathbf{G}\right)}}$$

The columns of the $\mathbf{G}$ matrix are nearly orthogonal. This may be used to simplify the marginal posterior for ω. The matrix $\mathbf{G}^{\mathrm{T}}\mathbf{G}$ is approximately equal to $N/2$ times the identity matrix. It is easy to show that:

$$\mathbf{d}^{\mathrm{T}}\mathbf{d} - \mathbf{d}^{\mathrm{T}}\mathbf{G}\left(\mathbf{G}^{\mathrm{T}}\mathbf{G}\right)^{-1}\mathbf{G}^{\mathrm{T}}\mathbf{d} \approx \sum_{i=1}^{N} d_i^2 - \frac{2}{N}\left(\sum_{i=1}^{N} d_i\cos(\omega t_i) + \sum_{i=1}^{N} d_i\sin(\omega t_i)\right)^2$$

Define

$$I(\omega) = \left(\sum_{i=1}^{N} d_i\cos(\omega t_i)\right)$$

$$Q(\omega) = \left(\sum_{i=1}^{N} d_i\sin(\omega t_i)\right)$$

then further approximations may be made to obtain:

$$\mathbf{d}^{\mathrm{T}}\mathbf{d} - \mathbf{d}^{\mathrm{T}}\mathbf{G}\left(\mathbf{G}^{\mathrm{T}}\mathbf{G}\right)^{-1}\mathbf{G}^{\mathrm{T}}\mathbf{d} \approx \sum_{i=1}^{N} d_i^2 - \frac{2}{N}I(\omega)^2 - \frac{2}{N}Q(\omega)^2$$

The Schuster periodogram is defined as

$$C(\omega) = \frac{1}{N}\left(I^2(\omega) + Q^2(\omega)\right)$$

so that the marginal density for the angular frequency may be expressed in terms of the Schuster periodogram as:

$$p(\omega \mid \mathbf{d}, \mathbf{I}) \propto \left[1 - \frac{2C(\omega)}{\sum_{i=1}^{N} d_i^2}\right]^{\frac{2-N}{2}}$$

Note that the term inside the square brackets is small (thus implying that the marginal density is large) if $2C(\omega) \approx \sum_{i=1}^{N} d_i^2$. This occurs if most of the data energy is concentrated around a single frequency ω.
The Schuster periodogram, (and hence the Discrete Fourier transform), is designed to determine the value of a single frequency in white Gaussian zero mean noise. Therefore it should really only be used on data that satisfies the single frequency model, and is not really designed for use in resolving two or more closely spaced frequencies, nor should it be used when the data is corrupted by non-Gaussian noise.
It is interesting to compare the discrete Fourier transform power spectrum and the Bayesian marginal density for a single frequency. If the frequency bin positions correspond with the sample points in the time series then the two procedures will be exactly equivalent. But what happens if the data contain multiple frequencies? If the number of data points, N, is large and the frequencies are such that

$$|\omega_i - \omega_j| >> \frac{2\pi}{N} \quad i \neq j$$

then the single model function method given above is sufficient.
However, if the condition is not satisfied, the model function G matrix must be written, for the case of two frequencies, as

$$\mathbf{G} = \begin{bmatrix} \cos(\omega_1 t_1) & \sin(\omega_1 t_1) & \cos(\omega_2 t_1) & \sin(\omega_2 t_1) \\ \cos(\omega_1 t_2) & \sin(\omega_1 t_2) & \cos(\omega_2 t_1) & \sin(\omega_2 t_1) \\ \cos(\omega_1 t_3) & \sin(\omega_1 t_3) & \cos(\omega_2 t_1) & \sin(\omega_2 t_1) \\ \vdots & \vdots & \vdots & \vdots \\ \cos(\omega_1 t_N) & \sin(\omega_1 t_N) & \cos(\omega_2 t_1) & \sin(\omega_2 t_1) \end{bmatrix}$$

The analysis then continues as before and it is necessary to determine the maximum of $p(\omega_1, \omega_2|\mathbf{d}, M)$ with respect to the two frequencies. The extension to the general case of Q frequencies is evident.

9.3.1.2 General Linear Changepoint Detector

A second application of the general linear model is to the linear changepoint detector [2, 3, 14] which is a matrix based detector that has analytically solvable marginalisation integrals when used with the Gaussian noise assumption. It is a very flexible method that can be adapted to a variety of changepoint detection problems. The central concept behind the linear changepoint detector is that an observed time sequence is modelled by different models at different points in time.

The models are known but the time instants at which the models change are not known. The various models may be expressed as the linear combination of some basis functions. For data with a single changepoint from one linear model to another, the problem may be formulated as:

$$d_i = \begin{cases} \sum_{q=1}^{Q} \alpha_q \, g_q(i) + e_i & \text{if } i \leq m \\ \sum_{q=1}^{Q} \beta_q \, g_q(i) + e_i & \text{otherwise} \end{cases}$$

where $g_q(i)$ is the value of a time-dependent model function evaluated at time t_i. This may be expressed in the general linear model form:

$$\mathbf{d} = \mathbf{G}\,\mathbf{b} + \mathbf{e}$$

The model matrix contains the known basis function terms $g_q(i)$ and the unknown changepoint m. Equation (9.6) may be used to express the posterior probability density function for the changepoint as:

$$p\,(m \mid \mathbf{d}, M) \propto \frac{\left[\mathbf{d}^{\mathbf{T}}\mathbf{d} - \mathbf{d}^{\mathbf{T}}\mathbf{G}\left(\mathbf{G}^{\mathbf{T}}\mathbf{G}\right)^{-1}\mathbf{G}^{\mathbf{T}}\mathbf{d}\right]^{\frac{-(N-Q)}{2}}}{\sqrt{\det\left(\mathbf{G}^{\mathbf{T}}\mathbf{G}\right)}}$$

The changepoint detector has been illustrated for a single changepoint but clearly the development is general and can be extended to an arbitrary number of changepoints. However the computational load can rapidly become excessive in that each evaluation of the posterior density requires that the model matrix $\mathbf{G}$ be evaluated. Some progress has been made in developing iterative methods to overcome this problem [13]

9.4 Numerical Bayesian Methods

In the last sections it has been shown how to conduct inference in situations where the integrals required to perform marginalisation have been analytically tractable. However, in more realistic situations the integrations have to be performed numerically and this requirement has lead to a full scale investigation of such integration methods. It is now clear that methods based on Monte Carlo techniques hold the greatest scope. Two areas of Bayesian analysis that require evaluation of integrals are:

- Evidence evaluation:

$$P(D|M_i) = \int P(D|\mathbf{w}, M_i)P(\mathbf{w}|M_i)d\mathbf{w}$$

- Marginalisation:

$$p(\mathbf{w}_1|\mathbf{D}) = \int_{\mathbf{w}_2} p(\mathbf{w}_1, \mathbf{w}_2|\mathbf{D})\, d\mathbf{w}_2$$

9.4.1 The Monte Carlo Method

The marginalisation and evidence integrals each have the form:

$$I = \int f(\mathbf{x})\, p(\mathbf{x})\, d\mathbf{x}$$

where $f(\mathbf{x})$ is a function of $\mathbf{x}$ and $p(\mathbf{x})$ is the probability density function.
The integral may be interpreted as the expected or mean value of the function $f(\mathbf{x})$.
The principle of the Monte Carlo method is:

1. Draw N random variates from a random generator with probability density $p(\mathbf{x})$

$$\mathbf{x}_i \leftarrow p(\mathbf{x}) \qquad i \in \{1, 2, \ldots, N\}$$

2. Compute the average:

$$I \approx \frac{1}{N} \sum_{i=1}^{N} f(\mathbf{x})$$

A fundamental requirement of the Monte-Carlo technique is that of drawing samples from a multi-variate probability density function. For some particular density functions this is possible but in general it is not possible to achieve this directly and it is necessary to use Markov chains. Integration techniques using sampling methods based on Markov chains were first developed for applications in statistical physics. Two branches of development originated in the 1950s. The classic paper by Metropolis *et al* introduced what is now known as the *Metropolis algorithm.* This method was popularized for Bayesian applications, along with its variant the Gibbs sampler, by the influential papers of Geman and Geman, [4], who applied it to image processing, and Gelfand and Smith, [5], who demonstrated its application to Bayesian problems in general.
A Markov chain is defined to be a series of random variables $x_1, x_2, \ldots x_N$ such that the influence of random variables $x_1, x_2, \ldots x_N$ on the value of x_{N+1} is mediated by the value of x_N alone:

$$p\,(x_{N+1} \mid x_1, x_2, \ldots x_N) = p\,(x_{N+1} \mid x_N)$$

The random variables have a common range known as the *state space* of the Markov chain. State space variables may be either discrete or continuous. The set of discrete space variables may be finite or countably infinite. In most cases[2] the set of state space variables is identical to the parameter space.
A Markov chain begins with an initial distribution for x_0, and thereafter the distribution of x_i is determined by the transition probabilities between states:

$$p_{N+1}\,(x) = \sum_{\{x'\}} p_N\,(x')\, T_N\,(x', x)$$

Denote the initial distribution as $p_0\,(x)$. In continuous state spaces the summation is replaced by an integral.

---

[2]The Hybrid Monte Carlo state space is a superset of parameter space.

The function T_N is known as a ***base transition.*** In some Markov chains the base transition does not depend on N and the Markov chain is said to be homogeneous (e.g. the global Metropolis algorithm). In other examples that we encounter, each base transition only affects one component of the state space vector $\mathbf{x}$ at a time and the base transitions are rotated in sequence (e.g. Local Metropolis algorithm, Gibbs sampler).
The principle underlying the use of the Gibbs sampler is that one can break down the problem of drawing samples from a multivariate density into one of drawing successive samples from densities of smaller dimensionality. In its usual form the Gibbs sampler draws samples from univariate densities.

9.4.2 Gibbs Sampler

Assume that the parameter space consists of k components $\{a_1, a_2, a_3 \ldots a_k\}$. The components are initialized to starting values $\{a_1^0, a_2^0, a_3^0 \ldots a_k^0\}$ and the Gibbs sampler proceeds by drawing random variates from conditional densities in a cyclical iterative pattern as follows:

First iteration:

$$\begin{aligned} a_1^1 &\leftarrow p(a_1 \mid a_2^0\, a_3^0 \ldots a_{k-1}^0\, a_k^0) \\ a_2^1 &\leftarrow p(a_2 \mid a_3^0\, a_4^0 \ldots a_k^0\, a_1^1) \\ a_3^1 &\leftarrow p(a_3 \mid a_4^0\, a_5^0 \ldots a_1^1\, a_2^1) \\ &\vdots \\ a_k^1 &\leftarrow p(a_k \mid a_1^1\, a_2^1 \ldots a_{k-2}^1\, a_{k-1}^1) \end{aligned}$$

Second iteration:

$$\begin{aligned} a_1^2 &\leftarrow p(a_1 \mid a_2^1\, a_3^1 \ldots a_{k-1}^1\, a_k^1) \\ a_2^2 &\leftarrow p(a_2 \mid a_3^1\, a_4^1 \ldots a_k^1\, a_1^2) \\ a_3^2 &\leftarrow p(a_3 \mid a_4^1\, a_5^1 \ldots a_1^2\, a_2^2) \\ &\vdots \\ a_k^2 &\leftarrow p(a_k \mid a_2^2\, a_1^2 \ldots a_{k-2}^2\, a_{k-1}^2) \end{aligned}$$

n^{th} **iteration:**

$$\begin{aligned} a_1^n &\leftarrow p(a_1 \mid a_2^{n-1}\, a_3^{n-1} \ldots a_{k-1}^{n-1}\, a_k^{n-1}) \\ &\vdots \end{aligned}$$

where $p(a_i \mid a_j\ j \neq i)$ denotes the conditional density of the ith component of the vector $\mathbf{a}$. The superscript number denotes the current iteration. Note that as soon as a variate is drawn, then it is inserted immediately into the conditional probability density function, and it remains there until it is substituted in the next iteration.

At the end of the jth iteration the sample $\{a_1^j, a_2^j, a_3^j \ldots a_{k-1}^j, a_k^j\}$ is a sample from the joint density. In common with other Markov chain approaches the Gibbs sampler requires an initial transient period to converge to equilibrium. How much of the initial series is affected by the initial state is difficult to ascertain, but some literature is available on the subject. This initial period of length M is known as the "burn in" and it varies in length depending on the problem. One should always discard the first M samples.

9.4.3 Metropolis Subchains

If the conditional densities are easy to sample from then the Gibbs sampler will be easy to implement. In many cases some of the densities are not of a simple standard form. In such cases one can resort to using the Metropolis algorithm as a means of drawing samples from the density. This form of Gibbs sampler may be regarded as a hybrid strategy between the true Gibbs sampler and the local Metropolis algorithm, where "Gibbs steps" for sampling conditional density are interleaved between "Metropolis steps" for sampling difficult conditional densities.

The Metropolis-Hastings algorithm is an extremely flexible method for producing a random sequence of samples from a given density. The Metropolis algorithm was originally introduced by Metropolis *et al* for computing the properties of substances composed of interacting molecules. This algorithm has been used extensively in statistical physics; its use being so widespread that in some circles the term has become almost synonymous with Monte Carlo work. A generalization introduced by Hastings, [6], and extended by Peskun, [7], is presented below. The standard form of simulated annealing algorithm uses the Metropolis algorithm, with the result that a large proportion of the optimization community do not distinguish between simulated annealing and the Metropolis algorithm. The Gibbs sampler can be viewed as a special case of the Metropolis-Hastings algorithm.

9.4.4 The General Algorithm

In this section we describe the Metropolis-Hastings algorithm in its more general form as described by Hastings. It is assumed that the density is given in a functional form $y = P(x)$, where $x \in \Phi$. The density does not need to be normalized. The algorithm explores the parameter space Φ by means of a random walk. Suppose X_i is the ith element of such a random walk. It is proposed that the next variate in the random sequence be Y_i which is produced by adding a random perturbation ζ to X_i:

$$Y_i = X_i + \zeta$$

where ζ is drawn from the proposal density $S(\zeta)$.

There are two possibilities for the choice of the next variate in the random sequence:

- $X_{i+1} = Y_i$ **Accept** the proposed random variate.
- $X_{i+1} = X_i$ **Reject** the proposal and repeat X_i.

The probability of accepting Y_i instead of the current value is given by the Metropolis-Hastings[3] acceptance function:

$$A(X_i, Y_i) = \min(1, Q(X_i, Y_i))$$

where

$$Q(X_i, Y_i) = \frac{P(Y_i)\, T(Y_i \mid X_i)}{P(X_i)\, T(X_i \mid Y_i)}$$

[3] Alternatively, one can use the Boltzmann acceptance function but this is less efficient.

The conditional probability density $T(Y_i \mid X_i)$ is identical to the proposal density $S(\zeta)$. Therefore $T(X_i \mid Y_i)$ is given by the probability density $S(-\zeta)$. Let ϵ be a uniform random variate drawn over the range [0, 1]. If the condition[4]:

$$P(Y_i)\,T(Y_i \mid X_i) > \epsilon\, P(X_i)\,T(X_i \mid Y_i)$$

holds then the next term X_{i+1} in the random sequence is $X_{i+1} = Y_i$, otherwise we have $X_{i+1} = X_i$.

Clearly, if the proposal density $S(\zeta)$ is symmetrical about the origin then the acceptance probability above is given by

$$Q(X_i, Y_i) = \frac{P(Y_i)}{P(X_i)}$$

Note that the probability of acceptance is independent of the probability density $S(\zeta)$. This is the original form of the Metropolis algorithm.

9.5 The Interpolation of Missing Samples for Audio Restoration

The aim of this section is to outline a method for restoring missing samples in digital audio signals. The section of audio signal in question is modelled as a stationary autoregressive process, and missing samples are imputed using the Gibbs sampler, [8].

Clicks are a familiar problem in audio gramophone signals, [9], and take the form of sudden unexpected bursts of impulsive noise with random but finite duration. These bursts of noise have numerous causes such as dirt, electrical interference or mechanical damage to the storage medium. The original signal is often effectively lost. Several methods of detecting clicks have been devised, with the best approaches being model based. Once a click has been detected the *suspect* samples are removed and replaced by interpolation. Figure 9.1 illustrates the problem in which it is required to replace the missing data $\mathbf{x}_u$ using a signal model conditioned on the known data $\mathbf{x}_{k1}$ and $\mathbf{x}_{k2}$.

Missing (ie. unknown) data:

$$[x(m)\ x(m+1)\ \ldots x(m+L-1)]^T = \mathbf{x}_u$$

Observed (ie. known) data

$$[x(1)\ x(2)\ \ldots x(m-1)]^T = \mathbf{x}_{k1}$$

$$[x(m+L)\ x(m+L+1)\ \ldots x(N)]^T = \mathbf{x}_{k2}$$

Define the Augmented data vector containing both observed and missing data:

$$\mathbf{x} = [\mathbf{x}_{k1}^{\mathbf{T}}\ \mathbf{x}_u^{\mathbf{T}}\ \mathbf{x}_{k2}^{\mathbf{T}}]^{\mathbf{T}}$$

[4]A small computational saving can be made if one recognizes that $\epsilon \leq 1$. In other words, if $p(Y_i)T(Y_i \mid X_i) > p(X_i)T(X_i \mid Y_i)$ then Y_i is automatically accepted as the next variate in the random sequence, so the random variate ϵ need not be generated.

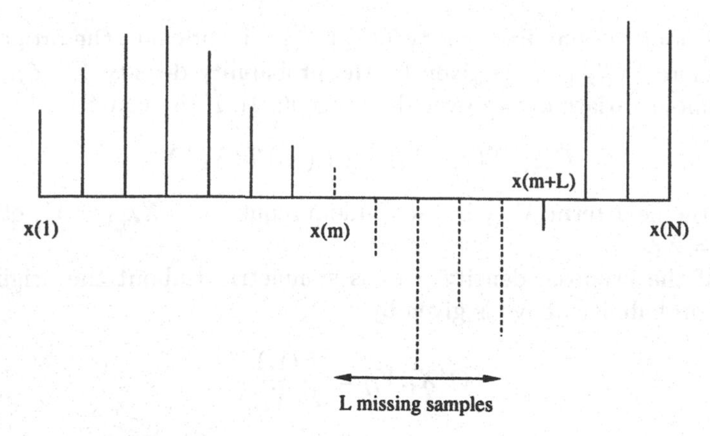

FIGURE 9.1. Audio signal with missing data

Audio signals are, in general, well-modelled as Autoregressive processes:

$$x(n) = \sum_{p=1}^{Q} a_p\, x(n-p) + e(n)$$

This expression may be written in matrix form in two ways as follows:

$$\mathbf{e} = \mathbf{G_a}\,\mathbf{x} \tag{9.7}$$

or

$$\mathbf{e} = \mathbf{x} - \mathbf{G_x}\,\mathbf{a} \tag{9.8}$$

In Equation (9.7) the matrix contains the AR model parameters and the data appears as a vector whereas in Eq. (9.8) the data appears in the matrix and the AR parameters are in vector form. The reason for using the two forms is algebraic convenience in what follows.
Consider the excitation energy $\mathbf{e}^T\,\mathbf{e}$.
From Equation (9.7):

$$\begin{aligned}\mathbf{e}^{\mathbf{T}}\,\mathbf{e} &= \mathbf{x}^T\,\mathbf{G}_a{}^T\,\mathbf{G}_a\,\mathbf{x} \\ &= [\mathbf{x}_{k1}^{\mathbf{T}}\;\mathbf{x}_u^{\mathbf{T}}\;\mathbf{x}_{k2}^{\mathbf{T}}]\,\mathbf{G}_{\mathbf{a}}^{\mathbf{T}}\,\mathbf{G}_{\mathbf{a}}\,[\mathbf{x}_{k1}^{\mathbf{T}}\;\mathbf{x}_u^{\mathbf{T}}\;\mathbf{x}_{k2}^{\mathbf{T}}]^{\mathbf{T}}\end{aligned}$$

Partitioning the matrix and combining the observed data $\mathbf{x}_{k1}$ and $\mathbf{x}_{k2}$ into a single vector $\mathbf{x}_k$ gives:

$$\mathbf{e}^{\mathbf{T}}\,\mathbf{e} = [\mathbf{x}_k^{\mathbf{T}}\,\mathbf{x}_u^{\mathbf{T}}]\begin{bmatrix}\mathbf{A} & \mathbf{B}\\ \mathbf{B}^{\mathbf{T}} & \mathbf{D}\end{bmatrix}\begin{bmatrix}\mathbf{x}_k\\ \mathbf{x}_u\end{bmatrix}$$

where $\mathbf{x}_k^{\mathbf{T}} = [\mathbf{x}_{k1}^{\mathbf{T}}\;\mathbf{x}_{k2}^{\mathbf{T}}]$.

$$\therefore\; \mathbf{e}^{\mathbf{T}}\,\mathbf{e} = \mathbf{x}_k^{\mathbf{T}}\,\mathbf{A}\,\mathbf{x}_k + 2\mathbf{x}_k^{\mathbf{T}}\,\mathbf{B}\,\mathbf{x}_u + \mathbf{x}_u^{\mathbf{T}}\,\mathbf{D}\,\mathbf{x}_u \tag{9.9}$$

which is a quadratic function of the unknown data vector $\mathbf{x}_u$.
An alternative expression for the excitation energy can be obtained from Eq. (9.8) as follows:

$$\mathbf{e}^{\mathbf{T}}\,\mathbf{e} = (\mathbf{x} - \mathbf{G_x}\,\mathbf{a})^{\mathbf{T}}(\mathbf{x} - \mathbf{G_x}\,\mathbf{a})$$

$$\therefore\; \mathbf{e}^{\mathbf{T}}\,\mathbf{e} = \mathbf{x}^{\mathbf{T}}\,\mathbf{x} + 2\mathbf{x}^{\mathbf{T}}\,\mathbf{G_x}\,\mathbf{a} + \mathbf{a}^{\mathbf{T}}\,\mathbf{G}_{\mathbf{x}}^{\mathbf{T}}\mathbf{G_x}\,\mathbf{a} \tag{9.10}$$

which is a quadratic function of the AR parameter vector $\mathbf{a}$.
Assume that the AR excitation $e(n)$ is a zero-mean i.i.d. Gaussian process $N(0,\sigma^2)$, then:

$$p(\mathbf{e}) = \prod_{n=1}^{N} p(e(n))$$

$$\therefore p(\mathbf{e}) = (2\pi\sigma^2)^{-\frac{N}{2}} \exp\left(-\frac{\mathbf{e}^{\mathbf{T}}\mathbf{e}}{2\sigma^2}\right)$$

$$\therefore p(\mathbf{x} \mid \mathbf{a}, \sigma) = (2\pi\sigma^2)^{-\frac{N}{2}} \exp\left(-\frac{\mathbf{e}^{\mathbf{T}}\mathbf{e}}{2\sigma^2}\right) \tag{9.11}$$

which is the likelihood function for the augmented data.
If the missing data $\mathbf{x_u}$ are fixed then one may maximize the likelihood with respect to the AR parameter vector $\mathbf{a}$ by means of Eqs. (9.11) and (9.10) to give:

$$\hat{\mathbf{a}} = \left(\mathbf{G}_{\mathbf{x}}^{\mathbf{T}}\mathbf{G}_{\mathbf{x}}\right)^{-1}\mathbf{G}_{\mathbf{x}}^{\mathbf{T}}\,\mathbf{x}$$

If the AR parameters are fixed then one may maximize the likelihood with respect to the unknown data $\mathbf{x}_u$ by means of Eqs. (9.11) and (9.9) to give:

$$\hat{\mathbf{x}}_u = -\mathbf{D}^{-\mathbf{T}}\mathbf{B}^{\mathbf{T}}\mathbf{x}_k$$

Vaseghi, [10], describes a method for jointly estimating the AR parameters and the missing data by successively maximizing the likelihood with respect to $\mathbf{a}$ and then with respect to $\mathbf{x}_u$. The procedure is iterated until the values converge at a maximum of the likelihood function $p(\mathbf{x} \mid \mathbf{a}, \sigma)$.

The EM algorithm may also be used although it works in a slightly different way from the above method. Upon each iteration, the interpolant $\hat{\mathbf{x}}_u(i+1)$ may be computed *linearly* from the previous iterate $\hat{\mathbf{x}}_u(i)$, essentially leapfrogging the AR step. The resultant interpolant maximizes the predictive density $p(\mathbf{x}_u \mid \mathbf{x}_k)$ – not the likelihood – so the results of using the EM algorithm are slightly different from those of the ML procedure.
The problem may also be considered in a Gibbs Sampling framework which is, as discussed earlier, a Markov chain based Monte Carlo sampling scheme for simulating jointly distributed random variates and which may be used for sampling the joint density $p(\mathbf{x}_u, \mathbf{a}, \sigma \mid \mathbf{x}_k)$.
The joint probability for the unknown variables may be written as:

$$p(\mathbf{x}_u, \mathbf{x}_k \mid \mathbf{a}, \sigma) \equiv p(\mathbf{x} \mid \mathbf{a}, \sigma)$$

$$\therefore p(\mathbf{x}_u, \mathbf{a}, \sigma \mid \mathbf{x}_k) = \frac{p(\mathbf{x} \mid \mathbf{a}, \sigma)\, p(\mathbf{a}, \sigma)}{p(\mathbf{x}_k)}$$

The Gibbs sampler requires the individual conditional densities.

$$p(\mathbf{x}_u \mid \mathbf{x}_k, \mathbf{a}, \sigma) = \frac{p(\mathbf{x}_u, \mathbf{a}, \sigma \mid \mathbf{x}_k)}{p(\mathbf{a}, \sigma \mid \mathbf{x}_k)}$$

$$= \frac{p(\mathbf{x}_u, \mathbf{a}, \sigma \mid \mathbf{x}_k)}{\int p(\mathbf{x}_u, \mathbf{a}, \sigma \mid \mathbf{x}_k)\, d\mathbf{x}_k}$$

which is a multivariate Gaussian density in $\mathbf{x}_u$.

$$p(\mathbf{a} \mid \sigma, \mathbf{x}_k, \mathbf{x}_u) = \frac{p(\mathbf{x}_u, \mathbf{a}, \sigma \mid \mathbf{x}_k)}{\int p(\mathbf{x}_u, \mathbf{a}, \sigma \mid \mathbf{x}_k)\, d\mathbf{a}}$$

which is a multivariate Gaussian density in $\mathbf{a}$.

$$p(\sigma \mid \mathbf{x}_k, \mathbf{x}_u, \mathbf{a}) = \frac{p(\mathbf{x}_u, \mathbf{a}, \sigma \mid \mathbf{x}_k)}{\int p(\mathbf{x}_u, \mathbf{a}, \sigma \mid \mathbf{x}_k)\, d\sigma}$$

which leads to a square-root inverted gamma density in σ.
It is straightforward to sample from these density functions so the Gibbs' Sampler may be used to sample from the joint posterior density.
Assume a starting position $(\mathbf{x}_u^0, \mathbf{a}^0, \sigma^0)$ in parameter space. Consider,

$$\begin{array}{lcl}
\mathbf{x}_u^1 & \leftarrow & p\,(\mathbf{x}_u \mid \mathbf{a}^0, \sigma^0, \mathbf{x}_k) \\
\mathbf{a}^1 & \leftarrow & p\,(\mathbf{a} \mid \sigma^0, \mathbf{x}_u^1, \mathbf{x}_k) \\
\sigma^1 & \leftarrow & p\,(\sigma \mid \mathbf{x}_u^1, \mathbf{a}^1, \mathbf{x}_k) \\
\mathbf{x}_u^2 & \leftarrow & p\,(\mathbf{x}_u \mid \mathbf{a}^1, \sigma^1, \mathbf{x}_k) \\
\mathbf{a}^2 & \leftarrow & p\,(\mathbf{a} \mid \sigma^1, \mathbf{x}_u^2, \mathbf{x}_k) \\
\vdots & & \vdots \\
\sigma^i & \leftarrow & p\,(\sigma \mid \mathbf{x_u}^i, \mathbf{a}^i, \mathbf{x}_k)
\end{array}$$

where the notation $\mathbf{x}_u^i \leftarrow p\,(\mathbf{x}_u \mid \mathbf{a}^{i-1}, \sigma^{i-1}, \mathbf{x}_k)$ means that a random sample $\mathbf{x}_u^i$ is drawn from the conditional probability density function of $\mathbf{x_u}$ with all other parameters fixed at the given values. The principle of the method is that the distribution of the random deviates $(\mathbf{x}_u^i, \mathbf{a}^i, \sigma^i)$ in parameter space asymptotically approaches the joint density. Exact details of how to draw random deviates from these density functions are given in [2]. The Gibbs sampling approach is only feasible once the very specialized structure of the matrices is taken into account. Matrix operations include the use of Levinson's algorithm, band LU decomposition and the Gaxpy-Cholesky decomposition, see [2] for further details and also results on real audio signals.
There is an important difference between the results obtained from the sampling approach and approaches based on Maximum Likelihood (ML). The sampling method provides an estimate of the missing audio data which is based on a *typical* section of AR model excitation $\mathbf{e}$ whereas the ML method provides a method which is based on the lowest excitation energy $\mathbf{e}^{\mathbf{T}}\,\mathbf{e}$. The implication of this is that for relatively long sections of missing data, the ML methods provide a signal estimate which decreases in amplitude towards the centre of the missing data gap. Another way of looking at this is that the excitation is assumed to be Gaussian and the "most probable" Gaussian signal is zero but this is certainly not typical of what one would observe. The result is that, for long sections of missing data, the sampling approach provides estimates which are perceptually superior to those obtained from the ML approach.

9.6 Conclusions

In this paper we have introduced the ideas associated with Bayesian inference applied to model selection and parameter estimation. The concept of model 'evidence'

was discussed and the technique of eliminating ***nuisance parameters*** by integration has been introduced and applied to frequency estimation and generalised change-point detection. The requirement to carry out many of the necessary integrals using numerical methods has been developed and Markov Chain Monte Carlo (MCMC) methods have been introduced.
These numerical techniques, and the Gibbs sampler in particular, have been applied to the problem of interpolation of missing data in an audio signal application.
The last few years have witnessed a massive development in Bayesian numerical techniques and these are just starting to be used within the signal processing community. There are many exciting areas that are waiting to be addressed by these techniques and the next few years should see many such applications.

9.7 References

[1] R.T. Cox. Probability, frequency and expectation. *Amer. J. Phys.*, 14:1-13, 1946.

[2] J.J.K. Ó Ruanaidh and W.J. Fitzgerald. *Numerical Bayesian Methods applied to Signal Processing.* Springer-Verlag, 1995.

[3] W.J. Fitzgerald., J.J.K. O' Ruanaidh, and J.A. Yates. Generalised Change-point Detection. *Cambridge University Engineering Department Tech. Report*, CUED/F-INFENG/TR187, 1994.

[4] S. Geman and D. Geman. Stochastic Relaxation, Gibbs Distributions, and the Bayesian Restoration of Images. *IEEE Trans. on Pattern Analysis and Machine Intelligence*, PAMI-6:721-741,1984.

[5] A.E. Gelfand and A.F.M. Smith. Sampling Based Approaches to Calculating Marginal Densities. *Journal of the American Statistical Association*, 85, 410:398-409,1990.

[6] W.K. Hastings. Monte Carlo sampling methods using Markov Chains and their applications. *Biometrika*, 57:97-109, 1970.

[7] P.H. Peskun. Guidelines for choosing the transition matrix in Monte Carlo methods using Markov chains. *J. Computational Physics*, 40:327-344, 1981.

[8] J.J.K. Ó Ruanaidh and W.J. Fitzgerald. Interpolation of missing samples for audio restoration. *Electronics Letters*, 30, 8:622, 1994.

[9] P.J.W. Rayner and S.J. Godsill. The detection and correction of artefacts in degraded gramophone recordings. *Proc. IEEE ASSP Workshop on Applications of Signal Processing to Audio and Acoustics*, 1991.

[10] S.V. Vaseghi. Algorithms for the restoration of archived gramophone recordings. PhD thesis, 1988, Cambridge University Engineering Department, England

[11] H. Jeffreys. *Theory of Probability*, Oxford University Press, 1961

[12] J. Bernado and A.F.M. Smith. *Bayesian Theory*, John Wiley, 1994.

[13] J.J.K. Ó Ruanaidh, W.J. Fitzgerald, and K.J. Pope. Recursive location of a discontinuity in a time series. *Proc. Internat. Conf. on Acoustics, Speech and Signal Processing ICASSP'94*, IV:513-516, 1994.

[14] R.W. Tennant and W.J. Fitzgerald. Detection of phase changes in BPSK and QPSK using the Bayesian general linear detector. *Proc. First European Conference on Signal Analysis and Prediction ECSAP-97*, 149-152, 1997.

10

Regularized Signal Identification Using Bayesian Techniques

Anthony Quinn[1]

ABSTRACT
Techniques which identify parametric models for a time series are considered in this chapter. When the model structure is unknown, the key issue becomes one of *regularizing* the inference, namely, discovering a model which explains the data well, but avoids excessive complexity. Strict model selection criteria—such as Akaike's, Rissanen's and the evidence approach—are contrasted with full Bayesian solutions which allow parameters to be estimated *in tandem*. In the latter paradigm, *marginalization* is the key operator allowing model complexity to be assessed and penalized naturally via integration over parameter subspaces. As such, it is an important alternative (or adjunct) to 'subjective' penalization via the choice of prior. These various strategies are considered in the context of model order determination for both harmonic and autoregressive signals, and it is emphasized that effective and numerically efficient identification algorithms result even in the case of uniform priors, if judicious integration of parameters is undertaken.

10.1 Introduction

Signal identification may be defined as the task of inferring a suitably parameterized model to explain a set of observations ('data'). The 'explanation' may be in terms of the model's ability to fit the data, or to predict it. This definition embraces the detection, model selection and estimation tasks which are often considered separately in the literature [2]. Signal identification is important in areas as disparate as econometrics [16] and image segmentation [13], and continues to be a very active area of research [4, 8].

A mathematical setting for signal identification is given in Section 10.2, and approaches which infer model structure at the expense of parameter inference are reviewed in Section 10.3. The two principal approaches to Bayesian regularization are presented in Section 10.4, namely, via the prior, and via partial or complete integration over the competing hypotheses (i.e. marginalization). Identification using non-informative priors is emphasized, and justification for the uniform parameter prior is argued in detail. An important class of signal models is presented in Section 10.5, and two subclasses are then considered, namely, harmonic (Section 10.6) and autoregressive models (Section 10.7). Conclusions follow in Section 10.8.

---

[1]Department of Electronic and Electrical Engineering, University of Dublin, Trinity College, Dublin 2, Ireland, E-mail: aquinn@ee.tcd.ie

10.2 The Signal Identification Problem

Let $\mathbf{d} = (d_1, \ldots, d_N)^T \in \underline{\overline{C}}^N$ be a length N data record which is to be explained in terms of one of several parametric model hypotheses, $\mathcal{I}_k$. Let the ***superhypothesis***, $\mathcal{I}$, be the set of all such hypotheses:

$$\mathcal{I} = \{\mathcal{I}_1, \ldots, \mathcal{I}_M\} \tag{10.1}$$

M is assumed finite, so that $\mathcal{I}$ constitutes a ***closed*** hypothesis space. If the hypotheses, $\mathcal{I}_k$, are ***parametric***, then each $\mathcal{I}_k$ seeks to explain $\mathbf{d}$ in terms of a set of parameters, $\mathbf{\Theta}_k = (\Theta_{k1}, \ldots, \Theta_{kp_k})^T$, whose realization is $\boldsymbol{\theta}_k \in \underline{\overline{\Theta}}_k \subset \underline{\overline{C}}^{p_k}$. If signal identification is to be accomplished, one of the $\mathcal{I}_k$s will be the (parameter-free) null hypothesis. From a Bayesian perspective, both $K = k$ and $\mathbf{\Theta}_k = \boldsymbol{\theta}_k$ are ***probabilistic parameters (p.p.s)*** (a phrase more expressive than 'random variable', to describe attributes of the sample space [10]). In this chapter, $\mathcal{I}$ and $\mathcal{I}_k$ are Jeffreys' notation [5], denoting the proposition that the superhypothesis, and the kth parametric model, are true respectively. If the parametric signal identification problem has been well posed, then the elementary propositions '$\mathbf{\Theta}_K = \boldsymbol{\theta}_k$' are mutually exclusive, $\forall k$, $\forall \boldsymbol{\theta}_k$. This, in turn, implies that $\underline{\overline{\Theta}} = \bigcup_{k=1}^{M} \underline{\overline{\Theta}}_k$ is a partition of the sample space, $\underline{\overline{\Theta}}$. $\mathcal{I}_k$ may then be defined more precisely as $\mathcal{I}_k \Leftrightarrow \underline{\overline{\Theta}}_k$, so that

$$\mathcal{I} = \mathcal{I}_1 \vee \mathcal{I}_2 \vee \cdots \vee \mathcal{I}_M \tag{10.2}$$

where '$\vee$' denotes logical OR, and the $\mathcal{I}_k$s are mutually exclusive. Equation (10.2) permits a normalizable ***model prior***, $\mathrm{p}(\mathcal{I}_k \mid \mathcal{I})$, to be posited, if desired. Equivalently stated, $K = k$ is a consistently defined p.p.

In contrast, the signal spaces, $\underline{\overline{S}}_k$, implied by the $\mathcal{I}_k$s are not disjoint, and may, in fact, be nested: $\underline{\overline{S}}_k \subset \underline{\overline{S}}_{k+1}$, $k = 1, \ldots, M-1$. Such a situation arises if $\mathbf{\Theta}_{k+1} = (\mathbf{\Psi}_{k+1}^T, \mathbf{\Phi}_{k+1}^T)^T$, $\mathbf{\Phi}_{k+1} = \mathbf{\Theta}_k$, and

$$\underline{\overline{S}}_{k+1}\Big|_{\mathbf{\Psi}_{k+1} \in \underline{\overline{\Psi}}_{k+1}^0} = \underline{\overline{S}}_k \tag{10.3}$$

for some non-empty (but possibly singleton) set, $\underline{\overline{\Psi}}_{k+1}^0$. The models in Sections 10.6 and 10.7 manifest this behaviour. In conclusion, $\mathcal{I}_k \Rightarrow \underline{\overline{S}}_k$, but $\underline{\overline{S}}_k \not\Rightarrow \mathcal{I}_k$; i.e. $\underline{\overline{\Theta}}$ and $\underline{\overline{S}} = \bigcup_{k=1}^{M} \underline{\overline{S}}_k$ are not equipollent sets.

The goal of ***full*** signal identification is to infer $\mathbf{\Theta}_K$ using an appropriate criterion which avoids overfitting $\mathbf{d}$:

$$\widehat{\boldsymbol{\theta}_k} = \arg\min_{k} \min_{\boldsymbol{\theta}_k} C(\boldsymbol{\theta}_k \mid \mathbf{d}, \mathcal{I}) \tag{10.4}$$

A vast range of criterion functions, $C(\cdot)$, has been proposed to satisfy Ockham's Razor (i.e. the 'closeness of fit vs. simplicity' trade-off, which is the philosophical concept underlying regularization). Reviews may be found in [4, 6, 10].

10.3 Strict Model Selection

In this chapter, the task of inferring K (strict model selection),

$$\hat{k} = \arg\min_{k} C'(\mathcal{I}_k \mid \mathbf{d}, \mathcal{I}) \tag{10.5}$$

via some model-dependent, but parameter-independent, criterion, $C'(\cdot)$, is distinguished from the full identification task defined by (10.4). The literature tends to treat strict model selection in isolation from parameter estimation, in which case, the latter is accomplished as a ***subsequent*** task for the kth model [2], yielding an inference $\hat{\boldsymbol{\theta}}_{\hat{k}}$. This is the ***modular*** approach to identification [10]. Between the two extremes ((10.4) and (10.5)) lie criteria for partial identification of $\boldsymbol{\Theta}_K$, as will be seen in Sections 10.4.3 and 10.6.

10.3.1 Penalized Maximum Likelihood

Consider the case where each model in $\mathcal{I}_k$ declares a predictive density for observations $\mathbf{D} = \mathbf{d}$:

$$\mathcal{I}_k : \quad \mathbf{D}|\boldsymbol{\theta}_k \sim \mathrm{p}(\mathbf{d} \mid \boldsymbol{\theta}_k, \mathcal{I}_k) \tag{10.6}$$

($\mathcal{I}_k$, though tautological, is included in the argument of $\mathrm{p}(\cdot)$ for clarity). With $\mathbf{d}$ observed, and $\boldsymbol{\theta}_k$ unknown, the predictive density in (10.6) is conveniently interpreted as a non-measure function, $l(\boldsymbol{\theta}_k \mid \mathbf{d}, \mathcal{I}_k)$, of $\boldsymbol{\theta}_k$, known as the ***Likelihood Function (LF)*** [5, 10]. An approach to (strict) model selection is then to maximize $l(\cdot)$, augmented by a ***penalty function***, $\mathcal{P}_\square[p_k]$, which is monotonically increasing with the number of parameters, p_k, in the kth model, but insensitive to variations of $\boldsymbol{\theta}_k$:

$$\hat{k} = \arg\min_{k} \left[-L(\hat{\boldsymbol{\theta}}_k \mid \mathbf{d}, \mathcal{I}_k) + \mathcal{P}_\square[p_k] \right] \tag{10.7}$$

$L(\cdot) = \ln l(\cdot)$ (to within an additive constant), and $\hat{\boldsymbol{\theta}}_k = \arg\min_{\boldsymbol{\theta}_k} [-L(\boldsymbol{\theta}_k \mid \mathbf{d}, \mathcal{I}_k)]$ is the ***Maximum Likelihood (ML)*** estimate of $\boldsymbol{\Theta}_k$. If, ultimately, $\hat{\boldsymbol{\theta}}_{\hat{k}}$ is chosen (i.e. the ML estimate of the selected model), then the scheme may be considered 'unified' (10.4), since $\mathcal{P}_\square[p_k]$ is not a function of $\boldsymbol{\theta}_k$; i.e. $\hat{\boldsymbol{\theta}}_{\hat{k}} = \widehat{\boldsymbol{\theta}_k}$ (10.4).

In this chapter, all the signal hypotheses, $\mathcal{I}_k$, will adopt the additive noise model:

$$\mathcal{I}_k : \quad \mathbf{d} = \mathbf{g}_k(\boldsymbol{\theta}_k) + \mathbf{e}, \qquad \mathbf{e} \sim \mathrm{p}_E(\mathbf{e} \mid \boldsymbol{\beta}, \mathcal{I}) \tag{10.8}$$

Note that the prior on $\mathbf{e}$ is assumed independent of k, which is reasonable if information about the noise is available independently of any assumptions about the signal, $\mathbf{g}_k(\cdot)$. Normalizability of $\mathrm{p}_E(\cdot)$ often implies that

$$\mathrm{p}_E(\mathbf{e} \mid \boldsymbol{\beta}, \mathcal{I}) = f(\|\mathbf{e}\|_\square) \tag{10.9}$$

where $f(\cdot)$ is monotonically decreasing, and $\|\cdot\|_\square$ is the norm implied by $\mathrm{p}_E(\cdot)$.

For example, the (complex) Gaussian noise pdf—to be explored in this chapter—is $\mathrm{p}_E(\mathbf{e} \mid \boldsymbol{\Sigma}, \mathcal{I}) = \mathcal{N}(\mathbf{0}, \boldsymbol{\Sigma}) \propto \exp\left(-\mathbf{e}^H \boldsymbol{\Sigma}^{-1} \mathbf{e}\right)$, where $\mathbf{e}^H \boldsymbol{\Sigma}^{-1} \mathbf{e}$ is a squared Mahalanobis norm, $\|\mathbf{e}\|^2_{\boldsymbol{\Sigma}^{-1}}$. From (10.7) and (10.8):

$$\hat{k} = \arg\min_{k} \left[\|\mathbf{d} - \mathbf{g}_k(\hat{\boldsymbol{\theta}}_k)\|^2_{\boldsymbol{\Sigma}^{-1}} + \mathcal{P}_\square[p_k] \right] \tag{10.10}$$

It can readily be appreciated from (10.10) that, in the absence of the penalty term, the criterion seeks to fit $\mathbf{d}$ as closely as possible under the norm definition imposed by $\mathrm{p}_E(\cdot)$. Thus, Ockham's Razor is ignored (i.e. the unpenalized criterion is unregularized).

Consistent with Jeffreys' [5] suggestion that the complexity of a model should be defined as its number of degrees of freedom (i.e. functionally independent parameters), most of the criteria of the type (10.7) employed in current signal analysis involve penalty terms which are proportional to p_k. Perhaps the most popular is Rissanen's ***Mimimum Description Length (MDL)*** criterion [11], which, in asymptotic form, yields $\mathcal{P}_{\mathrm{MDL}}[p_k] = \frac{1}{2} p_k \ln N$, and corresponds to asymptotic forms of several other criteria [6]. It is deduced by minimizing the number of binary digits required to code $\mathbf{d}$ under each $\mathcal{I}_k$. A generalization is found in the compact coding ideas of Wallace and Freeman [15]. In fact, combined information- and estimation-theoretic techniques have yielded new approaches in ***both*** areas. A related criterion—Akaike's Information Criterion (AIC) [1]—proposes $\mathcal{P}_{\mathrm{AIC}}[p_k] = p_k$ [6].

The Bayesian criterion of the type (10.5) is based on the marginal ***a posteriori*** model inference:

$$\hat{k} = \arg\max_{k} \mathrm{p}(\mathcal{I}_k \mid \mathbf{d}, \boldsymbol{\alpha}, \mathcal{I}) \tag{10.11}$$

Such strategies will be considered in the next Section. However, it is relevant to note here that asymptotic approximations are usually required in order to evaluate the integrations over $\overline{\Theta}_k$ which generate $\mathrm{p}(\cdot)$ in (10.11). For example, a second order Taylor expansion around $\hat{\boldsymbol{\theta}}_k$ yields a penalized ML criterion of the kind (10.7) [4].

10.4 Bayesian Signal Identification

Adopting the Bayesian perspective - which accepts probability as the consistent calculus for quantifying and manipulating degrees of belief in hypotheses - the signal identification task becomes one of eliciting the following ***a posteriori (AP)*** inference:

$$\begin{aligned} \mathrm{p}(\boldsymbol{\theta}_k \mid \mathbf{d}, \boldsymbol{\beta}, \boldsymbol{\gamma}, \mathcal{I}) &\propto (\boldsymbol{\theta}_k \mid \mathbf{d}, \boldsymbol{\beta}, \mathcal{I}_k)\mathrm{p}(\boldsymbol{\theta}_k \mid \boldsymbol{\gamma}, \mathcal{I}) & (10.12) \\ &\propto (\boldsymbol{\theta}_k \mid \mathbf{d}, \boldsymbol{\beta}, \mathcal{I}_k)\mathrm{p}(\boldsymbol{\theta}_k \mid \boldsymbol{\gamma}_k, \mathcal{I}_k)\mathrm{p}(\mathcal{I}_k \mid \mathcal{I}) & (10.13) \end{aligned}$$

In (10.12), $\mathrm{p}(\boldsymbol{\theta}_k \mid \boldsymbol{\gamma}, \mathcal{I})$ is the prior over all competing hypotheses $\boldsymbol{\Theta}_K = \boldsymbol{\theta}_k \in \overline{\Theta}$. It is a mixture of the priors, $\mathrm{p}(\boldsymbol{\theta}_k \mid \boldsymbol{\gamma}_k, \mathcal{I}_k)$, over $\overline{\Theta}_k$. $\boldsymbol{\gamma}$ is the set of ***all*** known parameters, $\boldsymbol{\gamma}_k$, of the prior laws, taken together, and $\mathrm{p}(\mathcal{I}_k \mid \mathcal{I})$ is the ***model prior*** (Section 10.2). Note that, since (10.2) implies a partition, the proposition $\boldsymbol{\Theta}_K = \boldsymbol{\theta}_k$ (denoted simply by $\boldsymbol{\theta}_k$ in (10.12) and (10.13)) is equivalent to $(K = k) \wedge (\boldsymbol{\Theta}_k = \boldsymbol{\theta}_k)$, where $\wedge$ denotes logical AND. AP inference for signal identification then becomes a task of distributing probability over the superspace, $\overline{\Theta} = \bigcup_{k=1}^{M} \overline{\Theta}_k$, of $\mathcal{I}$, in the light of observations $\mathbf{d}$. This is expressed by (10.12) and (10.13).

A certainty equivalent [10] in the form of a point estimate is usually required. In common with the majority of the literature, ***Maximum Joint a Posteriori (MAP)*** optimization is adopted in this chapter:

$$\widehat{\boldsymbol{\theta}_k} = \arg\max_{k} \max_{\boldsymbol{\theta}_k} \mathrm{p}(\boldsymbol{\theta}_k \mid \mathbf{d}, \boldsymbol{\beta}, \boldsymbol{\gamma}, \mathcal{I}) \tag{10.14}$$

This is a Bayesian criterion of the type (10.4). Without special assumptions concerning the priors, (10.14) will not constitute a regularized inference.

10.4.1 Prior-Regularized Inference

Consider the following prior assignments for (10.13):

$$\mathrm{p}(\mathcal{I}_k \mid \mathcal{I}) = \frac{1}{M} \tag{10.15}$$

$$\mathrm{p}(\boldsymbol{\theta}_k \mid \gamma_k, \mathcal{I}_k) \propto \exp\left[-\beta_k(\boldsymbol{\theta}_k \mid \gamma_k, \mathcal{I}_k)\right] \tag{10.16}$$

From (10.8), (10.13), (10.14), (10.15), and (10.16), and assuming, for ease of notation, that $f(x) = \exp(-x)$ in (10.9), then

$$\widehat{\boldsymbol{\theta}_k} = \arg\min_k \min_{\boldsymbol{\theta}_k} \left[\|\mathbf{d} - \mathbf{g}_k(\boldsymbol{\theta}_k)\|_{\Box} + \beta_k(\boldsymbol{\theta}_k \mid \gamma_k, \mathcal{I}_k)\right] \tag{10.17}$$

Careful choice of $\beta_k(\cdot)$ in (10.16) can therefore yield a regularized criterion. A vast literature exists (for a review, see [10]) on suitable priors for specific problems. Examples include Gaussian parameter priors (i.e. $\beta_k(\boldsymbol{\theta}_k \mid \mathbf{T}_k, \mathcal{I}_k) = \boldsymbol{\theta}_k^H \mathbf{T}_k^{-1} \boldsymbol{\theta}_k$, where $\mathbf{T}_k$ is the ***a priori*** covariance matrix, for $\boldsymbol{\Theta}_k$), to encourage parameter inferences of small Mahalanobis norm, and Markovian priors, important in image segmentation [13] for encouraging connectivity between inferred texture labels.

10.4.2 Non-Informative Prior Assignment

If conservative inference is to be effected, the need exists to adopt non-informative priors over the parameter spaces, $\overline{\Theta}_k$. Such priors also improve the ***reportability*** [3] of the resulting inference, and any estimates based upon it. A powerful technique for assigning non-informative priors is discussed in Section 10.5.1. However, an immediate and compelling choice remains the improper uniform prior (more correctly interpreted as Lebesgue measure for $\boldsymbol{\Theta}_K$):

$$\mathrm{p}(\boldsymbol{\theta}_k \mid \boldsymbol{\gamma}, \mathcal{I}) = \mathrm{p}(\boldsymbol{\theta}_k \mid \mathcal{I}) = const., \quad \forall k \tag{10.18}$$

Substituting (10.18) into (10.12), and noting that the dependence on hyperparameters, $\boldsymbol{\gamma}$, disappears, then $\mathrm{p}(\boldsymbol{\theta}_k \mid \mathbf{d}, \boldsymbol{\beta}, \mathcal{I}) \propto l(\boldsymbol{\theta}_k \mid \mathbf{d}, \boldsymbol{\beta}, \mathcal{I}_k)$. This choice is in sympathy with (i) the approach taken by Bayes (and Laplace) to the quantification of ignorance (and indifference) [10], (ii) the Maximum Entropy Method (Section 10.5.1) when there are no mean constraints [12], and (iii) the modelling assumptions implicit in (unpenalized) likelihood-based inference.

It has been argued in the literature (e.g. [2]) that proper priors - i.e. ones that are normalizable over *each* $\overline{\Theta}_k$ - are necessary for regularized inference, if the dimension of $\overline{\Theta}_k$ varies with k. The uniform prior can imply infinitely greater prior support for higher order models. To see this, assume that $\overline{\Theta}_k = [0, L)^k \subset \overline{\mathcal{R}}^k$, a k-dimensional hypercube of side L. Then, from (10.18), define

$$r = \frac{\mathrm{p}(\mathcal{I}_{k+1} \mid \mathcal{I})}{\mathrm{p}(\mathcal{I}_k \mid \mathcal{I})} = \frac{\int_{\overline{\Theta}_{k+1}} \mathrm{p}(\boldsymbol{\theta}_{k+1} \mid \mathcal{I}) d\boldsymbol{\theta}_{k+1}}{\int_{\overline{\Theta}_k} \mathrm{p}(\boldsymbol{\theta}_k \mid \mathcal{I}) d\boldsymbol{\theta}_k} = L \tag{10.19}$$

Marginalization (to be considered in Section 10.4.3) has been employed in the third term above, to yield the necessary total probabilities. From (10.19), $r \to \infty$ as $\overline{\Theta}_k \to \overline{\mathcal{R}}^k$. This objection is rejected on four counts, as follows, where, without loss of generality, it is assumed that p_k is a monotonically increasing function of k:

(i) $\mathcal{I}_{k+1}$ embraces an uncountably larger number of model hypotheses, $\Theta_{k+1} = \theta_{k+1}$, than $\mathcal{I}_k$. For example, in the nested case (Section 10.2), $\overline{S}_{k+1} \supset \overline{S}_k$. Thus, although nesting of $\overline{\Theta}_k$ does not follow, the choice $\mathrm{p}(\mathcal{I}_k \mid \mathcal{I}) < \mathrm{p}(\mathcal{I}_{k+1} \mid \mathcal{I})$ is both reasonable, and consistent with prior indifference.

(ii) Again, in the nested case, it is desirable that a prior assignment, consistent with (10.3), should satisfy the following:

$$\int_{\overline{\Theta}_k} \mathrm{p}(\theta_k \mid \mathcal{I}) d\theta_k = \left[\int_{\overline{\Theta}_k} \mathrm{p}(\psi_{k+1}, \theta_k \mid \mathcal{I}) d\theta_k \right]_{\psi_{k+1} \in \overline{\Psi}^0_{k+1}} \tag{10.20}$$

The uniform prior (10.18) is a solution of (10.20); a normalizable prior will, in general, fail to satisfy (10.20). It might be argued that a constraint,

$$\mathrm{p}(\psi_{k+1}, \theta_k \mid \mathcal{I})\Big|_{\psi_{k+1} \in \overline{\Psi}^0_{k+1}} = 0$$

could be imposed ***a priori***. In fact, this constitutes an anti-Ockham prior, and severely curtails the Ockham behaviour of the resulting marginal AP inference, as explained in [7].

(iii) If a rectangular model prior, $\mathrm{p}(\mathcal{I}_k \mid \mathcal{I})$, is to be achieved (i.e. $r = 1$ in (10.19)) without infinite penalties on transitions $\Theta_k \to \Theta_{k+1}$ ***a priori***, then non-Lebesgue measure is necesary on $\overline{\Theta}$. Such a choice is difficult to defend in realistic signal identification problems.

(iv) It will be seen in Section 10.5.2, for a large class of signal models - and in Sections 10.6 and 10.7 in specific cases - that highly effective Ockham behaviour is elicited in the AP inference using the uniform parameter prior.

In conclusion, a prior which is normalizable over each $\overline{\Theta}_k$ constitutes a strong complexity-penalizing prior, by (i). True prior indifference is expressed by using a uniform or diffuse prior over $\overline{\Theta}$, thereby satisfying (10.20). This approach to prior assignment reflects the natural partition of the sample space into elementary outcomes, $\Theta_K = \theta_k$, and avoids the need for hyperparameters, γ (10.13). Complexity penalization is then achieved using the technique in the next Section.

10.4.3 Marginalization-Regularized Inference

In the probability calculus, parameters of a hypothesis can be integrated out of the inference (10.12). Letting $\Theta_k = (\Psi_k^T, \Phi_k^T)^T$ as before, then the inference for each Ψ_k is obtained by integrating over the space, $\overline{\Phi}_k$, of Φ_k [10]:

$$\mathrm{p}(\psi_k \mid \mathbf{d}, \beta, \gamma, \mathcal{I}) = \int_{\overline{\Phi}_k} \mathrm{p}(\theta_k \mid \mathbf{d}, \beta, \gamma, \mathcal{I}) d\phi_k \tag{10.21}$$

Lebesgue measure is assumed. If the choice, $\Phi_k = \Theta_k$, is made, then the marginal AP model inference in (10.11) results, with $\alpha = (\beta^T, \gamma^T)^T$.

Although commonly associated with the notion of 'eliminating nuisance parameters' [2], the primary importance of marginalization is that it yields an inference for a parameter ***subspace***, $\overline{\Theta}_k\big|_{\psi_k}$, whose implied signal subspace is $\overline{S}_k\big|_{\psi_k} = \{\mathbf{g}_k(\psi_k, \phi_k),$

$\forall \phi_k \in \overline{\Phi}_k\} \subset \overline{\underline{\mathcal{C}}}^N$ [7, 10]. The structure of the latter (including the number of basis functions, their degree of linear dependence, etc.) is therefore naturally embraced by (10.21). What results is an *objective* measure of signal complexity [7, 10]. This induces regularized inferences of the kind (10.7) and (10.17) for propositions $\Psi_K = \psi_k$, without the need to employ potentially unreportable priors (10.16).

Note, from (10.21), that marginalization over ever larger subsets, Φ_k, of p.p.s, Θ_k, concentrates the inference into ever smaller numbers of parameters, Ψ_k. In so doing, the complexity of ever larger signal subspaces, $\underline{S}_k|_{\psi_k}$, is measured, but inferability of Φ_k is lost. This trade-off represents an 'uncertainty principle' for probabilistic inference. Nevertheless, the practice - referred to as the 'evidence framework' [2], or the 'marginal profile policy' [10] - persists of inferring a model, $\hat{k}$, via (10.11), and then inferring a set of parameters, $\hat{\theta}_{\hat{k}}$, for the selected model. The true MAP estimate, $\widehat{\theta_k}$ (10.14), is not guaranteed under this policy [10].

10.5 The General Non-Linear Signal Model

Consider the wide class of signals of the type (10.8), under the constraint

$$\mathbf{g}_k(\boldsymbol{\theta}_k) = \mathbf{G}_k(\boldsymbol{\omega}_k)\mathbf{b}_k \tag{10.22}$$

$\mathbf{G}_k(\cdot) \in \overline{\underline{\mathcal{C}}}^{N\times k}$ is a matrix of k N-dimensional basis vectors, each non-linearly parameterized in $\boldsymbol{\omega}_k \in \overline{\underline{\mathcal{R}}}^{r_k}$, and linearly combined via $\mathbf{b}_k \in \overline{\underline{\mathcal{C}}}^k$. Thus, $\mathcal{I}_k$ (10.8) and (10.22) constitutes a nonlinear regression analysis of $\mathbf{d}$, with unknown parameters $\boldsymbol{\Theta}_k = (\boldsymbol{\omega}_k^T, \mathbf{b}_k^T)^T \in \overline{\underline{\mathcal{C}}}^{p_k}$, $p_k = 2k + r_k$, and $\boldsymbol{\beta}$ assumed known. (Noise parameters may be handled similarly to the signal parameters, if they are unknown [9, 10]). Specific cases include harmonic (Section 10.6), wavelet, AR (Section 10.7) and linear regression models. The criteria described in Section 10.3.1 may be employed. A number of Bayesian strategies will be considered next, however.

10.5.1 *Entropic Prior*

A prior density may be elicited by maximizing its entropy subject to any testable mean constraints [2, 5, 10, 16]. Optimality properties have been established for this procedure [14]. The following entropic prior can be deduced [8, 10] for the parameters of the GNLS model with a constraint, γ^2, on the power of the signal:

$$p(\boldsymbol{\omega}_k, \mathbf{b}_k \mid \gamma, \mathcal{I}_k) \propto \exp\left[-\frac{k}{N\gamma^2}\mathbf{b}_k^H \mathbf{G}_k^H \mathbf{G}_k \mathbf{b}_k\right] \tag{10.23}$$

Its behaviour will be considered in Section 10.6.

10.5.2 *Ockham Parameter Inference (OPI)*

As an alternative to (10.23), analysis now proceeds using only the uniform parameter prior (10.18) for $\boldsymbol{\Theta}_K$. Substituting the latter, as well as (10.8) and (10.22), into (10.12), letting $\mathbf{e} \sim \mathcal{N}(\mathbf{0}, \boldsymbol{\Sigma})$, and integrating over the space, $\overline{\underline{\mathcal{C}}}^k$, of each $\mathbf{b}_k$ (which can be accomplished analytically under these assumptions), then the *Maximum Marginal a Posteriori (MaAP)* estimate for $\boldsymbol{\omega}_k$ is found to be [7, 10]

$$\widehat{\boldsymbol{\omega}_k} = \arg\min_k \min_{\boldsymbol{\omega}_k} \left\{ \|\mathbf{d} - \mathbf{G}_k \hat{\mathbf{b}}_k\|^2_{\boldsymbol{\Sigma}^{-1}} - \mathcal{D}_k(\boldsymbol{\omega}_k) \right\} \tag{10.24}$$

$$\mathcal{D}_k(\boldsymbol{\omega}_k) = -\ln|\mathbf{G}_k^H \boldsymbol{\Sigma}^{-1} \mathbf{G}_k| + k \ln \pi \tag{10.25}$$

Lebesgue measure is assumed. $\hat{\mathbf{b}}_k(\boldsymbol{\omega}_k) = (\mathbf{G}^H \boldsymbol{\Sigma}^{-1} \mathbf{G})^{-1} \mathbf{G}^H \boldsymbol{\Sigma}^{-1} \mathbf{d}$ is the ML estimate of $\mathbf{b}_k$ at each $\boldsymbol{\omega}_k$ [10]. (10.25) is the *Ockham Parameter Inference (OPI)* [7, 8, 10], and it arises in (10.24) naturally from integration (10.21). Comparing with (10.10), $-\mathcal{D}_k(\boldsymbol{\omega}_k) = \mathcal{P}_{\text{MaAP}}([k], \boldsymbol{\omega}_k)$ is a penalty function with integer argument, k, and continuous argument $\boldsymbol{\omega}_k$. The function $\mathcal{C}_k(\boldsymbol{\omega}_k) = |\mathbf{G}_k^H \boldsymbol{\Sigma}^{-1} \mathbf{G}_k|$ has been axiomatically deduced to be the objective complexity measure for the proposition $\boldsymbol{\Omega}_K = \boldsymbol{\omega}_k$, namely, the one implying the signal subspace $\underline{\mathcal{S}}_k|_{\boldsymbol{\omega}_k}$ [10]. The advantages of employing (10.24) include the following: (i) subjective priors (10.16), and penalty terms (10.7) and (10.10) requiring subsidiary desiderata (such as a minimization of description length), are unnecessary; (ii) true marginal MAP (i.e. MaAP) estimates, $\widehat{\boldsymbol{\omega}_k}$, are inferred jointly with the model structure, which is not the case for the strict selection criteria (10.5), (10.7), (10.10), and (10.11); (iii) marginalizations are analytical, whereas numerical approximations are necessary to evaluate (10.11) under (10.8) and (10.22); (iv) the OPI (10.25) regularizes direct parameter estimation (when $M = 1$ (10.1)), so that estimation thresholds can be avoided [7, 10]. It is emphasized that the additive structure (10.24) is a consequence of the modelling assumptions, specifically the use of a Gaussian law. Notwithstanding this, complexity-penalizing terms will arise whenever marginalization is undertaken [7], since the ***structure*** of competing signal subspaces is assessed (Section 10.4.3) with this operator.

10.6 Order Determination and Frequency Estimation for a Harmonic Model

Consider the problem of fitting a complex, discrete-time signal, $d[n]$ with a set of complex exponentials, whose digital frequencies, ω_i, are harmonically constrained, i.e. $\omega_i = i\omega_0$:

$$\mathcal{I}_k: \quad d[n] = \sum_{i=1}^{k} b_{ki} e^{ji\omega_0 n} + e[n], \qquad \begin{array}{l} k = 1, \ldots, M \\ n = 0, \ldots, N-1 \end{array} \tag{10.26}$$

This belongs to the GNLS class defined by (10.8) and (10.22), where $d_i = d[i-1]$, $i = 1, \ldots, N$, are the elements of the data vector, $\mathbf{d}$, and $(\mathbf{G}_k)_{il} = e^{jl\omega_0(i-1)}$. Let $N = 32$, and $\mathbf{e} \sim \mathcal{N}(\mathbf{0}, \mathbf{I}_N)$ (complex). Thus, $p_k = 2k + 1$, yielding the MDL and AIC penalty terms in Fig. 10.1. Note that the terms are ***constant*** with respect to ω_0 for any fixed k. Hence, they fail to regularize ***estimation*** of ω_0. Substituting (10.26) into (10.23), and integrating over $\mathbf{b}_k$, the marginal entropic prior for ω_0 is obtained for each $\mathcal{I}_k$. It has been shown [8] to be of the same form as the OPI (10.25), which is plotted in Fig. 10.2. Singularities at $\omega_0 = 0$ encourage lower order models. Note that, unlike the terms in Fig. 10.1, the one in Fig. 10.2 can be employed

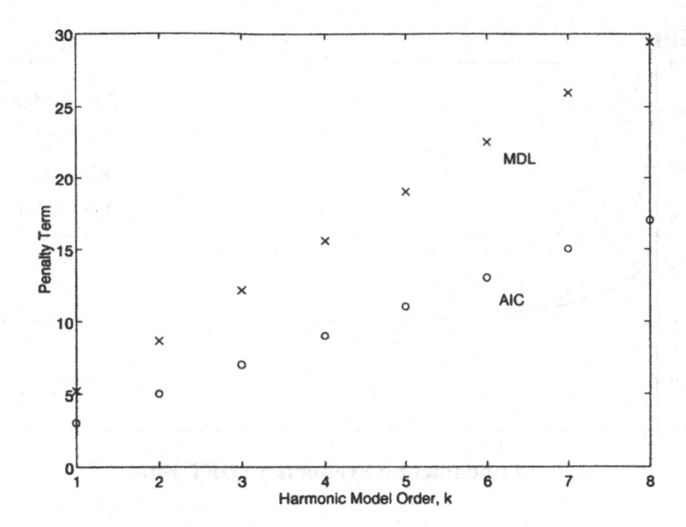

FIGURE 10.1. Variation of penalty terms with order of the harmonic model

successfully in alternative-free inference (i.e. when the model structure is *known*) to regularize estimation of ω_0. Accurate sub-threshold estimation is then possible. Detailed study of terms such as the one in Fig. 10.2 may be found in [7, 10].

10.7 Order Determination for Autoregressive (AR) Models

Consider a superhypothesis, $\mathcal{I}$, which seeks to explain a *real* time series, $d[n]$, in terms of a linear autoregressive model:

$$\mathcal{I}_k: \quad d[n] = -\sum_{i=1}^{k} b_{ki} d[n-i] + e[n], \qquad \begin{array}{l} k = 1, \ldots, M \\ n = 0, \ldots, N-1 \end{array} \tag{10.27}$$

The $N-M$ equations for $n = M, \ldots, N-1$ may be gathered into vector-matrix form, to yield the following:

$$\mathbf{d} = -\mathbf{Y}_k \boldsymbol{a}_k + \mathbf{e} \tag{10.28}$$

where $\mathbf{d} = (d[M], \ldots, d[N-1])^T \in \overline{\mathcal{R}}^{N-M}$, $\mathbf{Y}_k \in \overline{\mathcal{R}}^{(N-M)\times k}$ is a Toeplitz matrix with ilth element $d[M+i-1-j]$, and $\mathbf{e}\left(\in \overline{\mathcal{R}}^{N-M}\right) \sim \mathcal{N}(\mathbf{0}, \mathbf{I}_{N-M})$. Thus, (10.28) belongs to the GNLS class (10.22), with a basis function matrix dependent on the data themselves, and with $\boldsymbol{\omega}_k = \emptyset$. Hence, it is linearly parameterized. It may be shown by factorization (i.e. the chain rule of probability) [9] that

$$l(\boldsymbol{a}_k \mid \mathbf{d}, \mathbf{d}_0, \mathcal{I}_k) = \frac{1}{(2\pi)^{(N-M)/2}} \exp\left[-\frac{1}{2}\|\mathbf{d} + \mathbf{Y}_k \boldsymbol{a}_k\|_2^2\right] \tag{10.29}$$

where $\mathbf{d}_0 = (d[0], \ldots, d[M-1])^T$ are the initial conditions for $\mathcal{I}_M$, and, therefore, for $\mathcal{I}_k$, $k = 1, \ldots, M$. $\|\cdot\|_2$ denotes the Euclidean norm. Recalling that $\boldsymbol{a}_k$ is local to $\mathcal{I}_k$, then, from Bayes' theorem, the AP inference for $\boldsymbol{a}_k$ is

$$\mathrm{p}(\boldsymbol{a}_k \mid \mathbf{d}, \mathbf{d}_0, \mathcal{I}) = \frac{l(\boldsymbol{a}_k \mid \mathbf{d}, \mathbf{d}_0, \mathcal{I}_k)\mathrm{p}(\boldsymbol{a}_k \mid \mathcal{I})}{\mathrm{p}(\mathbf{d} \mid \mathbf{d}_0, \mathcal{I})} \tag{10.30}$$

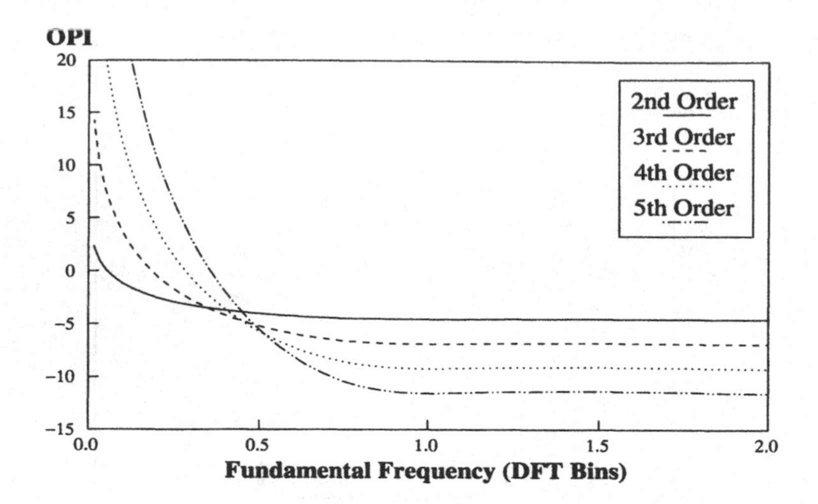

FIGURE 10.2. Ockham Parameter Inference (OPI) for ω_0 in the harmonic model

Adopting a uniform parameter prior across all hypotheses (10.18), and recognizing that $\mathrm{p}(\mathbf{d} \mid \mathbf{d}_0, \mathcal{I})$ is independent of $\boldsymbol{a}_k$, then

$$\mathrm{p}(\boldsymbol{a}_k \mid \mathbf{d}, \mathbf{d}_0, \mathcal{I}) \propto l(\boldsymbol{a}_k \mid \mathbf{d}, \mathbf{d}_0, \mathcal{I}_k)$$

Integration over the space, $\overline{\mathcal{A}}_k$, of $\boldsymbol{a}_k$ yields the total probability for $\mathcal{I}_k$ (Section 10.4.3). Since the conditional proposition, $(\mathcal{I}_k \mid \mathbf{d}, \mathbf{d}_0, \mathcal{I})$, implies the space of all signals spanned by the columns of $-\mathbf{Y}_k$ (10.28), a regularized inference will result. Letting $\overline{\mathcal{A}}_k = \overline{\mathcal{R}}^k$, and assuming Lebesgue measure, then the required integrations may be achieved analytically [9], to yield

$$\mathrm{p}(\mathcal{I}_k \mid \mathbf{d}, \mathbf{d}_0, \mathcal{I}) \propto (2\pi)^{\frac{k}{2}} \left|\mathbf{Y}_k^T \mathbf{Y}_k\right|^{-\frac{1}{2}} \exp\left[\frac{1}{2} \mathbf{d}^T \mathbf{Y}_k (\mathbf{Y}_k^T \mathbf{Y}_k)^{-1} \mathbf{Y}_k^T \mathbf{d}\right] \tag{10.31}$$

Note that the integration over $\overline{\mathcal{R}}^k$ implies non-zero prior probability for unstable models in each $\mathcal{I}_k$. If these are to be rejected ***a priori***, a bounded rectangular prior is necessary, leading to bounded integration of (10.30).

Maximization of (10.31) then yields the following MaAP criterion for model order determination:

$$\hat{k} = \arg\min_k \left[\frac{1}{2} \|\mathbf{d} + \mathbf{Y}_k \hat{\boldsymbol{a}}_k\|_2^2 + \frac{1}{2} \ln\left|\mathbf{Y}_k^T \mathbf{Y}_k\right| - \frac{k}{2} \ln 2\pi\right] \tag{10.32}$$

$\hat{\boldsymbol{a}}_k = -\mathbf{Y}_k^{\#} \mathbf{d}$, where $\mathbf{Y}_k^{\#}$ denotes the Moore-Penrose pseudo-inverse of $\mathbf{Y}_k$. Thus, $\hat{\boldsymbol{a}}_k$ is the ML estimate of the parameters of the kth model (10.27). It is equal to (i) the Least Squares estimate, because of the Gaussian assumption, and (ii) the Yule-Walker estimate [6], resulting from the autocorrelation method, because $\mathbf{Y}_k^T \mathbf{Y}_k$ is proportional to a matrix of sample autocorrelations [9]. Comparing with (10.7), the marginalization-induced penalty term is

$$\mathcal{P}_{\mathrm{MaAP}}[k] = \frac{1}{2} \ln\left|\mathbf{Y}_k^T \mathbf{Y}_k\right| - \frac{k}{2} \ln 2\pi \tag{10.33}$$

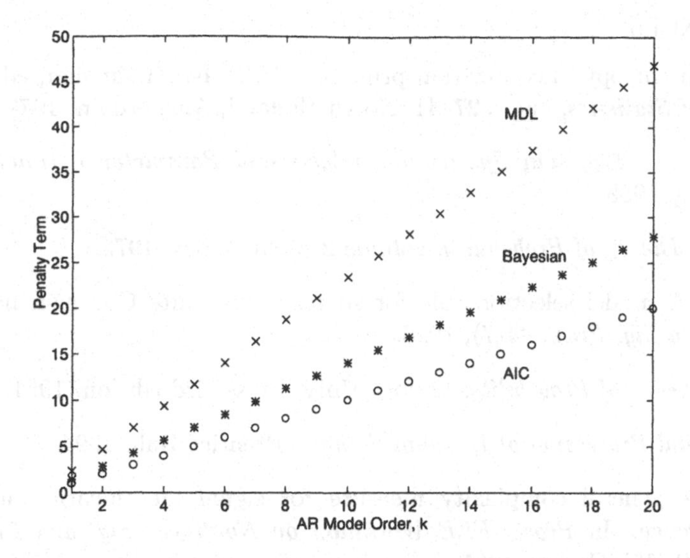

FIGURE 10.3. Marginalization (MaAP), AIC and MDL penalty terms for AR model order determination

This is a data-dependent, parameter-free Ockham inference, playing an equivalent rôle, in AR model selection, to the OPI in model selection with parameter estimation (10.25). (10.33) can be derived from fundamentals as an Ockham Prior [7, 10], but has entered (10.32) objectively via integration. (10.33) is plotted in Fig. 10.3 for a 4th order AR process with $N = 128$ and $M = 20$.

The Bayesian penalty lies between the AIC penalty (i.e. k) and the MDL penalty (i.e. $(k/2)\ln(N-M)$) in this case. However, because of its stochastic nature, Monte Carlo techniques are necessary to characterize the penalty term in (10.32). Sampling distributions are currently being developed. This work—and associated performance figures such as mean-squared error—will be reported shortly [9].

10.8 Conclusions

In this chapter, regularization of the parametric signal identification problem was explained. Non-flat priors were considered, but the non-informative uniform prior over the composite parameter space, $\underline{\Theta}$, was shown to optimally express prior indifference for signal identification. Marginalization was employed as an objective means of measuring and criticizing the complexity of rival signal spaces. Cases of global and partial marginalization were presented, each inducing Ockham terms. A key advantage of marginalization is that the number of parameters to be inferred may be chosen freely, yielding an Ockham term which is a function of those parameters. In this manner, the marginal inference is sensitive to parameter-dependent complexity variations, a feature not observed in criteria such as MDL, AIC or evidence. Frequency and order determination for the harmonic model, as well as order determination for the AR model, were presented. In future work [9], regularized inference of ***subsets*** of AR parameters will be considered. In time series classification, a feature vector of predefined dimension can be extracted in this manner.

10.9 References

[1] H. Akaike. On entropy maximization principle. In P. R. Krishnaiah, editor, *Applications of Statistics*, pages 27–41. North-Holland, Amsterdam, 1977.

[2] G. L. Bretthorst. *Bayesian Spectrum Analysis and Parameter Estimation.* Springer-Verlag, 1989.

[3] B. de Finetti. *Theory of Probability*, volume 2. John Wiley, 1975.

[4] P.M. Djurić. A model selection rule for sinusoids in white Gaussian noise. *IEEE Trans. on Sig. Proc.*, 44(7), 1996.

[5] H. Jeffreys. *Theory of Probability.* Oxford Univ. Press, 3rd edition, 1961.

[6] B. Porat. *Digital Processing of Random Signals.* Prentice-Hall, 1994.

[7] A. Quinn. A general complexity measure for signal identification using Bayesian inference. In *Proc. IEEE Workshop on Nonlinear Sig. and Image Process. (NSIP '95)*, Greece, 1995.

[8] A. Quinn. Novel parameter priors for Bayesian signal identification. In *Proc. IEEE Int. Conf. on Acoust., Sp., Sig. Proc. (ICASSP)*, Munich, 1997.

[9] A. Quinn. New results for autoregressive signal identification. In *IEEE Int. Conf. on Acoust., Sp., Sig. Proc. (ICASSP)*, 1998. In preparation.

[10] A.P. Quinn. Bayesian Point Inference in Signal Processing. Ph.D. Thesis, Cambridge University Engineering Dept., 1992.

[11] J. Rissanen. Modeling by shortest data description. *Automatica*, 14, 1978.

[12] R.D. Rosenkrantz, editor. *E. T. Jaynes: Papers on Probability, Statistics and Statistical Physics.* D. Reidel, Dordrecht-Holland, 1983.

[13] O. Schwartz and A. Quinn. Fast and accurate texture-based image segmentation. In *Proc. 3rd IEEE Int. Conf. on Image Proc.*, Lausanne, 1996.

[14] J.E. Shore and R.W. Johnson. Axiomatic derivation of the principle of maximum entropy and the principle of minimum cross-entropy. *IEEE Trans. on Inf. Th.*, IT-26(1), 1980.

[15] C.S. Wallace and P.R. Freeman. Estimation and inference by compact coding. *J. Roy. Statist. Soc. B*, 49(3), 1987.

[16] A. Zellner. *An Introduction to Bayesian Inference in Econometrics.* John Wiley and Sons, 1971.

11

System Identification

Lennart Ljung[1]

ABSTRACT

In this contribution we give an overview and discussion of the basic steps of System Identification. The four main ingredients of the process that takes us from observed data to a validated model are: (1) The data itself, (2) The set of candidate models, (3) The criterion of fit and (4) The validation procedure. We discuss how these ingredients can be blended to a useful mix for model-building in practice.

11.1 Introduction

The process of going from observed data to a mathematical model is fundamental in science and engineering. In the control area this process has been termed "System Identification" and the objective is then to find dynamical models (difference or differential equations) from observed input and output signals. Its basic features are however common with general model building processes in statistics and other sciences.

System Identification has been an active research area for more than thirty years. It has matured and many of the techniques have become standard tools in control and signal processing engineering. The "mainstream approach" is described e.g. in [8] and [19]. Over the past few years there has been a significantly renewed interest in the area with topics like "unknown-but-bounded" disturbances, [17, 11], set membership techniques [4, 14] subspace techniques [15], H_∞-identification [16, 6], worst case analysis [5, 10], as well as how to deal with unmodelled dynamics [13].

The procedure is characterized by four basic ingredients:

1. The observed data
2. A set of candidate models
3. A criterion of fit
4. Validation

We shall in the sequel discuss these items more closely.

[1]Department of Electrical Engineering, Linköping University, S-581 83 Linköping, Sweden, E-mail: Ljung@isy.liu.se

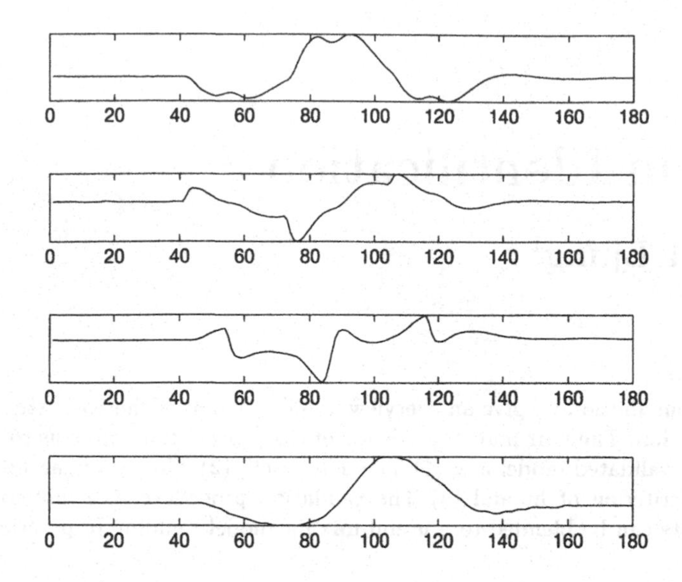

FIGURE 11.1. Results form test flights of the new Swedish aircraft JAS- Gripen, developed by SAAB Military Aircraft AB, Sweden. From the top a) Pitch rate. b) Elevator angle. c) Canard angle. d) Leading edge flap

11.2 The Data

The area of system identification begins and ends with real data. Data required to build models and to validate models. The result of the modelling process can be no better than what corresponds to the information contents in the data.

Let us take a look at two data sets:

Example 1 An unstable aircraft. *Figure 11.1 shows some results from test flights of the new Swedish aircraft JAS-Gripen, developed by SAAB Military Aircraft AB, Sweden. The problem is to use the information in these data to determine the dynamical properties of the aircraft for fine-tuning regulators, for simulations, and so on. Of particular interest are the aerodynamical derivatives.*

Example 2 Vessel dynamics. *See Figure 11.2. The problem is to determine the residence time in the buffer vessel. The pulp spends about 48 hours total in the process, and knowing the residence time in the different vessels is important in order to associate various portions of the pulp with the different chemical actions that have taken place in the vessel at different times. (The κ-number is a quality property that in this context can be seen as a marker allowing us to trace the pulp.)*

11.3 The Set of Models: Model Structures

The single most important step in the identification process is to decide upon a model structure, i.e., a set of candidate models. In practice typically a whole lot of them are tried out and the process of identification really becomes the process of evaluating and choosing between the resulting models in these different structures.

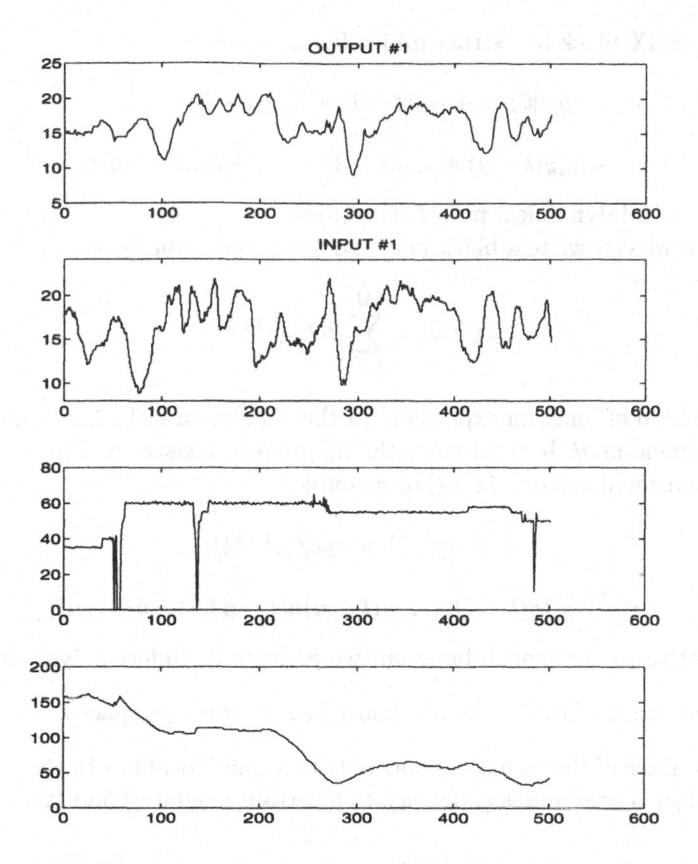

FIGURE 11.2. From the pulp factory at Skutskär, Sweden. The pulp flows continuously through the plant via several buffer tanks. From the top a) The κ-number of the pulp flowing into a buffer vessel. b) The κ-number of the pulp coming out from the buffer vessel. c) Flow out from the buffer vessel. d) Level in the buffer vessel

It is natural to distinguish between three types of model structures:

1. Black-box structures
2. Structures from physical modelling
3. Structures from semi-physical modelling

11.3.1 Black-Box Structures

A black-box structure is one where the parameterization in terms of a parameter vector θ is chosen so that the family of candidate models covers as "many common and interesting" ones as possible. No particular attention to the actual application is then paid. For a linear system (a linear mapping from past data to future ones) we could for example think of choosing the parameters as the impulse response coefficients, of a finite impulse response model

$$\hat{y}(t|\theta) = \sum_{k=1}^{M} \theta_k u(t-k) \tag{11.1}$$

Here $u(t)$ is the input to the process and $\hat{y}(t|\theta)$ is the model's predicted or "guessed" output at time t. The actual output will be $y(t)$. More common in control applications is the ARX black box structure for linear systems:

$$\hat{y}(t|\theta) = -a_1 y(t-1) - a_2 y(t-2) - \dots$$
$$-a_n y(t-n) + b_1 u(t-1) + \dots + b_m u(t-m) \tag{11.2}$$

"the mother of all dynamical model structures".

In general we can write a black box structure conceptually as

$$\hat{y}(t|\theta) = \sum_{k=1}^{M} \theta_k h_k(z^{t-1}) \tag{11.3}$$

i.e. as some kind of function expansion. In the general case the basis functions $\{h_k\}$ may also depend on θ. In most cases the h_k are also constrained to be functions of a fixed dimensional vector $\varphi(z^t)$ (For example

$$h_k(z^{t-1}) = h_k(\varphi(z^{t-1}))$$
$$\varphi(z^t) = (y(t-1), \dots, y(t-n) u(t-1), \dots, u(t-n))$$

It is instructive to distinguish between two principally different basis functions:

- Global: Each of the h_k have support in the whole φ- space
- Local: Each of the h_k has support only in a small local box in the φ-space.That is, $h_k(\varphi)$ is zero unless φ belongs to a certain neighborhood (that depends on k).

Among black-box structures that use global basis functions are all the usual linear black box models, Volterra series expansions and so on.

The local basic functions models can be visualized as a multidimensional table: The φ-space has been split up into a number of boxes. A new observation $\varphi(t)$ then falls into one of these boxes, the one corresponding to say h_k, and the predicted output is then taken as θ_k (or possibly interpolated, taking into account few neighboring boxes). The sizes and locations of the boxes can be determined with the aid of estimation data. The extreme case is when the boxes are determined so that exactly one data point $\varphi(t)\, t = 1 \dots, N$ has fallen in each box: this is the so called nearest neighbor approach [22]. All this is well established in the statistical literature under names of "non-parametric regression" and "density estimation" [20, 3].

Neural network model structures, e.g. [12], represent a spectacular revival of these techniques. So called radial basis networks correspond to localized bases (where the "boxes" overlap like Gaussian distribution functions), while the feed-forward sigmoid network formally would use global basis functions (although the "dynamic effects" really are localized). Fuzzy modelling [7] is again an example of localized basis functions with interpolation rules, which are inherited by the "membership function".

It is worth stressing that these new techniques of neural net modelling and fuzzy identification represent useful realizations of non-linear black box modelling with some particular new structures, but at the same time they definitely fall into a very old and classical framework of estimation techniques (See, e.g. [9, 2]).

11.3.2 Structures from Physical Modelling

When we have physical insight into the properties of the system to be identified, it is natural to exploit this: *"Don't estimate what you already know!"* Basically we then write down those physical laws and relationships that describe the system. Most often they are then summarized in a state space form where the parameters are unknown physical constants in the description. The identification process is then to estimate these constants.

11.3.3 Semi-Physical Model Structures

The logical route to utilize available physical knowledge may be quite laborious. It is then tempting to instead try some simple black-box structures, such as the ARX model (11.2) *("Try Simple Things First")*. This is quite OK, but it should in any case be combined with physical insight. Here is a toy example to illustrate the point:

"Suppose we want to build a model for how the voltage applied to an electric heater affects the temperature of the room. Physical modelling entails writing down all equations relating to the power of the heater, heat transfer, heat convection and so on. This involves several equations, expressions and unknown heat transfer coefficients and so on. A simple black-box approach would instead be to use, say the ARX-model (11.2) with u as the applied voltage and y the room temperature. But that's too simple! A moment's reflection reveals that it's the heater power rather than the voltage that gives the temperature change. Thus use (11.2) with $u=$ squared voltage and $y=$ room temperature"

I would like to coin the term ***semi-physical modelling*** for introducing non-linear transformation of the raw measurement, based on high-school physics and common sense. The transformed measurements are then used in black-box structures such as the ARX structure.

Clearly semi-physical modelling is in frequent use. It is however also true that many failures of identification are indeed to be blamed on not applying this principle.

11.3.4 Hybrid Structures

Of particular current interest is to conceive model structures that are capable of dealing both with dynamic effects, described by differential/difference equations and with logical constraints, "the ifs and the buts" of the system. Not so many concrete results have yet been obtained in this area, but quite intense work is going on now. We may point to some work on using three models and pattern recognition for these hybrid model structures [21, 18].

11.4 The Criterion of Fit

The system identification problem really is a variant of the following archetypical problem in science and human learning: "We are shown a collection of vector pairs

$$z^N = \{[y(t); \varphi(t)], \quad t = 1, 2 \ldots N\}$$

Call this "the training set". We are then shown a new value $\varphi(N+1)$ and are asked to name a corresponding value $y(N+1)$"

The variable t could be thought of as time, but could be anything. The vectors $y(t)$ and $\varphi(t)$ may take values in any sets (finite sets or subsets of $\Re^n$ or anything else) and the dimension of $\varphi(t)$ could very well depend on t (and could be unbounded). The formulation covers most kinds of classification and model building problems.

How to solve this problem? The mathematical modelling approach is to construct a function $\hat{g}_N(t, \varphi(t))$ based on the "training" set, and to use this function for pairing $y(t)$ to new $\varphi(t)$:

$$\hat{y}(t) = \hat{g}_N(t, \varphi(t)) \tag{11.4}$$

Where do we get the function g from?

Basically, we have to follow the process described in the previous section, and carry out the search for g in a family of functions that is parameterized in terms of a finite number of parameters, i.e., a model structure. We thus match the measured value $y(t)$ against the candidate $g(t,\, \theta,\, \varphi(t)) = \hat{y}(t|\theta)$

$$y(t) \sim \hat{y}(t|\theta)$$

How shall we proceed to match these two? There are essentially two approaches. We can assume that the variables are related by

$$y(t) = \hat{y}(t|\theta) + v(t) \tag{11.5}$$

where $v(t)$ is the effect of ***unmeasured inputs*** that influence the system

The unmeasured input v is usually thought of as "disturbances and noise". Clearly we need some sort of assumptions about the character of v, in order to proceed to find a good value of θ, based on the information in z^t. There are two basic approaches to such assumptions.

- *Non-probabilistic*: Constrain the set of possible signals $\{v(t)\}$ in some way, like

$$|v(t)| \leq C \quad \forall t \tag{11.6}$$

  In general we may write for the "allowed" disturbances:

$$v \in V(\theta) \tag{11.7}$$

- *Probabilistic:* Assign probabilities to the different possible $\{v(t)\}$ sequences. That is, describe $\{v(t)\}$ as a random process with known or parameterized probability distribution:

$$v \text{ has pdf } p_v(\cdot, \theta) \tag{11.8}$$

A Pragmatic Approach

A more pragmatic approach to estimating the dynamics of a system is simply to postulate a ***predictor model structure***, i.e. look for a description of the observed data within a family of models

$$\hat{y}(t|\theta) = g_t(z^t, \theta) \tag{11.9}$$

where the prediction of $y(t)$ is denoted by $\hat{y}(t|\theta)$. The function g is based on observations available at time $t-1$,

$$z^t = [y(t-1), u(t-1), \ldots, y(0), u(0)] \tag{11.10}$$

and is an arbitrary (differentiable) function of these data and of the parameter vector θ. The actual output will then differ from the prediction by an error $e(t)$

$$y(t) = \hat{y}(t|\theta) + e(t) \tag{11.11}$$

We then seek that value of θ that has the best track record in achieving good predictions

$$\hat{\theta}_N = \arg\min_{\theta} \frac{1}{N} \sum_{t=1}^{N} \ell(t, \theta, \epsilon(t, \theta)) \tag{11.12}$$

$$\epsilon(t, \theta) = y(t) - \hat{y}(t|\theta) \tag{11.13}$$

It is clear that by invoking a probabilistic framework, i.e. by assigning a pdf to $\{e(t)\}$ in (11.11) the pragmatic estimate (11.12) can be seen as an ML estimate.

It is also clear that by choosing

$$\ell(t, \theta, x) = \begin{cases} 0 & |x| \leq C \\ \infty & |x| > C \end{cases} \tag{11.14}$$

the method (11.12) will pick out those θ which are consistent with the assumption $|e(t)| \leq C$ in (11.10) for all $0 \leq 1 \leq N$. Thus the non-probabilistic approach also fits into (11.12).

Most "traditional" control oriented descriptions of System Identification follow this mixture of pragmatic and probabilistic approaches. See e.g. [8] and [19].

11.5 Back to Data: The Practical Side of Identification

It follows from our discussion that the most essential element in the process of identification – once the data have been recorded – is to try out various model structures, compute the best model in the structures; using (11.12), and then validate this model. Typically this has to be repeated with quite a few different structures before a satisfactory model can be found.

While one should not underestimate the difficulties of this process, I suggest the following simple procedure to get started and gain insight into the models.

1. Find out a good value for the delay between input and output, e.g. by using correlation analysis.

2. Estimate a fourth order linear model with this delay using part of the data, and simulate this model with the input and compare the model's simulated output with the measured output over the whole data record. In MATLAB language this is simple,

```
z = [y u];
compare(z,arx(z(1:200,:),[4 4 1]));
```

If the model/system is unstable or has integrators, use prediction over a reasonable large time horizon instead of simulation.

Now, either of two things happen:

- *The comparison "looks good".* Then we can be confident that with some extra work – trying out different orders, and various noise models – we can fine tune the model and have an acceptable model quite soon. Let me add here that I am amazed by the large amount of applications that fall into this category.

- *The comparison "does not look good".* Then we must do further work. There are three basic reasons for the failure.

  1. *A good description needs higher order linear dynamics.* This is actually in practice the least likely reason, except for systems with mechanical resonances. One then obviously has to try higher order models or focus on certain frequency bands by band pass filtering.

  2. *There are more signals that significantly affect the output.* We must then look for what these signals might be, check if they can be measured and if so include them among the inputs. Signal sources that cannot be traced or measured are called "disturbances" and we simply have to live with the fact that they will have an adverse effect on the comparisons.

  3. *Some important non-linearities have been overlooked.* We must then resort to semi-physical modelling to find out if some of the measured signals should be subjected to non-linear transformations. If no such transformations suggest themselves, one might have to try some non-linear black-box model, like a neural network.

Clearly, this advice does not cover all the art of identification, but it is the best half page summary of the practical process of identification that I can offer.

Example 3 *Aircraft dynamics*

Let us try the recipe on the aircraft data in figure 11.1! Picking the canard angle only as the input, estimating a fourth order model based on the data points 90 to 180, gives figure 11.3. (We use 10-step ahead prediction in this example since the models are unstable – as they should be, JAS has unstable dynamics in this flight case). It does not "look good". Let us try alternative 2: More inputs. We repeat the procedure using all three inputs in figure 11.1. That is, the model is computed as

```
arx([y u1 u2 u3], [4 4 4 4 1 1 1])
```

on the same data set. The comparison is shown in Figure 11.4. It "looks good". By further fine-tuning, as well as using model structures from physical modelling, only slight improvements are obtained.

Example 4 *Buffer vessel dynamics*

Let us now consider the pulp process of Figure 11.2. We use the κ-number before the vessel as input and the κ- number after the vessel as output. The delay is preliminarily estimated to 12 samples. Our recipe, where a fourth order linear model is estimated using the first 200 samples and then simulated over the whole record gives figure 11.5. It does not look good.

Some reflection shows that this process indeed must be non-linear (or time-varying): the flow and the vessel level definitely affect the dynamics. For example, if the

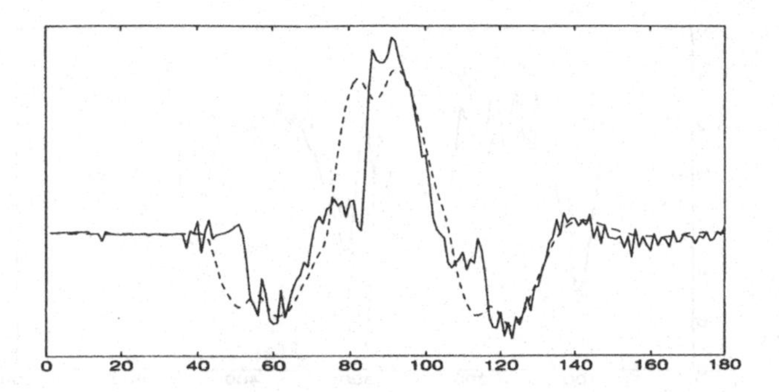

FIGURE 11.3. Dashed line: Actual Pitch rate. Solid line: 10 step ahead predicted pitch rate, based on the fourth order model from canard angle only

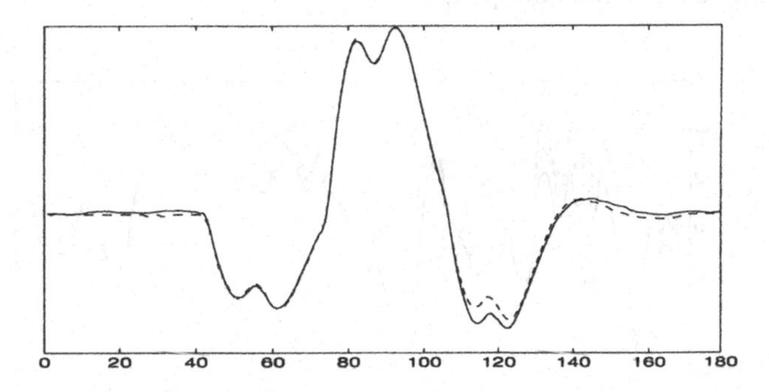

FIGURE 11.4. As figure 11.3 but using all three inputs

flow was a plug flow (no mixing) the vessel would have a dynamics of a pure delay equal to vessel volume divided by flow.

Let us thus resample the date accordingly, i.e. so that a new sample is taken (by interpolation from the original measurement) equidistantly in terms of integrated flows divided by volume. In MATLAB terms this will be

```
z = [y,u]; pf = flow./level;
t =1:length(z)
newt =
table1([cumsum(pf),t],[pf(1)sum:(pf)]' );
newz = table1([t,z], newt);
```

We now apply the same procedure to the resampled data. This gives figure 11.6. This "looks good". Somewhat better numbers can then be obtained by fine-tuning the orders.

11.6 Conclusions

The area of process identification is one where real practical application and rather advanced mathematical tools and perspectives meet. The meeting place is really the software into which many years' research has been packaged. There are now

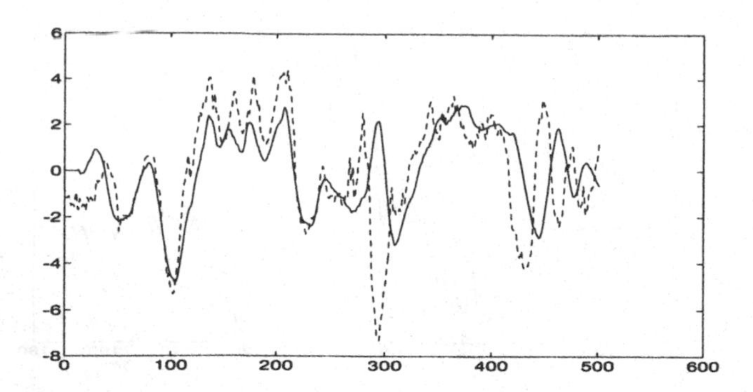

FIGURE 11.5. Dashed line: κ-number after the vessel, actual measurements. Solid line: Simulated κ-number using the input only and a fourth order linear model with delay 12, estimated using the first 200 data points

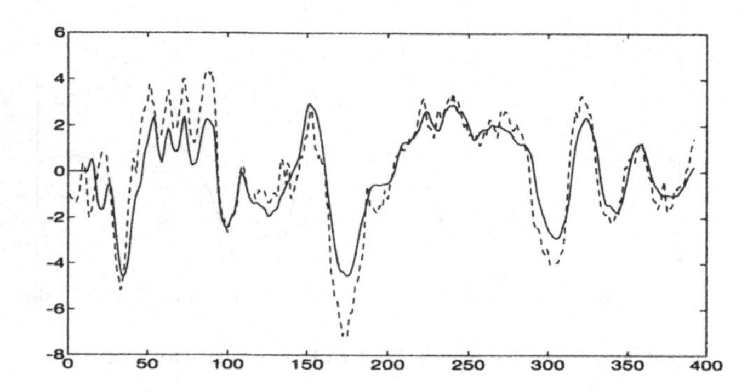

FIGURE 11.6. Same as figure 11.5 but applied to resampled data

many successful such packages commercially available. They have become standard tools in many industrial applications. This again stresses that it is the engineer's perspective that is the ultimate one in this area.

Acknowledgments: I am happy to acknowledge many years' financial support from the National Swedish Board for Technical Development, and the Swedish Research Council for Engineering Sciences.

I am also grateful to SAAB Military Aircraft AB for allowing me to show the test flight data from JAS-Gripen and to Skutskär Paper and Pulp Mill for the process data. The analysis of the latter data has been carried out by Torbjörn Andersson and Predrag Pucar [1].

11.7 References

[1] T. Andersson and P. Pucar. Estimation of residence time in continuous flow systems with dynamics. *Journal of Process Control*, 5:9–17, February 1995.

[2] A.R. Barron. Statistical properties of artificial neural networks. In *Proceedings of the 28th IEEE Conference on Decision and Control*, pages 280–285, 1989.

[3] L. Devroye and L. Györfi. *Non-parametric density estimation.* Wiley, New York, 1985.

[4] E. Fogel. System identification via membership set constraints with energy constrained noise. *IEEE Trans on Automatic Control*, AC-24:615–622, 1979.

[5] L. Guo and P. Khargonekar. A class of algorithms for identification in h_∞. *Automatica*, 1992.

[6] A.J. Helmicki. Control oriented system identification: A worst case / deterministic approach in h_∞. *IEEE Transactions on Automatic Control*, 36(October):1163–1176, 1991.

[7] C.W. Ku and Y.Z. Lu. Fuzzy model identification and self-learning. *IEEE Trans. on SMC*, 17, 1987.

[8] L. Ljung. *System Identification - Theory for the User*. Prentice-Hall, Englewood Cliffs, N.J., 1987.

[9] L. Ljung and T. Söderström. *Theory and Practice of Recursive Identification*. MIT press, Cambridge, Mass., 1983.

[10] P. Mäkilä and J. Partington. Robust approximation and identification in H_∞. pages 70–76. Proc. American Control Conference, 1991.

[11] M. Milanese and G. Belforte. Estimations theory and uncertainty intervals evaluation in the presence of unkown but bounded errors: Linear families of models and estimators. *IEEE Trans. on Automatic Control*, AC-27:408–414, 1982.

[12] K.S. Narendra and K. Parathasarathy. Identification and control of dynamical systems using neural networks. *IEEE Trans. Neural Networks*, 1:4–27, 1990.

[13] B.M. Ninness. *Stochastic and Deterministic Modeling*. PhD thesis, Dept. of Electrical Engineering, University of Newcastle, NSW, Australia, August, 1993.

[14] J. Norton. Identification and application of bounded parameter models. *Automatica*, 23:497–507, 1987.

[15] P. Van Overschee and B. De Moor. N4SID: subspace algorithms for the identification of combined deterministic-stochastic systems . *Automatica, (Special Issue)*, 30(1):75–93, 1994.

[16] P. Parker and R. Bitmead. Adaptive frequency response estimation. In *Proc. Conference on Decision and Control*, pages 348–353, 1987.

[17] F.C. Schweppe. Recursive state estimation - unknown but bounded errors and system inputs. *IEEE Trans. on Automatic Control*, (37):22–28, 1968.

[18] A. Skeppstedt, L. Ljung, and M. Millnert. Construction of composite models from observed data. *Int. J. Control*, 55(1):141–152, 1992.

[19] T. Söderström and P. Stoica. *System Identification*. Prentice-Hall Int., London, 1989.

[20] C.J. Stone. Consistent non-parametric regression (with discussion). *Ann. Statist.*, 5:595–645, 1977.

[21] J.E. Strömberg, F. Gustafsson, and L. Ljung. Trees as black-box model structures for dynamical systems. In *Proc. 1st European Control conference (ECC'91)*, pages 1175–1180, Grenoble, France, 1991.

[22] T.M. Cover and P.E. Hart. Nearest neigbor pattern classification. *Trans. IEEE Info. Theory*, IT-13:21–27, 1967.

12

ARMAX Model Identification with Unknown Process Order and Time-Varying Parameters

S. D. Likothanassis[1]
E. N. Demiris[2]

ABSTRACT
Identification of a system, in which the generating mechanisms are unknown, has been a central issue in the field of signal processing for many years. The aim of this paper is two-fold: first to investigate and present a method for simultaneously selecting the order and identifying the time-varying parameters of an AutoRegressive Moving Average model with eXogenous input (ARMAX), and second to evaluate the method via computer experiments. The proposed algorithm is based on the reformulation of the problem in the standard state space form and the subsequent implementation of a bank of Kalman filters, each fitting a different order model. Then the problem is reduced to selecting the true model, using the well-known multi-model partitioning theory for general (not necessarily Gaussian) data pdf's. Simulations illustrate that the proposed method selects the correct model order and identifies the model parameters in a sufficiently small number of iterations, even when the true model order does not belong to the bank of Kalman filters. Furthermore, the method is adaptive, in the sense that it can successfully track changes in the model structure in real time. Finally, the algorithm can be implemented in parallel and a VLSI implementation is also feasible.

12.1 Introduction

Adaptive filtering has been an active area of digital signal processing for a long time because adaptive techniques have found applications in various fields and that the advances in VLSI technology together with the decreasing cost of hardware and digital signal processors have made it possible to implement complex algorithms at a reasonable cost. Selecting the correct order and estimating the parameters of a model, such as ARMAX, are fundamental in linear prediction, system identification, and spectral analysis. The problem of fitting an ARMAX model to a given time

[1]Department of Computer Engineering and Informatics, University of Patras, Patras 26500, and Computer Technology Institute (C.T.I.), Patras 26100, Greece, E-mail: Likothan@cti.gr
[2]Department of Computer Engineering and Informatics, University of Patras, Patras 26500, Greece, E-mail: Demiris@ceid.upatras.gr

series has attracted much attention because it arises in a variety of applications, such as adaptive control, speech analysis and synthesis, radar, sonar, seismology, and biomedical engineering. Furthermore, the effect of varying process parameters on system performance is of crucial importance, specifically when the desired system performance alters and deteriorates because of disturbances of the system's parameters, whenever the design is based on nominal parameter values. Therefore it is desirable to incorporate parameters' variation into system design.

An ARMAX model (a generalization of the AR, ARX and ARMA models), can be represented as follows:

$$A(q)y(t) = B(q)u(t) + C(q)e(t) \tag{12.1}$$

or

$$y(t) = \sum_{i=1}^{n_a} a_i y(t-i) + \sum_{j=1}^{n_b} b_j u(t-j) + \sum_{k=1}^{n_c} c_k e(t-k) + e(t) \tag{12.2}$$

where N is the number of the noisy measurements of the discrete time process $y(t)$, $u(t)$ represents the exogenous input, $e(t)$ is a zero-mean white noise process, with variance R, not necessarily Gaussian, $n = (n_a, n_b, n_c)$ is the order of the predictor, and $a_i : \ i = 1, \ldots, n_a, \ b_j : \ j = 1, \ldots, n_b, \ c_k : \ k = 1, \ldots, n_c$, are the predictor coefficients.

Clearly the problem is two-fold: one has to select the order of the predictor and then to compute the predictor coefficients. Perhaps the most crucial part of the problem is the former.

Several information theoretic criteria have been proposed for the model order selection task. The most well known of the proposed solutions for this problem include the Final Prediction Error (FPE), Akaike's Information Criterion (AIC) proposed by Akaike [1, 2, 3], the Minimum Description Length (MDL) Criterion proposed by Schwartz [16] and Rissanen [14], and a new approach based on the MDL criterion [11]. Most of the techniques computed by the above criteria are based on the assumption that the data are Gaussian and on asymptotic results. Hence, strictly speaking, their applicability is limited only to the Gaussian case.

In this paper, the method presented for simultaneous ARMAX model order selection and identification is based on adaptive multi-model partitioning filters [7] - [9]. The method is not restricted to the Gaussian case. It is applicable to on-line/adaptive operation and is computationally efficient. Furthermore, it can be realized in a parallel processing fashion, a fact which makes it amenable to VLSI implementation. A similar method for AR model order selection is found in [6].

The main points and the organization of this paper are the following: In Section 12.2, the ARMAX model order selection and identification problem is reformulated, the adaptive multi-model partitioning algorithm is briefly presented, and its application to the specific problem is discussed. In Section 12.3, a summary of the variables used to formulate the solution to the ARMAX model order selection and identification problem as well as the algorithm are presented. In Section 12.4, simulation examples and figures are presented that demonstrate the performance of the method. Finally Section 12.5, summarizes the conclusions.

12.2 Reformulation of the ARMAX Model Order Selection and Identification Problem

Let us further define the vector of coefficients as follows:

$$\vartheta(t) = [a_1(t), \ldots, a_{n_a}(t), b_1(t), \ldots, b_{n_b}(t), c_1(t), \ldots, c_{n_c}(t)]^\top \tag{12.3}$$

where $0 \le t \le N$ (N denotes the number of samples). Notice that the coefficients a_i, b_j, and c_k have been replaced by $a_i(t)$, $b_j(t)$, and $c_k(t)$, respectively, to reflect the possibility that the coefficients are subject to random perturbations. This fact can be modeled by assuming that

$$\vartheta(t+1) = \vartheta(t) + w(t), \quad t = 1, 2, \ldots, N \tag{12.4}$$

where $w(t) = [w_1(t), w_2(t), \ldots, w_{n_a+n_b+n_c}(t)]$ is a zero-mean white noise process, with variance Q, not necessarily Gaussian (we assume that $e(t)$ and $w(t)$ are independent). Equation (12.2) can be written as

$$y(t) = h^\top(t)\vartheta(t) + e(t) \tag{12.5}$$

where

$$h^\top(t) = [y(t-1)\ldots y(t-n_a)u(t-1)\ldots u(t-n_b)e(t-1)\ldots e(t-n_c)] \tag{12.6}$$

Equations (12.4) and (12.5) are in the standard state space form, so the results of the state space estimation techniques can be readily used to estimate the values of the $\vartheta(t)$'s. This formulation is well known and can be found in [4], for example.

Remarks:

1. We have to assign values to the variances of the processes $w(k)$ and $e(k)$ denoted by Q and R, respectively. Assessing the values of Q and R is not always an easy task. If R is not readily obtainable, it can be estimated by a technique described in [15]. The effect of estimating R via this technique is investigated in [13]. Again, Q can be estimated by a technique described in [13].
2. We assume that an a priori mean of the vector $\vartheta(0)$ can be set to zero when no knowledge about their values is available before any measurements are taken (we notice that is the most likely case). The usual choice of the initial variance of the vector $\vartheta(0)$, denoted by P_0 is $P_0 = mI$, where m is a large integer.
3. We assume that measurements of $y(t)$ and $u(t)$ are set to zero for $k < 0$, a technique well known as "prewindowing".

The ARMAX model identification problem is now stated as follows: Given a set of observations, $y(t)$ and $u(t)$, where $0 \le t \le N$, from an ARMAX(n_a, n_b, n_c) process, we have to determine the unknown parameter vector:

$$v = [n_a,\ n_b,\ n_c,\ \vartheta(t),\ Q,\ R] \tag{12.7}$$

Now let us assume that the order n or the parameters (n_a, n_b, n_c) are unknown and what we know is only that the true order satisfies the condition $n_0 \le n \le n_{MAX}$. Then it is clear that the true model is one of a family of models described by relationships (12.4) and (12.5). The true model is specified by the true value of the parameters (n_a, n_b, n_c). Then the problem is to select the correct model among various "candidate" models. In other words, we have to design an optimal estimator (in the minimum variance sense), when uncertainty is incorporated in the signal model. The solution to this problem has been given by the multi-model partitioning theory [7] - [9] described in the following.

The multi-model partitioning algorithm operates on the following discrete model:

$$\vartheta(t+1) = F(t+1, t|n)\vartheta(t) + w(t) \tag{12.8}$$

$$z(t) = h^{\top}(t|n)\vartheta(t) + e(t) \tag{12.9}$$

where $n = (n_a, n_b, n_c)$ is the unknown parameter, in our case the model order. This parameter is assumed to be a random variable with known or assumed a priori pdf $p(n|0) = p(n)$. The optimal MMSE estimate of $\vartheta(t)$ is given by

$$\hat{\vartheta}(t|t) = \int_n \hat{\vartheta}(t|t;n)p(n|t)dn \tag{12.10}$$

where $\hat{\vartheta}(t|t;n)$ is the conditional MMSE state vector estimate obtained by the corresponding Kalman filter matched to the model with parameter value n and initialized with initial conditions $\hat{\vartheta}(0|0;n)$ and $P(0|0;n)$. The model-conditional pdf $p(n|t)$ is given by

$$p(n|t) = \frac{L(t|t;n)}{\int_n L(t|t;n)p(n|t-1)dn} p(n|t-1) \tag{12.11}$$

where

$$L(t|t;n) = |P_z(t|t-1;n)|^{-\frac{1}{2}} exp\{-\frac{1}{2}\|\tilde{z}(t|t-1;n)\|P_z^{-1}(t|t-1;n)\}$$

and

$$P_z(t|t-1;n) = \hat{h}^{\top}(t|n)P(t|0;n)\hat{h}(t|n) + R(t)$$

$$\tilde{z}(t|t-1;n) = z(t) - \hat{h}^{\top}(t|n)F(t+1,t|n)\hat{\vartheta}(t|t-1;n)$$

$$\hat{h}^{\top}(t|n) = [y(t-1)\dots y(t-n_a)u(t-1)\dots u(t-n_b)\epsilon(t-1)\dots\epsilon(t-n_c)]$$

$$\epsilon(t) = z(t) - \hat{h}^{\top}(t|n)\hat{\vartheta}(t|t-1;n) \tag{12.12}$$

The important feature of the adaptive multi-model partitioning algorithm [7] - [9] is that all the filters needed to implement it can be independently realized. A wide range of applications require high-speed, real-time digital signal processing. Minimization of processing time is efficiently achieved with parallel processing. The solution adopted in this paper for the ARMAX identification problem is presented in Fig. 12.1. The parallelism in this block diagram emphasizes the ability to implement the algorithm in parallel, thus saving enormous computational time [10].

Remarks:

1. Equations (12.10), (12.11) pertain to the case where the pdf of n is continuous in n. When this is the case, one is faced with the need for a nondenumerable infinity of Kalman filters for the exact realization of the optimal estimator. The usual approximation performed to overcome this difficulty is somehow to

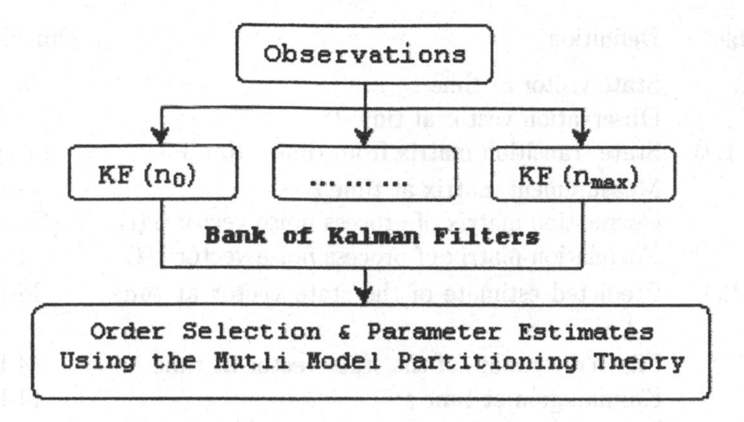

FIGURE 12.1. Solution structure

approximate the pdf of n by a finite sum [17]. Fortunately, in our case the sample space is naturally discrete, so that no approximations are necessary, and the estimator is indeed optimal. Since this is the case, the integrals in (12.10), (12.11) must be replaced by summations running over all possible values of (n_a, n_b, n_c).

2. If we rewrite Equations (12.4), (12.5) and define the state vector as a vector containing all model coefficients, then the model is easily extended to the multivariate case [12].

12.3 Summary of Relationships between Kalman Filter and Multi-Model Partitioning Algorithm

Let us assume that $M = n_a + n_b + n_c$, $\vartheta(t)$ denotes the M-dimensional parameter state vector and $z(t)$ denotes the observed data of the system, both measured at time t. Thus the system model described by Equations (12.8) and (12.9) is represented in the form of a signal-flow graph in Fig. 12.2.

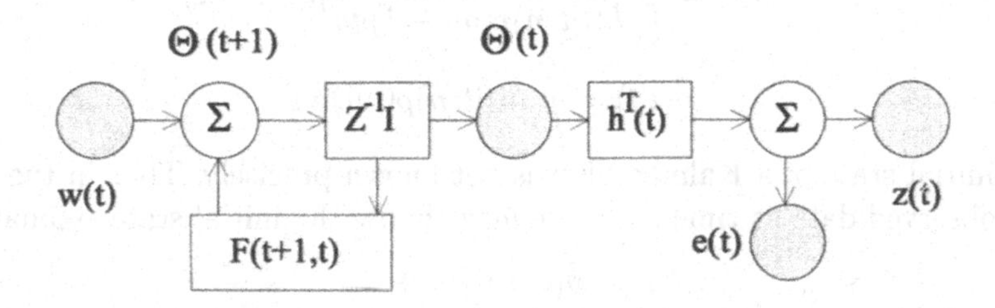

FIGURE 12.2. Signal-flow-graph representation of the system model

Following is a summary of the variables used to formulate the solution to ARMAX model identification:

Variable	Definition	Dimension
$\vartheta(t)$	State vector at time t	M-by-1
$z(t)$	Observation vector at time t	1-by-1
$F(t+1,t)$	State transition matrix from time t to $t+1$	M-by-M
$\hat{h}^{\top}(t)$	Measurement matrix at time t	1-by-M
QI	Correlation matrix of process noise vector $w(t)$	M-by-M
R	Correlation matrix of process noise vector $e(t)$	1-by-1
$\hat{\vartheta}(t+1\|t)$	Predicted estimate of the state vector at time $t+1$	M-by-1
$\hat{\vartheta}(t\|t)$	Filtered estimate of the state vector at time t	M-by-1
$G(t)$	Kalman gain at time t	M-by-1
$\epsilon(t)$	Innovations vector at time t	1-by-1
$P(t+1\|t)$	Correlation matrix of the error in $\hat{\vartheta}(t+1\|t)$	M-by-M
$P(t\|0)$	Correlation matrix of the error in $\hat{\vartheta}(t\|t)$	M-by-M

The multi-model partitioning algorithm including the Kalman filter and the initial conditions is presented to facilitate the reader's understanding:

Computation $t = 0, 1, 2, 3, \ldots$: For each n do,

$$G(t|n) = F(t+1,t|n)P(t|t-1;n)\hat{h}(t|n)\left[\hat{h}^{\top}(t|n)P(t|t-1;n)\hat{h}(t|n) + R\right]^{-1}$$

$$\epsilon(t|n) = z(t) - \hat{h}^{\top}(t|n)\hat{\vartheta}(t|t-1;n)$$

$$\tilde{z}(t|t-1;n) = z(t) - \hat{h}^{\top}(t|n)F(t+1,t|n)\hat{\vartheta}(t|t-1;n)$$

$$\hat{\vartheta}(t+1|t;n) = F(t+1,t|n)\hat{\vartheta}(t|t-1;n) + G(t|n)\epsilon(t|n)$$

$$\hat{\vartheta}(t|t;n) = F(t+1,t|n)^{-1}\hat{\vartheta}(t+1|t;n)$$

$$P(t|0;n) = P(t|t-1;n) - F(t+1,t|n)^{-1}G(t|n)\hat{h}^{\top}(t|n)P(t|t-1;n)$$

$$P(t+1|t;n) = F(t+1,t|n)P(t|0;n)F^{\top}(t+1,t|n) + QI$$

$$P_z(t|t-1;n) = \hat{h}^{\top}(t|n)P(t|0;n)\hat{h}(t|n) + R(t)$$

$$L(t|t;n) = |P_z(t|t-1;n)|^{-\frac{1}{2}}exp\{\frac{-1}{2}\|\tilde{z}(t|t-1;n)\|\ P_z^{-1}(t|t-1)\}$$

EndFor

$$p(n|t) = \frac{L(t|t;n)}{\int_n L(t|t;n)p(n|t-1)dn}p(n|t-1)$$

$$\hat{\vartheta}(t|t) = \int_n \hat{\vartheta}(t|t;n)p(n|t)dn$$

The initial state of a Kalman filter is not known precisely. Thus in the absence of any observed data at time $t = 0$, we may choose the initial state estimate as

$$\hat{\vartheta}(0|0;n) = 0$$

and its correlation matrix as

$$P(0|0;n) = cI,\ c = \text{large positive constant}$$

and assuming that all the filters have equal model-conditional pdfs at time $t = 0$,

$$p(n|0) = \frac{1}{F},\ F = \text{the number of the Kalman filters in the bank}$$

12.4 Examples and Figures

The proposed algorithm has been tested extensively in several simulation experiments. All experiments were conducted 100 times (100 Monte Carlo runs). For examples 1 and 2 the data record is of length 500, and the data generating process at time instant $t = 0$, is given by

$$y(0) = 1.8y(-1) - 0.9y(-2) + 1.0u(-1) - 1.2u(-2) + 0.4e(-1) + e(0)$$

Based on this model ($n = (2,2,1)$), an ARMAX system like that of (12.4) and (12.5) was simulated with a MA process as an exogenous input given by

$$u(t) = 2.76c(t-1) - 3.81c(t-2) + 2.65c(t-3) - 0.92c(t-4) + c(t)$$

where $c(t)$ was a zero-mean white noise process with variance equal unity. This signal is labelled as "EXG".

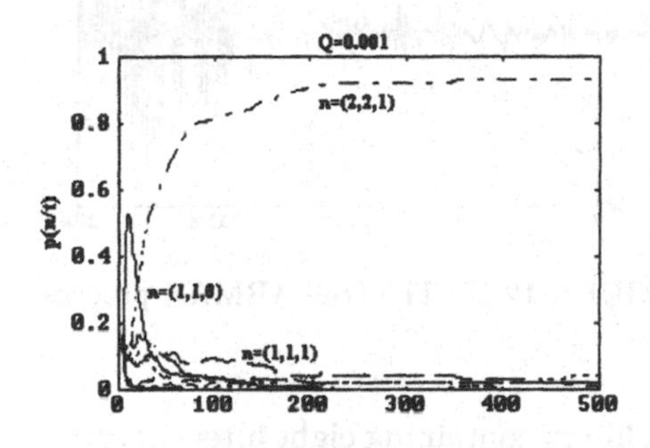

FIGURE 12.3. The evolution of the a posteriori probabilities

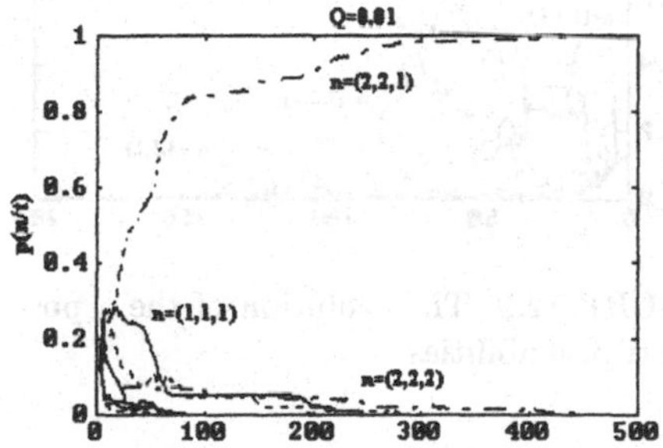

FIGURE 12.4. The evolution of the a posteriori probabilities

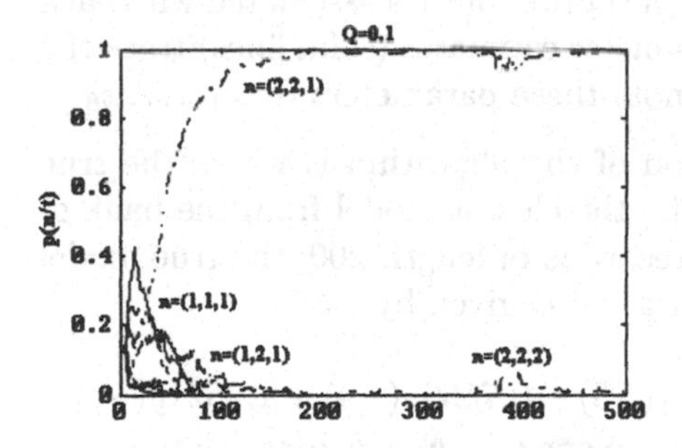

FIGURE 12.5. The evolution of the a posteriori probabilities

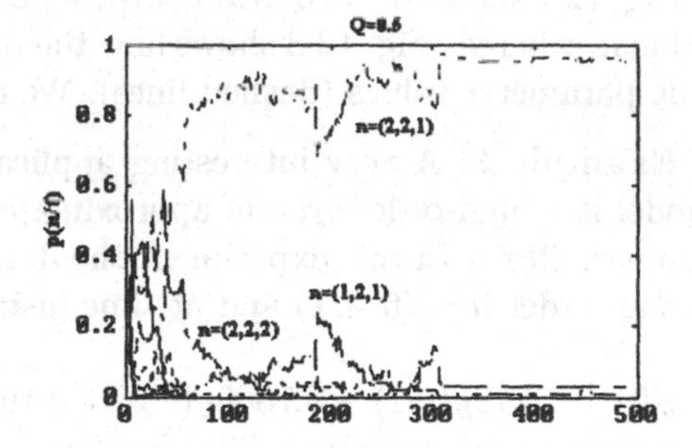

FIGURE 12.6. The evolution of the a posteriori probabilities

Example 1: We realized a bank of Kalman filters containing eight filters of order $(1,1,1), (1,1,2), (2,1,1), (2,2,1), (2,1,2), (1,2,1), (1,2,2)$, and $(2,2,2)$, respectively. We ran the algorithm four times, using a time-varying parameter vector with four different values of the noise variance $w(k)$: $Q = 0.001, Q = 0.01, Q = 0.1,$ *and* $Q = 0.5$, respectively.

Figures 12.3, 12.4, 12.5, and 12.6 show the evolution of the a posteriori probabilities associated with each value n for $Q = 0.001, Q = 0.01, Q = 0.1$, and $Q = 0.5$, respectively.

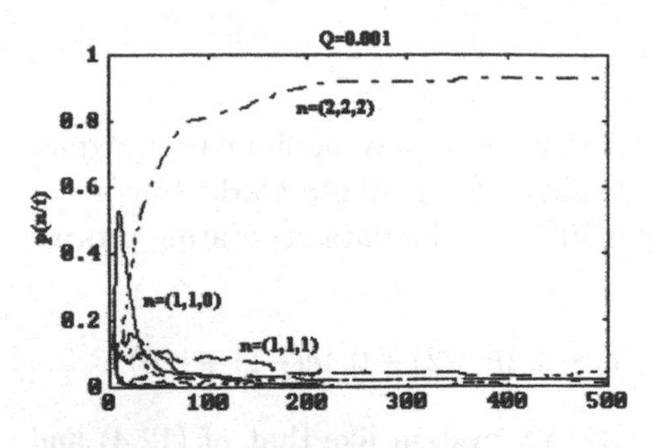

FIGURE 12.7. The evolution of the a posteriori probabilities

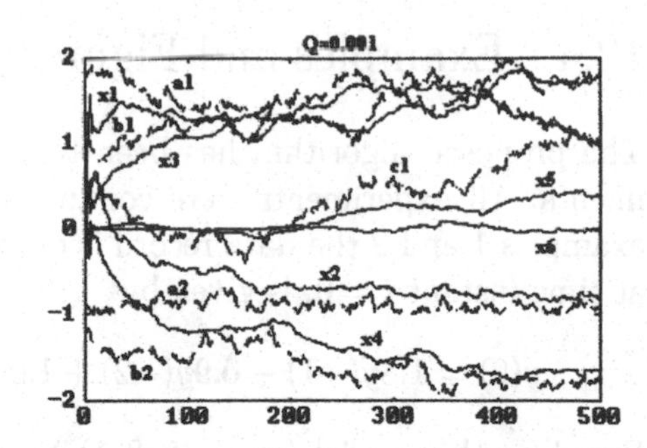

FIGURE 12.8. Parameter estimates (solid) and true parameters (dashed)

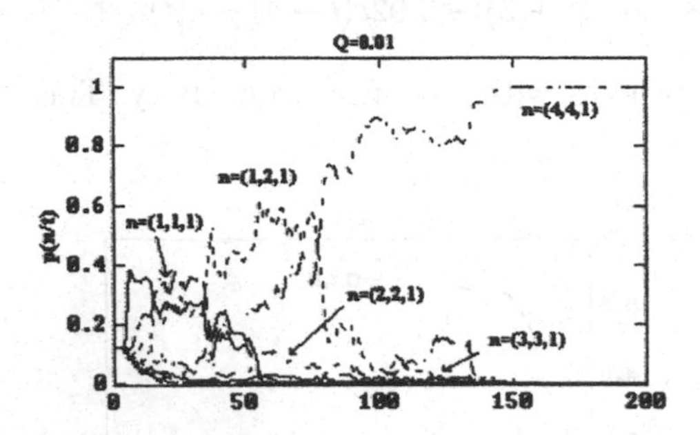

FIGURE 12.9. The evolution of the a posteriori probabilities

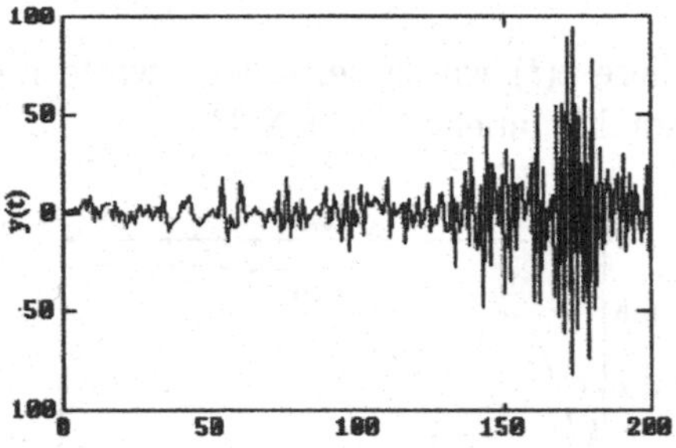

FIGURE 12.10. The true ARMAX process

Example 2: We realized a bank of Kalman filters containing eight filters of order $(1,1,0),(1,1,1),(2,1,1),(1,1,2),(2,1,2),(1,2,1),(1,2,2), and (2,2,2)$, respectively. We ran the algorithm adding a noise with variance $Q = 0.001$ to the parameter's vector, according to the relationship (12.4).

Fig. 12.7 shows the evolution of the a posteriori probabilities associated with each value n, whereas Fig. 12.8 shows how the parameter estimates (solid lines) track the true parameter values (dashed lines). We denote these parameters by $x_1, \ldots, x_8$.

Example 3: A very interesting application of the algorithm is where the true model is a high-order system approximated by the closest model from the bank of Kalman filters. In this experiment the data record is of length 200, the true model has an order $n = (6,4,1)$ and at time instant $t = 0$ is given by

$$y(0) = 1.279y(-1) - 0.7805y(-2) + 0.1635y(-3) - 0.7566y(-4) + 1.0621y(-5)$$
$$-0.7821y(-6) - 0.2997u(-1) - 0.4147u(-2) - 0.2794u(-3) + 0.4973u(-4) + e(0)$$

where the exogenous input $u(t)$ is the "EXG" signal. The bank of Kalman filters contains eight filters of orders $(0,1,1),(1,1,1),(1,2,1)$, $(2,2,1)$, $(2,3,1)$, $(3,3,1)$, $(3,4,1)$, and $(4,4,1)$, respectively, and the variance of the noise added to the parameter's vector is $Q = 0.01$.

Fig. 12.9 shows the evolution of the a posteriori probabilities associated with each value n, Fig. 12.10 shows the true ARMAX process $y(t)$, Fig. 12.11 shows the error vector $E(t)$ expressed as $E(t) = y(t) - \hat{y}(t)$, and Fig. 12.12 shows the estimated ARMAX process.

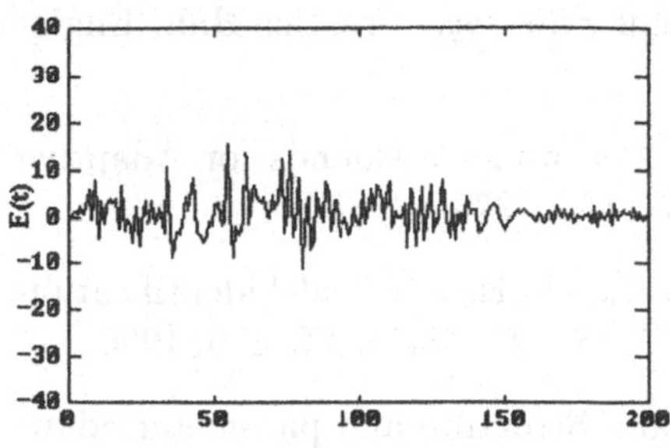

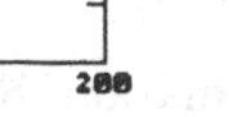

FIGURE 12.11. The error vector E(t)

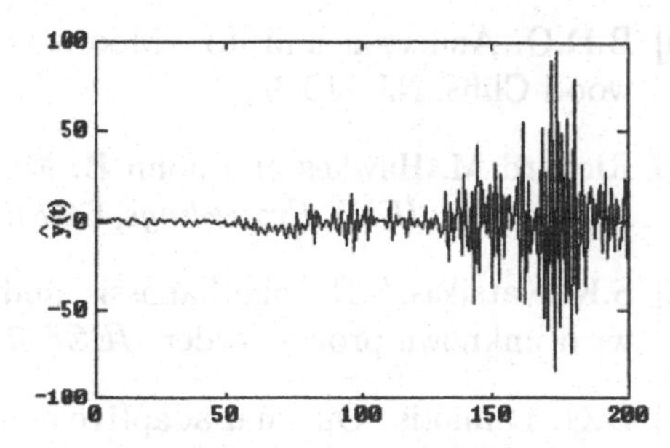

FIGURE 12.12. The estimated ARMAX process

12.5 Conclusions

The experiments above indicate that (a) when the true model is one of the models included in the bank of Kalman filters, the proposed method selects the correct model order and identifies the time-varying parameters in a few steps, even when we consider a noise variance, such as $Q = 0.5$, in the random-walk of the parameter vector; (b) when the true model is not included in the bank of Kalman filters the proposed algorithm selects the closest model, in our experiment the model with order $n = (2, 2, 2)$, and successfully identifies the time-varying parameters as the sixth parameter tends to zero after a few iterations. We note that the closest model is that which minimizes the Kullback information criterion [5]; (c) in the special case when the true order n satisfies the condition $n > n_{max}$, our method successfully approximates a high order system by a low order system (in our experiment by the closest system that is the model with order $(n_a, n_b, n_c) = (4, 4, 1)$). It has been shown in [5] that this fact is very interesting in a control system such as in antiaircraft systems.

As shown by the experimental results, the a posteriori probability for the true model tends to unity as the rest of the probabilities tend asymptotically to zero. Thus, when the process order changes during the operation, the proposed algorithm can detect this change and then can select the new order and consequently identify the values of the parameters with an accuracy dependent on the noise level.

The superiority of this method is that it works in real time and successfully track changes of the parameter vector. Furthermore, the algorithm can be implemented in parallel. A VLSI implementation is also feasible.

12.6 References

[1] H. Akaike. Fitting autoregressive models for prediction. *Ann. Inst. Of Stat Math*, 21:243–347, 1969.

[2] H. Akaike. Information theory and an extension of the maximum likelihood principle. *In Proc. 2nd Int. Symp. Inform. Theory. Budapest, Hungary: Akademia Kiado*, pp. 267–281, 1973.

[3] H. Akaike. A new look at the statistical model identification. *IEEE Trans. Automat. Conr.*, 26:1–18, 1977.

[4] B.D.O. Anderson and J.B. Moore. *Optimal Filtering.* Prentice-Hall, Englewood Cliffs, NJ,, 1979.

[5] Richard M. Hawkes and John B. Moore. Performance Bounds for Adaptive Estimation. *IEEE Proceedings*, 64(8):1143–1150, 1976.

[6] S.K. Katsikas, S.D. Likothanassis, and D.G. Lainiotis. AR model identification with unknown process order. *IEEE Trans. A.S.S.P.*, 38(5):872–876, 1990.

[7] D.G. Lainiotis. Optimal adaptive estimation: Structure and parameter adaptation and parameter adaptation. *IEEE Trans. Automat. Contr.*, AC-16:160–170, 1971.

[8] D.G. Lainiotis. Partitioning: A unifying framework for adaptive systems I: Estimation. *Proc. IEEE*, 64:1126–1143, 1976.

[9] D.G. Lainiotis. Partitioning: A unifying framework for adaptive systems II: Estimation. *Proc. IEEE*, 64:1182–1198, 1976.

[10] D.G. Lainiotis, S.K. Katsikas, and S.D. Likothanassis. Adaptive deconvolution of seismic signals: Performance, computational analysis, parallelism. *IEEE Trans. Acoust., Speech, Signal Processing*, 36(11):1715–1734, 1988.

[11] Gang Liang, D. Mitchell Wilkes, and James A. Cadzow. ARMA Model Order Estimation Based on the Eigenvalues of the Covariance Matrix. *IEEE Trans. On Signal Processing*, 41(10):3003–3009, 1993.

[12] S.D. Likothanassis and S.K. Katsikas. Multivariable AR model identification and control. *In Proc. IASTED Int. Symp. Modeling, Identification and Control*, Grindelwald, Switzerland, pp. 248-252, 1987.

[13] A.K. Mahalanabis and S. Prasad. On the application of the fast kalman algorithm to adaptive deconvolution of seismic data. *IEEE Trans. Geosci. Remote Sensing*, GE-21(4):426–433, 1983.

[14] J. Rissanen. Modeling by shortest data description. *Automatica*, 14:465–471, 1978.

[15] S.K. Katsikas. Optimal algorithms for geophysical signal processing. Ph. D. Thesis, University of Patras, Dept. of Computer Engineering and Informatics, Greece, 1986.

[16] G. Schwarz. Estimation of the dimension of a model. *Ann. Stat.*, 6:461–464, 1978.

[17] R.L. Shengbush and D.G. Lainiotis. Simplified parameter quantization procedure for adaptive estimation. *IEEE Trans. Automat. Contr.*, AC-14:424–425, 1969.

13 Using Kautz Models in Model Reduction

Albertus C. den Brinker[1]
Harm J.W. Belt[2]

ABSTRACT
A method is presented for model reduction. It is based on the representation of the original model in an (exact) Kautz series. The Kautz series consists of orthogonalized exponential sequences. The Kautz series is non-unique: it depends on the ordering of the poles. The ordering of the poles can be chosen such that the first terms contribute most to the overall impulse response of the original system in a quadratic sense. Having a specific ordering, the reduced model order, say n, can be chosen by considering the energy contained in a truncated representation. The resulting reduced order model is obtained simply by truncation of the Kautz series at the nth term. Since only a selection of the poles already present in the original model is made, the numerical problems associated with the calculation of the optimal poles are avoided. The model order reduction method is illustrated by two examples. In the first example the Kautz model order reduction method is compared with the balanced model order reduction technique. In the second example, the Kautz model reduction method is combined with Prony's method to estimate exponential sequences from a noisy data set.

13.1 Introduction

In many applications it is convenient to have an adequate description of a linear system by an IIR filter of lowest possible order (according to some error criterion). Often, a quadratic norm is used to compare the original system with its lower-order approximation. Taking an ARMA model description [10], the aim is to determine the poles and zeros (or numerator and denominator polynomials) of the lower-order model. It is well known that using a quadratic norm the error surface usually contains many local minima and that especially the determination of the poles turns out to be numerically ill-conditioned problem. Furthermore, the model order has to be chosen beforehand and thus if the best model within this class is not adequate, the whole estimation procedure has to be repeated.

For these reasons, among others, several suboptimal optimization procedures were developed. One of these is Prony's method starting from the description of the sys-

[1]Faculteit der Elektrotechniek, Technische Universiteit Eindhoven, P.O. Box 513, NL-5600 MB Eindhoven, Netherlands, E-mail: A.C.d.Brinker@ele.tue.nl
[2]Philips Research Laboratories, Prof. Holstlaan 4, NL-5656 AA Eindhoven, Netherlands, E-mail: Belt@natlab.research.philips.com

tem in terms of the impulse response [9], but it turns out that this method generally overestimates the damping terms of the poles [3]. Another well-known technique is balanced model order reduction (BMOR) [11], which starts from the state-space description. The BMOR is also suboptimal in the sense of a quadratic norm. Although this method is appealing from several points of view, it is also associated with numerical problems as a consequence of the required matrix decompositions and inversions.

The numerical problems associated with model reduction can be easily illustrated. Suppose we have a function which is the sum of two exponential sequences $A_1 p_1^k$ and $A_2 p_2^k$. In order that model reduction is possible at all, the poles should be in each others proximity, i.e. a large part of the sequence p_1^k can be modelled by p_2^k (and vice versa). (We exclude the trivial case that one of the exponential sequences contains hardly any energy at all.) A reduced model involves finding a sequence Bq^k such that the original function is 'best' represented by this simpler representation. However, finding q is a numerically difficult task: small variations in q around its optimal value typically have little effect on the error. Instead of trying to solve this problem, we avoid it by exploiting the property that p_1^k, p_2^k and its optimal lower-order model q^k have so much in common. In essence, the technique proposed is the following. Since p_1^k offers already a fair description of p_2^k (and vice versa) we simply select q as that p_i which models most of the energy of both sequences and calculate B accordingly. (Note that the excluded trivial case where one of the exponential sequences contains relatively little energy also fits in nicely with such a procedure and in that case the poles do not need to be located in each others proximity).

Clearly, such technique is suboptimal in a quadratic sense: we do not optimize over the poles, but make a selection on an already given set of poles. The essential part of the procedure is an ordering algorithm, which has been simplified by using the orthogonalization of exponential sequences as given by a Kautz series. Furthermore, it is noted that the order selection and model calculation are performed simultaneously and that if the original model is stable, so is the reduced model.

In the next section the Kautz functions (orthogonalized exponential sequences) are introduced. In the two subsequent sections the ordering of the poles and the actual model reduction are discussed. In Section 13.5 some extensions are considered. Section 13.6 contains an example of the proposed model reduction technique. In Section 13.7 it is argued that the proposed model reduction technique when combined with Prony's method is capable of estimating damped sinusoids from a data sequence. Section 13.8 illustrates this by an example.

13.2 Kautz Filters

Suppose we have a discrete-time model of large order N given by its transfer function

$$H(z) = \sum_{n=1}^{N} \frac{R_n z}{z - p_n} = \frac{\sum_{n=1}^{N} B_n z^n}{\prod_{n=1}^{N} (z - p_n)}, \tag{13.1}$$

with p_n the poles and R_n the residues ($R_n \neq 0$). We assume that $H(z)$ is a stable filter $|p_n| < 1$. We introduce P as the set of poles, i.e., $P = \{p_n; n = 1, \cdots, N\}$, and we assume that $p_n \neq p_m$ for $n \neq m$. Thus the impulse response is a sum of exponential sequences $h(k) = \sum_{n=1}^{N} R_n p_n^k$. The transfer function $H(z)$ is written

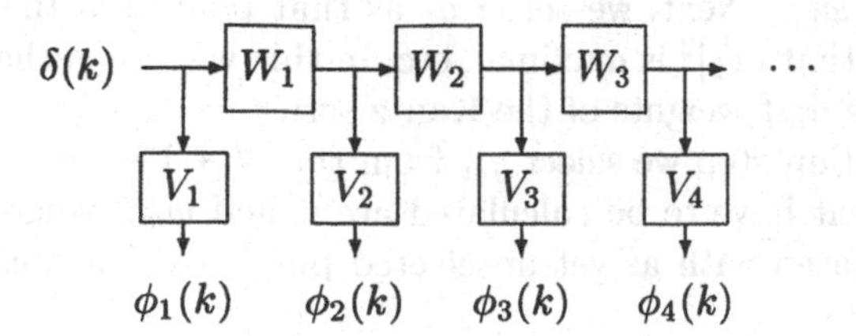

FIGURE 13.1. A Kautz filter

as a Kautz series

$$H(z) = \sum_{n=1}^{N} w_n \Phi_n(z) \tag{13.2}$$

with

$$\Phi_n(z) = \sqrt{1 - q_n q_n^*}\, \frac{z}{z - q_n} \prod_{l=1}^{n-1} \frac{1 - z q_l^*}{z - q_l} \tag{13.3}$$

the z-transforms of the Kautz functions $\phi_n(k)$ [4, 6], where * denotes conjugation and $\{q_n; n = 1, \cdots, N\} = P$. This Kautz series is of finite order and exact. The Kautz filter is shown in Fig. 13.1 where $W_n(z) = (1 - z q_n^*)/(z - q_n)$ and $V_n(z) = \sqrt{1 - q_n q_n^*}\, z/(z - q_n)$. Note that the nth Kautz function is a function of the parameters $q_1, \cdots, q_n$.

The Kautz functions are orthonormal, i.e., $\sum_{k=0}^{\infty} \phi_m(k)\phi_n^*(k) = \delta_{m,n}$ with $\delta_{m,n}$ the Kronecker delta. As a consequence of the orthonormality, the weights w_n can be determined by

$$w_n = \sum_{k=0}^{\infty} h(k)\phi_n^*(k) = \frac{1}{2\pi j} \oint H(z)\Phi_n^*(1/z^*)\, \frac{dz}{z}$$

where $j = \sqrt{-1}$ and the contour integral is taken counter-clockwise around the unit circle. The Kautz series is non-unique: there are $N!$ orderings of the set of poles P resulting in allowable q_i's. In fact, (13.2) can be regarded as $N!$ possible Gram-Schmidt orthogonalization procedures.

We note that the energy E contained in the impulse response can be calculated in the time domain, in the frequency domain using Parseval's theorem, or from the expansion coefficients in the Kautz series:

$$E = \sum_{k=0}^{\infty} h(k)h^*(k) = \frac{1}{2\pi} \int_{<2\pi>} H(e^{j\theta})H^*(e^{j\theta})d\theta = \sum_{n=1}^{N} |w_n|^2$$

13.3 Ordering of the Poles

13.3.1 General case

In order to be able to perform the model reduction we can choose the ordering of the poles q_n such that the most energy is contained in the first weights. This is done in a recursive way. First, we select the first pole q_1 as that pole p_n which yields maximum energy $|w_1|^2$. Next, we select q_2 as that pole from the remaining set of poles $P\backslash\{q_1\}$ such that $|w_2|^2$ is maximal, etc. In this way we gather as much energy as possible into the first weights of the Kautz series.

In the mth iteration step we select q_m from the $N+1-m$ yet unselected poles p_n. The weights that have to be calculated are called w_{mn} where n only takes on those values associated with as yet unselected poles p_n. The weights w_{mn} can be calculated by

$$w_{mn} = \frac{1}{2\pi j} \oint H(z)\Phi_m^*(1/z^*)\frac{dz}{z} = \sqrt{1-p_n p_n^*} \sum_{l=1}^{N} \frac{R_l}{1-p_l p_n^*} \prod_{i=1}^{m-1} \frac{p_l - q_i}{1-p_l q_i^*} \tag{13.4}$$

where Φ_m is a function of the already selected poles $q_1, \cdots, q_{m-1}$ plus an additional pole p_n from the as yet unselected poles $P\backslash\{q_1, \cdots, q_{m-1}\}$. We choose w_m as that value of w_{mn} with maximal absolute value and q_m as the associated p_n. In order to obtain a unique Kautz series it is required that this maximum is unique. We assume that in practical situations this is always the case.

13.3.2 Reduction of Real Models

The algorithm for the ordering of the poles requires a unique maximum for the possible weights that are calculated in each iteration step. In view of symmetry, this condition will not be met in the case of a real model having one or more complex-conjugated pole pairs. Furthermore, if one starts from a real system, one usually wants a reduced model that is real as well. In order to attain this, the pole selection procedure has to be adapted in order to guarantee that complex-conjugated pole pairs occur sequentially. Therefore, we suggest an adaptation of the pole selection scheme for a real model such that complex-conjugated pole pairs are kept together. This reduces the number of possible Gram-Schmidt orthogonalization procedures from $N!$ to $(N_r + N_c)!$ where N_r and N_c are the number of real poles and complex-conjugated pole pairs, respectively, and $N = N_r + 2N_c$.

We propose the following adaptation:

1. Select a first-order section with a real-valued pole if
   a. the absolute value of the weight associated with this pole is largest of all (as yet unselected) real poles,
   b. the squared absolute value of this weight is larger than the sum of squared absolute values of the weights of two additional first-order sections for any (as yet unselected) complex-conjugated pole pair.
2. If there is no real pole satisfying both previous conditions, we select two additional first-order sections either with a complex-conjugated pole pair or with two real-valued poles. The selection is as before based on a maximum of additional energy.

13.4 Model Reduction

We now have the Kautz series with the desired ordering of the poles. We can plot the modelled energy $E(n)$ defined as

$$E(n) = \sum_{m=1}^{n} |w_m|^2$$

or the relative modelled energy $E_r(n) = E(n)/E$ as a function of n and select the appropriate order as the minimum of n for which $E(n)/E > 1 - \epsilon$ where ϵ is the admissible relative deviation in squared norm between the original model and its lower-order approximation. The reduced-order model is then defined as the nth order model $H_n(z) = \sum_{i=1}^{n} w_i \Phi_i(z)$ where Φ_i are governed by the ordered poles. We then have deleted the $N - n$ exponential sequences (poles) which, after projection on the remaining set of exponential sequences (poles), contribute least to the overall impulse response according to a quadratic norm.

If the pole ordering outlined in Section 13.3.2 is used, not every order n of the reduced system yields a real system. In order for the reduced model to be real, truncation in between a complex-conjugated pole pair ($q_{n+1} = q_n^*$) must be prohibited.

If a desired level of accuracy $E_r(n) = E(n)/E$ is known beforehand then the ordering process can be terminated whenever this accuracy is reached. Obviously, this makes the proposed procedure faster.

13.5 Extensions

The procedure can in principle be adapted to the case of multiple occurring poles. However, we lose degrees of freedom in the ordering process of the poles. If all the poles are equal, as for instance starting with an FIR model, there is no flexibility at all. Therefore, we restrict ourselves to the case of simple poles. The other argument is of course that in practice this will be the usual case.

The transfer function $H(z)$ given by (13.1) is a restricted Nth order rational function. The general case can also be treated in the previous way using the following procedure. We write

$$H(z) = \frac{\sum_{n=0}^{N} B_n z^n}{\prod_{n=1}^{N}(z - p_n)} = C + \frac{\sum_{n=0}^{N-1} B'_n z^n}{\prod_{n=1}^{N}(z - p_n)} \tag{13.5}$$

The constant C represents the direct feedthrough. The second part can be used in the model reduction scheme presented before using one-time delayed versions of the Kautz functions given in (13.3).

The presented ordering procedure is numerically well-conditioned; it requires no matrix decomposition, matrix inversions nor other numerically cumbersome methods. However, should one start with a transfer function given by

$$H(z) = \frac{\sum_{n=0}^{N} B_n z^n}{\sum_{n=0}^{N} A_n z^n}$$

one must find the roots of the denominator polynomial. This may give numerical problems. Two remarks can be made. First of all the procedure can also be applied by starting with selecting poles q_i from a set P' not identical to the exact poles given by P. Even the dimensions of the sets P and P' need not be equal. The only restriction would be that a very accurate Kautz model of the original system is available by using set P'. This leads us to the second remark that we can start from the roots of the denominator polynomial even if in itself this procedure might be associated with numerical inaccuracies. In order to obtain a stable model however, it is required that the roots obtained from the polynomial are within the unit circle.

The method presented previously can be adapted to weighted quadratic error criteria. Consider the error criterion (norm)

$$\frac{1}{2\pi j}\oint F(z)F^*(1/z^*)W(z)\frac{dz}{z}$$

where the contour integral is taken around the unit circle and the weighting function $W(z)$ is a nonnegative real number for $|z| = 1$. Similar to the Kautz functions, a set of orthogonal functions can be constructed for this norm. For instance, taking $W(z) = z/((z - \xi)(1 - z\xi))$ with $-1 < \xi < 1$, it can be easily shown that the functions

$$\Psi_m(z) = (1 - \xi z^{-1})\Phi_m(z)$$

are orthogonalized exponential sequences under the given window.

13.6 Example

As an example, we generated a 30th-order model ($N = 30$) by the random model generator of MatLab. This routine gives a numerator and denominator polynomial which we will call $A(z)$ and $B(z)$, respectively. We excluded those models having multiple poles and those that are only marginally stable (poles at the unit circle). Furthermore, we removed all initial pure delays in the impulse response by shifting the coefficients in B such that $B_N \neq 0$ [see (13.1)]. The motivation is that $B_{N-k} = 0$ for $k = 0, \cdots, K < N - 1$ implies that $h(k) = 0$ for $k = 0, \cdots, K$ and thus that the impulse response $h(k)$ is in fact a function defined on the interval $[K+1, \infty)$ rather than on $[0, \infty)$ as is the case for the Kautz functions.

From the numerator and denominator polynomials of the generated high-order model, the poles and residues were calculated yielding a representation according to (13.1). The generated model is real and therefore the procedure outlined in Section 13.3.2 was used. The squared error divided by the energy of the original impulse response of two typical examples is shown in Fig. 13.2. The allowed truncation orders (i.e., yielding real reduced models) are indicated by the crosses.

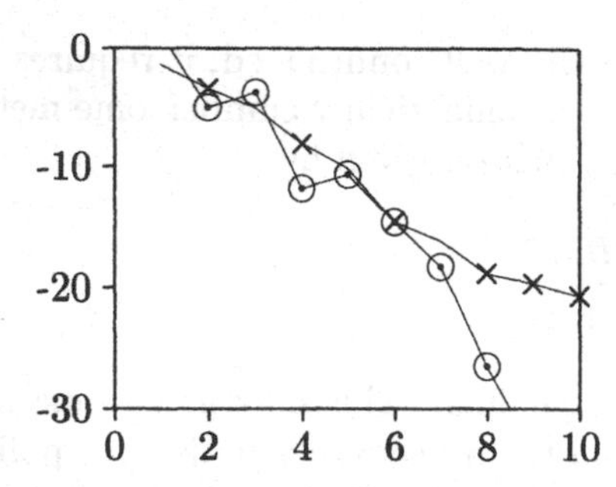

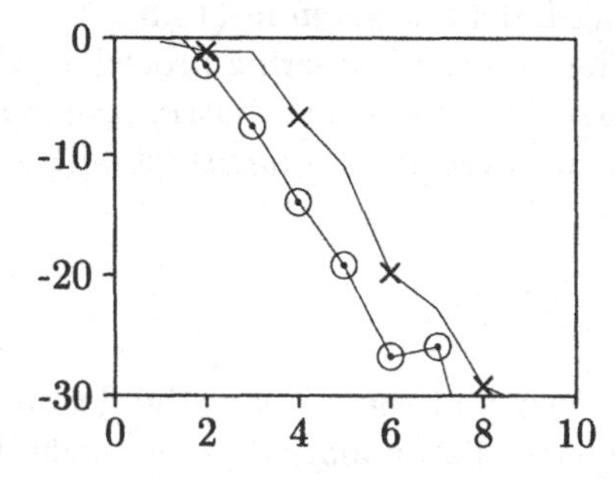

FIGURE 13.2. Two typical examples of the results of model reduction in terms of the relative quadratic error criterion (in dB) as a function of the reduced order for the Kautz model (crosses) and the balanced model order reduction (circles)

For a comparison, the original system was converted to a state space model and the balanced model order reduction (BMOR) scheme [11] was used to estimate reduced models of different orders. The results thereof are also shown in Fig. 13.2 in terms of the quadratic error. For the random models generated by MatLab of order around 30 and a given accuracy of the reduced model in terms of the quadratic criterion, the required model order for the BMOR is typically one less than for the reduced Kautz model. However, for higher model orders, the BMOR more often than not failed. The matrices calculated in the BMOR routine became ill-conditioned and the required Cholesky decomposition was no longer possible as a consequence of numerical inaccuracies.

13.7 Modelling Impulse Response Data

The model reduction technique discussed in the previous section starts with a description of a large-scale system described by its poles and residues according to (13.1). A common case encountered in modelling is also where noisy data of the impulse response of an unknown system is available. The Kautz model can in that case also give a good estimate of the model underlying the data when combined with Prony's method. Such a procedure is described and illustrated in the following.

Suppose we have at our disposal the data $x(k)$, $k = 0, \cdots, M-1$, which are noisy data representing an impulse response. The impulse response corresponds to a system having a rational transfer function $F(z)$. Thus

$$x(k) = f(k) + n(k)$$

where $n(k)$ is a zero-mean noise sequence and $f(k)$ is the impulse response with $f(k) = \mathcal{Z}^{-1}\{F(z)\}$ and $F(z)$ having poles p_i, $i = 1, \cdots, N$. Our aim is to estimate $F(z)$ (thus $f(k)$) from $x(k)$. A well-known technique is Prony's method [9].

Prony's method is a classical and straightforward approximation to the cumbersome least-squares solution. It is well known that if we use Prony's method with the exact order of the unknown system (in our case N), then the damping parameters in the poles tend to be overestimated [9, 3]. If we use Prony's method with an order L substantially higher than N than we model not only the system but also part of the noise. In fact, having $M = 2L$ the complete data $x(k)$ is absorbed in the calculated numerator and denominator polynomial representing Prony's model without any loss of information (although numerical problems may arise in the reconstruction of the data from the calculated numerator and denominator polynomials). It turns out that if L is substantially larger than N, a subset of the poles implicitly given by the denominator polynomial are close to the original poles [2, 8].

Using Prony's method with $L > N$ we obtain a set of poles from which the poles describing (mainly) the impulse response have to be separated from those modelling the noise. A possible way to do this in the case of noise-free data or moderate noise levels is using time reversal of the data [9, 7]. In the next example this procedure is combined with the Kautz reduction technique.

13.8 Example

As an example we consider the acoustical impulse response shown in Fig. 13.3. We prefer this example since it is not synthetic data, i.e., it is a priori unknown whether the data can be described by a rational transfer function, and furthermore, we are ignorant of the order (if any) of the underlying model as well as of the noise characteristics.

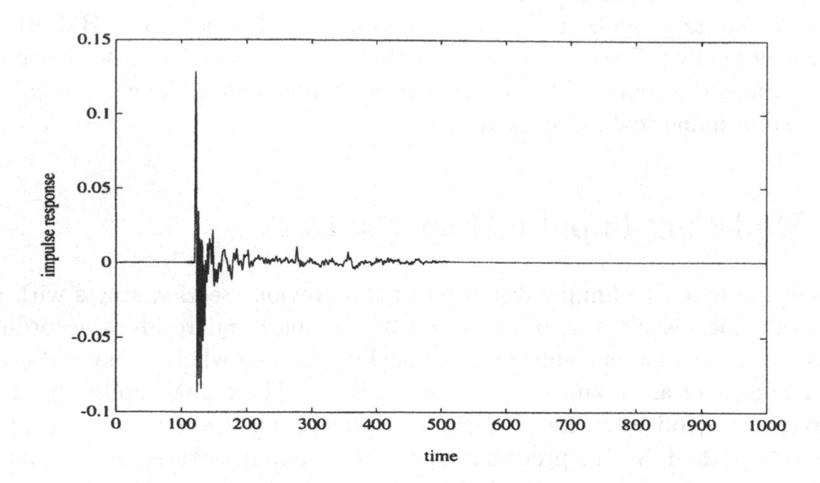

FIGURE 13.3. Example of an acoustical impulse response

To model the data, we eliminated the delay and took 480 samples (122 – 601), and denoted these by $y(n)$, $n = 0, \cdots, M-1$ with $M = 480$. Next, the poles had to be determined before we could enter the Kautz model reduction scheme. This was done by estimating the backward prediction polynomial $B(z^{-1})$ according to

$$Yb = -c \tag{13.6}$$

where

$$\begin{aligned}
Y &= \begin{pmatrix} y(1) & y(2) & \cdots & y(L) \\ y(2) & y(3) & \cdots & y(L+1) \\ \vdots & & & \vdots \\ y(M-L) & y(M-L+1) & \cdots & y(M-1) \end{pmatrix} \\
b &= (B(1), B(2), \cdots, B(L))^t \\
c &= (y(0), y(1), \cdots, y(M-L-1))^t
\end{aligned}$$

Next, we take the inverse of the roots of the polynomial $B(z^{-1}) = 1 + \sum_{l=1}^{L} B(l)z^{-l}$. In [7] it is shown that if the data is noise free, if the underlying system consists of N damped harmonics and if $L > N$, then this procedure would yield N poles inside the unit circle and $L - N$ poles outside the unit circle. Those within the unit circle are the desired poles. If there is not too much noise, the previous observation still holds as can be shown experimentally. Of course, in that case the poles within the unit circle are no longer exactly those of the underlying model, and furthermore, there may occasionally occur an extra pole within the unit circle (this depends critically upon the noise level). Also, in the case of noise the estimated poles are generally closer to the original ones for larger L.

In [8] it was suggested to calculate the singular value decomposition and truncate

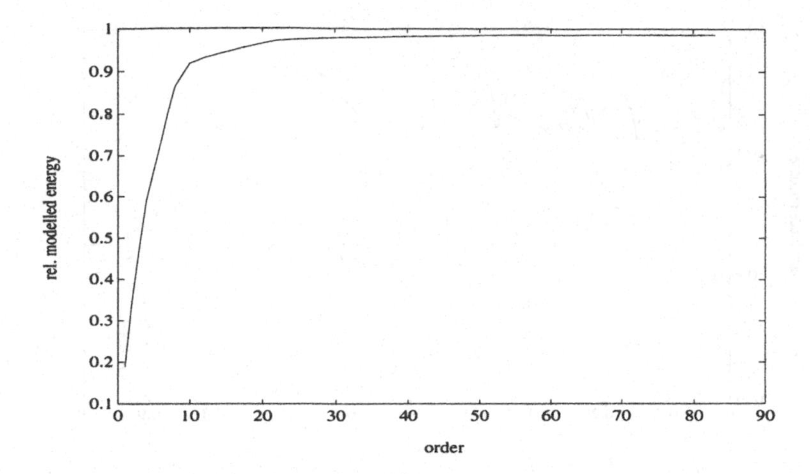

FIGURE 13.4. Relative modelled energy $E_r(n)$ as a function of the model order n

the SVD solution by setting

$$b^t = -\sum_{m=1}^{L_t} \sigma_m^{-1}[u_m^h c]v_m \tag{13.7}$$

with $N < L_t < L$ and where σ_m are the singular values and v_m and u_m are the eigenvectors of Y^hY and YY^h, respectively. Of course, if N is unknown, it is hard to select an appropriate value for L_t.

We took for L a large value ($L = 160$) and used both methods, i.e. with ($L_t = 80$) and without truncation. The standard method gave 83 stable poles, whereas the truncated SVD method yielded only 36 stable poles.

With these poles we entered the Kautz model reduction scheme and ordered the poles as discussed in Section 13.3.2 with the difference that the weights were calculated in the time domain. The ordering process uses

$$w_{mn} = \sum_{k=0}^{M-1} y(k)(a_{m-1} * f_n)(k) \tag{13.8}$$

instead of (13.4) where a_{m-1} is the $m-1$st order allpass section given by the already selected poles $q_1, \cdots, q_{m-1}$, i.e.

$$\mathcal{Z}\{a_{m-1}(k)\} = \prod_{l=1}^{m-1} \frac{1 - zq_l^*}{z - q_l}$$

and $f_n(k) = \sqrt{1 - p_n p_n^*}\, p_n^k$ where p_n is one of the as yet unselected poles. We note that in order for the orthogonality to hold, summation in (13.8) should extend in fact to ∞. This truncated version (13.8) is allowed if $y(k) = 0$ for $k > M$.

The results of the Kautz ordering are given in Fig. 13.4 in the form of the relative modelled energy

$$E_r(n) = \sum_{m=1}^{n} |w_m|^2/E \tag{13.9}$$

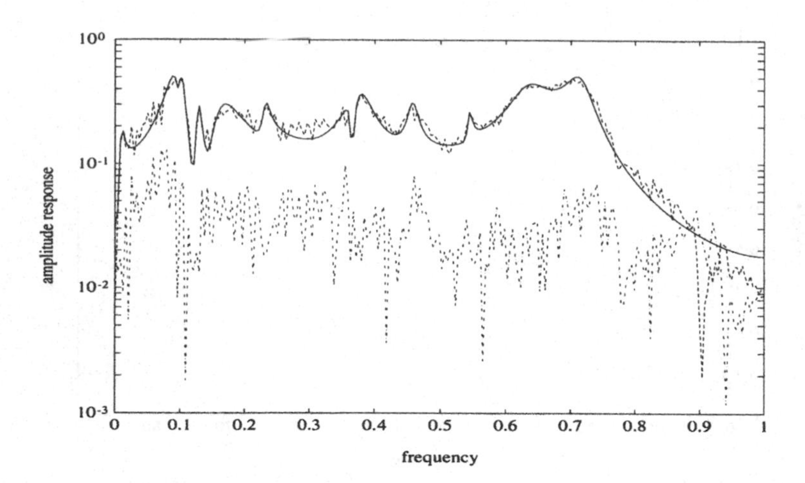

FIGURE 13.5. Amplitude characteristic of the original impulse response (dashed line), the model (solid line), and the difference (dash-dotted line)

with E the energy in the impulse response. We observe a fast initial increase of the modelled energy as a function of the reduced order and a subsequent slowly levelling-off. The plot for the truncated SVD method practically coincides with the one shown here.

In order to select the order we used the Minimum Description Length defined by $\arg\{\min_n J(n)\}$ with

$$J(n) = M \log_{10}(1 - E_r(n)) + 2n \log_{10} M \tag{13.10}$$

This measure suggested to take an order of 24 using the truncated SVD method (order of 30 for the standard method). Using 24 as the order we obtained $E_r = 0.9741$ and 0.9736 using the poles of the truncated SVD method and the standard method, respectively.

In Fig. 13.5 we have plotted the frequency characteristic by plotting the absolute value of the Fourier transforms of the original impulse response, its approximation, and of the difference between the original and its approximation. As was expected from the fact that E_r is so close to unity, there is only a small unmodelled part and in a course way, the unmodelled part varies along with the model part [12].

Another example using synthetic data can be found in [5]. There, we simply used a high-order Prony method to estimate the poles.

The results of the method do critically depend on the first step: the estimation of the poles by Prony's method or its alternatives [8] used here. In order that accurate estimates of the poles can be retrieved, the data must satisfy the following conditions:

- The number of data M. Depending on the decay rates of the poles and the noise level, M must be chosen such that beyond M $y(k) = 0$ or that the data is buried in the noise. If M is chosen too large, the tail of the data contains only information about the noise and no information concerning the impulse response.
- The order L. Prony's method splits the data into two parts. The second part contains the data $L+1, \cdots, M$. All the poles must be 'visible' in this second

part. This implies that when choosing L too large, all the poles with a fast decay are not traceable from the data by Prony's method.

- Coloured noise. In view of the previous remark, if the effective length of the autocorrelation function of the noise is much smaller than L, then presumably such noise will not be treated differently from white noise.

13.9 Conclusions

We have presented a method for model reduction using Kautz filters. The method is a simple non-iterative procedure which does not involve the numerical pitfalls associated with methods requiring matrix decompositions or inversions. The method gives a simultaneous order and parameter estimate and is evaluated in terms of a quadratic error criterion. The resulting reduced-order model is always stable.

By an example we have shown that the method gives results comparable to those using balanced model order reduction and we observed that the method still works for large model orders where the balanced model order reduction typically fails as a consequence of numerical inaccuracies.

It was also shown that the method can be used to estimate exponential sequences from a noisy data set when used in combination with Prony's method. For this latter case, there exist numerous other methods including Steiglitz-McBride [14], cyclo-correlation based methods [1], and methods using higher-order statistics [13]. The current method has as yet to be compared to these other methods.

13.10 References

[1] K. Abed-Meraim, A. Belouchrani, A. Mansour, and Y. Hua. Parameter estimation of exponentially damped sinusoids using second order statistics. In *Proc. EUSIPCO-96, Eighth European Signal Processing Conference*, pp. 2025–2028, Trieste, Italy, 1996.

[2] M.L. Van Blaricum and R. Mitra. A technique for extracting the poles and residues of a system directly from its transient response. *IEEE Trans. Antennas Propag.*, 23:777–781, 1975.

[3] M.L. Van Blaricum and R. Mitra. Problems and solutions associated with Prony's method for processing transient data. *IEEE Trans. Antennas Propag.*, 26:174–182, 1978.

[4] P.W. Broome. Discrete orthonormal sequences. *Journal of the Association for Computing Machinery.* 12(2):151–168, 1965.

[5] A.C. den Brinker and H.J.W. Belt. Model reduction by orthogonalized exponential sequences. In *Proc. ECSAP-97, First Europ. Conf. on Signal Analysis and Prediction*, pp. 121–124, Prague, 1997.

[6] W.H. Kautz. Transient synthesis in the time domain. *IEEE Trans. Circuit Theory*, 1:29–39, 1954.

[7] R. Kumaresan. On the zeros of the linear prediction-error filter for deterministic signals. *IEEE Trans. Acoust. Speech Signal Process.*, 31:217–220, 1983.

[8] R. Kumaresan and D.W. Tufts. Estimating the parameters of exponentially damped sinusoids and pole-zero modeling in noise. *IEEE Trans. Acoust. Speech Signal Process.*, 30:833–840, 1982.

[9] S.L. Marple Jr. *Digital Spectral Analysis with Applications*, Prentice-Hall, Inc., Englewood Cliffs, N.J., 1987.

[10] R.N. McDonough and W.H. Huggins. Best least-squares representation of signals by exponentials. *IEEE Trans. Autom. Contr.*, 13:408–412, 1968.

[11] B.C. Moore. Principal component analysis in linear systems: controllability, observability, and model reduction. *IEEE Trans. Autom. Contr.*, 26:17–32, 1981.

[12] B. Ninness and F. Gustafsson. A unifying construction of orthonormal bases for system identification. *IEEE Trans. Autom. Contr.*, 42:515–521, 1997.

[13] C.K. Papadopoulos and C.L. Nikias. Parameter estimation of exponentially damped sinusoids using higher order statistics. *IEEE Trans. Acoust. Speech Signal Process.*, 38:1424–1436, 1990.

[14] K. Steiglitz and L.E. McBride. A technique for the identification of linear systems. *IEEE Trans. Autom. Contr.*, 10:461–464, 1965.

14

Condition Monitoring Using Periodic Time-Varying AR Models

Andrew C. McCormick[1]
Asoke K. Nandi[1]

ABSTRACT
Model based approaches to vibration monitoring can provide a means of detecting machine faults even if data is only available from the machine in its normal condition. Previous work has assumed stationarity in the vibration signals and exploited only time-invariant models. The cyclostationary nature of rotating machine vibrations can be exploited by using periodic time-varying autoregressive models to model the signal better than time-invariant models, and this can improve the fault detection performance. Experimental data collected from a small rotating machine set subjected to several bearing faults was used to compare time-varying and time-invariant model based systems.

14.1 Introduction

The analysis of vibration data for monitoring the condition of machinery requires the use of signal processing algorithms to extract condition dependent features from the signal. From such features, the condition can be classified using a variety of dsicriminant analysis techniques ranging from thresholding to artificial neural networks [5]. These approaches do however require that data are available from the machine under faulty conditions which may arise.

By using autoregressive (AR) techniques to model the vibrations [2], a change in the machine condition can be detected by a change in the statistics of the error between the predicted signal and the measured signal. As the machine's condition deteriorates, through wear or damage, the model and signal will not match so well and the error will increase. The error time series for a signal matching the model will be minimum variance, zero mean, i.i.d. white noise, however if the signal changes, the variance may increase, the signal may become correlated and non-linear components may appear in the vibrations affecting the mean and the higher-order statistics of the error. Therefore fault detection involves determining whether significant changes have occured in any of these statistical features.

Time-invariant AR models are the optimal linear models of stationary signals, but the periodic nature of rotating machinery gives rise to vibration signals which

[1]Department of Electronic and Electrical Engineering, University of Strathclyde, 204 George Street, Glasgow, G1 1XW, United Kingdom, E-mail: asoke@eee.strath.ac.uk

are cyclostationary [4] in nature, and these can be better modeled using periodic-time-varying models [7].

14.2 Cyclostationarity in Vibration Signals

Rotating machinery is often run at a constant speed for significant lengths of time giving a periodic input to the vibrations. In many cases, this periodicity exhibits its effect only as periodic signals at harmonics of the rotation frequency. However in some machinery, especially if faults develop, second order cyclostationary effects may arise, where there is some non-linear combination of periodic and random signals. This could arise from amplitude modulation when the load being driven by the machine is varying randomly or through effects which affect the vibrations for only part of the rotation cycle. Such vibrations could arise if a machine goes out of balance causing the shaft to rub against some obstruction for part of its cycle. Another potential cause of cyclostationary vibrations are the short periodic impulses of random noise caused by bearing faults. An indication of the cyclostationarity of a signal can be obtained by comparing the ratio of signal which is stationary to that which is periodically correlated. Cyclostationarity in the signal can be computed using the the cyclic-autocorrelation:

$$R_{xx}^{\alpha}(\tau) = E\{x(t-\frac{\tau}{2})x(t+\frac{\tau}{2})e^{-j2\pi\alpha t}\} \tag{14.1}$$

Cyclostationary signal components affect this function at positions where $\alpha \neq 0$ and therefore the cyclostationarity of a signal can be measured by comparing the stationary components along the $\alpha = 0$ axis with the cyclostationary components. The degree of cyclostationarity [8] indicates whether a signal is cyclostationary at a particular frequency α and can be computed for discrete time signals using:

$$DCS^{\alpha} = \frac{\sum_{\tau} |R_{x}^{\alpha}(\tau)|^2}{\sum_{\tau} |R_{x}^{0}(\tau)|^2} \tag{14.2}$$

14.3 Periodic AR Models

If a stationary time series can be modeled as a pth-order linear time-invariant autoregressive process,

$$y(n) = \sum_{k=1}^{p} a(k)y(n-k) + \eta(n) \tag{14.3}$$

then a model can be identified and used as a one-step-ahead predictor:

$$\hat{y}(n) = \sum_{k=1}^{p} a(k)y(n-k) \tag{14.4}$$

If the signal is created by a linear stationary process, then a linear time-invariant model will provide the minimum mean square error in prediction. If the vibrations are cyclostationary, a time-invariant model will not be the optimal predictor of the next value of the signal. A periodic-time-variant model may fit the signal better, e.g.

$$y(n) = \sum_{k=1}^{p} h_k(n \,(\mathrm{mod}\; T))y(n-k) + \eta(n) \tag{14.5}$$

For sampled signals, this model requires that T is an integer number of samples. For vibration signals, this cannot be guaranteed and therefore a frequency domain representation, where the coefficients $h(n\,(\mathrm{mod}\; T))$ are expanded into their Fourier series, gives a periodic autoregressive model with cyclic components existing at harmonics of the rotation frequency.

$$\begin{aligned} y(n) &= \sum_{k=1}^{p} a(k,0)y(n-k) \\ &+ \sum_{k=1}^{p}\sum_{\alpha=1}^{M} a(k,\alpha)\cos(\frac{2\pi\alpha n}{T})y(n-k) \\ &+ \sum_{k=1}^{p}\sum_{\alpha=1}^{M} b(k,\alpha)\sin(\frac{2\pi\alpha n}{T})y(n-k) + \eta(n) \end{aligned} \tag{14.6}$$

This representation requires two decisions to be made about the structure of the model. Both the number of previous inputs used in determining the next value and the number and frequencies of the periodic sections need to be chosen. To ensure that the model is not over-determined, some form of complexity regularization such as the Akaike Information Criterion (AIC) [1] or something similar [6, 3] can be used. This however only balances reductions in the error with the number of parameters in the model and cannot help decide at which frequencies parameters should be, without an exhaustive comparison of every possible model structure.

Since the vibration signals are produced by a physical mechanical process any AR model of the sampled signal will only be an approximation and therefore the reasons for restricting the model complexity will either be due to computational limitations or to ensure that the model generalizes the signal.

Therefore the algorithm used does not attempt to find an optimal model structure, but tries find a model which is an improvement on the time-invariant models. The number of previous inputs was determined from a time-invariant AR model with the order chosen using the AIC. Sections were then added one at a time with the frequencies chosen as those most significant in the cyclic-autocorrelation. The coefficients of each section were trained using a gradient descent optimization algorithm with weight update every sample. After training for a fixed number of iterations (chosen through trial and error) the new prediction error was calculated and if this new error gave a reduction in the AIC, the section was included as part of the model.

When a model has been determined, it can then be used for detecting faults by comparing the prediction error with that obtained from the training data. Since the error from the training data is random, it is necessary to compare the statistics of the error for both known and unknown data. The simplest statistics which can be compared are either the variance or the second order moment (mean square value) of the error. It may also however be useful with some faults to look at the mean, the skewness or kurtosis of the errors. Since the error should be uncorrelated for the training data, a feature which measures how closely the autocorrelation of the error approximates a delta function may also be useful.

The mean, variance, skewness and kurtosis can be estimated using the following definitions, where the expectation operation $E\{.\}$ is performed using time averaging.

$$\mu = E\{e(n)\} \tag{14.7}$$

$$\sigma^2 = E\{(e(n) - \mu)^2\} \tag{14.8}$$

$$\gamma_3 = \frac{E\{(e(n) - \mu)^3\}}{\sigma^3} \tag{14.9}$$

$$\gamma_4 = \frac{E\{e^4(n)\} - 4\gamma_3\sigma^3\mu - 6\sigma^2\mu^2 - 3\sigma^4 - \mu^4}{\sigma^4} \tag{14.10}$$

The autocorrelation, $R_{ee}(\tau) = E\{e(n)e(n+\tau)\}$, can be used to produce a feature which measures how uncorrelated the error is:

$$\rho = \left(\sum_{\tau} \frac{R_{ee}^2(\tau)}{R_{ee}^2(0)} \right) - 1 \tag{14.11}$$

This feature will be zero for a completely uncorrelated error time series and will increase as the correlation increases. The formalism in Equations (14.8) to (14.11) assumes that the prediction error, $e(n)$, is stationary and this is valid as long as the data is consistent with the model in Equation (14.7). It is understood that this cannot be guarranteed a priori, although the result in Section 14.4 appear to indicate that this may not be a problem.

To compare these features for known and unknown conditions, it was assumed that the estimates of the features were Gaussian distributed. A threshold was set, at a chosen distance from the mean value of the feature estimated from the known condition data. If an estimated feature from the unknown condition test data was further than this distance from the mean then the condition was detected as faulty. If the Gaussian assumption is correct then by choosing a threshold of 2.576 units of standard deviations of the feature should falsely classify only 1% of the known condition data as faulty. The false alarm rate can be reduced by increasing the threshold although this will reduce the detection probability. Since several features are estimated, it is possible to use these jointly to increase the sensitivity to change by deciding that the condition is faulty if any of the features lie outside a threshold. The model based fault detection scheme is illustrated in Figure 14.1 which details how the fault is detected from the vibration signal.

14.4 Experimental Results

To evaluate objectively the use of periodic time-varying models for fault detection, vibration data was collected from the following machine test rig. The machine was subjected to several different bearing faults with the vibration and once per revolution signals recorded to allow accurate determination of rotation speed and position.

The machine consisted of a rotating shaft driven by a d.c. electric motor through a flexible coupling. The shaft was held in place by two rolling element bearings,

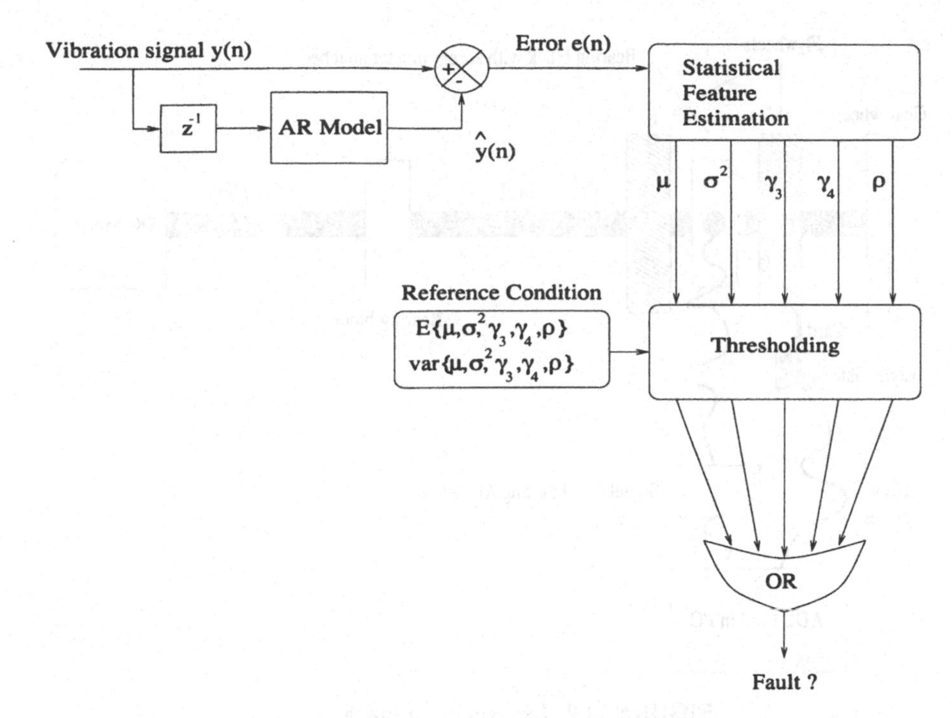

FIGURE 14.1. Fault detection from vibration signal

each with 8 balls of diameter 6.35mm, and with ball pitch diameter of 28.5mm. Two metal flywheels were used to load the shaft either side of one of the bearings, and a gear-wheel was attached to the end of the shaft. A diagram of the machine is shown in Figure 14.2.

One of the bearings was replaced by bearings which had the following defects: damage to the inner race, damage to the outer race and damage to one of the rolling elements and damage to the cage which holds the elements. Vibrations were also recorded for a second bearing which had not been damaged. This produced vibration signals for a total of six conditions: 1) Reference (Training) Normal Condition; 2) Second (Validation) Normal Condition; 3) Inner Race Fault Condition; 4) Outer Race fault Condition; 5) Rolling Element Fault Condition; 6) Broken Cage Fault Condition.

An accelerometer was attached to this bearing to measure vibrations and a once per revolution signal was obtained by attaching a small piece of card to one of the flywheels which passed through a light gate every revolution. Vibrations and the once per revolution pulse were sampled for 40 seconds at 24kHz using a Loughborough Sound Instruments DSP Card fitted to a Viglen 33MHz 486PC. These were recorded with the rotation period of the machine set to approximately 23ms for the machine in its normal condition and with each faulty bearing attached.

Model based systems were implemented using both linear time-invariant and time-varying models. An AR model was determined for an undamaged bearing, from 4 seconds of vibration data, using the Yule- Walker equations and the model order was chosen using the Akiake Information Criterion (AIC). This AR model was tested to evaluate its success at fault detection. It was used as the time-invariant section of a periodic AR model, and additional sections were added one at a time, with the coefficients being determined using the gradient descent algorithm. The

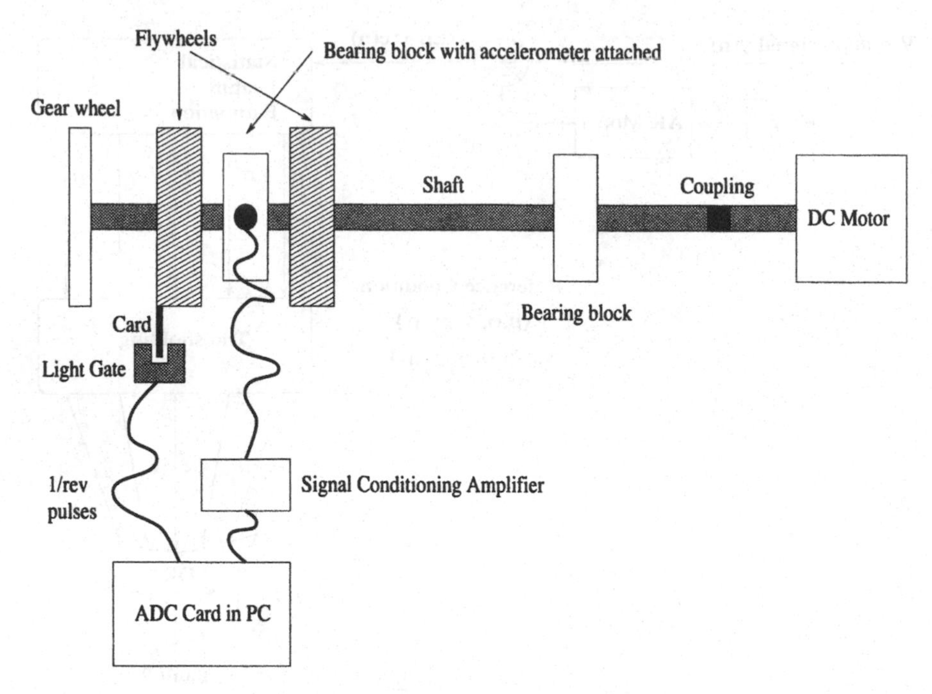

FIGURE 14.2. Experimental machine set

sections were added if they reduced the AIC.

For evaluation purposes, the prediction error time series were calculated from the vibrations recorded under each machine condition. These were then divided into 100 segments and features were calculated in each segment. Features calculated using the training data were used to set thresholds. Using these thresholds, each segment was classified as faulty or not-faulty for each individual feature and for all the features in combination.

Table 14.1 shows the percentage of test segments classified as faulty using each of the features individually and jointly using all the features (Π), using thresholds set at the 99% confidence interval.

This produced a very high number of false alarms indicating that the threshold has been set too tight and throwing doubts on the assumption of Gaussianity. By increasing the threshold to twenty times the standard deviation, the percentage of false alarms is much reduced, as is shown in Table 14.2. The probability of false alarm is reduced to 2% for all normal conditions but the probability of detection

	Bearing condition					
	1	2	3	4	5	6
μ	12	43	100	100	94	14
σ^2	6	100	100	100	100	91
γ_3	21	41	100	89	100	9
γ_4	5	15	100	45	100	4
ρ	8	99	100	100	100	99
Π	29	100	100	100	100	100

TABLE 14.1. Percentage of segments detected as faulty using time-invariant model and 99% threshold

is down to 76%. The system fails to detect the cage fault. The table also indicates that it might be possible to use the variance feature on its own.

	Bearing condition					
	1	2	3	4	5	6
μ	0	0	66	97	0	0
σ^2	0	2	100	100	100	4
γ_3	0	4	100	40	100	2
γ_4	0	0	100	2	100	4
ρ	0	0	86	0	100	0
Π	0	4	100	100	100	4

TABLE 14.2. Percentage of segments detected as faulty using time-invariant model and 20σ threshold

Table 14.3 shows the percentage of segments classified as faulty using the 99% confidence interval threshold for the periodic-time-varying model. This shows a similar pattern to Table 14.1 where there were a large number of false alarms.

	Bearing condition					
	1	2	3	4	5	6
μ	9	69	100	100	86	2
σ^2	5	100	100	100	100	100
γ_3	20	40	100	80	100	31
γ_4	5	13	100	34	100	0
ρ	6	98	100	100	100	100
Π	28	100	100	100	100	100

TABLE 14.3. Percentage of segments detected as faulty using time-varying model and 99% threshold

	Bearing condition					
	1	2	3	4	5	6
μ	0	0	48	5	2	0
σ^2	0	0	100	100	100	0
γ_3	0	4	94	32	100	0
γ_4	0	1	100	2	100	0
ρ	0	0	81	0	100	100
Π	0	4	100	100	100	100

TABLE 14.4. Percentage of segments detected as faulty using time-varying model and 20σ threshold

Table 14.4 shows the percentage of segments classified using the much higher threshold of 20 times the standard deviation. In this case the joint classification produces a 100% fault detection rate and a 2% false alarm rate. This table also illustrates the need to monitor more than one feature as both the variance and the feature based on the autocorrelation are required to detect all faults. The other features appear less essential although large changes in the skewness or kurtosis can be observed for both the inner race and rolling element faults.

It is clear that the periodic-time-varying model based system performs better than the time-invariant system. However using the Gaussian assumption to set the threshold gave poor results and therefore some other means of deciding where the thresholds should be set needs to be investigated, looking in more detail at the features to obtain a better indication of their distribution under normal circumstances.

14.5 Conclusions

The application of periodic time-variant AR models to the problems of fault detection and diagnosis in machinery has been demonstrated. This has shown that it is possible to detect machine faults with a system which is given knowledge of the normal machine condition only. Comparison was made between time-invariant and time-varying systems and it was found that the time-varying model based system performed better, detecting cage fault which the time-invariant model based system failed to detect.

Acknowledgments: The authors wish to thank Weir Pumps Ltd., of Glasgow for the loan of the experimental machine set, and Mr Lindsay Jack for his assistance in collecting the experimental data. The authors also wish to acknowledge the financial assistance of the EPSRC and DERA, Winfrith.

14.6 References

[1] H. Akaike. A new look at the statistical model identification. *IEEE Transactions on Automatic Control*, 19:716–723, 1974.

[2] D.C. Baillie and J. Mathew. A comparison of autoregressive modeling techniques for fault diagnosis of rolling element bearings. *Mechanical Systems and Signal Processing*, 10(1):1–17, 1996.

[3] J.R. Dickie and A.K. Nandi. AR modelling of skewed signals using third-order cumulants. *IEE Proceedings-Part VIS*, 142(2):78–86, 1995.

[4] W.A. Gardner. Exploitation of spectral redundancy in cyclostationary signals. *IEEE Signal Processing Magazine*, 8(2):14–36, 1991.

[5] A.C. McCormick and A.K. Nandi. Real time classification of rotating shaft loading conditions using artificial neural networks. *IEEE Transactions on Neural Networks*, 8(3):748–757, 1997.

[6] A.K. Nandi and J.A. Chambers. New lattice realisation of the predictive least-squares order selection criterion. *IEE Proceedings-F*, 138(6):545–550, 1991.

[7] M. Pagano. Periodic and multiple autoregression. *Annals of Statistics*, 6:1310–1317, 1978.

[8] G.D. Zivanovic and W.A. Gardner. Degrees of cyclostationarity and their application to signal detection and estimation. *Signal Processing*, 22:287–297, 1991.

15

Rayleigh Quotient Iteration for a Total Least Squares Filter in Robot Navigation

Tianruo Yang[1]
Man Lin[1]

ABSTRACT
Noisy sensor data must be filtered to obtain the best estimate of the robot position in robot navigation. The discrete Kalman filter, usually used for predicting and detecting signals in communication and control problems has become a common method for reducing the effect of uncertainty from the sensor data. However, due to the special domain of robot navigation, the Kalman approach is very limited. Here we propose the use of a Total Least Squares Filter which is solved efficiently by the Rayleigh quotient iteration method. This filter is very promising for very large amounts of data and from our experiments we can obtain more precise accuracy faster with cubic convergence than with the Kalman filter.

15.1 Introduction

The discrete Kalman filter [7] usually used for predicting and detecting signals in communication and control problems has become a common method for reducing the effect of uncertainty from the sensor data. Due to the fact that most functions in applications are non-linear, the extended Kalman filter, which linearizes the function by taking a first-order Taylor expansion, is introduced and this linear approximation is then used as the Kalman filter equation [1, 8].

In robot navigation, several problems often occur when we apply either the Kalman or the extended Kalman filter. An underlying assumption in any least squares estimation is that the entries in the data matrix are error-free [2, 6] which means that the time intervals at which measurements are taken are exact. But in many actual applications, sampling error, human errors, and instrument errors may preclude the possibility of knowing the data matrix exactly. In some cases, the errors in data matrix can be at least as great as the measurement errors. At this moment, applying the Kalman filter gives very poor results. And also the linearization process of the extended Kalman filter can introduce significant error into the problem [2]. The extended Kalman is not guaranteed to be optimal or even to converge because it needs a very good initial estimate of the solution. In some

[1]Linköping University, Department of Computer Science, 581 83 Linköping, Sweden,
E-mails: {tiaya, linma}@ida.liu.se

cases, it can easily fall into a local minimum when this initial guess is poor which is the situation faced by robot navigators.

Just recently, Yang proposed two ways, namely, the Krylov subspace method [17] and the Lanczos bidiagonalization process with updating techniques to solve successfully the Total Least Squares Filter.

In this paper, we propose a Rayleigh quotient iteration method for the Total Least Squares Filter (TLS) which does not require numerous measurements to converge because the camera images in robot navigation are time-consuming and take the errors in the data matrix into consideration. Recently Boley and Sutherland described a Recursive Total Least Squares Filter (RTLS) which is very easy to update [2]. In some ways, that is still a time-consuming algorithm. Here we apply the Rayleigh quotient iteration process which is more computationally attractive to solve the total least squares problem. The experiments indicate that this approach achieves greater accuracy with promising computational costs and converge cubically.

The paper is organized as follows. In Section 15.2, we briefly describe the Total Least Squares (TLS) filter. The Rayleigh quotient iteration method with local and global convergence are presented in Section 15.3. Finally we offer some comments and remarks.

15.2 Total Least Squares Filter

Given an over-determined system of equation $Ax = b$, the TLS problem in its simplest form is to find the smallest perturbation to A and b that makes the system of equations compatible. Specifically, we find an error matrix E and vector r such that for some vector x

$$\min_{E,r} \|(E,\ r)\|_F, \qquad (A+E)x = b+r$$

The vector x corresponding to the optimal (E, r) is called the TLS solution.

The most common algorithms for computing the TLS solution are based on the Singular Value Decomposition (SVD), a computationally expensive matrix decomposition [5]. A very complete survey of computational aspects and analysis of the TLS problem is given by Van Huffel and Vandewalle in [15]. Recently Van Huffel [14] presented some iterative methods based on inverse iteration and Chebyshev acceleration, which compute the basis of a singular subspace associated with the smallest singular values. Their convergence properties, the convergence rate, and the operation counts per iteration step are analyzed so that they are highly dependent on the gap of singular values. Also some recursive TLS filters have been developed for application in signal processing [3, 18].

15.3 Rayleigh Quotient Iteration Method

Here we proceed to solve this filter by inverse iteration where we shift with the Rayleigh quotient

$$\rho(x) = \frac{(x^T A^T - b^T)(Ax - b)}{(1 + x^T x)} = \frac{r^T r}{1 + x^T x} \tag{15.1}$$

Hence the Rayleigh quotient approximation is simply $\sigma = \|r\|_2/\|(x^T - 1)\|_2$. In this iteration method, if x_{k-1}, is the current iteration, we compute the next approximation x_k and λ_k by solving

$$\begin{pmatrix} J_1 & A^T b \\ b^T A & \gamma \end{pmatrix} \begin{pmatrix} x_k \\ -1 \end{pmatrix} = \lambda_k \begin{pmatrix} x_{k-1} \\ -1 \end{pmatrix} \tag{15.2}$$

where

$$J_1 = A^T A - \sigma^2 I, \quad \gamma = b^T b - \sigma^2, \quad \sigma^2 = \rho(x_{k-1})$$

The solution is obtained from

$$\begin{pmatrix} J_1 & A^T b \\ 0 & \tau \end{pmatrix} \begin{pmatrix} x_k \\ -1 \end{pmatrix} = \lambda_k \begin{pmatrix} x_{k-1} \\ -w^T x_{k-1} - 1 \end{pmatrix} \tag{15.3}$$

where

$$\tau = (b - Aw)^T b - \sigma^2, \quad J_1 w = A^T b$$

The first block equation is $J_1 x_k = A^T b + \lambda_k x^{(k-1)}$. To improve the numerical stability, we write $A^T b = A^T r_{k-1} + J_1 x_{k-1} + \sigma^2 x_{k-1}$, where $r_{k-1} = b - Ax_{k-1}$. Then

$$x_k = x_{k-1} + w + \lambda_k d, \quad w = z + \sigma^2 d$$

where

$$J_1 z = A^T r, \qquad J_1 d = x_{k-1}$$

ALGORITHM: Rayleigh Quotient Iteration

$x = x_{LS}$; $\quad r = b - Ax$;
for $\quad k = 1, 2, \ldots \quad$ do
$\quad \sigma^2 = r^T r / (1 + x^T x)$;
$\quad f_1 = -A^T r - \sigma^2 x$;
$\quad f_2 = -b^T r + \sigma^2$;
$\quad \rho = (f_1^T f_1 + (f_2)^2)^{1/2}$;
$\quad$ solve $\quad (A^T A - \sigma^2 I) z = A^T r$;
$\quad$ solve $\quad (A^T A - \sigma^2 I) d = x$;
$\quad w = z + \sigma^2 d$;
$\quad s = s - Aw$;
$\quad \eta = (s^T b - \sigma^2)/(x^T(x + w) + 1)$;
$\quad x = x + w + \eta d$;
$\quad r = s - \eta(Ad)$;
endfor

15.3.1 Local Convergence

Since the matrix $A^T A$ is real and symmetrical, it holds that if a singular vector is known to precision ϵ, the Rayleigh approximates the corresponding singular value to the precision ϵ^2. This leads to *cubic convergence* for the Rayleigh quotient iteration.

When the Rayleigh sequence x_k converges to z, the behavior is best described in terms of $\phi_k = \angle(x_k, z)$, the error angle. It turns out that, as $k \to \infty$, $\phi_k \to 0$ to the third order which ensures that the number of correct digits in x_k triples at each step for k large enough and this often means $k > 2$.

The current iterate x_k can be written in terms of ϕ_k as

$$x_k = z \cos \phi_k + u_k \sin \phi_k$$

where $u_k^T z = 0$ and $\|u_k\| = 1 = \|z\|$.

We can state the result formally. The analysis is similar to that for the power method [9].

Theorem 15.3.1 *Assume that the Rayleigh sequence x_k converges to the eigenvector z of $(x^T, -1)^T$ associated with the eigenvalue σ_{n+1}^2. As $k \to \infty$, the error angles ϕ_k satisfy*

$$\lim |\phi_{k+1}/\phi_k^3| \leq 1$$

15.3.2 Global Convergence

The best computable measure of accuracy of an eigenpair is its residual vector. The key fact, not appreciated until 1965, is that, however poor the starting vector, the residual norms always decrease.

Theorem 15.3.2 *For the Rayleigh quotient iteration, the residual norms*

$$\rho_k = \|t_k\|_2, \qquad t_k = \begin{pmatrix} -A^T r_k - \sigma_k^2 x_k \\ -b^T r_k + \sigma_k^2 \end{pmatrix}$$

always decrease, $\rho_{k+1} \leq \rho_k$, for all k.

Rayleigh quotient iteration is used to improve a given approximate singular vector, but in general we cannot say to which singular vector it will converge. It is also possible that some unfortunate choice of the starting vector will lead to endless cycling. However, it can be shown that such cycles are unstable under perturbation, so this will not occur in practice. The following theorem is a minor variation based on Kahan's original argument.

Theorem 15.3.3 *Let x_k be the Rayleigh sequence generated by any unit vector x_0. As $k \to \infty$,*

1. *σ_k^2 converges, and either*
2. *the approximate eigenpair converges cubically, or*
3. *$x_{2k} \to x_+$, $x_{2k+1} \to x_-$, linearly, where x_+ and x_- are the bisectors of a pair of eigenvectors whose eigenvalues have mean $\rho = \lim_k \sigma_k$. In this situation it is unstable under perturbation of x_k.*

15.3.3 New Convergence Results

In [12], Szyld presents a method for finding selected eigenvalue-eigenvector pairs of the generalized problem $Ax = \lambda Bx$, where A is symmetric and B is positive-definite.

The method finds an eigenvalue in a given interval or determines that the interval is free of eigenvalues while computing an approximation to the eigenvalue closest to that interval. His method consists of inverse and Rayleigh quotient iteration steps. The convergence is studied and it is shown how an inclusion theorem gives one of the criteria for switching from inverse to Rayleigh quotient iteration. The existence of an eigenvalue in the desired interval is guaranteed when this criterion is fulfilled. We will modify this approach to our eigenvalue problem in

$$\begin{pmatrix} A^TA & A^Tb \\ b^TA & b^Tb \end{pmatrix}\begin{pmatrix} x \\ -1 \end{pmatrix} = \sigma_{n+1}^2\begin{pmatrix} x \\ -1 \end{pmatrix}$$

which can be expressed as $\hat{A}\hat{x} = \sigma_{n+1}^2\hat{x}$.

A combination of inverse and Rayleigh quotient iterations of the following form will be used:

$$\begin{aligned}(\hat{A} - \mu I)y_{s+1} &= x_s \\ \omega_{s+1} &= (y_{s+1}^T y_{s+1})^{-1/2} \\ x_{s+1} &= \omega_{s+1}y_{s+1}\end{aligned} \tag{15.4}$$

Here $\mu = \gamma$ in the case of inverse iteration, and $\mu = \mu_s = x_s^T A x_s$ for the Rayleigh quotient case. It is well known that inverse iteration converges linearly to the eigenvalue closest to γ, whereas Rayleigh quotient iteration exhibits cubic convergence. However, the latter method has the drawback that it can converge to different eigenvalues depending on the starting vector. We should establish a criterion to guarantee that the method converges to the desired eigenvalue and not to any other.

Let $(\sigma_i^2, \hat{x}_i)$ be the eigenpair of the matrix $\hat{A}$. For convenience of notation we assume that the eigenvalues are numbered according to this order: $0 < \sigma_{n+1}^2 \leq \sigma_n^2 \leq \ldots \leq \sigma_1^2$. At each step of the inverse iteration, there are two residuals that can be computed, $q_s = \hat{A}x_s - \gamma x_s$, with respect to the shift γ, and $p_s = \hat{A}x_s - \mu_s x_s$ with respect to the Rayleigh quotient. They are related by the equation

$$p_s = q_s + (\gamma - \mu_s)x_s$$

We study the convergence of q_s and obtain the following lemma based on that in [12].

Lemma 15.3.4 *One important relationship between p_s and q_s is*

$$\|p_s\|_2^2 = \|q_s\|_2^2 - (\gamma - \mu_s)^2 \tag{15.5}$$

which shows that

$$\|p_s\|_2 \leq \|q_s\|_2, \quad |\gamma - \mu_s| \leq \|q_s\|_2 \tag{15.6}$$

and

$$\|q_s\|_2 \to |\gamma - \sigma_{n+1}^2| \quad as \quad s \to \infty \tag{15.7}$$

It turns out that the convergence sequence is actually monotonically decreasing; thus we have the following lemma.

Lemma 15.3.5 *The residual norms of the inverse iteration with respect to the shift γ are monotonically decreasing, i.e., $\|q_{s+1}\|_2 \leq \|q_s\|_2$.*

Given x_s and $\mu = \mu_s$, Rayleigh quotient iteration consists of computing x_{s+1} from equation (15.4). The convergence of Rayleigh quotient iteration is cubic in all practical cases (see [9, 10] and the references given therein for a full discussion). Wilkinson [16], Ruhe [11] and Szyld [13] proved the following results:

$$\omega_s \leq \|p_s\|_2, \qquad |\mu_s - \mu_{s+1}| \leq \|p_s\|_2 \tag{15.8}$$

$$\|p_{s+1}\|_2 \leq \|p_s\|_2, \qquad |\mu_s - \lambda_i| \leq \|p_s\|_2 \tag{15.9}$$

where λ_i is the eigenvalue closest to μ_s, $|\mu_s - \mu_{s+1}| \to 0$, and in all practical cases, $\|p_s\|_2 \to 0$.

Now we can turn to our main condition which is used to guarantee the convergence of the Rayleigh quotient iteration method to the smallest desired singular value.

Theorem 15.3.6 *If $\|r_{LS}\|_2^2 \leq \sigma_n^2 - \sigma_{n+1}^2$, then the Rayleigh quotient iteration converges to σ_{n+1}^2.*

Proof: From the definition of $\|q_1\|_2$, if we consider the shift $\gamma = 0$, then

$$\begin{aligned} \|q_1\|_2 &= \|(A\ \ b)^T(A\ \ b)\begin{pmatrix} x_{LS} \\ -1 \end{pmatrix}\|_2 \\ &= \|r_{LS}\|_2^2 \end{aligned}$$

Since the convergence sequence is actually monotonically decreasing, i.e., $\|q_{s+1}\|_2 \leq \|q_s\|_2$, it follows that $\|q_s\|_2 \leq \|r_{LS}\|_2^2$ and from (15.6)

$$\|p_s\|_2 \leq \|r_{LS}\|_2^2, \qquad |\mu_s| \leq \|r_{LS}\|_2^2 \tag{15.10}$$

Also from (15.7) and the lemma, we obtain $\sigma_{n+1}^2 \leq \|q_s\|_2$. Combining with $\|q_s\|_2 \leq \|r_{LS}\|_2^2$, we know that σ_{n+1}^2 lies in the interval $(-\|r_{LS}\|_2^2, \|r_{LS}\|_2^2)$. If $\|r_{LS}\|_2^2 \leq \epsilon$ where $\epsilon = \sigma_n^2 - \sigma_{n+1}^2$, then σ_{n+1}^2 is the only singular value in that fixed interval, and further with (15.10), then the Rayleigh quotient iteration algorithm converges only to the desired singular value σ_{n+1}^2.

15.4 Experimental Results

To compare the performance with the Kalman filter and the Recursive Total Least Squares Filter (RTLS) in practice, we ran our Rayleigh quotient iteration approach for the Total Least Squares Filter algorithm for a set of experiments, including one with a physical mobile robot and camera, as suggested in [2].

Due to limited space, we selected only one experiment as an example where we simulated a simple robot navigation problem typical of that faced by an actual mobile robot [1, 4, 8]. The robot has identified a single landmark in a two-dimensional environment and knows the landmark location on a map. It does not know its own position. It moves in a straight line with a known uniform velocity. Its goal is to estimate its own starting position relative to the landmark by measuring the visual

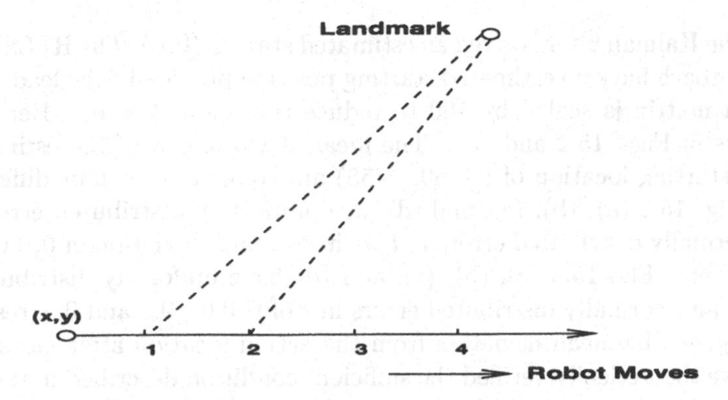

FIGURE 15.1. Diagram of measurement

angle between its heading and the landmark. Measurements are taken periodically as it moves. Fig. 15.1 gives a simple diagram of the problem. Assume that the landmark is located at (0,0), that the y coordinate of the robot's starting position does not change as the robot moves, and that the robot knows which side of the landmark it is on. To map this robot-based system to the ground coordinate system, it suffices to know only the robot's compass heading from an internal compass. We will follow the simple way described in [2] of knowing the compass heading independently.

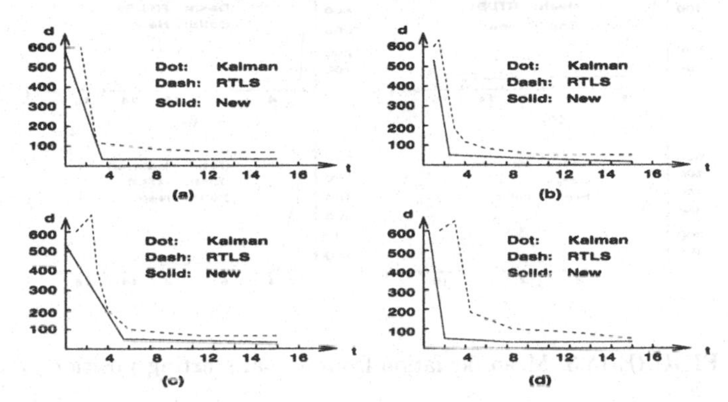

FIGURE 15.2. Mean deviation to actual starting position I

In the experiment, we assumed that the y coordinate of the robot path was negative, that robot velocity v was 20 per unit of time and that measurements of α were taken at unit time intervals. At any time t_i, we know that

$$\cot(\alpha_i) = \frac{x + t_i * v}{y}$$

where (x, y) is the robot starting position and α_i is the angle of the robot heading to the landmark. Random errors with a uniform distribution are added to the angle measures and a normally distributed random error is added to the time measurement. We can formulate the problem so that the data matrix and the measurement vector contained error are as follows:

$$A_i = [1 \ \ -\cot(\alpha_i)], \ x_i = [x^T \ y^T]^T, \ b_i = -t_i * v$$

where, at time t_i, A_i is the data matrix, b_i is the measurement vector, and x_i is the estimated state vector consisting of the coordinates (x, y) of the robot starting position. The Kalman filter is given an estimated start of $(0, 0)$. The RTLS algorithm and our approach have no estimated starting position provided. The leading column of the data matrix is scaled by 100 to reduce the allowed errors. Here we show some results in Figs. 15.2 and 15.3. The mean deviations d of the estimates from the actual starting location of $(-460, -455)$ are compared for four different error amounts. Fig. 15.2 (a), (b), (c), and (d) have uniformly distributed errors in α of $\pm 2°$ and normally distributed errors in t with standard deviation of $0, 0.05, 0.1$ and 0.5, respectively. Fig. 15.3 (a), (b), (c), and (d) have uniformly distributed errors in α of $\pm 4°$ and normally distributed errors in t of $0, 0.05, 0.1$ and 0.5, respectively. Table 15.1 gives the mean deviation from the actual location after measurements. Here we have theoretically verified the sufficient condition described in the previous section. However, for those which cannot satisfy the above sufficient condition, we still need to do much more work. For the experiments, the new approach, namely the Rayleigh quotient iteration method converges more quickly than the RTLS filter which is also faster than the classical Kalman filter when errors are introduced in both the measurement vector and the data matrix. Also we can see clearly that the new approach can achieve a closer estimate of the actual location than the RTLS and Kalman filters as well.

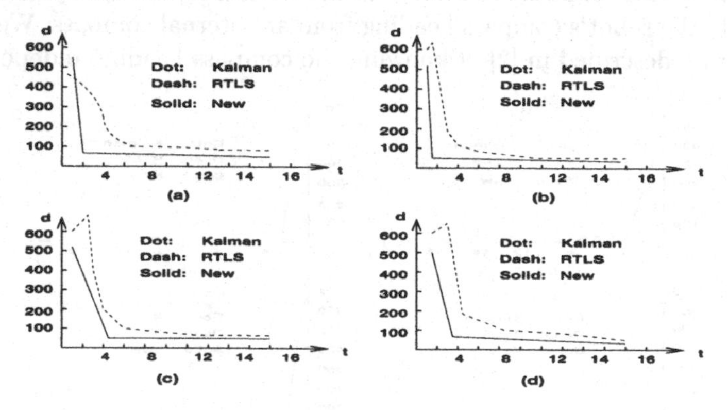

FIGURE 15.3. Mean deviation from actual starting position II

Error(α)	Error(t)	0	0.05	0.1	0.5
$\pm 2°$	Kalman	32.47	20.27	24.54	28.56
	RTLS	20.24	15.90	24.81	23.75
	New	19.45	13.32	21.68	19.37
$\pm 4°$	Kalman	21.01	31.80	34.63	36.67
	RTLS	10.11	24.97	32.13	31.39
	New	8.42	18.65	30.36	26.37

TABLE 15.1. Mean deviation from actual location

15.5 Conclusion

In this paper, we propose a new approach, namely, the Total Least Squares Filter solved by the Rayleigh quotient iteration method. This filter is very suitable for large amounts of data with relatively few readings and makes very few assumptions about the data or the problem solved. Compared with the classical Kalman filter, we take the error term of data matrix into consideration and do not take care whether the initial guess guarantees that the solution is optimal without falling into a local minimum when the initial solution is poor, which is very typical in robot navigation. We apply it to robot navigation, and the experiments indicate that this is a successful approach. Further research should focus on the convergence of this approach because here we give only a sufficient condition which is usually very difficult to verify to guarantee global convergence.

15.6 References

[1] N. Ayache and O.D. Faugeras. Maintaining representations of the environment of a mobile robot. *IEEE Transaction on Robotics and Automation*, 5(6):804–819, 1989.

[2] D.L. Boley and K.T. Sutherland. Recursive total least squares: An alternative to the discrete Kalman filter. Technical Report CS-TR-93-32, Department of Computer Science, University of Minnesota, April 1993.

[3] C.E. Davila. Efficient recursive total least squares algorithm for FIR adaptive filtering. *IEEE Transactions on Signal Processing*, 42(2):268–180, 1994.

[4] H. Durrant-White, E. Bell, and P. Avery. The design of a radar-based navigation system for large outdoor vehicles. In *Proceedings of 1995 International Conference on Robotics and Automation*, IEEE, pp. 764–769, 1995.

[5] G.H. Golub and C.F. Van Loan. An analysis of the total least squares problem. *SIAM Journal on Numerical Analysis*, 17:883–893, 1980.

[6] G.H. Golub and C.F. Van Loan. *Matrix Computations.* 2nd edition, Johns Hopkins University Press, Baltimore, 1989.

[7] S. Haykin. *Adaptive Filter Theory.* 2nd edition, Prentice-Hall, Engelwood Cliffs, NJ, 1991.

[8] A. Kosaka and A.C. Kak. Fast vision-guided mobile robot navigation using model-based reasoning and prediction of uncertainties. *CVGIP: Image Understanding*, 56(3):271–329, 1992.

[9] B.N. Parlett. *The Symmetric Eigenvalue Problem.* Prentice-Hall, Englewood Cliffs, NJ, 1980.

[10] B.N. Parlett and W. Kahan. On the convergence of a practical QR algorithm. In A.J.H. Morrell, editor, *Information Processing 68*, North-Holland, Amsterdam, 1:114–118, 1969.

[11] A. Ruhe. Computation of eigenvalues and eigenvectors. In V.A. Baker, editor, *Sparse Matrix Techniques*, Lecture Notes in Mathematics 572, Springer-Verlag, pp. 130–184, 1977.

[12] D. Szyld. Criteria for combining inverse and Rayleigh quotient iteration. *SIAM Journal on Numerical Analysis*, 25(6):1369–1375, 1988.

[13] D.B. Szyld. *A two-level iterative method for large sparse generalized eigenvalue calculations.* Ph.D. thesis, Department of Mathematics, New York University, New York, 1983.

[14] S. Van Huffel. Iterative algorithms for computing the singular subspace of a matrix associated with its smallest singular values. *Linear Algebra and its Applications*, 154/156:675–709, 1991.

[15] S. Van Huffel and J. Vandewalle. *The Total Least Squares Problem: Computational Aspects and Analysis*, volume 9 of *Frontiers in Applied Mathematics.* SIAM, Philadelphia, 1991.

[16] J. H. Wilkinson. Inverse iteration in theory and practice. *Symposium on Mathematics*, 10:361–379, 1972.

[17] T. Yang and M. Lin. Iterative total least squares filter in robot navigation. In Proceedings of *IEEE 1997 International Conference on Acoustics, Speech, and Signal Processing (ICASSP-97)*, Munich, Germany, 1997.

[18] K.B. Yu. Recursive updating the eigenvalue decomposition of a covariance matrix. *IEEE Transactions on Signal Processing*, 39(5):1136–1145, 1991.

16

Wavelet Use for Noise Rejection and Signal Modelling

Aleš Procházka[1]
Martina Mudrová[1]
Martin Štorek[1]

ABSTRACT
Wavelet functions form an essential tool of modern signal analysis and modelling in various engineering, biomedical and environmental systems. The paper is devoted to selected aspects of construction of wavelet functions and their properties enabling their use for time-scale signal analysis and noise rejection. The main part of the paper presents the basic principles of signal decomposition in connection with fundamental signal components detection and approximation. The methods are verified for simulated signals at first and then used for real signal analysis representing air pollution inside a given region of North Bohemia covered by several measuring stations. The following part of the chapter is devoted to comparison of linear and non-linear signal modelling and prediction for original and de-noised signals. The paper emphasizes the algorithmic approach to methods of signal analysis and prediction and it presents the use of wavelet functions both for signal analysis and modelling.

16.1 Introduction

Signal analysis is one of the essential parts of digital signal processing in various information and control systems and is often based upon different discrete transforms. The wavelet transform forms in many cases a very efficient alternative to the short-time Fourier transform enabling signal analysis with varying time-frequency resolution. Its application further includes signal components detection and signal decomposition, data compression and noise rejection both for one-dimensional and two-dimensional signals [14, 15, 24]. The use of wavelet and radial basis functions as transfer functions of neural networks for signal modelling and prediction is presented in various papers as well [3, 9, 11, 21, 27]. Problems closely related to this topic include signal segmentation [6, 19], restoration, statistical signal processing [13] and efficient optimization methods [7, 20].

The theory of wavelet transforms is studied in many papers and books published recently [2, 15, 22, 23, 25] describing also the relations between the wavelet trans-

[1]Institute of Chemical Technology, Department of Computing and Control Engineering, Technická 1905, 166 28 Prague 6, Czech Republic,
E-mails: {Ales.Prochazka, Martina.Mudrova, Martin.Storek}@vscht.cz

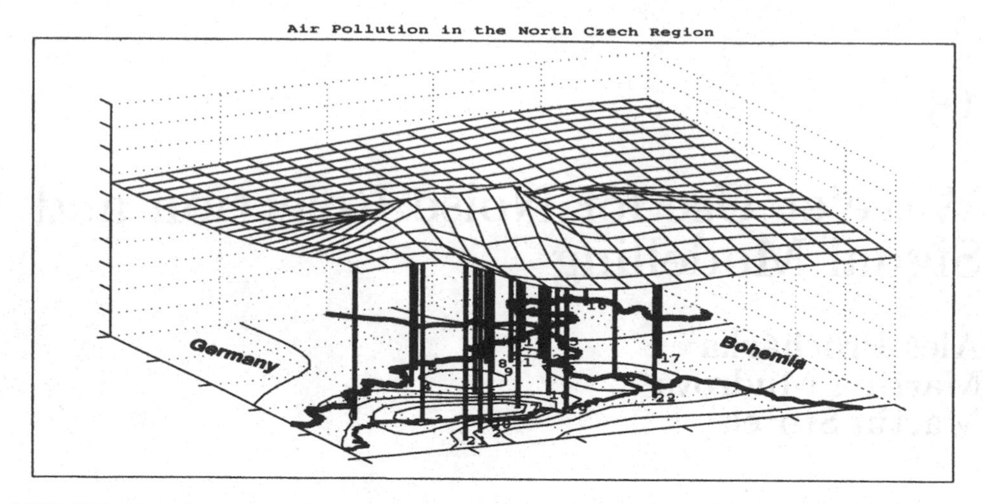

FIGURE 16.1. Air pollution in North Bohemia extrapolated from simultaneous measurements of 22 stations at a selected time

form and the short-time Fourier transform [10]. The analysis and the construction of wavelet functions both in the analytical and in the recurrent forms with description of their properties are important results of these studies.

The chapter is devoted to basic principles of signal analysis and decomposition by wavelet transform in connection with rejection of noise components forming an additive part of a given sequence. Results of these studies are applied to the processing of a given multichannel signal representing air pollution in North Bohemia. An example of such a signal is given in Fig. 16.1 using measurements at a selected time from 22 measuring stations in this region for signal interpolation and extrapolation. The goal of such a signal processing is to analyse noisy data to provide information about air pollution in the whole region and to predict the situation several sampling periods ahead. General methods of signal de-noising and prediction studied in the chapter are closely related to some other principles described in [1, 12, 18].

16.2 Signal Analysis and Wavelet Decomposition

Wavelet transforms (WT) provides an alternative to the short-time Fourier transform (STFT) for non-stationary signal analysis and both STFT and WT result in signal decomposition into two-dimensional function. Its independent variables stand for time and frequency in case of the STFT and they represent time and scale (closely related to frequency) in case of the WT. The basic difference between these two transforms is in the construction of the window function which has a constant length in case of the STFT (including rectangular, Blackman and other window functions) while in case of the WT wide windows are applied for low frequencies and short windows for high frequencies to ensure constant time-frequency resolution. Local and global signal analysis can be combined in this way.

In case of a continuous (non-stationary) signal $x(t)$ the continuous WT coefficients are given by relation

$$C(a,b) = \frac{1}{\sqrt{a}} \int_{-\infty}^{\infty} x(t)\, W(\frac{t-b}{a})\, dt \tag{16.1}$$

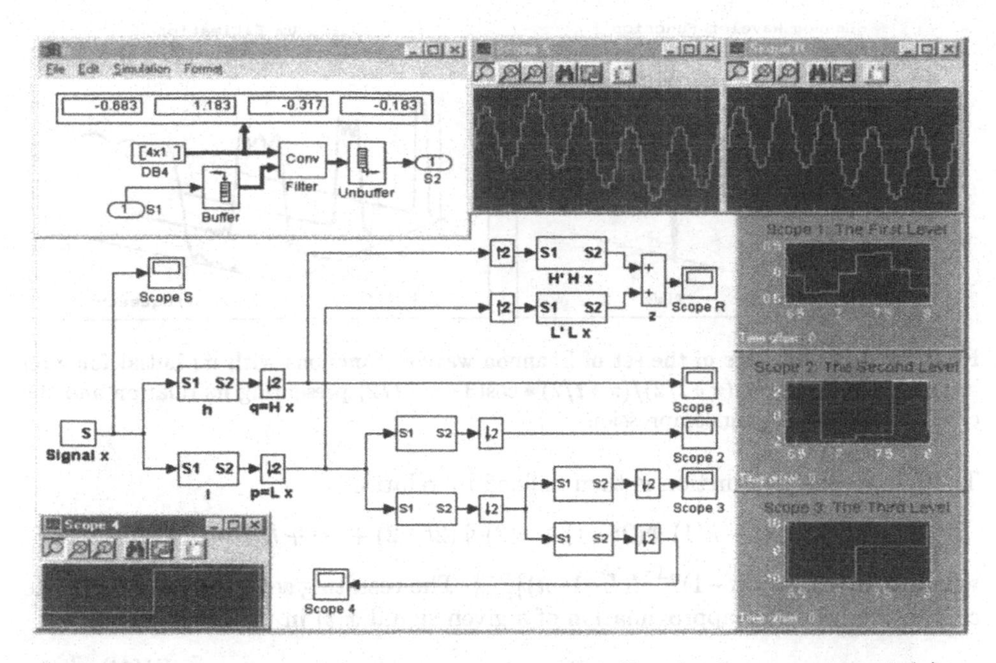

FIGURE 16.2. The Simulink model of the pyramidal filter bank structure used (a) to evaluate and to present wavelet transform coefficients for a given sequence $\{x(n)\}_{n=0}^{N-1}$ and complementary low-pass and high-pass filters at selected levels (b) to reconstruct the given sequence of values

where $W(t)$ is the basic (mother) wavelet function and values a and b stand for its translation and dilation respectively. There are strict criteria for the choice of such a basic wavelet function given by a requirement of the transforms reversibility [10, 14]. The most frequently used wavelets include Daubechies, Shannon, Morlet, Gaussian and harmonic functions [10, 15].

The discrete WT assumes the evaluation of wavelet coefficients for discrete parameters of dilation $a = 2^m$ and translation $b = k\, 2^m$ using only the initial (mother) wavelet function $W(t)$ localized at a particular position. The dilated and translated version in the form

$$W_{m,k}(t) = \frac{1}{\sqrt{a}}\, W\left(\frac{1}{a}\,(t-b)\right) = \frac{1}{\sqrt{2^m}}\, W\left(2^{-m}t - k\right) \tag{16.2}$$

is then used for signal analysis and description. Two basic approaches include definition of wavelet functions either in the analytical form or by recurrent equations. A very commonly used Daubechies wavelet function of order N is defined through the scaling function $\Phi(t)$ resulting from equation

$$\Phi(t) = l(0)\;\Phi(2t) + l(1)\;\Phi(2t-1) + l(2)\;\Phi(2t-2) + \cdots + l(L-1)\;\Phi(2t-L+1) \tag{16.3}$$

for coefficients $\{l(n)\}_{n=0}^{L-1}$. The (iterative) solution of this equation is based upon the relation

$$\Phi_j(t) = l(0)\;\Phi_{j-1}(2t) + l(1)\;\Phi_{j-1}(2t-1) + \cdots + l(L-1)\;\Phi_{j-1}(2t-L+1)$$

for $j = 1, 2, \cdots$ and the given starting Haar function

$$\Phi_0(t) = \begin{cases} 1 & \text{for } t \in \langle 0, 0.5) \\ 0 & \text{in other cases} \end{cases} \tag{16.4}$$

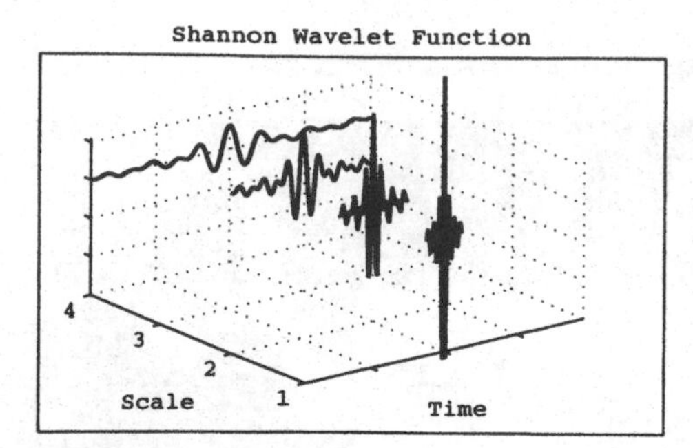

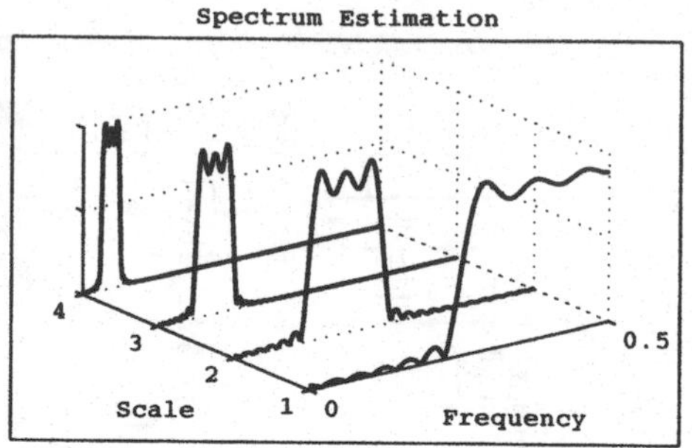

FIGURE 16.3. Analysis of the set of Shannon wavelet functions with its initial function in the form $W(t) = \sin(\pi * t/2)/(\pi * t/2) * \cos(3 * \pi * t/2)$ presenting its dilation and the corresponding spectrum compression

The initial wavelet function is then defined by relation

$$W(t) = h(0)\ \Phi(2t) + h(1)\ \Phi(2t-1) + h(2)\ \Phi(2t-2) + \cdots + h(L-1)\ \Phi(2t-L+1)$$

where $\{h(n)\}_{n=0}^{L-1} = \{(-1)^{n+1} l(L-1-n)\}_{n=0}^{L-1}$. The resulting set of wavelet functions can be then used for approximation of a given signal $x(t)$ in the form [15]

$$x(t) = c(0)\Phi(t) + c(1)W(t) + [c(2)\ c(3)] \begin{bmatrix} W(2t) \\ W(2t-1) \end{bmatrix} + [c(4)\ c(5)\ c(6)\ c(7)] \begin{bmatrix} W(4t) \\ W(4t-1) \\ W(4t-2) \\ W(4t-3) \end{bmatrix} + \ldots$$

for the set of wavelet transform coefficients $\{c(n)\}$.

The wavelet transform coefficients for a given column vector $\{x(n)\}_{n=0}^{N-1}$ may be evaluated efficiently using Mallat's pyramidal structure as shown in Fig. 16.2. This model enables the study of signal decomposition and reconstruction based upon the convolution between values of the given sequence and filter coefficients as shown in detail for block l in the upper-left corner of the Fig. 16.2 using Daubechies wavelet of order 4 in this case. From the signal processing point of view this algorithm assumes the use of the half band low-pass scaling sequence $\{l(n)\}_{n=0}^{L-1}$ to generate the corresponding wavelet sequence $\{h(n)\}_{n=0}^{L-1}$ which is convolved with the signal $\{x(n)\}$ and subsampled by two. Introducing decomposition matrices

$$\mathbf{L}_{N/2,N} = \begin{bmatrix} l(1) & l(0) & 0 & 0 & \cdots \\ l(3) & l(2) & l(1) & l(0) & \cdots \\ \cdots & \cdots & \cdots & \cdots & \cdots \\ 0 & \cdots & l(L-1) & \cdots & l(0) \end{bmatrix} \tag{16.5}$$

$$\mathbf{H}_{N/2,N} = \begin{bmatrix} h(1) & h(0) & 0 & 0 & \cdots \\ h(3) & h(2) & h(1) & h(0) & \cdots \\ \cdots & \cdots & \cdots & \cdots & \cdots \\ 0 & \cdots & h(L-1) & \cdots & h(0) \end{bmatrix} \tag{16.6}$$

it is possible to define the separation process for resulting signals of the first level decomposition $\mathbf{p} = \mathbf{L}\,\mathbf{x}$ and $\mathbf{q} = \mathbf{H}\,\mathbf{x}$. Signal reconstruction may be achieved by introducing the transposed matrices $\mathbf{L}_{N,N/2}^T$ and $\mathbf{H}_{N,N/2}^T$ as shown in Fig. 16.2 and the reconstructed signal $\mathbf{z}$ is given by

$$\mathbf{z} = (\mathbf{L}^T * \mathbf{L} + \mathbf{H}^T * \mathbf{H}) * \mathbf{x} = \mathbf{T}^T * \mathbf{T} * \mathbf{x} \tag{16.7}$$

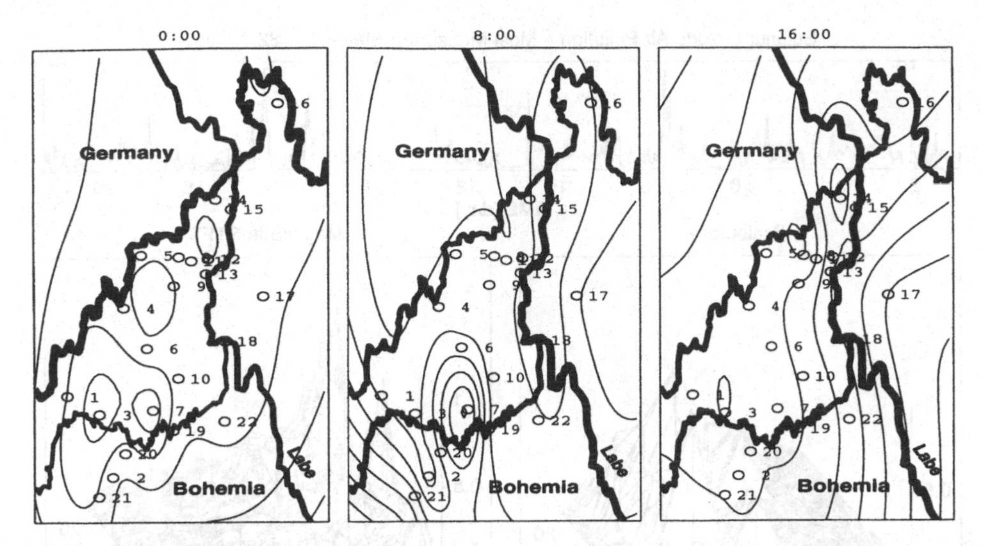

FIGURE 16.4. One day evolution of air pollution in North Bohemia extrapolated from 22 measuring stations in this region

The requirement for perfect signal reconstruction gives the necessary and sufficient condition for matrix $\mathbf{E} = \mathbf{T}^T\mathbf{T}$ to be the identity matrix. Matrix $\mathbf{T}$ defined by the original matrices $\mathbf{L}$ and $\mathbf{H}$ in the form

$$\mathbf{T} = \begin{bmatrix} \cdots & & & \cdots & \cdots & & & \cdots \\ l(L-1) & l(L-2) & l(L-3) & \cdots & l(0) & 0 & 0 & \cdots \\ h(L-1) & h(L-2) & h(L-3) & \cdots & h(0) & 0 & 0 & \cdots \\ 0 & 0 & l(L-1) & l(L-2) & \cdots & \cdots & l(0) & \cdots \\ 0 & 0 & h(L-1) & h(L-2) & \cdots & \cdots & h(0) & \cdots \\ \cdots & & & \cdots & \cdots & & & \cdots \end{bmatrix}$$

must be orthonormal in this case. The whole process of signal reconstruction from the first level of wavelet decomposition is illustrated in Fig. 16.2.

Various wavelet and scaling sequences can be used for signal analysis defined above. Their dilation correspond to spectrum compression according to Fig. 16.3. The most common choice include Daubechies wavelets even though their frequency characteristics give only an approximation to band-pass filters. On the other hand harmonic wavelets introduced in [15] can have broader application in many engineering problems owing to their very attractive spectral properties.

16.3 Real System Description

The region of North Bohemia represents a heavily industrial area with problems of air pollution which is monitored by several measuring stations to enable extrapolation and prediction of selected pollutants. As some measuring devices can be corrupted and resulting signal can be affected by the additive noise it is necessary to interpolate missing values first and then to reject signal noise components. The autoregressive (AR) method introduced in [6] can be applied very efficiently for restoration of missing values in the first stage of signal processing. The map with

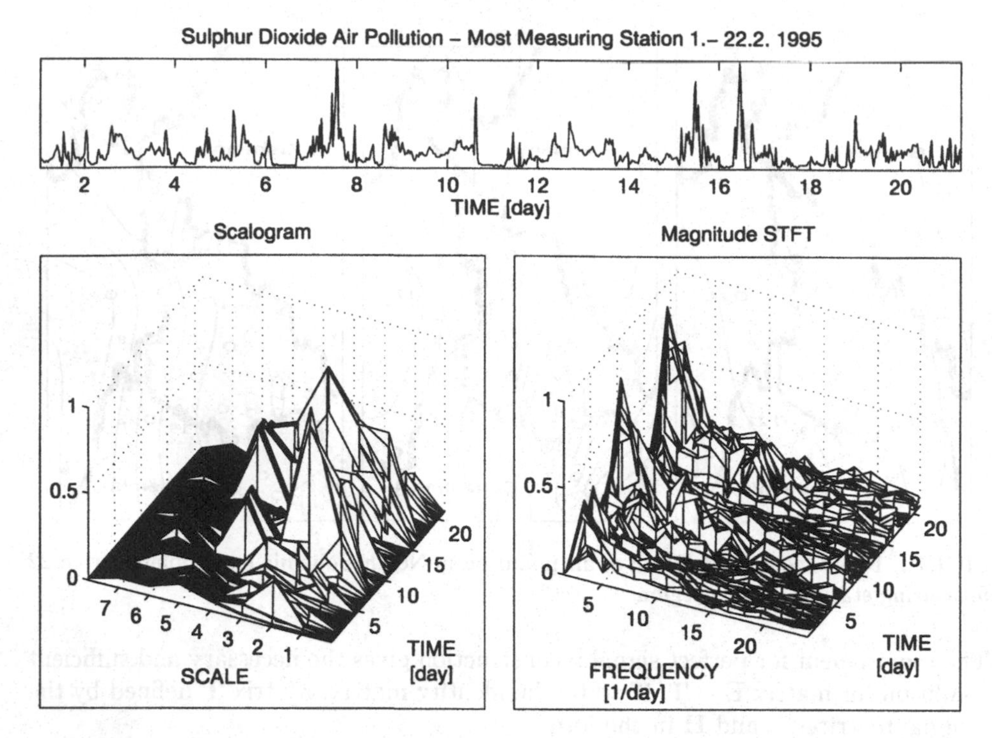

FIGURE 16.5. Spectrogram and scalogram of a given data segment representing air pollution measured in the city of Most during a three weeks period

the evolution of air pollution during a chosen day is presented in Fig. 16.4. The following preliminary analysis of a selected time series is given in Fig. 16.5 comparing results obtained by the STFT and WT. Owing to its definition the Wavelet transform provides better localization of signal peaks in this case thus enabling the use of an appropriate level of wavelet decomposition for the precise detection of extreme values of air pollution.

16.4 Signal Noise Rejection

Signal decomposition by a given set of wavelet functions into separate levels or scales results in the set of wavelet transform coefficients that can be used for signal compression, signal analysis, and in the case that the coefficients are not modified, they allow perfect signal reconstruction. In the case when only selected levels of signal decomposition are used, or wavelet transform coefficients are processed, it is possible to extract signal components or to reject its undesirable parts. Using the threshold method introduced by [24, 4] it is further possible to reject noise and to improve signal to noise ratio.

The de-noising algorithm assumes that the signal has low frequency components and that it is corrupted by additive Gaussian white noise with its power much lower than that of the analyzed signal. The whole method consists of the following steps:

- Signal decomposition using a chosen wavelet function up to the selected level and evaluation of wavelet transform coefficients

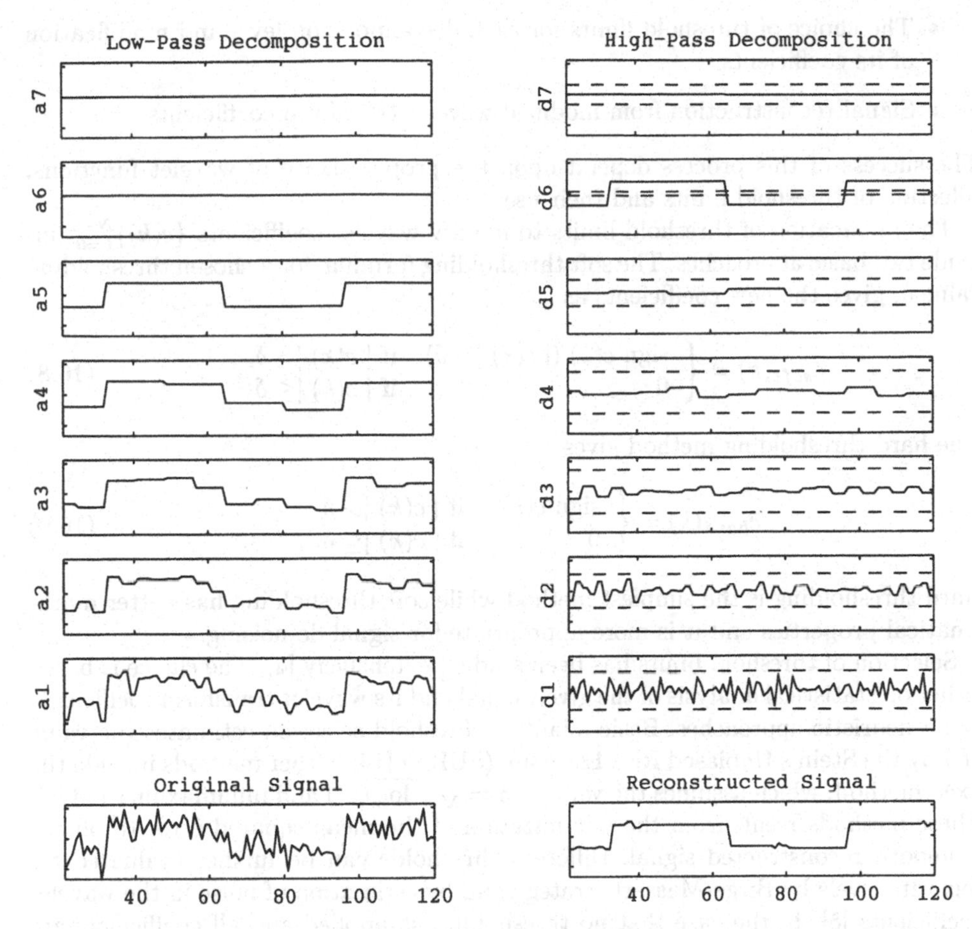

FIGURE 16.6. The use of the Haar wavelet function for decomposition of the given simulated signal $x(n) = 1$ for $n = N/4, \cdots, N/2-1, 3N/4, \cdots, N-1$ and $x(n) = 0$ for other values of n ($N = 128$) with an additive noise component and signal reconstruction for given threshold limits

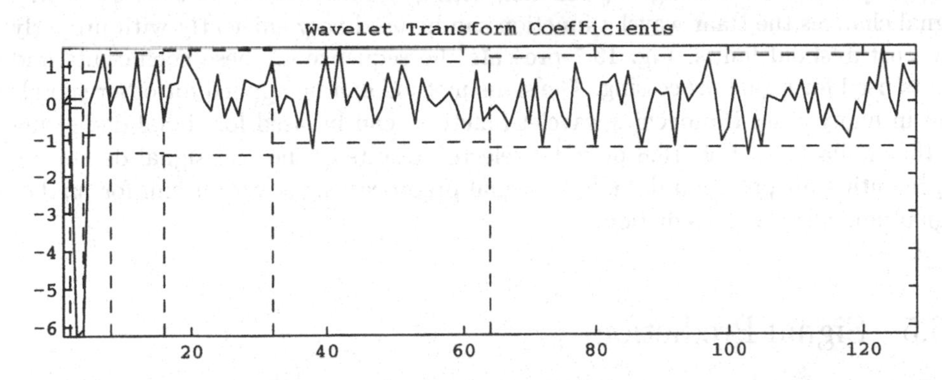

FIGURE 16.7. Wavelet transform coefficients $\{c(k)\}_{k=0}^{N-1}$ for the given signal $\{x(n)\}_{n=0}^{N-1}$ with its length $N = 128$ and selected threshold limits used for signal de-noising presenting segments of coefficients belonging to is decomposition level $l = D, D-1, \cdots, 1$ where $D = log_2(N)$

- The choice of threshold limits for each decomposition level and modification of its coefficients
- Signal reconstruction from modified wavelet transform coefficients

The success of this process depend upon the proper choice of wavelet functions, selection of threshold limits and their use.

The application of threshold limits to modify wavelet coefficients $\{c(k)\}_{k=0}^{N-1}$ include two basic approaches. The soft thresholding formula, for a chosen thresholding value δ, gives the new coefficients as

$$\bar{c}_{soft}(k) = \begin{cases} \text{sign } c(k) \ (\mid c(k) \mid -\delta) & \text{if } \mid c(k) \mid > \delta \\ 0 & \text{if } \mid c(k) \mid \leq \delta \end{cases} \tag{16.8}$$

The hard thresholding method gives

$$\bar{c}_{hard}(k) = \begin{cases} \text{sign } c(k) & \text{if } \mid c(k) \mid > \delta \\ 0 & \text{if } \mid c(k) \mid \leq \delta \end{cases} \tag{16.9}$$

Hard thresholding is the simplest method while soft thresholding has better mathematical properties and it is more appropriate for signal de-noising.

Selection of threshold limits has been studied extensively [4]. The choice is based either on statistical analysis of the given signal and its wavelet transform coefficients or on heuristic approaches. Basic adaptive threshold selection estimates the value of δ by the Stein's Unbiased Risk Estimate (SURE) [14]. Other methods include the fixed method, which assumes the value of $\delta = \sqrt{2 \, \log{(N)}}$ and minimax method [4]. These methods result from the minimization of the mean-squared error to obtain a smooth reconstructed signal. Different thresholds can be further evaluated for separate levels by Birge-Massart strategy and by estimation of noise in the wavelet coefficients [5]. In the case that no thresholding is applied and all coefficients are used it is possible to obtain perfect reconstruction of the original signal $\{x(n)\}_{n=0}^{N-1}$. Properly chosen wavelet functions and threshold levels enable efficient extraction of low pass signal components and improvement of signal to noise ratio.

Fig. 16.6 together with Fig. 16.7 provides results of signal de-noising for a simulated sequence representing step function with an additive noise. In the case of step signal changes the Haar wavelet function can be used very efficiently with properly chosen threshold limits. Fig. 16.7 presents the sequence of these coefficients and limits used for signal processing. The same method can be applied for other signals and in many cases Daubechies wavelet functions can be used for signal de-noising in the similar way. Fig. 16.8 provides selected results of the real signal de-noising representing air pollution data. Such signal preprocessing is very useful for further signal modelling and prediction.

16.5 Signal Prediction

The problem of parameter estimation is very common in signal analysis, classification, prediction and system identification as well. In the simplest case it is possible to use the autoregressive (AR) model [8, 16, 17] of order N to describe a signal $\mathbf{x} = [x(0), x(1), \cdots, x(M+N-1)]$ by a vector of coefficients $\mathbf{a} = [a(0), \cdots, a(N-1)]'$.

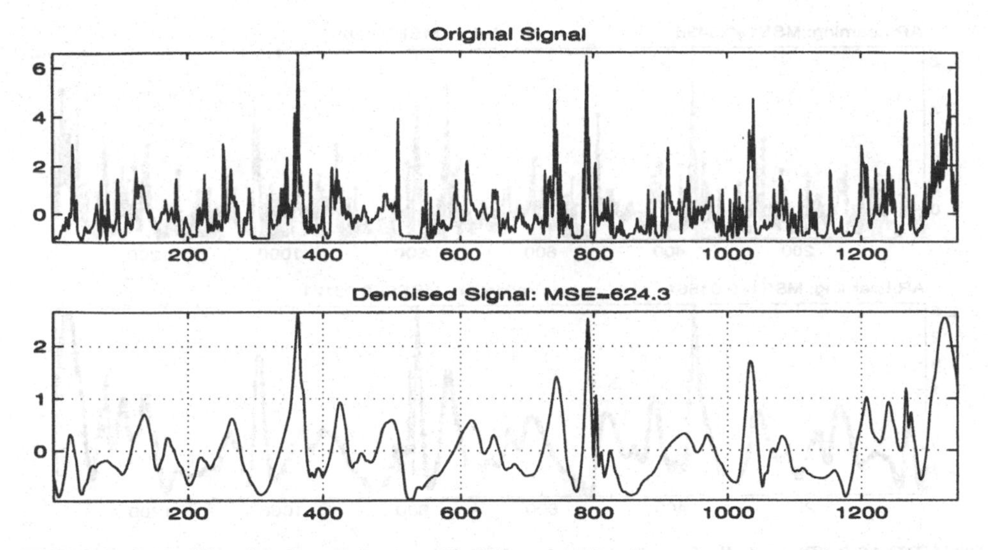

FIGURE 16.8. Wavelet de-noising applied for real signal representing a chosen segment of air pollution sequence with the sampling period of $T_s = 3$ hours processed by the Daubechies wavelet function of the 8th order

The problem of one step ahead prediction results in the solution of M equations of N variables $\mathbf{A}_{M,N}\,\mathbf{a}_{N,1}=\mathbf{y}_{M,1}$ in the form

$$\begin{bmatrix} x(N-1) & x(N-2) & \cdots & x(0) \\ x(N) & x(N-1) & \cdots & x(1) \\ & & \cdots & \\ x(N+M-2) & x(N-M-3) & \cdots & x(M-1) \end{bmatrix} \begin{bmatrix} a(0) \\ a(1) \\ \cdots \\ a(N-1) \end{bmatrix} = \begin{bmatrix} x(N) \\ x(N+1) \\ \cdots \\ x(N+M-1) \end{bmatrix} \quad (16.10)$$

The AR model order N can be estimated from the signal autocorrelation function as the number of its significant values [12]. Then it is possible to apply the singular value decomposition (SVD) for the linear regression model order selection and the following QR factorization for the selection of independent variables in the regression problem which make a fundamental contribution to the quality of the mathematical model. Such subset selection, described in [12, 26], is used in various applications of system identification and control and it allows the simplification of the autoregressive model especially in case of seasonal time series with significant periodic components.

Other methods of signal prediction are based on adaptive systems and non-linear models using artificial neural networks (ANN) in many cases. Let us assume the ANN with R inputs, $S1$ nodes of its first layer with non-linear (sigmoidal or wavelet) transfer function $F1$ and one output node with the linear transfer function. Applying the input column vector $\mathbf{P}_{R,1}$ to such ANN it is possible to evaluate its output

$$\mathbf{Y} = \mathbf{W2}{*}F1\,(\mathbf{W1} * \mathbf{P} + \mathbf{B1}) + \mathbf{B2} \quad (16.11)$$

where matrices $\mathbf{W1}_{S1,R}, \mathbf{W2}_{1,S1}$ stand for neural network coefficients and vectors $\mathbf{B1}_{S1,1}, \mathbf{B2}_{1,1}$ represent biases.

To achieve better of signal prediction it is useful in many cases to apply signal de-noising in the first step before the chosen prediction method is applied. Selected results of AR modelling are presented in Fig. 16.9 comparing that achieved both

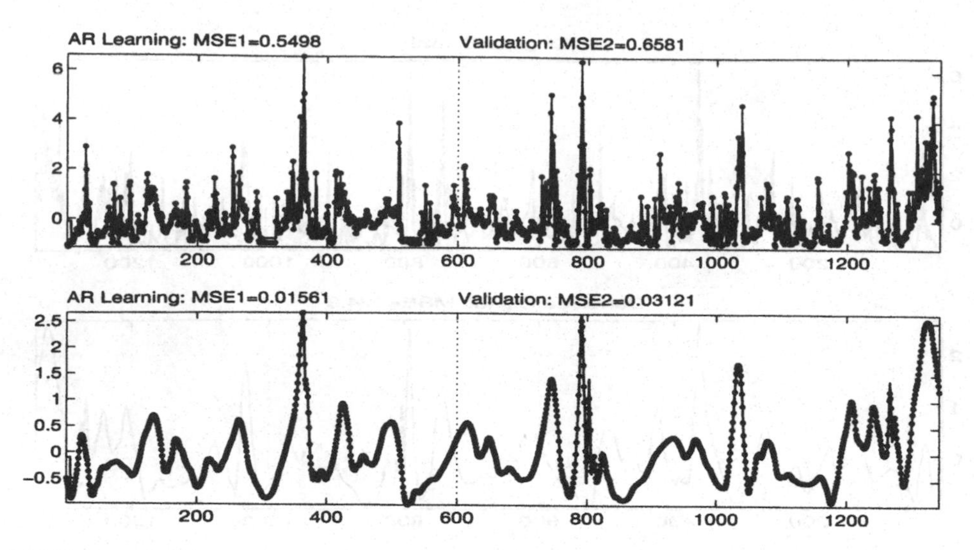

FIGURE 16.9. The prediction of a real signal with its sampling period $T_s = 3$ hours long for the original and de-noised signal using the autoregressive model of the 7th order

Type	*Original Signal*		*De-noised Signal*	
	Learning	*Validation*	*Learning*	*Validation*
AR Model	0.5498	0.6581	0.0156	0.0312
Tansig NN	0.4970	0.7728	0.0096	0.0606
Wavelet NN	0.5412	0.6598	0.0132	0.0328

TABLE 16.1. Mean square error of three hours air pollution prediction at the city of Most measuring station

for original signals and de-noised signals using wavelet signal decomposition described above. The mean square errors both in the learning and validation parts are summarized in Table 16.1 for various models and for original and de-noised signals. The experiments indicate that with signal de-noising in the preprocessing, similar prediction errors are achieved with AR models as are achieved with selected non-linear structures.

16.6 Conclusion

The contribution describes selected properties of the wavelet transform compared with that of the short-time Fourier transform. The main part of the paper presents

- the ability of the WT to decompose a given signal
- wavelet coefficient modification for noise rejection
- signal reconstruction

Methods and algorithm presented in the paper are applied for processing of a real signal representing air pollution to interpolate its missing values, to increase its signal to noise ratio and to predict its behaviour. Linear and non-linear methods are compared to find the most efficient method for given real signal modelling. Further problems include missing values restoration, analysis of wavelet de-noising methods, neural network transfer functions choice and their coefficients optimization for signal prediction.

Acknowledgments: The paper has been supported by the grant No. 201/94/0130 of the Grant Agency of Czech Republic. All real signals were kindly provided by the Czech Hydrometeorological Institute in Prague.

16.7 References

[1] P. Arena, S. Baglio, L. Fortuna, and G. Nunnari. Neural Networks to Predict Ozone Pollution in Industrial Areas. In *VIIIth European Signal Processing Conference EUSIPCO-96*. European Association for Signal Processing, 1996.

[2] I. Daubechies. The Wavelet Transform, Time-Frequency Localization and Signal Analysis. *IEEE Trans. Inform. Theory*, 36:961–1005, 1990.

[3] B. Delyon, A. Juditsky, and A. Benveniste. Accuracy Analysis for Wavelet Approximation. *IEEE Trans. on Neural Networks*, 6(2):332–348, 1995.

[4] D.L. Donoho. De-Noising by Soft-Thresholding. *IEEE Trans. on Information Theory*, 41(3):613–627, 1995.

[5] D.L. Donoho and I.M. Johnstone. Ideal Spatial Adaptation by Wavelet Shrinkage. *Biometrica*, pp. 425–485, 1994.

[6] W. Etter. Restoration of Discrete-Time Signal Segments by Interpolation Based on the Left-Sided and Right-Sided Autoregressive Parameters. *IEEE Trans. Signal Processing*, 44(5):1124–1135, 1996.

[7] M.T. Hagan and M.B. Menhaj. Training Feedforward Networks with the Marquardt Algorithm. *IEEE Trans. Neural Networks*, 5(6):989–993, 1994.

[8] M.H. Hayes. *Statistical Digital Signal Processing and Modeling.* John Wiley and Sons, Inc., New York, 1996.

[9] S. Haykin. *Neural Networks.* IEEE Press, New York, 1994.

[10] Y.T. Chan. *Wavelet Basics.* Kluwer Academic Publishers, Boston, 1995.

[11] E.S. Chng, S. Chen, and B. Mulgrew. Gradient Radial Basis Function Networks for Nonlinear and Nonstationary Time Series Prediction. *IEEE Trans. on Neural Networks*, 7(1):190–194, 1996.

[12] P.P. Kanjilal. *Adaptive Prediction and Predictive Control.* Peter Peregrinus Ltd., IEE, U.K., 1995.

[13] S.M. Kay. *Fundaments of Statistical Signal Processing.* Prentice Hall, 1993.

[14] M. Misiti, Y. Misiti, G. Oppenheim, and J.M. Poggi. *Wavelet Toolbox.* The MathWorks, Inc., Natick, Massachusetts 01760, 1996.

[15] D.E. Newland. *An Introduction to Random Vibrations, Spectral and Wavelet Analysis.* Longman Scientific & Technical, Essex, U.K., third edition, 1994.

[16] B. Porat. *A Course in Digital Signal Processing.* John Wiley and Sons, Inc., New York, 1997.

[17] J.G. Proakis and D.G. Manolakis. *Digital Signal Processing.* Prentice Hall, 1996.

[18] A. Procházka, M. Mudrová, and J. Fiala. Non-Linear Time Series Modelling and Prediction. *Neural Network World,* 6(2):215–221, 1996.

[19] A. Procházka, M. Sláma, and E. Pelikán. Bayesian Estimator Use in Signal Processing. *Neural Network World,* 6(2):209–213, 1996.

[20] A. Procházka and V. Sýs. Application of Genetic Algorithms for Wavelet Networks Signal Modelling. In *VIIth European Signal Processing Conference EUSIPCO-94,* pp. II/1078–II/1081. European Association for Signal Processing, 1994.

[21] A. Procházka and V. Sýs. Time Series Prediction Using Genetically Trained Wavelet Networks. In *Neural Networks for Signal Processing 3 - Proceedings of the 1994 Workshop,* pp. 195–203. IEEE, 1994.

[22] O. Rioul and M. Vetterli. Wavelets and Signal Processing. *IEEE SP Magazine,* pp. 14–38, 1991.

[23] G. Strang. Wavelets and Dilation Equations: A brief introduction. *SIAM Review,* 31(4):614–627, 1989.

[24] G. Strang and T. Nguyen. *Wavelets and Filter Banks.* Wellesley-Cambridge Press, 1996.

[25] M. Vetterli. Wavelets and Filter Banks: Theory and Design. *IEEE Trans. Signal Processing,* 40(9):2207–2232, 1992.

[26] M. Štorek and A. Procházka. Neural Networks in Signal Prediction. In *The First European Conference on Signal Analysis and Prediction ECSAP-97,* pp. 228–231. ICT Press, 1997.

[27] B.A. Whitehead and T.D. Choate. Cooperative-Competitive Genetic Evolution of Radial Basis Functions Centers and Widths for Time Series Prediction. *IEEE Trans. Neural Networks,* 7(4):869–880, 1996.

17

Self-Reference Joint Echo Cancellation and Blind Equalization in High Bit-Rate Digital Subscriber Loops

Iván A. Pérez-Álvarez[1]
José M. Páez-Borrallo[2]
Santiago Zazo-Bello[3]

ABSTRACT
The strong growth in DS1 (1.544 Mbit/sec) has led to recent intense research in high bit-rate digital subscriber lines (HDSL). Channel equalization and echo cancellation are required for a dual-duplex HDSL communication, but these systems have to be trained before they start to transmit the data. In this work we present a new scheme that permits the initialization (or re-initialization after an impulsive disturb) of a full-duplex HDSL data transmission system without using a known training data sequence at both ends of the line. It also estimate the impulse responses of both echo paths and channels jointly. As an example, we show the performance of the new proposed structure with a practical channel.

17.1 Introduction

During 1980's, the digital subscriber lines (DSL) were developed for ISDN basic rate access, applying adaptive equalization and echo cancellation techniques to the subscriber loop for first time. Adaptive filtering techniques are particularly suitable for loop applications because of the wide range of different channel characteristics encountered due to different cable makeups. The adaptive filters used in the DSL can track substantial changes in the channel through time. For example, due to changes in cable ambient temperature.

Nowadays, it is a fact the large market demand for full-duplex high-bit rate services on wire loops. The strong growth in DS1 (1.544 Mbit/sec) service, coupled with continuing speed and complexity advances in VLSI, has led to recent intense

[1]Dpto. de Señales y Comunicaciones, Universidad de Las Palmas de Gran Canaria, Campus Universitario de Tafira, 35017 Las Palmas, Spain, E-mail: ivan@masdache.teleco.ulpgc.es
[2]Dpto. de Señales, Sistemas y Radiocomunicaciones, ETS Ing. Telecomunicación, Universidad Politécnica de Madrid, Ciudad Universitaria s/n, 28040 Madrid, Spain, E-mail: paez@gaps.ssr.upm.es
[3]Universidad Alfonso X El Sabio, Villanueva de la Cañada, 28691 Madrid, Spain, E-mail: szazo@uax.es

research in high bit-rate digital subscriber lines (HDSL). Other solutions as fiber to the curp (FTTC) and fiber to the home (FTTH) technologies are the future solutions for the subscribers [19]. In order to achieve 1.544 Mbit/sec throughput, the length of the HDSL lines must be shorter than the DSL length, that is why the objective is the carrier serving area (CSA) which is carried out by 65-70% of local loop plant [1]. The solution adopted to deliver 1.544 Mbit/sec over CSA loops is based on the use of two loop pairs operating in duplex fashion. This architecture, referred to as 'dual-duplex', reduces the required throughput of a CSA loop from 1.544 Mbit/sec to around 800 Kbit/sec.

In order to allow each individual loop to operate in a full-duplex fashion, channel equalization and echo cancellation are required for a dual-duplex HDSL communication. But, before actual data are transmitted, the echo canceller and the equalizer have to be trained to mitigate leakage of the locally transmitted signal (echo) and to cope with intersymbol interference (ISI) caused by channel distortion. The reduction of this system initialization time is of great importance in the development of HDSL transmission techniques over the two-wire lines, and a part of the efficiency of the overall system depends on the communication's initial setup time.

Different authors have studied the techniques for joint adaptation of echo canceller and equalizer. Mueller [11] and Falconer [4] were the first to work in this field, but the convergence rates of their methods are very slow. Salz [18] and Cioffi [3] introduced least square based echo-cancellation algorithms achieving a factor improvement of 5 to 10 in the convergence speed over the classical LMS algorithm. Marcos and Macchi [10] analyze a transceiver structure for full-duplex data transmission and compare the performance for the coupled and decoupled structures (echo canceller and equalizer). More recently, Long and Ling [9] presented an echo canceller fast training scheme which estimates the coefficients by sending a special periodic sequence and correlating a segment of the sequence with the real echo samples.

The main drawbacks that we can find in those methods are, in general, that the use of the LMS algorithms during convergence can result in unacceptably long training, and especially if retraining is required. In case of faster converging techniques, like in [18, 3] and [9], some applications have serious drawbacks in the form of unacceptably high computational load. And, what is most important from the point of view of this work, in all these cases the start-up methods are operated in the half-duplex transmission mode.

To solve this problems, Chen [2, 7] has recently proposed a new method to reduce the system initialization time in high-speed full-duplex data transmission, which can simultaneously estimate the impulse response of echo paths and channels at both ends. To achieve this, during the start-up period, two mutually orthogonal periodic sequences are designed and used to co-estimate the near echo, the far echo and the channel response.

In this work we present a new scheme, for multilevel PAM baseband line code, that permits the initialization (or re-initialization after an impulsive disturb) of a full-duplex HDSL data transmission system without using a known training data sequence at both ends of the line. It also estimate the impulse responses of both echo paths and channels jointly. After an initial handshaking procedure, no especial sequences need to be sent by the transmitters at both ends, whereas the receivers can estimate the echo and channels paths using the self-reference schemes proposed in [14, 16].

The structure of this chapter is depicted as follows: firstly, in Section 17.2, we will review the main problems that we can find in the HDSL transceivers, with specific subsections to study the different problems that the error references used to drive the adaptive filters have. In Section 17.3, we will consider the study of a new identification filter and its properties, and we will see how some of the important problems mentioned in the previous section are overcome. In Section 17.4, we will show how this filter can be applied to two well known communications subsystems, echo and ISI cancellers. In Section 17.5, we present a new structure for a HDSL transceiver with joint cancellation and blind equalization which makes use of the subsystems presented in the previous section. We will finish this chapter with two last sections, one devoted to present some simulations of this transceiver in the CSA environments (Section 17.6) and another one devoted to conclusions (Section 17.7).

17.2 Review of the Main Problems in the HDSL Transceivers

In this section we are going to review the main problems that we can find in the HDSL transceivers. The nature of these problems are different. Some of them are characteristic of full-duplex communications and others appear with the use of dispersive channels. A different kind of problem come with the start-up procedures, which can introduce an important delay at the beginning of the communication. This problem can be overcome if we choose the appropriate technique, as we will see in the next sections. Finally other set of problems are derived from the type of error reference used to drive the different adaptive filters in the transceiver. To study these last problems we dedicate two specific subsections, one of them related with non-zero error reference and another with the zero error reference.

Fig. 17.1 shows a simplified version of a HDSL transceiver. In it, we find the classical adaptive subsystems, the echo canceller and the equalizer, as well as the main signals involved in the transmitting and receiving processes. We have the baseband transmitted signal represented by the near-end symbols, $\{I_N\}$. This signal goes to the line through the hybrid subsystem, which makes the connections between 2-wire and 4-wire transmission facilities. Due to the variation of loop makeups, the hybrid is an imperfectly matched electrical bridge. Therefore, part of the transmitted signal will appear at the receiving circuit as near echo, $\{N_e\}$. This is the **first problem** if we want to work in a full-duplex environment and an echo canceller is required to cancel this echo.

The far-end baseband signal $\{I_f\}$ is received through the 2-wire line contaminated by Gaussian noise, $\{N\}$, and intersymbol interference, $\{ISI\}$. Other impairments, out of the scope of this work, are manifested in form of other noises, like impulse noise, near-end crosstalk (NEXT) and far-end crosstalk (FEXT). The existence of $\{ISI\}$ is the **second problem** and an equalizer is required to remove it.

The **third problem** that we want remark here, although it has been viewed in the introduction section, is the procedure to train the different adaptive filters during the start-up time and the need of using special sequences to do it. The usual method to train the different adaptive filters present in the transceiver is to dedicate an initial time to work in half-duplex fashion. Thus, in a first stage, one transceiver is only dedicated to transmit a known sequence while the other one listen to it.

This situation allows the proper adjustment of the coefficients of the echo canceller in the transmitter transceiver, and the coefficients of the equalizer for the receiver transceiver. After that, in a second stage, the process is inverted and the rest of the adaptive filters are trained. Finally, in a third stage, both transceivers work in full-duplex fashion where, among several other things, they are dedicated to refine their adaptive filters. This whole process, which takes places at the beginning of a transmission or in a retraining stage takes an important amount of time that is desirable to reduce as much as possible.

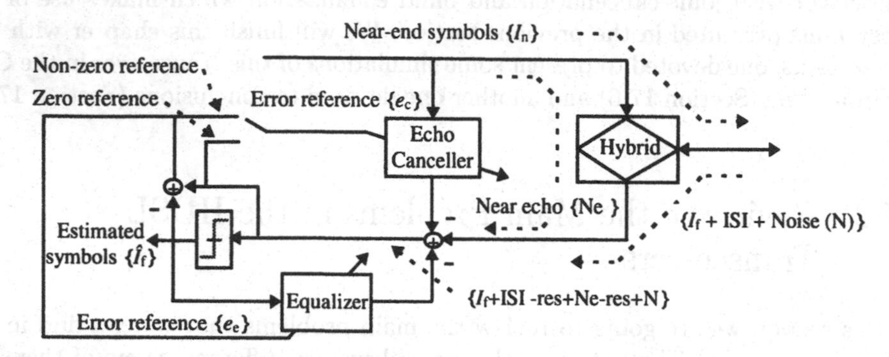

FIGURE 17.1.

17.2.1 *The Masking Effect on the Error Reference: The Zero Reference Case*

The signals generated by the echo canceller and the equalizer, which are estimates of $\{N_e\}$ and $\{ISI\}$ respectively, are subtracted from the incoming signal (Fig. 17.1). As a result of this operation we obtain a new signal composed by the sum of far-end signal $\{I_f\}$, residual echo $\{N_e\text{-}res\}$, residual intersymbol interference $\{ISI\text{-}res\}$ and noise $\{N\}$. Some classical adaptive echo cancellation schemes [4] (called zero-reference schemes) use this signal as the error reference for their adaptive algorithms. Therefore, the error reference used by the echo canceller is

$$e_c = I_f + N_e\text{-}res + ISI\text{-}res + N \tag{17.1}$$

The main problem in this case is that the true echo reference, N_e-*res*, is contaminated by other signals which are considered by the adaptive algorithm like interference signals.

In particular, the presence of the interference symbols, in this case the far-end symbols $\{I_f\}$, "mask" the true echo reference. This effect takes relevance in advanced stages of convergence, or from other point of view, when the residual error drops below the interference signal level, which masks the "natural" error reference of the adaptive algorithm, i.e. the residual echo. This masking effect forces the algorithm to slow down its speed of convergence towards the steady state value.

Another consequence of the presence of the interference symbols is that the probability density function (*pdf*) of the error reference used by the adaptive algorithm is a multimodal function, depending on the number of levels of the PAM far-end signal. A particular case for binary signal, [-1,+1], has been represented in Fig. 17.2, with an interference to error rate (IER) of 5 dB and where the $\{ISI\text{-}res\}$ and $\{N\}$ components have been neglected. In this case the *pdf* is a bimodal function.

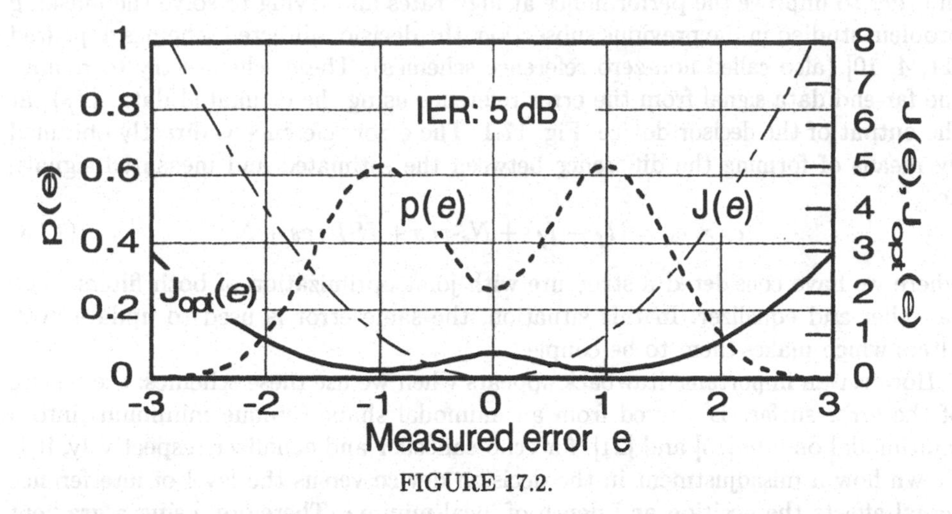

FIGURE 17.2.

In a general way, the cost function for adaptive filters used in standard data communications systems is based on some absolute moment of the observed error, e.g. $|e|$ for LMA or e^2 for LMS. They are not always the best measurement we can perform since their *pdf* may not fit, depending on the convergence state, with the minima position's of the cost function. In [8] it is shown that the optimal criterion or error cost function for a recursive algorithm coincides also with that of the maximum likelihood estimate (MLE), i.e.

$$J_{opt}[e] = -\log[p(e)] \tag{17.2}$$

where $p(e)$ is the *pdf* of the observed error. This situation is represented in Fig. 17.2, for the binary signal case, where the solid line represents the J_{opt}, whereas the dashdot line represents the cost function for the LMS, $J[e] = e^2$. In this situation, when the IER starts to be high ($|I_f| \gg |N_e\text{-}res|$), the minimum of the cost function for the LMS algorithm has little information about the true error, N_e-*res*. This minimum is placed in front of a minimum of the *pdf* and consequently the frequency of observing a proper error reference diminishes. Observe how the optimal error criterion (17.2) deviates from the quadratic one, e^2, as long as the IER increases. In other words, the signal component in the observed error is masking more and more the residual error during the convergence period.

To avoid this undesired behaviour during an important period of convergence time, some direct schemes based on the improvement of error reference can be found in the literature. Among others we point up e.g. a) adding an uncorrelated dithering signal to the error reference [6], b) training periods in a half-duplex start-up phase, or c) removing part of the incoming signal from the error reference using estimations of the far-end symbols, decision-directed schemes [11]. Because of its importance, we are going to review in the next subsection the problems that this scheme has.

17.2.2 The Risk of a Multimodal Error Surface: The Non-Zero Reference Case

In order to improve the performance at high rates and trying to solve the masking problem studied in the previous subsection, the decision-directed schemes appeared [11, 4, 10] (also called non-zero reference schemes). These schemes try to remove the far-end data signal from the error reference using the estimated data, $\{\hat{I}_f\}$, at the output of the decisor device, Fig. 17.1. The error reference is directly obtained by means of forming the difference between the estimated and measured signals, i.e.

$$e_c = e_e = (I_f - \hat{I}_f) + N_e\text{-}res + ISI\text{-}res + N \tag{17.3}$$

where we have considered a structure with joint optimization of both filters, echo canceller and equalizer. In this situation, the same error is used to update both filters which makes them to be coupled.

However an important drawback appears when we use these schemes: the nature of the error surface is altered from an unimodal shape (unique minimum) into a multimodal one. In [15] and [21], for echo canceller and equalizer respectively, it is shown how a misadjustment in the decision device versus the level of interference signal affects the position and depth of local minima. Therefore, using a gradient search of the minima for these non-globally convex error surfaces, at least as it is known for the quadratic algorithm in the self-adaptive equalization techniques, there exists some risk of convergence towards a local minimum for some unconsidered situations: i.e., when a fixed adaptation step is used, the eye pattern of the incoming data signal is not sufficiently open, or when an incorrect initial tap settings is taken.

It is our objective to introduce a new adaptive filter that overcomes the problems already presented in this section. In particular, those related with error reference and with start-up period.

17.3 The New Identification Filter and its Properties

Fig. 17.3 shows an identification scheme which differs from the classical one, [5], in two aspects: the presence of the interference symbols $\{I\}$ in the residual error $\{\varepsilon\}$, and the modification performed on the standard error reference, $\{e = I + \varepsilon\}$, before being used by the adaptive algorithm to be transformed in a new error reference $\{e_n\}$.

In order to overcome the problem of the interferent signal studied in Section 17.2, we can exploit some statistical characteristics of interferent signal. That is, we can use the discrete nature of the interference signal in order to self-cancellate the masking effect in the current error reference $\{e\}$ by using a delayed version of itself. To provide a better understanding of the proposed solution, we will first study the easier case: the binary case.

17.3.1 A Simplified Situation: The Binary Case

As we have mentioned above, our objective is try to gain advantage of the symmetrical 2-level nature of the interference signal to improve the error reference that

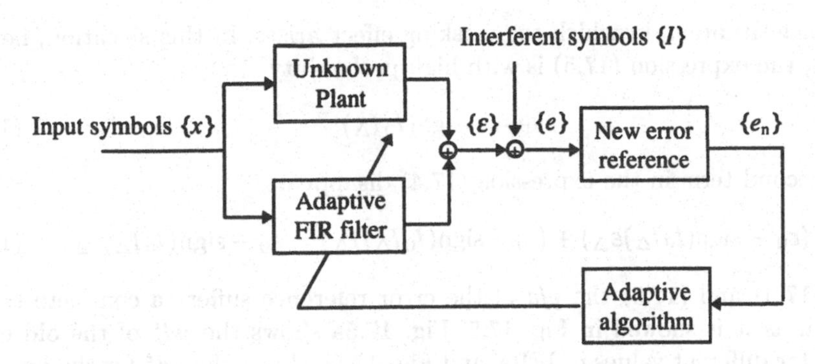

FIGURE 17.3.

drives the adaptive algorithm. The idea, first introduced in [6], extended in [12], and full developed and analyzed in [13], is supported by the fact that the interferent signal, sampled at symbol rate with the correct sampling phase, takes only two possible and opposite values, i.e. [-1,+1]. Therefore, a delayed (not too retarded) error reference sample might be a good replica of the current sample level if we don't take into account its sign. This new situation offers us the possibility of cancelling, if the sign problem is solved, the masking effect in the old error reference and, thus, providing a new and cleaner error reference. In Fig. 17.4 we show the internal structure of the "New error reference" box of Fig. 17.3.

In mathematical terms the new error reference for binary case is

$$e_n = e_0 \oplus e_\Delta = (\varepsilon_0 \oplus \varepsilon_\Delta) + (I_0 \oplus I_\Delta) \tag{17.4}$$

where the subindexes 0 and Δ mean current and delayed sample respectively, and $\oplus$ is the sign control operation represented in Fig. 17.4, which commands the delayed sample sign.

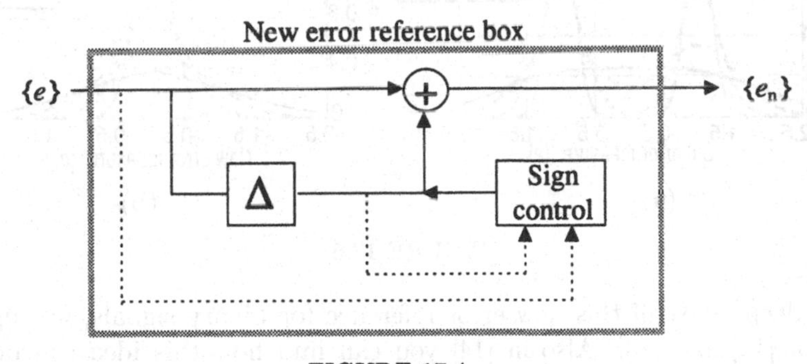

FIGURE 17.4.

The sign control operation must exploit the knowledge about the sign in the current and delayed sample to select the appropriate sign (Fig. 17.4). This can be achieved using the next expression

$$\oplus = -\text{sign}(e_0 e_\Delta) \tag{17.5}$$

which is a stochastic sign that forces the algorithm to self-cancellate the effect of the previous signal independently of its sign. This happen if we are in an advanced convergence situation, in which the masking effect arises. In this situation, i.e. for high IER, the expression (17.5) is with high probability

$$\oplus \approx -\text{sign}(I_0 I_\Delta) \tag{17.6}$$

and the second term in the expression (17.4) disappears

$$e_n \approx (\varepsilon_0 - \text{sign}(I_0 I_\Delta)\varepsilon_\Delta) + (I_0 - \text{sign}(I_0 I_\Delta) I_\Delta) = \varepsilon_0 - \text{sign}(I_0 I_\Delta)\varepsilon_\Delta \tag{17.7}$$

With (17.4) and (17.5) the ***pfd*** of the error reference suffers a complete transformation, as it is showed in Fig. 17.5. Fig. 17.5a shows the ***pdf*** of the old error reference for different values of IERs, and Fig. 17.5b shows the ***pdf*** for the new one with the same IERs.

A set of important conclusions can be obtained from these figures. The first one is that the new error reference transforms the ***pdf*** from a bimodal function to a symmetric and unimodal one. The second one is that this new ***pdf*** is nearly Gaussian. These two conclusions are very important. They mean that the new error is a better reference than the old one to be used by a cost function of the type $J(e) = |e|^k$, and in particular by $J(e) = e^2$ (LMS). It is better from the point of view that we saw in Section 17.2.1. The third conclusion is that this behaviour, symmetry and unimodality of ***pdf***, remains not only for high IERs but for moderate IERs also. Finally, other important property must be pointed up in this self-cancelling algorithm. It doesn't have the far-end level signal problem that we saw in Section 17.2.2 for the non-zero reference algorithms. It overcomes this problem obtaining the signal level information from the own signal, in its delayed sample.

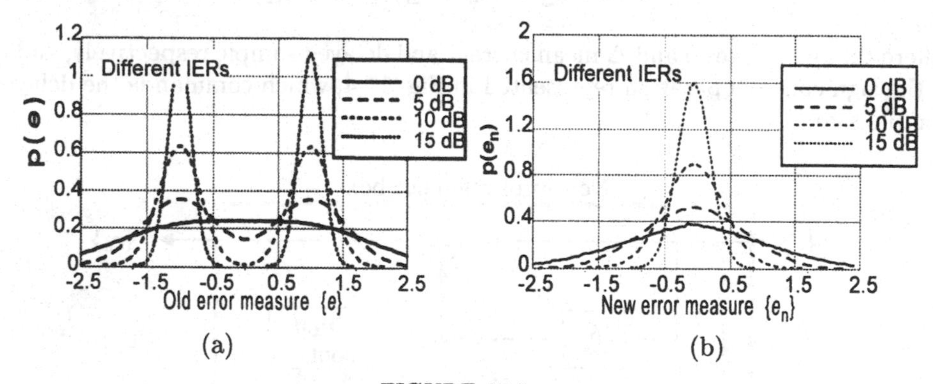

FIGURE 17.5.

For a deep study of this new error reference for binary signals, we suggest the lecture of [12] and [13]. Also in [14] you can find how this idea can be applied successfully to constant modulus signals (PSK) in the case of blind equalizers.

17.3.2 A Generalized Error Reference for M-PAM Signals

The main drawback of the new error reference presented in the previous subsection, (17.4) and (17.5), is the degradation of algorithm's performance as the number of PAM levels increase. Let us observe that unless I_0 and I_Δ have identical amplitudes the masking effect is not removed, or from another point of view, the corresponding ***pdf*** is still multimodal. For a 4-PAM, or 2B1Q line code, the ***pdf*** of $\{e\}$ is modified from a four-modal function to a three-modal one, with a zero-level reference probability equal to 50%. For a 8-PAM the *pdf* is modified from an eight-modal function into a seven-modal one, with a zero-level reference probability of 25%. And so on. This leads to the next conclusion: if the number of levels increases the probability of the zero-level in the new reference decreases, and also the probability of unmasking the interference signal too.

To avoid this degradation in the error criteria, in [16] was proposed a new error reference which uses a delayed and **weighted** version of the error observation, $\{e\}$. This new error reference is

$$e_n = \left\{ e_i / \min_i \left(|e_i| \right) \right\} \quad \text{where} \quad e_i = e_0 - \beta_i \text{sign}(e_0 e_\Delta) e_\Delta \tag{17.8}$$

where e_Δ is a delay version of error, e_0 is the current one and factors β_i

$$\beta_i = m_i / l_i \quad \text{with} \quad m_i, l_i = 1, 3, ..., M-1 \tag{17.9}$$

all possible amplitude ratios for the corresponding M-PAM data. For instance, for a 4-PAM, $\beta_i = \{1/3, 1, 3\}$. In (17.8), we have neglected the absolute time dependency of the signals, but we remark the relative delay in the subindexes, 0 and Δ.

17.3.2.1 *Pdf* of the New Error Reference

In this new error reference, factor β_i weights the delayed error signal to adjust its level to the current one. That is, in an advanced stage of convergence

$$e_i \approx \text{sign}(|I_0|)(|I_0| - \beta_i |I_\Delta|) + (\varepsilon_0 - \text{sign}(I_0 I_\Delta) \beta_i \varepsilon_\Delta) \tag{17.10}$$

where with high probability, $\beta_i = |I_0 / I_\Delta|$, and thus the far-end signal is cancelled, the residual error is unmasked and the error reference becomes

$$e_n \approx \varepsilon_0 - \text{sign}(I_0 I_\Delta) \beta_i \varepsilon_\Delta = \varepsilon_0 - (I_0 / I_\Delta) \varepsilon_\Delta \tag{17.11}$$

With this new error reference, we obtain a complete transformation of the error *pdf*. As an example of what we have stated, we present in Fig. 17.6 the $\{e\}$ and $\{e_n\}$ *pdf* for a 4-PAM signal and IERs in the range of -40dB to 40dB. Fig. 17.6b shows the $\{e_n\}$ *pdf* for the new error. Observe the unimodal character of all curves for $\{e_n\}$, even in the case of large IERs, opposite to the case of $\{e\}$ where we can see the multimodal nature of the *pdf* for high IERs (when $|I| \gg |\varepsilon|$).

The same conclusions obtained for the binary case can be applied with this new error reference, with a very important improvement. When the residual error $\{\varepsilon\}$ is still large enough compared with the interferent signal (earliest state of convergence, when $|I| \ll |\varepsilon|$), the new error reference losses its characteristic Gaussian shape, but maintaining the unimodal character. This is due to the fact that expression $\text{sign}(e_0 e_\Delta)$ will be, with high probability, equal to $\text{sign}(\varepsilon_0 \varepsilon_\Delta)$ and the effective error reference (17.8) becomes in this stage

$$e_n = \left\{ e_i / \min_i \left(|e_i| \right) \right\} \quad \text{where} \quad e_i \approx |\varepsilon_0| - \beta_i |e_\Delta| \tag{17.12}$$

that is, e_n is just a linear combination of two Gaussian random variables. Observe in Fig. 17.6b that the corresponding *pdf* shapes for (17.12) are better approximated by exponential functions that Gaussian ones. This would recommend a sign-LMS algorithm (LMA) for this early convergence stage.

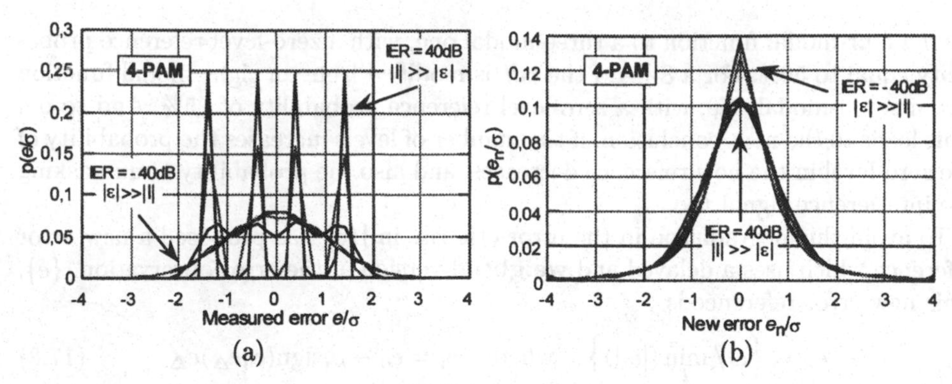

FIGURE 17.6.

17.3.2.2 Convergence Analysis in the Extreme Cases

The new adaptive algorithm takes the generic form of (see Fig. 17.3)

$$\mathbf{w}(n+1) = \mathbf{w}(n) + \mu g[e(n)]\mathbf{x}(n) \tag{17.13}$$

where $g[e] = e_n$ in the new algorithm (17.8) and $g[e] = e$ for the classical LMS. In any case, subtracting from (17.13) the optimal solution $\mathbf{w}^*$, multiplying (17.13) by its transpose and assuming an independent input sequence $\{x\}$, it is possible to obtain the following scalar convergence equation for the residual error variance [15, 13]

$$\sigma_\varepsilon^2(n+1) = \sigma_\varepsilon^2(n)\left[1 - \mu S_1(n) + \mu^2 S_2(n)\right] = \sigma_\varepsilon^2(n) P(\mu, n) \tag{17.14}$$

where in general

$$S_1 = 2\sigma_x^2 \frac{E\{\varepsilon(n) g[e(n)]\}}{\sigma_\varepsilon^2}, \quad S_2 = N\sigma_x^4 \frac{E\{g^2[e(n)]\}}{\sigma_\varepsilon^2} \tag{17.15}$$

and where $E\{\bullet\}$ denotes the expectation operation and N is the FIR filter length.

The factor $P(\mu, n) = 1 - \mu S_1(n) + \mu^2 S_2(n)$, a quadratic polynomial in μ, is the responsible for the algorithm's convergence and performance. In fact, it has to be bounded by $0 < P(\mu, n) < 1$ during the convergence period and it should equal to 1 in the steady state ($n = \infty$).

The new error reference introduces some difficulties in the convergence analysis of the adaptive algorithm since it depends on both, the current and delayed residual error, and it incorporates the minimum function. To simplify this analysis we will assume that the interference signal is not corrupted by other signals, i.e. $e = I + \varepsilon$, the residual error $\{\varepsilon\}$ is a Gaussian random processes with zero mean and independent of $\{x\}$, and the sequences $\{x\}$ and $\{I\}$ are i.i.d.

With these assumptions it is straightforward the following expression for the LMS algorithm, where $g[e] = e = I + \varepsilon$,

$$P(\mu, n) = 1 - 2\mu\sigma_x^2 + \mu^2 N\sigma_x^4\left(1 + \frac{E\{x^2\}}{\sigma_\varepsilon^2(n)}\right) \tag{17.16}$$

where in the steady state

$$P(\mu, \infty) \quad = \quad 1 = 1 - 2\mu\sigma_x^2 + \mu^2 N\sigma_x^4\left(1 + \frac{E\{x^2\}}{\sigma_\varepsilon^2(\infty)}\right) \Rightarrow$$

$$\Rightarrow \quad \mu = \frac{2\sigma_\varepsilon^2(\infty)}{N\sigma_x^2\left(\sigma_\varepsilon^2(\infty) + E\{x^2\}\right)} \approx \frac{2\sigma_\varepsilon^2(\infty)}{N\sigma_x^2 E\{x^2\}} \tag{17.17}$$

that is, the adaption step μ is directly proportional to the final residual error variance or inversely to IER $= E\{x^2\}/\sigma_\varepsilon^2(\infty)$.

To obtain a similar expression for our new algorithm, we split the problem into two:

1. Case of high IER (in advanced state of convergence)
   Considering e_n given in (17.11), $\sigma_\varepsilon^2 = \sigma_0^2 \approx \sigma_\Delta^2$ and statistical independence between ε_0 and ε_Δ, we can write

$$S_1 = \frac{2\sigma_x^2}{\sigma_\varepsilon^2}\left[E\{\varepsilon_0^2\} - E\left\{\frac{I_0}{I_\Delta}\right\}E\{\varepsilon_0\varepsilon_\Delta\}\right] = 2\sigma_x^2 \tag{17.18}$$

   and

$$\begin{aligned} S_2 &= \frac{N\sigma_x^4}{\sigma_\varepsilon^2}\left[E\{\varepsilon_0^2\} + E\left\{\left(\frac{I_0}{I_\Delta}\right)^2\right\}E\{\varepsilon_\Delta^2\} - 2E\{\varepsilon_0\varepsilon_\Delta\}\right] = \\ &= N\sigma_x^4\left[1 + E\{I_0^2\}E\left\{\frac{1}{I_\Delta^2}\right\}\right] \end{aligned} \tag{17.19}$$

   therefore, forcing $P(\mu, \infty) = 1$, we obtain for μ

$$\mu = \frac{2}{N\sigma_x^2\left(1 + E\{I_0^2\}E\left\{\frac{1}{I_\Delta^2}\right\}\right)} \tag{17.20}$$

   an adaption step **independent** on the final IER. In other words, in the hypothetical situation of absence of other additive interference, this algorithm will make the residual error variance converge to zero as long as the adaption step is below (17.20).

2. Case of low IER (in the initial state of convergence)
   In this situation the error reference is given by (17.12). For this error

$$\begin{aligned} S_1 &= \frac{2\sigma_x^2}{\sigma_\varepsilon^2}\left[E\{\varepsilon_0^2\} - \sum_i \beta_i \int_0^\infty \varepsilon_0 f_0(\varepsilon_0) \int_{a_i\varepsilon_0}^{b_i\varepsilon_0} \varepsilon_\Delta f_\Delta(\varepsilon_\Delta)\, d\varepsilon_\Delta d\varepsilon_0\right] = \\ &= 2\sigma_x^2\left[1 - \frac{2}{\pi}\sum_i \beta_i\left(\frac{1}{1+a_i^2} - \frac{1}{1+b_i^2}\right)\right] \end{aligned} \tag{17.21}$$

   where $a_i = (\beta_{i+1} + \beta_i)/2$ and $b_i = (\beta_{i-1} + \beta_i)/2$. For S_2 we obtain after some similar calculus

$$\begin{aligned} S_2 &= N\sigma_x^2\left[1 + \frac{2}{\pi}\sum_i \beta_i^2\left(\frac{1}{1+a_i^2} - \frac{1}{1+b_i^2} + \tan^{-1}(b_i) - \tan^{-1}(a_i)\right)\right. \\ &\quad \left. -\frac{4}{\pi}\sum_i \beta_i\left(\frac{1}{1+a_i^2} - \frac{1}{1+b_i^2}\right)\right] \end{aligned} \tag{17.22}$$

observe that, for this situation, S_1 and S_2 are again **independent** from the current IER. It means that the converging factor $P(\mu, n)$ takes two different and constant values at the beginning and at the end of the transient period. It implies, in fact, two different speeds of convergence and a transition region in between.

Fig. 17.7 shows this effect for a 4-PAM signal and normalized values of $N = 1$ and $\sigma_x^2 = 1$. Here we have chosen a μ which guarantees a IER of 40 dB in the LMS algorithm. Also, for comparation purpose only, the same μ has been chosen in the new algorithm since, otherwise, the new algorithm would allow to choose a higher adaptation step due to the minimum function in the new error reference (17.8).

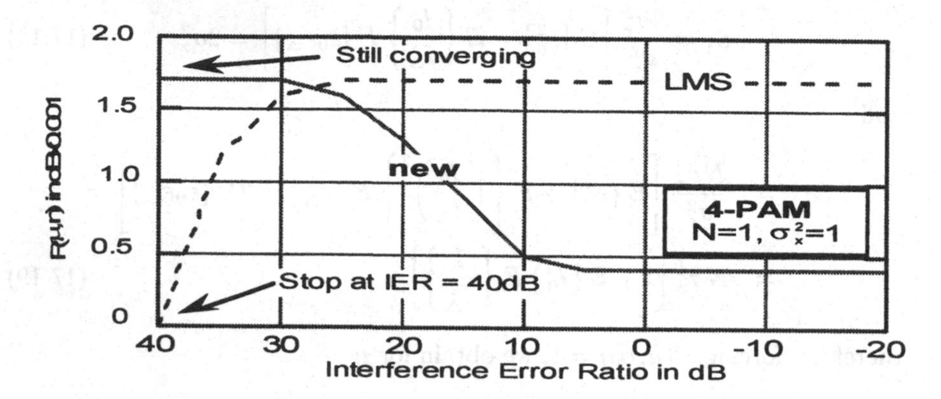

FIGURE 17.7.

Observe how the speed of convergence in the LMS algorithm goes to zero for IERs approaching 40 dB and how the corresponding curve for the new algorithm presents two stable and constant speed of convergences at the beginning and at the end of the analyzed IER region, with a smooth transition in between. We can also observe that the speed of convergence is higher for high IERs. This is mainly due to two facts: having a better, totally unmasked, error reference and the Gaussian shape of their *pdf*, which is optimal for the cost function chosen in (17.13), that is $J[e] = e_n^2$.

17.4 Applying the New Identification Filter to Communications Subsystems

The new algorithm presented in Section 17.3 and the structure of Fig. 17.3 can be applied to different subsystems in the field of communications. Thus, we can use it to identify the hybrid subsystem, as it was proposed in [16], and to identify the different parts, precursor and postcursor, of the channel as it was proposed in [17]. In other words, we can use the basic structure to implement the echo canceller and the ISI-canceller in the transceiver of a digital baseband communication system.

In the next two subsections we are going to see how the basic structure of the Fig. 17.3 carry out the implementation of these subsystems.

17.4.1 Echo Canceller

In this case, the structure presented in the Fig. 17.3 can be applied directly, being the only difference the meaning of the signals involved. Fig. 17.8 shows this subsystem where now the input signal is the near-end symbols $\{I_n\}$, the unknown plant is the hybrid, and the interferent signal is composed of far-end signal $\{I_f\}$, inter-symbol interference $\{ISI\}$ and Gaussian noise $\{N\}$. Finally, the adaptive FIR filter is the echo canceller.

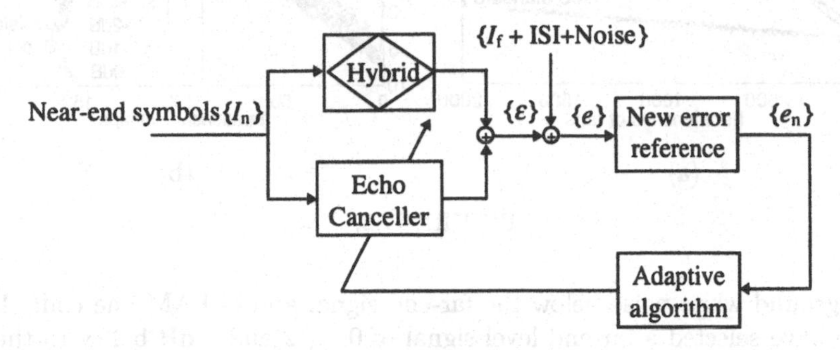

FIGURE 17.8.

Observe that the only difference with the classical structure of a transceiver presented in Fig. 17.1 is that in real situations the hybrid is a four ports subsystem, including the matching impedance port. In the model of Fig. 17.8 the hybrid is presented like a two-ports subsystem and the far-end signal, and the rest of the impairment signals, are added directly to the residual echo $\{\varepsilon\}$.

17.4.1.1 Comparisons with Zero and Non-Zero Reference Algorithms

Before we see the performance of the new identification filter integrated in a complex system as it is a transceiver (Section 17.6), it is interesting to dedicate a few lines to compare the new algorithm, in the simple echo canceller case, with the standard ones, and verify how some of the problems pointed up in Section 17.2 are overcome. We present two computer simulations to compare its results with the corresponding LMS (zero-reference) and the stochastic gradient decision-directed (non-zero reference) algorithms.

Fig. 17.9a displays the results for the new and LMS algorithms in an echo canceller subsystem of $N = 6$ coefficients, a hybrid with exponential response, working in an environment of -80dB of background white noise below the far-end signal and 8-PAM line code. The new algorithm has been started with four different initial settings and the adaptation step μ for LMS standard is set for a steady state of Signal to Echo Ratio of 35dB (SER = 35dB). A different and higher μ is used for our algorithm.

Observe the great difference in the convergence speed, due to the unmasking effect, and the lack of sensitivity to the initial settings in the new algorithm. Observe, also, the two different periods of convergence (2 slopes) before reaching the noise region (-80 dB). This is the expected behaviour of the convergence after the analysis carried out in Section 17.3.2.2 and the results obtained in Fig. 17.7.

Fig. 17.9b displays the results for the new and decision-directed algorithms in an echo canceller subsystem of $N = 6$ coefficients working in an environment of -35dB

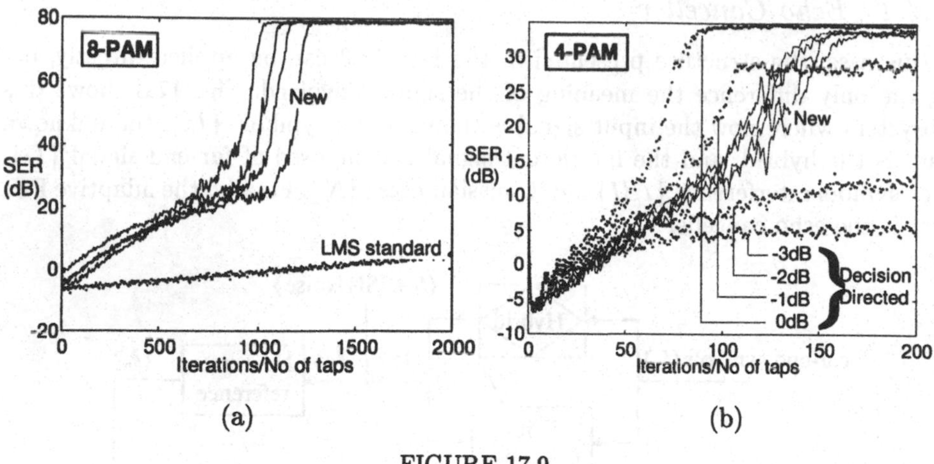

FIGURE 17.9.

of background white noise below the far-end signal and 4- PAM line code. In this case we have selected a far-end level signal of 0, 1, 2 and 3 dB below to the level expected by the decision device.

We can see the good performance of the new algorithm against the level misadjustment and the degradation of the decision-directed algorithm when the far-end level is more than 1 dB below what is expected by the decision device. Obviously the 0 dB case is an ideal situation. It is very difficult to achieve this in a real case because the CAG subsystem, which must be present to guarantee a correct level at the slicer's input, can not distinguish the far-end signal from the rest of the signals at the slicer's input, which is composed by far-end symbols, inter-symbol interference, residual echo and noise (see Fig. 17.1). Opposite to this behaviour, the new algorithm is insensitive to the far-end level signal as it was remarked in Sections 17.3.1 and 17.3.2.

17.4.2 ISI Canceller

For the ISI-canceller case the structure presented in Fig. 17.3 suffers some transformations. Thus, in Fig. 17.10 is presented the ISI-canceller structure based on the new identification filter.

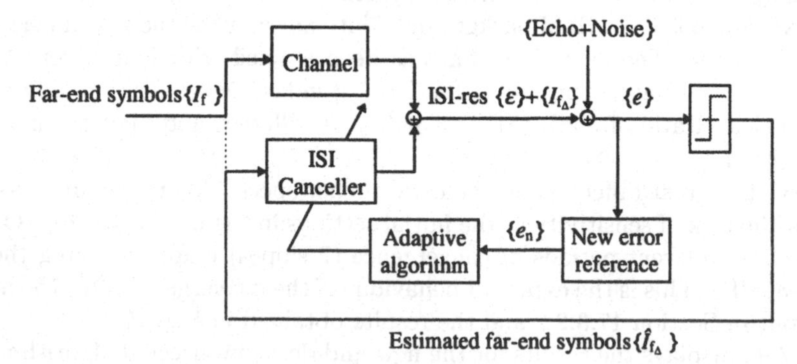

FIGURE 17.10.

In this case the input signal to the unknown filter, the channel, are the far-end symbols $\{I_f\}$. In an ideal situation, those symbols would be the input to the ISI-canceller too. The problem is that we have not access to these far-end symbols and, therefore, we must use the estimated far-end symbols $\{\hat{I}_{f\Delta}\}$ at the slicer's output. This is the main change in relation to the structure presented in Fig. 17.3, although other modifications will be introduced depending on the part of the channel we want to identify.

In a general case, we must suppose that we have a non-minimum phase channel. Therefore, the main symbol at the channel's output will be delayed a Δ value in relation with the channel input symbol. This situation means that the ISI-canceller input is also delayed with respect to the channel input the same Δ value and, therefore, the ISI-canceller can only identify the postcursor part of channel. In Fig. 17.11 we show an equivalent structure to Fig. 17.10, that help us to understand this in a better way.

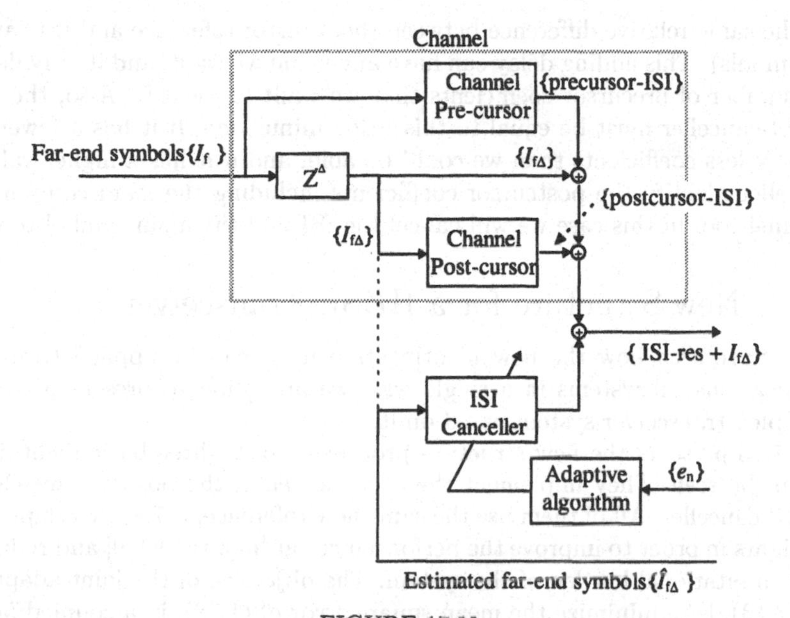

FIGURE 17.11.

Fig. 17.11 shows the channel splitter in two filters: the precursor channel and the postcursor channel. Let us observe how the ISI-canceller's input is synchronized with the postcursor filter's input. The precursor ISI is considered by the adaptive algorithm like just other disturbing signal. That is why it only identify the postcursor channel's coefficients.

To identify the precursor coefficients, we must advance somehow the input of our ISI-canceller in relation to the error reference. Obviously, we can not use the real far-end symbols, but we are allow to introduce a delay between channel's output and error reference, as it is showed in Fig. 17.12.

To understand this in a better way let us observe how, if we had have the real far-end symbols at the ISI-canceller's input, then the error reference would be delayed with respect them. This delay will be exactly Δ (see Fig. 17.10). Thus, as what we have are the symbols estimated at the slicer's output with a delay Δ in relation to the far-end symbols, we add a controlled delay in the error reference in order to

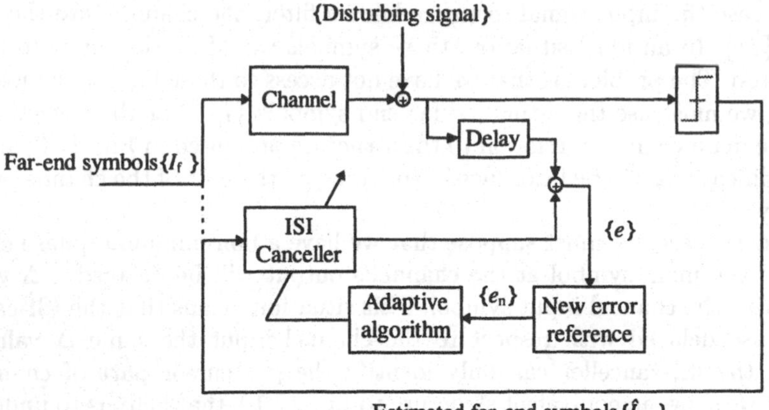

FIGURE 17.12.

obtain the same relative difference between them (error reference and ISI canceller input symbols). This adding delay can have any value we want, and it only depends on the number of precursor coefficients that we want to identify. Also, the length of the ISI-canceller must be equal to this value minus one. If it has a lower value we identify less coefficients than we could be able, and if it has a higher value the ISI-canceller identify the postcursor coefficients including the main component of the channel too. In this case we will cancel the ISI and the main symbol too.

17.5 A New Structure for a HDSL Transceiver

Once we have seen how the new identification filter can be applied to different communications subsystems in a single way, we are going to present a complete and complex transceiver system based on it.

Fig. 17.13 presents the new structure proposed. In it, three basic identification filters can be seen. They implement the echo canceller, the postISI canceller and the preISI canceller. All of them use the same new reference, (17.8), to adapt jointly their weights in order to improve the performances at high rates [10] and reduce the global computational burden of the system. The objective of the joint adaptation, using (17.13), is to minimize the mean square error of (17.8), in a coupled fashion.

All delay boxes have the same value (Δ) and, although the designer is free to choose any value, the optimum one is equal to the length of the preISI canceller plus one, as it was seen in Section 17.4.2.

In case of the echo canceller, the only difference respect to what we saw in Section 17.4.1 is the delay introduced in the vector signal of adaptive the algorithm (17.13). The objective is to synchronize this vector signal with the error reference which is also delayed. The same situation happens with the postISI canceller where we must introduce a delay box to synchronize the vector signal and the error reference. Finally, the preISI canceller has the same structure that it was discussed in Section 17.4.2.

As it has been presented in [16], and as we have seen here, the new identification filter of Fig. 17.3 does not need any especial training sequence to update his weights. That is the reason why, we can say that Fig. 17.13 implements a self-reference joint echo cancellation and blind equalization structure.

Finally, the constant presence of preISI and noise at the input of the slicer that

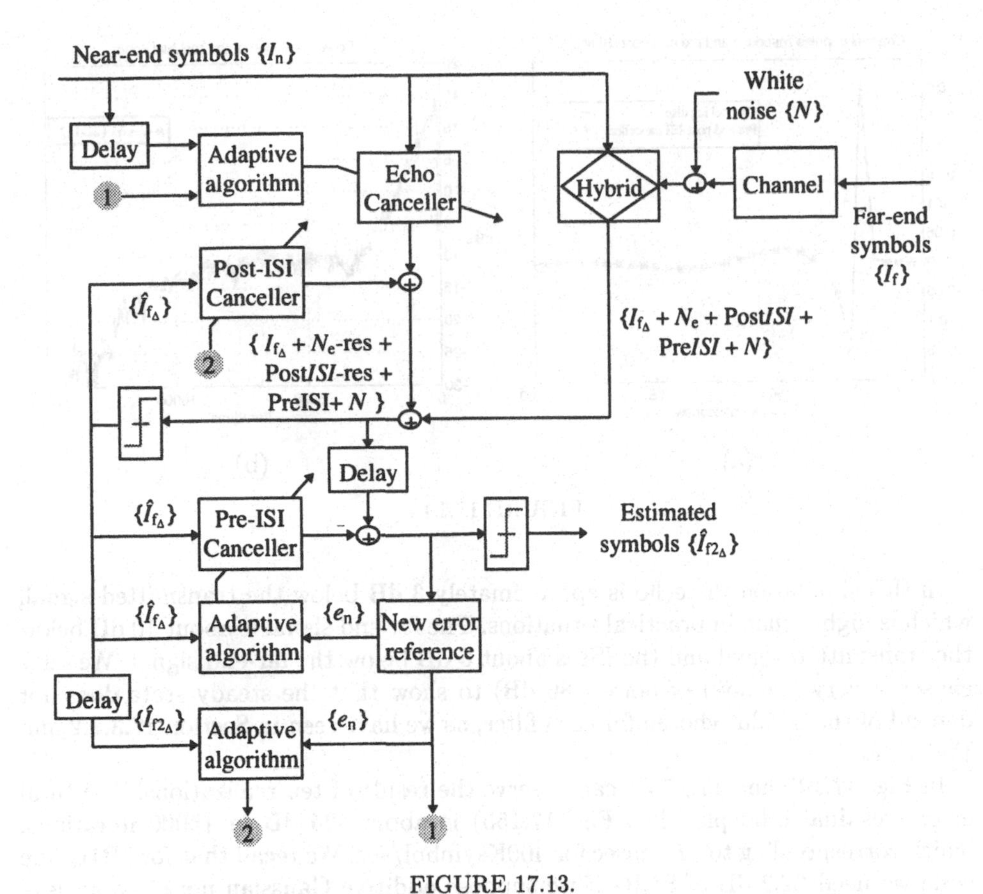

FIGURE 17.13.

feed the ISI canceller introduces the risk of many erroneous estimated symbols $\{\hat{I}_{f\Delta}\}$. However, in real cases, as we are going to see in next section, the anticausal part of HDSL loops proposed by the CSA models is minimal compared with the causal part, and the probability of a correct decision is high because the eye-pattern is enough opened. Other structures are possible, [17], but in general, the main problem that remains unsolved for future works is that the preISI always affects to some filters. That is the reason why the transceiver shown in Fig. 17.13 is valid only in moderate preISI environments.

17.6 Simulation in CSA Environment

In this section we present a computer simulation to show the performance of the new structure (Fig. 17.13) in HDSL loops. In Fig. 17.14a we show the impulse response of the channel which corresponds to a CSA cooper loop of 12Kft and 0.4mm of diameter with the effects of the line transformers included. We have chosen the same hybrid for the echo canceller that we chose in Section 17.4.1.1.

The line code chosen has been 4-PAM (2B1Q) and the received signal has been filtered by a highpass filter $(1 - Z^{-1})$ in order to mitigate the presence of the near zero pole introduced by the transformers [20]. The baud rate is 400Ksymbol/sec, or 800Kbit/sec, which corresponds with the dual-duplex solution to deliver 1.544 Mbit/sec over CSA loops with two loop pairs operating in duplex fashion.

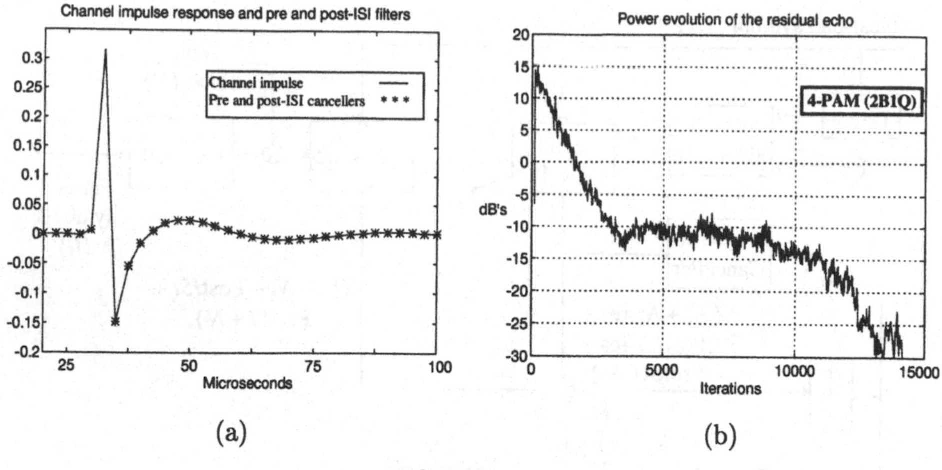

(a) (b)

FIGURE 17.14.

In this simulation the echo is approximately 3 dB below the transmitted signal, which is higher than in practical situations. The far end signal is about 10 dB below the transmitted signal and the ISI is about 6 dB below the far-end signal. We have chosen a very low level of noise (-80 dB) to show that the steady state does not depend of the μ value chosen for each filter, as we have seen in Section 17.3.2.2 and Fig. 17.7.

In Fig. 17.14b and 17.15 we can observe the result of ten realizations. The total error (residual echo plus ISI, Fig. 17.15b) is about -25 dB in 15000 iterations, which corresponding to 37.5 msec for 400Ksymbol/sec. We recall that for 2B1Q line code we need 22.2 dB of SNR (if we suppose additive Gaussian noise) to achieve a probability of error of 10^{-8} in the decision device. We can see that this new structure has a shorter training period than classical ones, than can spend several seconds. In Fig. 17.14b and Fig. 17.15a we can see the individual evolution of the echo canceller and the ISI canceller set. In Fig. 17.14a, we also show the coefficients

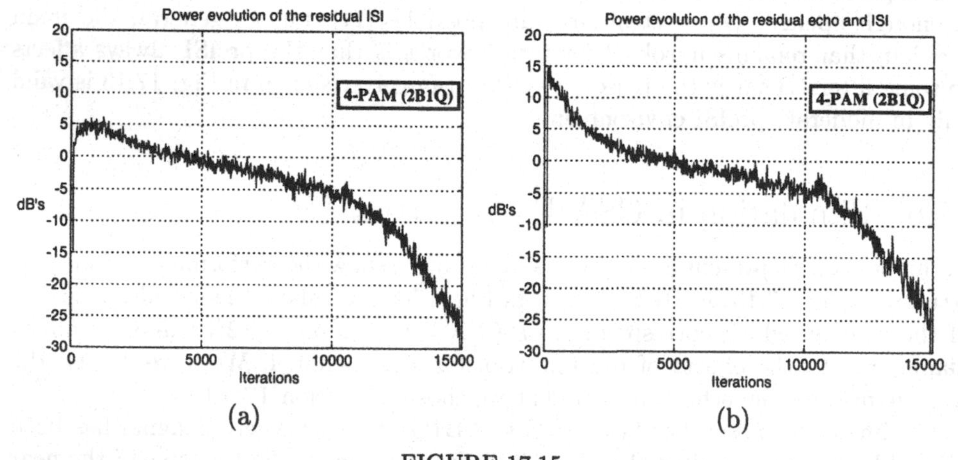

(a) (b)

FIGURE 17.15.

of the ISI-canceller filters at the time we stopped the simulation (15000 iterations). As we can see, the channel has been completely identified.

17.7 Conclusions

In this work a new and simple self-cancelling technique for adaptive filters using baseband signals (M-PAM) has been presented . We show, from the cost function point of view, that the new algorithm uses a better error criterion, close to the instantaneous optimal one, during all the transient period. The *pdf* corresponding to different stages of convergence shows an uniform behaviour which permits a high speed of convergence. In fact, we obtain two different phases of convergence, which are nearly constant with independence on the IER. It is also remarkable that it is insensitive to signal level and it seems to be insensitive for the initial settings.

We have shown how the self-cancellation technique can be applied in a global fashion. Simulation in HDSL environment point out a good performances and it is foreseeable to obtain the same behaviour in any channel with moderate precursor distortion. All these preliminary results provide us with a solid confidence about the properties of this new error reference. It focus our future work about the analytical study in the whole convergence range, look for other solutions for precursor disturbances and introduce other realistic signals (NEXT, Far-echo, etc.) in our simulations.

Acknowledgments: The authors would like to express their deep appreciation to Marta Herranz Luna and Javier Pérez Salvador for their valuable corrections and suggestions. Also to Juan Manuel Villar Navarro and Prof. Mariano García Otero for their helps with LaTeX and many other computer tools.

17.8 References

[1] S.V. Ahamed et al. Digital subscriber line (HDSL and ADSL) capacity of the outside loop plant. *IEEE JSAC*, 13(12):1540–1549, 1995.

[2] X. Chen et al. Simultaneous estimation of echo path and channel responses using full-duplex transmitted training data sequences. *IEEE Trans. on Inform. Theory*, 41(9):1409–1417, 1995.

[3] J.M. Cioffi. A fast echo caceller initialization method for CCITT V.32 modem. *IEEE Trans. on Comm.*, 38:629–638, 1990.

[4] D. D. Falconer. Adaptive reference echo cacellation. *IEEE Trans. on Comm.*, 30:2083–2094, 1982.

[5] S. Haykin. *Adaptive filter theory.* Prentice-Hall, Inc., 1991.

[6] A. Kanenasa et al. An echo canceller adaptation method for digital subscriber loop transmission. In *IEEE Proc. ICASSP'86*, vol. 2, pp. 957–960, Tokyo, Japan, 1986.

[7] W. Li et al. Full-duplex fast estimation of echo and channel responses in the presence of frequency offsets in both far echo and far signal. In *IEEE Proc. ICASSP'96*, vol. III, pp. 1715–1718, Atlanta, Georgia, U.S.A., 1996.

[8] L. Ljung et al. *Theory and practice of recursive identification.* The MIT Press, 1983.

[9] G. Long et al. Fast initialization of data-driven Nyquist in-band echo cancellers. *IEEE Trans. on Comm.*, 41:893–904, 1993.

[10] S. Marcos et al. Adaptive coupling of an echo canceller and an equalizer for digital transmission. In *Proc. EUSIPCO'88*, vol. IV, pp. 93–96, Grenoble, France, 1988.

[11] K.H.I. Mueller. Combining echo cacellation and decision feed-back equalization. *BSTJ*, 58:491–500, 1979.

[12] J.M. Páez-Borrallo et al. Adaptive LMA echo canceller in baseband data transmission with improved error reference. In *Proc. EUSIPCO'90*, vol. V, pp. 1947–1950, Barcelona, Spain, 1990.

[13] J.M. Páez-Borrallo et al. Self-improved error reference for adaptive echo cancellation in full-duplex baseband data communications. In *IEEE Proc. ICASSP'93*, vol. III, pp. 332–335, Minneapolis, U.S.A., 1993.

[14] J.M. Páez-Borrallo et al. Adaptive filtering in data communications with self improved error reference. In *IEEE Proc. ICASSP'94*, vol. III, pp. 65–68, Adelaide, Australia, 1994.

[15] J.M. Páez-Borrallo et al. Convergence analysis of decision-directed adaptive echo cancellers for baseband data transmision. *IEEE Trans. on Comm.*, 43:503–513, 1995.

[16] I.A. Pérez-Alvarez et al. A differential error reference adaptive echo canceller for multilevel PAM line codes. In *IEEE Proc. ICASSP'96*, vol. III, pp. 1707–1710, Atlanta, Georgia, 1996.

[17] I.A. Pérez-Alvarez et al. Self-reference joint echo cancellation and blind equalization in High Digital Subscriber Loops. In *Proc. ECSAP'97*, pp. 175–178, Prague, Czech Republic, 1997.

[18] J. Salz. On the start-up problem in digital echo cancellers. *BSTJ*, 62:1353–1364, 1983.

[19] A. Vicente-Castillo et al. Advanced on-line services: market situation and evolution. *Comunicaciones de Telefonica I+D, Magazin (English version)*, 7:5–19, 1996.

[20] G. Young. Reduced complexity decision feedback equalization for digital subscriber loops. *IEEE JSAC*, 9:810–816, 1991.

[21] S. Zazo-Bello. Study of new adaptive schemes for blind channel equalization based on Bussgang type algorithms. Ph.D. Thesis, Polytechnic University of Madrid, ETSIT, 1995.

Part III

SIGNAL PREDICTION

18

Predictability: An Information-Theoretic Perspective

Georges A. Darbellay[1]

ABSTRACT
The notion of the predictability of a random vector $\boldsymbol{Y}$ by another random vector $\boldsymbol{X}$ is defined. Its properties are presented, including its relationship to the notions of conditional predictability and linear predictability. The approach is purely nonparametric, in the sense that it is based solely on the data and makes no assumption on the nature of the predictive model. It enables the optimal selection of the input variables *before* building any model. An algorithm for estimating the predictability directly from the data is described. Two examples illustrate the use of these concepts.

18.1 Introduction

Prediction rests on the assumption of *dependence*. When attempting to make forecasts, we clearly assume that the future depends on the past. If a system can be studied with a deterministic model, it is then possible, at least in principle, to express this dependence in a functional form. For systems studied with the help of stochastic models, this dependence is essentially described by means of probability measures.

A deterministic model enables perfect predictions. In scientific investigations, however, one inevitably encounters the problem of precision, i.e. of the variability of the predictions. Imprecision of long term predictions is ubiquitous, and even classical mechanics is no exception [1]. This variability stems from different sources. One of them is the imperfect knowledge of the initial conditions. This issue becomes acute for the majority of non-linear systems, because of their highly sensitive dependence on the initial conditions. Even though such systems are governed by deterministic equations, the exact trajectory of the system becomes unpredictable past a certain time, a fact referred to as deterministic chaos. The higher the chaoticity of the system, the more unpredictable it is. Another cause is that it is often impossible to completely isolate a system from "external" or non-controllable influences, including any noise. This brings us to the question of the dimensionality of the system. Adding further variables to a model, provided that these variables can be identified, might increase the predictive power of the model, but extending a model is often not feasible. For high-dimensional systems one has to resort to some form

[1]Institute of Information Theory and Automation, Academy of Sciences,
Pod vodárenskou věží 4, 182 08 Prague 8, Czech Republic, E-mail: dbe@utia.cas.cz

of averaging. Then one is led to build stochastic models.

Stochasticity is not to be equated with structurelessness. A stochastic model allows to make statements about the evolution of a system, but now in a probabilistic language. Etymologically stochastic means "guessable", and a guess is also a prediction.

In this contribution we consider a measure of stochastic dependence and show how to use it as a measure of predictability. The theory is developed in the next section. In the third section, we present an algorithm to estimate the predictability directly from the data, *before* any model is built. In the fourth section we study the predictability of two time series. Our conclusions are summarised in the last section.

18.2 Theory

Stochastic dependence is a central concept of probability theory. We will consider two measures of dependence, upon which we will base our measures of predictability.

18.2.1 Information and Predictability

We start from the notion of stochastic *independence*. Let $\boldsymbol{X} = (X_1, ..., X_n)$ and $\boldsymbol{Y} = (Y_1, ..., Y_m)$ be two vectors of continuous random variables taking real values. The vector $\boldsymbol{X}$ can be regarded as the input of a system and the vector $\boldsymbol{Y}$ as its output. Each X_i can be understood as a predictor variable and each Y_j is to be predicted. The values taken by X_i and Y_j will be denoted by x_i and y_j respectively. Let $p_x(\boldsymbol{x})$ and $p_y(\boldsymbol{y})$ be the probability density functions of the random vectors $\boldsymbol{X}$ and $\boldsymbol{Y}$, and $p_{x,y}(\boldsymbol{x}, \boldsymbol{y})$ be their joint probability density function. The densities $p_x(\boldsymbol{x})$ and $p_y(\boldsymbol{y})$ will be referred to as the marginal densities. The random vectors $\boldsymbol{X}$ and $\boldsymbol{Y}$ are stochastically independent if and only if

$$p_{x,y}(\boldsymbol{x}, \boldsymbol{y}) = p_x(\boldsymbol{x}) p_y(\boldsymbol{y}) \qquad \forall\, \boldsymbol{x} \in \mathbb{R}^n, \forall\, \boldsymbol{y} \in \mathbb{R}^m \tag{18.1}$$

This condition is obviously very awkward to verify. It would be more practical to have a single number. Since this number will serve only for comparing whether one pair of random vectors is more or less dependent than another pair, its range is arbitrary. It is natural to choose *zero* for perfect independence and *one* for total dependence. One way of obtaining *zero* out of the definition above is to form the logarithm of the ratio of the joint probability with the product of the marginal probabilities

$$\ln \frac{p_{x,y}(\boldsymbol{x}, \boldsymbol{y})}{p_x(\boldsymbol{x}) p_y(\boldsymbol{y})} \tag{18.2}$$

which will be zero if Eq. (18.1) is satisfied and take its expectation

$$I(\boldsymbol{X}, \boldsymbol{Y}) = \int_{\mathbb{R}^d} p_{x,y}(\boldsymbol{x}, \boldsymbol{y}) \ln \frac{p_{x,y}(\boldsymbol{x}, \boldsymbol{y})}{p_x(\boldsymbol{x}) p_y(\boldsymbol{y})} d\boldsymbol{x} d\boldsymbol{y} \tag{18.3}$$

For notational convenience, we have introduced the dimension $d = n + m$. This number, $0 \leq I \leq \infty$, plays a central role in information theory, where it is called the mutual information [2]. To normalise it between *zero* and *one*, we use the following transformation

$$\rho = \sqrt{1 - e^{-2I}} \tag{18.4}$$

The reason for this choice is essentially that in this way ρ reduces to some well known measure of linear dependence when the joint probability density of $\boldsymbol{X}$ and $\boldsymbol{Y}$ is Gaussian, as will become apparent below in 18.2.2.

The number $\rho(\boldsymbol{X}, \boldsymbol{Y})$ captures the full dependence, both linear and non-linear, between $\boldsymbol{X}$ and $\boldsymbol{Y}$, and it can be interpreted as the ***predictability*** of $\boldsymbol{Y}$ by $\boldsymbol{X}$. This measure of predictability is based on probability distributions, i.e., those underlying the data, and it does ***not*** depend on the particular model used to predict $\boldsymbol{Y}$ from $\boldsymbol{X}$. From the properties of the mutual information, we can deduce that

- if $\rho(\boldsymbol{X}, \boldsymbol{Y}) = 0$, $\boldsymbol{X}$ contains no information on $\boldsymbol{Y}$, which means that $\boldsymbol{Y}$ cannot be predicted by means of $\boldsymbol{X}$.
- if $\rho(\boldsymbol{X}, \boldsymbol{Y}) = 1$, there exists a one-to-one relationship between $\boldsymbol{x}$ and $\boldsymbol{y}$, the values of taken by $\boldsymbol{X}$ and $\boldsymbol{Y}$. This is the limit case of determinism.
- when modelling the input-output pair $(\boldsymbol{X}, \boldsymbol{Y})$ for any model with input $\boldsymbol{X}$ and output $\boldsymbol{U} = f(\boldsymbol{X})$, where f is a measurable function, the predictability of $\boldsymbol{Y}$ by $\boldsymbol{U}$ cannot exceed the predictability of $\boldsymbol{Y}$ by $\boldsymbol{X}$, i.e.,

$$\rho(\boldsymbol{U}, \boldsymbol{Y}) \leq \rho(\boldsymbol{X}, \boldsymbol{Y}) \tag{18.5}$$

A good model would be close to the upper bound. The inequality (18.5) means that there cannot be more information in the model output $\boldsymbol{U}$ about $\boldsymbol{Y}$ than there is in the input data $\boldsymbol{X}$ about $\boldsymbol{Y}$.

18.2.2 Linear Predictability

For a d-dimensional Gaussian random vector $\boldsymbol{Z} = (\boldsymbol{X}, \boldsymbol{Y})$, the mutual information is expressed by

$$I(\boldsymbol{X}, \boldsymbol{Y}) = \frac{1}{2} \ln \frac{\det\Sigma_{xx} \det\Sigma_{yy}}{\det\Sigma} \tag{18.6}$$

Here $\Sigma = (\sigma_{ij})_{1 \leq i,j \leq d}$ is the $d \times d$ variance-covariance matrix of $\boldsymbol{Z}$, where $\sigma_{ij} = \int_{\mathbb{R}^d} (z_i - \mu_i)(z_j - \mu_j) p(\boldsymbol{z}) d\boldsymbol{z}$, with the expectations $\mu_i = \int z_i p(\boldsymbol{z}) d\boldsymbol{z}$. For $i \neq j$, σ_{ij} is a covariance. σ_{ii} is a variance. Σ_{xx} is the $n \times n$ variance-covariance matrix of $\boldsymbol{X}$ and Σ_{yy} is the $m \times m$ variance-covariance matrix of $\boldsymbol{Y}$. The symbol det denotes the determinant. It can be shown that for a Gaussian random vector $\boldsymbol{Z}$ the mutual information I depends solely on the coefficients of linear correlation

$$r_{ij} = r(X_i, Y_j) = \frac{\sigma_{ij}}{\sqrt{\sigma_{ii}\sigma_{jj}}} \tag{18.7}$$

For $n = m = 1$, i.e., $(\boldsymbol{X}, \boldsymbol{Y}) = (X, Y)$, (18.6) becomes

$$I(X, Y) = -\frac{1}{2} \ln(1 - r^2(X, Y)) \tag{18.8}$$

It follows from (18.4) that if (X, Y) is a Gaussian random vector, then $\rho(X, Y) = |r(X, Y)|$.

For $m = 1$, i.e. $(\boldsymbol{X}, \boldsymbol{Y}) = (\boldsymbol{X}, Y)$, (18.6) yields

$$\rho^2(\boldsymbol{X}, Y) = 1 - \frac{\det\Sigma}{\sigma_{yy}\,\det\Sigma_{xx}} \tag{18.9}$$

With the help of the Schur complement formula, this can be transformed into

$$\rho^2(\boldsymbol{X}, Y) = \frac{\boldsymbol{\sigma}_{xy}^{tr}\Sigma_{xx}^{-1}\boldsymbol{\sigma}_{xy}}{\sigma_{yy}} \tag{18.10}$$

where $\boldsymbol{\sigma}_{xy} = (\sigma_{x_1 y}, ..., \sigma_{x_n y})$ and where $\boldsymbol{\sigma}_{xy}^{tr}$ is the column vector obtained by transposition of $\boldsymbol{\sigma}_{xy}$. The left-hand side of (18.10) is nothing else than the square of the well known coefficient of multiple (linear) correlation [8].

For the general case in which $\boldsymbol{X}$ is a vector of n random variables and $\boldsymbol{Y}$ a vector of m random variables, (18.6) becomes

$$\rho^2(\boldsymbol{X}, \boldsymbol{Y}) = 1 - \frac{\det\Sigma}{\det\Sigma_{xx}\det\Sigma_{yy}} = 1 - \det(I_m - \Sigma_{yy}^{-1}\Sigma_{xy}\Sigma_{xx}^{-1}\Sigma_{xy}^{tr}) \tag{18.11}$$

Again we have used the Schur complement formula, and the decomposition

$$\Sigma = \begin{pmatrix} \Sigma_{xx}\Sigma_{xy}^{tr} \\ \Sigma_{xy}\Sigma_{yy} \end{pmatrix} \tag{18.12}$$

In Eq. (18.11), I_m denotes the $m \times m$ identity matrix. The m eigenvalues of $\Sigma_{yy}^{-1}\Sigma_{xy}\Sigma_{xx}^{-1}\Sigma_{xy}^{tr}$ are the canonical correlations as defined in multivariate statistical analysis (e.g. [9]).

In accordance with the considerations above, we define the ***linear predictability*** of $\boldsymbol{Y}$ by $\boldsymbol{X}$ as

$$\lambda(\boldsymbol{X}, \boldsymbol{Y}) = \sqrt{1 - \frac{\det\Sigma}{\det\Sigma_{xx}\,\det\Sigma_{yy}}} \tag{18.13}$$

where, again, Σ is the variance-covariance matrix of $(\boldsymbol{X}, \boldsymbol{Y})$, Σ_{xx} the variance-covariance matrix of $\boldsymbol{X}$ and Σ_{yy} the variance-covariance matrix of $\boldsymbol{Y}$. $\lambda(\boldsymbol{X}, \boldsymbol{Y})$ is a number between *zero* and *one*. It says how much of $\boldsymbol{Y}$ can be predicted from $\boldsymbol{X}$ with the help of a linear model. The definition (18.13) implies that if $(\boldsymbol{X}, \boldsymbol{Y})$ follows a Gaussian distribution, then $\rho(\boldsymbol{X}, \boldsymbol{Y}) = \lambda(\boldsymbol{X}, \boldsymbol{Y})$. In $\mathbb{R}^2$, $\lambda(X, Y) = |r(X, Y)|$.

18.2.3 Selection of Inputs

Predictability should increase as new explanatory variables are added. Let us introduce the input vectors $\boldsymbol{X}_{n-1} = (X_1, ..., X_{n-1})$ and $\boldsymbol{X}_n = (X_1, ..., X_{n-1}, X_n)$. From (18.4) and the chain rule for the mutual information

$$I(\boldsymbol{X}_n, \boldsymbol{Y}) = I(\boldsymbol{X}_{n-1}, \boldsymbol{Y}) + I(X_n, \boldsymbol{Y}|\boldsymbol{X}_{n-1}) \tag{18.14}$$

where | indicates conditioning, it follows that

$$\rho(\boldsymbol{X}_{n-1}, \boldsymbol{Y}) \leq \rho(\boldsymbol{X}_n, \boldsymbol{Y}) \tag{18.15}$$

since the mutual information is always positive and that

$$\rho^2(X_n, \boldsymbol{Y}|\boldsymbol{X}_{n-1}) = \frac{\rho^2(\boldsymbol{X}_n, \boldsymbol{Y}) - \rho^2(\boldsymbol{X}_{n-1}, \boldsymbol{Y})}{1 - \rho^2(\boldsymbol{X}_{n-1}, \boldsymbol{Y})} \geq 0 \tag{18.16}$$

The inequalities in (18.15) and (18.16) become equalities if and only if X_n does not provide any information that is not already contained in $\boldsymbol{X}_{n-1}$. The quantity $\rho^2(X_n, \boldsymbol{Y}|\boldsymbol{X}_{n-1})$, measures the conditional dependence between $\boldsymbol{Y}$ and X_n given $\boldsymbol{X}_{n-1}$. $\rho(X_n, \boldsymbol{Y}|\boldsymbol{X}_{n-1})$, can be interpreted as the increase in the predictability of $\boldsymbol{Y}$ as the new predictor variable X_n is added to the $n-1$ predictor variables $(X_1, ..., X_{n-1})$, and will be called the ***conditional predictability*** of $\boldsymbol{Y}$ by X_n given $\boldsymbol{X}_{n-1}$. The conditional predictability $\rho(X_n, \boldsymbol{Y}|\boldsymbol{X}_{n-1})$ will be zero if and only if the information provided by X_n is redundant. If one is interested in linear models only, then (18.16) becomes

$$\lambda^2(X_n, \boldsymbol{Y}|\boldsymbol{X}_{n-1}) = \frac{\lambda^2(\boldsymbol{X}_n, \boldsymbol{Y}) - \lambda^2(\boldsymbol{X}_{n-1}, \boldsymbol{Y})}{1 - \lambda^2(\boldsymbol{X}_{n-1}, \boldsymbol{Y})} \geq 0 \tag{18.17}$$

where the right hand side of the equation can be evaluated from (18.13). The positive number $\lambda^2(X_n, \boldsymbol{Y}|\boldsymbol{X}_{n-1})$ is nothing else but the square of the coefficient of partial (linear) correlation. The relationship (18.17) is well known in the theory of statistical regression [8]. It tells us whether the quality of the predictions of a linear model can be improved by adding the predictor variable X_n. (18.16) may be viewed as a non-linear generalisation of (18.17).

18.2.4 Comparing ρ and λ

Intuitively, one would like a measure of predictability to be larger than a measure of linear predictability only. Unfortunately, it is not always true that $\rho \geq \lambda$ [4]. Despite the fact that ρ appears to be an excellent generalisation of λ, since (18.16) generalises (18.17), it is clear that the difference $\rho - \lambda$ cannot be equated to the non-linear part of the predictability. However, we can say that a difference between ρ and λ, whether positive or not, signals the inadequacy of a linear model, on the grounds that linear correlations capture linear relationships. Furthermore, in the majority of cases, we do have $\rho(\boldsymbol{X}, \boldsymbol{Y}) \geq \lambda(\boldsymbol{X}, \boldsymbol{Y})$. The following example might help to see why. Consider two random variables X and Y whose values are distributed on a circle. For reasons of symmetry, it is clear that $\lambda(X, Y) = |r(X, Y)| = 0$. However $\rho(X, Y)$ is close to *one* because the relationship between X and Y is almost one-to-one. Section 18.4 provides some additional examples.

Our definition of predictability is model-independent in the sense that predictability can be estimated directly from the data ***before*** building any model. ***After*** a model of the input-output system $(\boldsymbol{X}, \boldsymbol{Y})$ has been built, with model output $\boldsymbol{U} = f(\boldsymbol{X})$, one would expect that $\boldsymbol{U}$ and $\boldsymbol{Y}$ are linearly correlated because for a good model $\boldsymbol{U}$ should track $\boldsymbol{Y}$. Thus they should co-vary. The linear predictability $\lambda(\boldsymbol{U}, \boldsymbol{Y})$ can be calculated from (18.13). In the majority of cases we conjecture that the following inequality holds:

$$\lambda(\boldsymbol{U}, \boldsymbol{Y}) \leq \rho(\boldsymbol{X}, \boldsymbol{Y}) \tag{18.18}$$

The predictability ρ is defined under far more general conditions than λ. It is not difficult to find examples when ρ is defined and λ is not [6]. One such example appears in 18.3.2.

18.3 Estimation from Data

In practice, one is often interested in building a model from some set of data. To this end, we have developed an algorithm for estimating the mutual information, and thus the predictability, from data [3]. The accuracy and the precision of this algorithmic estimator were tested on a range of distributions, two examples of which are given in 18.3.2.

18.3.1 Algorithm

The basic idea is to consider finite partitions of $\mathbb{R}^d$. A *finite partition* of $\mathbb{R}^d$ is any finite system of non-intersecting subsets of $\mathbb{R}^d$, whose sum is the whole of $\mathbb{R}^d$. In other terms, $\Gamma = \{C_1, ..., C_K\}$ is a finite partition of $\mathbb{R}^d$ if $C_i \subset \mathbb{R}^d$, $C_i \cap C_j = \emptyset$ for $i \neq j$, and $\cup_{i=1}^{K} C_i = \mathbb{R}^d$. In this work, partition will always mean finite partition. Each set C_i is called a cell of the partition Γ and K is the number of cells. A partition $\Lambda = \{D_l, l = 1, ..., L\}$ is a *refinement* of a partition $\Gamma = \{C_k, k = 1, ..., K\}$ if for each $D_l \in \Lambda$ there exists a $C_k \in \Gamma$ such that $D_l \subset C_k$. The sets D_l can be rearranged in such a way that $\Lambda = \{C_{kl}; k = 1, ..., K, l = 1, ..., L_k\}$, where $\bigcup_{l=1}^{L_k} C_{kl} = C_k$. A refinement will also be called a subpartition.

The most common type of partition is that of a product partition. Fig. 1(a) represents the generic type of a product partition. Formally, a partition $\{C_k\}$ is a product partition of $\mathbb{R}^d = \mathbb{R}^n \times \mathbb{R}^m$ if (i) every cell C_k can be written as $C_k = A_i \times B_j$ where (ii) $\{A_i\}$ is a partition of $\mathbb{R}^n$ and $\{B_j\}$ a partition of $\mathbb{R}^m$. A more general (and thus more flexible) type of partition is provided by the class of partitions with product cells. Such cells are formed by the product of d one-dimensional intervals, i.e., they are hypercubes in $\mathbb{R}^d$. From now on we will consider only such cells. A partition $\{C_k\}$ with product cells also satisfies (i) but not necessarily (ii). Now the A_i are simply n-dimensional hypercubes and the B_j are m-dimensional hypercubes. There are many different types, or subfamilies, of partitions with product cells. Fig. 1(b) illustrates the type of partitions we will use for our algorithm.

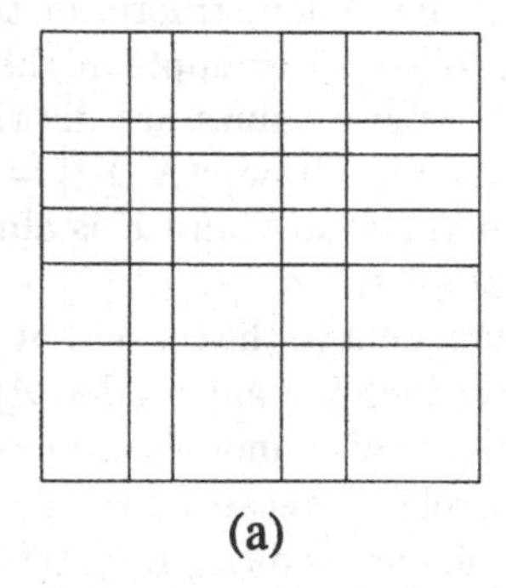

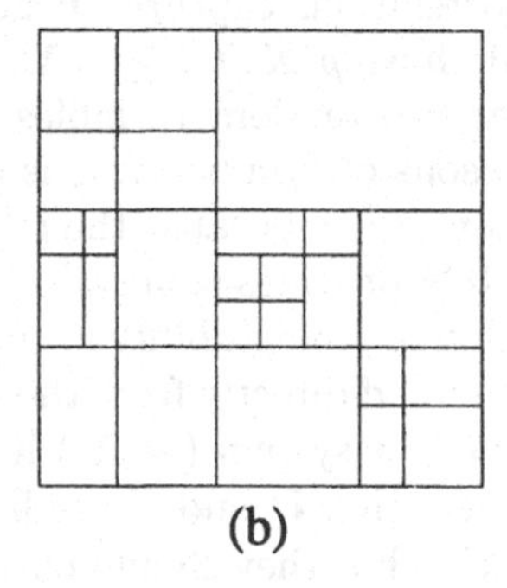

(a) (b)

FIGURE 18.1. Two examples of partitions in $\mathbb{R}^2$: (a) a product partition, (b) a type of partition with product cells

Let $P_{\boldsymbol{X}}$ be the probability distribution of $\boldsymbol{X}$, $P_{\boldsymbol{Y}}$ the probability distribution of $\boldsymbol{Y}$, and $P_{\boldsymbol{X},\boldsymbol{Y}}$ the joint probability distribution of $(\boldsymbol{X}, \boldsymbol{Y})$. This means that $P_{\boldsymbol{X},\boldsymbol{Y}}(C_k)$ is the probability that the values of $(\boldsymbol{X}, \boldsymbol{Y})$ are in the set C_k. $P_{\boldsymbol{X}}(C_k)$ will denote the probability that the values of $\boldsymbol{X}$ are in the orthogonal projection of C_k in $\mathbb{R}^n$, and $P_{\boldsymbol{Y}}(C_k)$ the probability that the values of $\boldsymbol{Y}$ are in the orthogonal projection of

C_k in $\mathbb{R}^m$. We recall that for continuous random variables the probability density functions (if they exist) are related to the probability distributions by

$$p_{x,y}(\boldsymbol{x},\boldsymbol{y})\Big|_{\boldsymbol{x}=\boldsymbol{a},\boldsymbol{y}=\boldsymbol{b}} = \lim_{\Delta_1,\ldots,\Delta_d\to 0} \frac{P_{\mathbf{X},\mathbf{Y}}(a_1-\Delta_1 < x_1 \le a_1,\ldots,b_m-\Delta_d < y_m \le b_m)}{\Delta_1\Delta_2\ldots\Delta_d} \tag{18.19}$$

and

$$p_x(\boldsymbol{x})\Big|_{\boldsymbol{x}=\boldsymbol{a}} = \lim_{\Delta_1,\ldots,\Delta_n\to 0} \frac{P_{\mathbf{X}}(a_1-\Delta_1 < x_1 \le a_1,\ldots,a_n-\Delta_n < x_n \le a_n)}{\Delta_1\Delta_2\ldots\Delta_n} \tag{18.20}$$

A similar relationship applies for $p_y(\boldsymbol{y})$.

Now we are equipped to express the mutual information in a form more suitable for its evaluation from data. Consider a fixed but arbitrary partition $\Gamma = \{C_k\}$ of $\mathbb{R}^d$. We may define the constants

$$c_k = \frac{P_{\mathbf{X},\mathbf{Y}}(C_k)}{P_{\mathbf{X}}(C_k)\,P_{\mathbf{Y}}(C_k)} \tag{18.21}$$

Then we can write

$$I(\mathbf{X},\mathbf{Y}) = \sum_k \int_{C_k} d\boldsymbol{x}d\boldsymbol{y}\; p_{x,y}(\boldsymbol{x},\boldsymbol{y}) \ln \frac{p_{x,y}(\boldsymbol{x},\boldsymbol{y})}{p_x(\boldsymbol{x})p_y(\boldsymbol{y})}$$

$$= \sum_k \int_{C_k} d\boldsymbol{x}d\boldsymbol{y}\; p_{x,y}(\boldsymbol{x},\boldsymbol{y}) \ln \frac{p_{x,y}(\boldsymbol{x},\boldsymbol{y})}{c_k\; p_x(\boldsymbol{x})p_y(\boldsymbol{y})} + \sum_k P_{\mathbf{X},\mathbf{Y}}(C_k) \ln \frac{P_{\mathbf{X},\mathbf{Y}}(C_k)}{P_{\mathbf{X}}(C_k)\,P_{\mathbf{Y}}(C_k)} \tag{18.22}$$

where

$$P_{\mathbf{X},\mathbf{Y}}(C_k) = \int_{C_k} d\boldsymbol{x}d\boldsymbol{y}\; p_{x,y}(\boldsymbol{x},\boldsymbol{y}) \tag{18.23}$$

The first sum in (18.22) can be made arbitrarily small. This is done by considering successive refinements of Γ. Using the notation C_{kl} is practical for the cells of the first refinement but not for further refinements. In order not to multiply the indices, we will relabel the cells of each successive refinement and call them C_k again. So, if we consider successive refinements,

$$\lim_{\Delta_1,\ldots,\Delta_d\to 0} c_k = \lim_{\Delta_1,\ldots,\Delta_d\to 0} \frac{P_{\mathbf{X},\mathbf{Y}}(\boldsymbol{a}^k-\boldsymbol{\Delta}_n{}^k < \boldsymbol{x} \le \boldsymbol{a}^k, \boldsymbol{b}^k-\boldsymbol{\Delta}_m{}^k < \boldsymbol{y} \le \boldsymbol{b}^k)}{P_{\mathbf{X}}(\boldsymbol{a}^k-\boldsymbol{\Delta}_n{}^k < \boldsymbol{x} \le \boldsymbol{a}^k)\,P_{\mathbf{Y}}(\boldsymbol{b}^k-\boldsymbol{\Delta}_m{}^k < \boldsymbol{y} \le \boldsymbol{b}^k)}$$
$$= \frac{p_{x,y}(\boldsymbol{x},\boldsymbol{y})}{p_x(\boldsymbol{x})p_y(\boldsymbol{y})}\Big|_{\boldsymbol{x}=\boldsymbol{a}^k,\boldsymbol{y}=\boldsymbol{b}^k} \tag{18.24}$$

where the cell C_k is the product

$$C_k = [a_1^k - \Delta_1^k, a_1^k] \times \ldots \times [a_n^k - \Delta_n^k, a_n^k] \times [b_1^k - \Delta_{n+1}^k, b_1^k] \times \ldots \times [b_m^k - \Delta_{n+m}^k, b_m^k]$$

In (18.24) the inequalities are meant component-wise, and the subscript k indicates the dependence on the cell C_k. We have used the obvious notation $\boldsymbol{\Delta}^k = (\boldsymbol{\Delta}_n{}^k, \boldsymbol{\Delta}_m{}^k) = (\Delta_1^k,\ldots,\Delta_n^k,\Delta_{n+1}^k,\ldots,\Delta_{n+m}^k)$. From (18.24) it follows that the argument of the logarithm in the first sum in (18.22) will tend to *one* as the partitions are made finer and finer. As a result, the sum in (18.22) will become arbitrarily small. Thus, we obtain

$$I(\boldsymbol{X},\boldsymbol{Y}) = \sum_k P_{\boldsymbol{X},\boldsymbol{Y}}(C_k) \ln \frac{P_{\boldsymbol{X},\boldsymbol{Y}}(C_k)}{P_{\boldsymbol{X}}(C_k)\, P_{\boldsymbol{Y}}(C_k)} \tag{18.25}$$

if the condition

$$\frac{P_{\boldsymbol{X},\boldsymbol{Y}}(C_{kl})}{P_{\boldsymbol{X}}(C_{kl})\, P_{\boldsymbol{Y}}(C_{kl})} = \frac{P_{\boldsymbol{X},\boldsymbol{Y}}(C_k)}{P_{\boldsymbol{X}}(C_k)\, P_{\boldsymbol{Y}}(C_k)} \qquad \forall l \quad \forall \{C_{kl}\} \tag{18.26}$$

is satisfied for every cell C_k. The two quantifiers mean that the equality in (18.26) must hold for every cell C_{kl} of every subpartition $\{C_{kl}\}$ of the cell C_k under consideration. Eq. (18.25) is a consequence of (18.26) and (18.24) implying $c_k = p_{x,y}(\boldsymbol{x},\boldsymbol{y})/p_x(\boldsymbol{x})p_y(\boldsymbol{y})$. We note that Eq. (18.25) and (18.26) are a new formulation of Dobrushin's information theorem [7], [3]. Eq. (18.26) expresses the key idea of our approach to the estimation problem: the partition should be fine enough so as to achieve *local independence* on every one of its cells.

To have a workable algorithm, we need a systematic procedure for constructing partitions and subpartitions. To partition a given d-dimensional cell, we will divide each one of its d edges (i.e., margins) into *two* equiprobable intervals. This scheme will be referred to as *marginal equiquantization*. At every partitioning step, a cell is partitioned into 2^d subcells. We use equiprobable intervals, rather than equidistant intervals, in accordance with the invariance of the mutual information under one-to-one transformations of its variables

$$I(\ (f_1(X_1),...,f_n(X_n)),\ (g_1(Y_1),...,g_m(Y_m))\ ) = I(\ (X_1,...,X_n),\ (Y_1,...,Y_m)\ ) \tag{18.27}$$

where the f_i and g_j denote these bijective component-wise transformations. The choice of equiprobable intervals guarantees that the marginal equiquantization scheme will produce the same sequence of partitions and thus the same mutual information, for all bijectively component-wise transformed variables.

Schematically, our algorithm can be formulated in the form of three rules.

(R0) Let $\mathbb{R}^d$ be the initial one-cell partition.

(R1) Every cell is to be partitioned by marginal equiquantization into 2^d subcells.

(R2) A cell is not partitioned any further if condition (18.26) is satisfied.

18.3.2 Estimator

We wish to estimate $I(\boldsymbol{X},\boldsymbol{Y})$ from a finite sample of N points $(\boldsymbol{x},\boldsymbol{y})$ in $\mathbb{R}^d = \mathbb{R}^n \times \mathbb{R}^m$. Probabilities can be estimated through frequencies, i.e., by dividing the number of points in a cell by the total number of points N. Formula (18.25) becomes

$$\hat{I}(\boldsymbol{X},\boldsymbol{Y}) = \frac{1}{N} \sum_{k=1}^{m} N_{\boldsymbol{X},\boldsymbol{Y}}(C_k) \ln \frac{N_{\boldsymbol{X},\boldsymbol{Y}}(C_k)}{N_{\boldsymbol{X}}(C_k) N_{\boldsymbol{Y}}(C_k)} + \ln N \tag{18.28}$$

The hat accent denotes the fact that $\hat{I}(\boldsymbol{X},\boldsymbol{Y})$ is an estimate obtained from a sample of $(\boldsymbol{X},\boldsymbol{Y})$. $N_{\boldsymbol{X},\boldsymbol{Y}}(C_k)$ is the number of points in cell C_k. The marginal numbers

$\rho=\lambda$	avg($\hat{\rho}$)	$\sqrt{\text{var}(\hat{\rho})}$	avg($\hat{\lambda}$)	$\sqrt{\text{var}(\hat{\lambda})}$	KS stat
0	0.0003	0.0032	0.0081	0.0062	0.5246
0.3	0.3002	0.0110	0.3007	0.0092	0.0132
0.6	0.6002	0.0070	0.6004	0.0063	0.0176
0.9	0.8999	0.0020	0.9001	0.0018	0.0150

TABLE 18.1. Simulation results for two-dimensional Gaussian distributions. Each line corresponds to a different value of the linear correlation coefficient r used to generate the Gaussian sample. These theoretical values of $r=\lambda=\rho$ are listed in the first column

$\rho=\lambda$	avg($\hat{\rho}$)	$\sqrt{\text{var}(\hat{\rho})}$	avg($\hat{\lambda}$)	$\sqrt{\text{var}(\hat{\lambda})}$	KS stat
0	0	0	0	0	0
0.3721	0.3733	0.0118	0.3720	0.0083	0.0503
0.6708	0.6724	0.0070	0.6711	0.0054	0.0659
0.9234	0.9222	0.0017	0.9235	0.0014	0.0535

TABLE 18.2. Simulation results for three-dimensional Gaussian distributions. Each line corresponds to a different value of the linear correlation coefficient $r_{gen}(=r_{12}=r_{13}=r_{23})$ used to generate the 3rd sample. The values of r_{gen} are 0, 0.3, 0.6, 0.9

$N_{\mathbf{X}}(C_k)$, respectively $N_{\mathbf{Y}}(C_k)$, are the numbers of points in the orthogonal projection of C_k in $\mathbb{R}^n$, respectively in the orthogonal projection of C_k in $\mathbb{R}^m$.

In practice the condition (18.26) cannot possibly be checked for every subpartition of the cell under investigation. To make matters tractable, we choose two subpartitions obtained through marginal equiquantization. Then the local independence condition (18.26) becomes

$$N_{\mathbf{X}}(C_{kl}) \approx N_{\mathbf{X}}(C_k)\frac{N_{\mathbf{X}_a}(C_{kl})N_{\mathbf{X}_b}(C_{kl})}{N_{\mathbf{X}_a}(C_k)N_{\mathbf{X}_b}(C_k)} \qquad l=1,...,2^{sd} \tag{18.29}$$

where $s=1,2$. The first subpartition has 2^d subcells, the second subpartition is obtained by partitioning the first subpartition and has 2^{2d} subcells. This approximation produces a workable estimator, as the test examples in the tables show.

For each of the two subpartitions in (18.29), the conditions are summarised into two $\mathcal{X}^2$ statistics, one statistic for each subpartition. This is done by summing over the subcells, i.e., the index l in (18.29). We found that a significance level between 1% and 3% for the $\mathcal{X}^2$ test works well. For small samples, that is to say less than 500 points, one may even choose a significance level of 5%. Higher (lower) significance levels usually result in overestimation (underestimation). In all the results reported in this work, a significance level of 3% has been chosen.

In the tables we report some tests performed on Gaussian and Burr distributions in $\mathbb{R}^2$ and $\mathbb{R}^3$. All calculations were obtained with samples of 10000 points drawn from the appropriate distribution. For each line in the tables, the averages and the standard deviations of the estimate of the predictability $\hat{\rho}$ and the estimate of the linear predictability $\hat{\lambda}$ were calculated over 100 samples, except for Table 18.1 where we used 1000 samples. For the calculation of λ the standard estimators of the variance and the covariance were used.

ρ	avg($\hat{\rho}$)	$\sqrt{\text{var}(\hat{\rho})}$	avg($\hat{\lambda}$)	$\sqrt{\text{var}(\hat{\lambda})}$	KS stat
0.0936	0.0854	0.0162	0.0781	0.0230	0.0632
0.5661	0.5638	0.0082	—	—	0.0336
0.7606	0.7602	0.0048	—	—	0.0507
0.9742	0.9739	0.0007	—	—	0.0427

TABLE 18.3. Simulation results for two-dimensional Burr distributions

ρ	avg($\hat{\rho}$)	$\sqrt{\text{var}(\hat{\rho})}$	avg($\hat{\lambda}$)	$\sqrt{\text{var}(\hat{\lambda})}$	KS stat
0.1266	0.1063	0.0207	0.1081	0.0332	0.0571
0.6417	0.6388	0.0077	—	—	0.0893
0.8138	0.8090	0.0047	—	—	0.0376
0.9817	0.9804	0.0005	—	—	0.0532

TABLE 18.4. Simulation results for three-dimensional Burr distributions

In the first column of the tables we listed the theoretical values of ρ. For Gaussian distributions they can be calculated from formula (18.6). For Burr distributions the formula can be found in [6] (ρ depends on only one parameter, which takes the values $10, 1, 0.5, 0.1$) starting from the top line in the tables. From the second and third columns it can be seen that the estimator $\hat{\rho}$ is accurate and fairly precise. Note that in Tables 18.3 and 18.4 the linear predictability is undefined for three cases out of four. The last column contains the values of the Kolmogorov-Smirnov(KS) statistic for testing the normality of $\hat{\rho}$. It is usually below the critical values, thus indicating that $\hat{\rho}$ and $\hat{I}$ follow a one-dimensional Gaussian distribution.

As the dimension d is increased, one has to be aware that the estimation of frequencies in high-dimensional spaces soon requires a very large number of data points. This is known as the curse of dimensionality, and one inevitably encounters it in all nonparametric (model-independent) approaches.

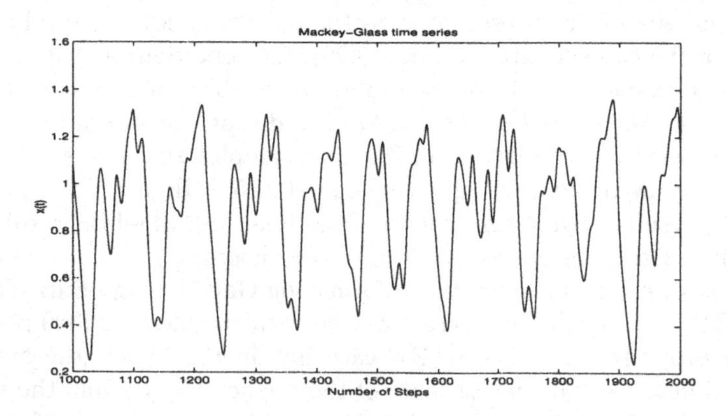

FIGURE 18.2. Section of the Mackey-Glass time series over 1000 steps

18.4 Analysis of the Predictability of Two Time Series

18.4.1 The Mackey-Glass Time Series

As a first illustration, we consider a well-known non-linear time series, namely the Mackey-Glass time series. It is generated by the time-delay differential equation

$$\frac{dx}{dt} = \frac{a\,x(t)}{1 + x^c(t-\tau)} - b\,x(t) \tag{18.30}$$

where the constants are often (as here) taken as $a = 0.2$, $b = 0.1$ and $c = 10$. The delay parameter τ determines the behavior of the time series, and for $\tau > 16.8$ the Mackey-Glass equation has a chaotic attractor. Here, we chose $\tau = 30$ and generated a time series from (18.30) with the time interval $\Delta t = 1$, a section of which is shown in Fig. 18.2.

In Fig. 18.3 we show the predictability of the output $x(t)$ over the next 100 steps, i.e., for $t = 1, ..., 100$, with the input $x(0)$, with the inputs $x(0)$ and $x(-20)$, and with the inputs $x(0)$, $x(-8)$, $x(-20)$ and $x(-28)$. These inputs were selected

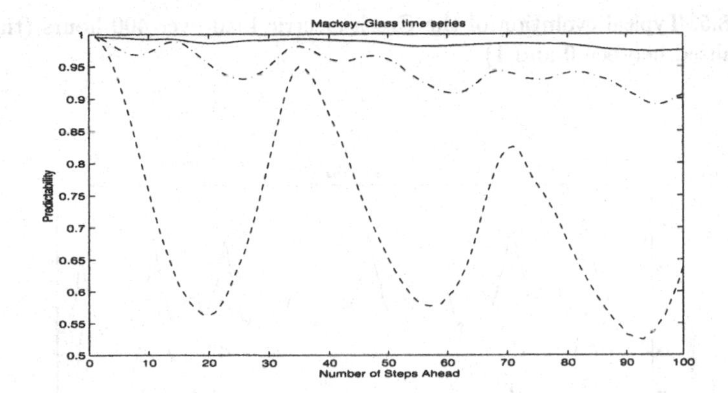

FIGURE 18.3. Predictability $\hat{\rho}(x(t), x(0))$ [dashed line], $\hat{\rho}(\ x(t),\ (x(0), x(-20))\ )$ [dashdot line], and $\hat{\rho}(\ x(t),\ (x(0),\ x(-8), x(-20), x(-28))\ )$ [full line], for $t = 1, ..., 100$

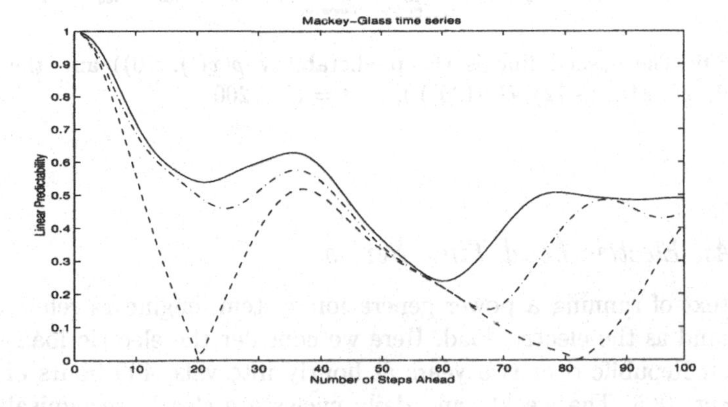

FIGURE 18.4. Linear predictability $\hat{\lambda}(x(t), x(0))$ [dashed line], $\hat{\lambda}(\ x(t),\ (x(0), x(-20))\ )$ [dashdot line], and $\hat{\lambda}(\ x(t),\ (x(0), x(-8), x(-20), x(-28))\ )$ [full line], for $t = 1, ..., 100$

so as to maximise the average predictability over the next 100 steps. The linear predictability is plotted in Fig. 18.4 for the same sets of inputs. The difference between the two figures speaks for itself. Note that $\hat{\rho}$ is always greater than $\hat{\lambda}$. For all calculations we used data sets with 100000 points.

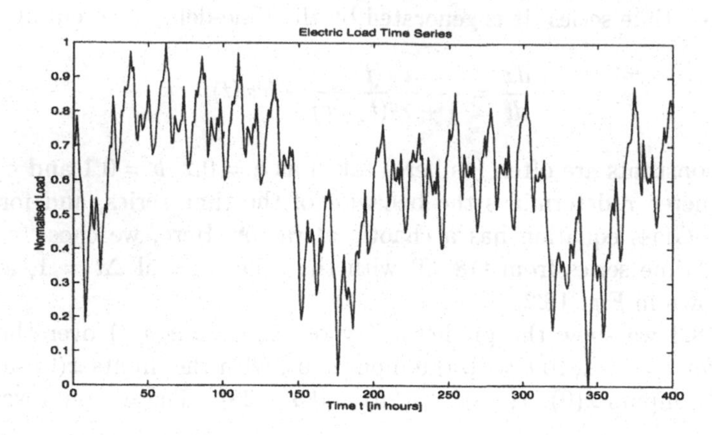

FIGURE 18.5. Typical evolution of the Czech electric load over 400 hours (the load has been normalised between 0 and 1)

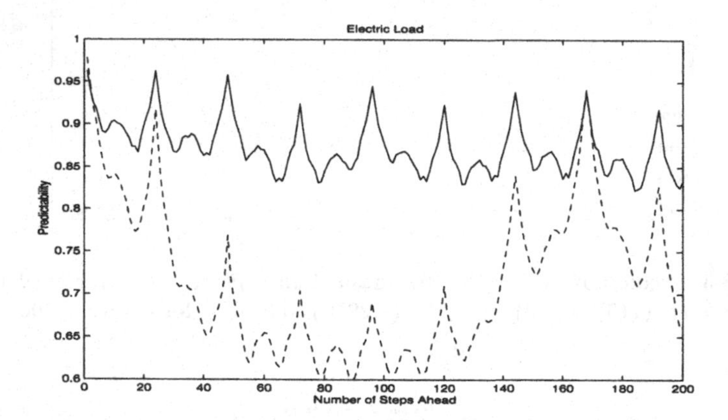

FIGURE 18.6. The dashed line is the predictability $\hat{\rho}(x(t), x(0))$ and the full line is $\hat{\rho}(\ x(t),\ (x(0), x(-24), x(-72), x(-120))\ )$, for $t = 1, ..., 200$

18.4.2 An Electric Load Time Series

In the context of running a power generation system, engineers refer to the electricity demand as the electric load. Here we consider the electric load time series of the Czech Republic over two years at hourly intervals, 400 hours of which are shown in Fig. 18.5. The weekly and daily cycles are clearly recognisable, though they are partially blurred by the random fluctuations produced by the diversity of consumers' needs.

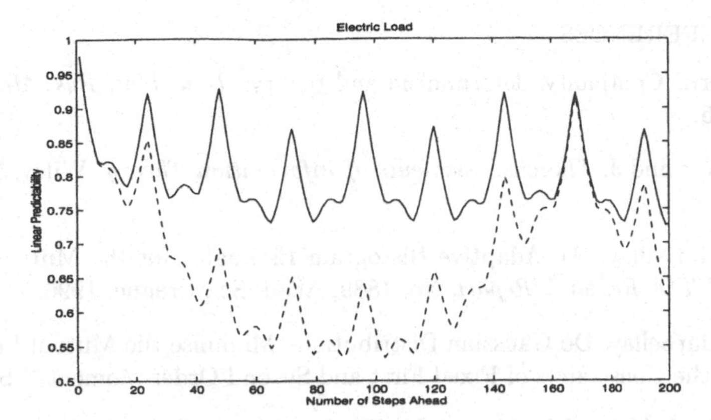

FIGURE 18.7. The dashed line is the linear predictability $\hat{\lambda}(x(t), x(0))$ and the full line is $\hat{\lambda}(\ x(t),\ (x(0), x(-24), x(-72), x(-120))\ )$, for $t = 1, ..., 200$

The predictability, respectively the linear predictability, of the next 200 hours is shown in Fig. 18.6, respectively Fig. 18.7. Again $\hat{\rho} \geq \hat{\lambda}$. However, at the daily peaks the differences between $\hat{\rho}$ and $\hat{\lambda}$ are negligible. With the help of the conditional predictability, we found that using other inputs but the daily peaks brings little improvement. Therefore it is doubtful that, for this time series, a non-linear model would significantly outperform a linear model. This was confirmed in a comparative study between ARMA and neural network models [5].

For all calculations we used data sets with 16000 points. In $\mathbb{R}^5$, we can notice that the predictability starts suffering from underestimation as is apparent in the first few steps where $\hat{\rho}(x(t), (x(0), x(-24), x(-72), x(-120))) < \hat{\rho}(x(t), x(0))$ which, in theory, is a contradiction. Nevertheless, with only 16000 points in $\mathbb{R}^5$, we still get meaningful results.

18.5 Conclusion

We have shown that a measure of stochastic dependence, the mutual information, can serve as a measure of predictability and that it is possible to estimate it directly from data before building any predictive model. This could prove very useful, as it answers three important questions before entering the often complex task of modelling the data under investigation. First, it provides an upper bound on the predictability that can be achieved with any model. This gives an idea of how much effort should or should not be spent in improving the current model(s). Second, by comparing the overall predictability with the linear predictability, one can assess whether a linear model is sufficient or whether a non-linear model is necessary. Third, the conditional predictability provides a criterion for the optimal selection of input variables.

Acknowledgments: **Financial support from the *Fonds National Suisse de la Recherche Scientifique* is gratefully acknowledged. The author wishes to thank J. Franek and M. Sláma for programming assistance, and to the *ČEZ* for providing the electric load data.**

18.6 References

[1] M. Born. Continuity, determinism and reality. *Dan. Mat. Fys. Medd.*, 30(2): 1, 1955.

[2] T. Cover and J. Thomas. *Elements of Information Theory.* Wiley, New York, 1991

[3] G.A. Darbellay. An Adaptive Histogram Estimator for the Mutual Information. *UTIA Research Report*, No. 1889, Acad. Sc., Prague, 1996.

[4] G.A. Darbellay. Do Gaussian Distributions Minimise the Mutual Information under the Constraints of Fixed First and Second Order Moments? Submitted.

[5] G.A. Darbellay and M. Slama. Electric Load Forecasting: Do Neural Networks Stand a Chance? Submitted.

[6] G.A. Darbellay and I. Vajda. Entropy Expressions for Continuous Multivariate Distributions. In preparation.

[7] R.L. Dobrushin. General formulation of Shannon's main theorem in information theory. *Uspekhi Mat. Nauk*, 14:3–104, 1959 (in Russian).

[8] C.R. Rao. *Linear Statistical Inference and Its Applications.* Wiley, New York, 1973.

[9] M.S. Srivastava and C.G. Khatri. *An Introduction to Multivariate Statistics.* Elsevier North Holland, New York, 1979.

19

Self-Adaptive Evolution Strategies for the Adaptation of Non-Linear Predictors in Time Series Analysis

André Neubauer[1]

ABSTRACT

The application of evolutionary computation techniques to the prediction of non-linear and non-stationary stochastic signals is presented - a task that arises, e.g., in time series analysis. Especially, the online adaptation of bilinear predictors with the help of a *multi-membered* (μ, λ) *- evolution strategy with self-adaptation of strategy parameters* is treated. Special emphasis is given to the tracking capabilities of this specific evolutionary algorithm in non-stationary environments. The novel modifications of the standard (μ, λ) - evolution strategy are detailed that are necessary to obtain a computationally efficient algorithm. Using the evolutionary adapted bilinear predictor as part of a *bilinear prediction error filter*, the proposed methodology is applied to estimating bilinear stochastic signal models. Experimental results are given that demonstrate the robustness and efficiency of the (μ, λ) - evolution strategy in this digital signal processing application.

19.1 Introduction

In digital signal processing, the prediction of stochastic signals is an important task with a variety of applications, e.g., in speech processing, system identification, and time series analysis. In the past, linear predictors were mainly used due to the availability of computationally efficient algorithms for the estimation of predictor parameters [6, 25].

A number of non-linear predictors were developed for stochastic signals with statistical characteristics that cannot adequately be handled by linear predictors. A classical example of recursive polynomial filters is the *bilinear* predictor that can be obtained via an extension of recursive linear predictors [10]. To cope with stochastic signals with time-varying statistical characteristics, the adaptation algorithm of the bilinear predictor must work in non-stationary environments.

In this respect, *evolutionary computation* techniques [4] are promising. *Evolutionary algorithms* [1, 20], i.e., *genetic algorithms* [7], *evolution strategies* [22, 23, 24], *evolutionary programming* [5], and *genetic programming* [8], are robust optimization methodologies that take their patterns from biological evolution and molecu-

[1]Siemens AG, Microelectronics Design Center, Wacholderstraße 7, 40489 Düsseldorf, Germany, E-mail: neubauer@ezmd.hl.siemens.de

lar genetics. For numerical parameter optimization problems, the ***Evolution Strategy*** (ES), developed in the sixties by Rechenberg and Schwefel at the Technical University of Berlin, is a powerful and robust optimization algorithm. As shown in [16] for adaptive linear predictors, this specific evolutionary algorithm can work in non-stationary environments.

A drawback of evolutionary algorithms proposed so far for adaptive filters [2, 3, 11, 13, 14, 15, 16, 19, 28] is the computational complexity inherent in the calculation of the appropriate error criterion. As the ***least squares*** error criterion is mainly used, the evaluation of each individual solution delivered by the ***evolutionary algorithm*** requires calculating a sum of squared errors over a time window of length L. A recursive implementation is more desirable for determining the error criterion that takes into account the history of the adaptive system.

Therefore, this paper proposes a modified multi-membered (μ, λ)-ES with self-adaptation of strategy parameters for the online adaptation of bilinear predictors, i.e., one generation corresponds to one time step of the stochastic signal. Providing each individual with an additional ***memory component***, the error criterion for an individual is recursively calculated with the help of its parents' memory. Experimental results for the prediction of non-stationary bilinear stochastic signals demonstrate the robustness and computational efficiency of this methodology.

This chapter is organized as follows. In the second section, non-linear and especially bilinear stochastic signal models are defined. The application of adaptive non-linear (bilinear) predictors to the analysis of non-linear (bilinear) stochastic signal processes is explained in the third section. The multi-membered (μ, λ)-ES with self-adaptation of strategy parameters and its modifications are detailed in the fourth section. The fifth section presents experimental results for the estimation of non-stationary bilinear stochastic signal models.

19.2 Non-Linear Stochastic Signal Models

Modern techniques in time series analysis for studying stochastic signals are based on parameterized stochastic signal models. A time-discrete ($t \in \mathbf{Z}$) stochastic signal process $\{x(t)\}$ is hereby assumed to be generated by a white Gaussian noise process $\{e(t)\}$ with mean $\mu_e = 0$ and variance σ_e^2 that provides the input signal of a non-linear system, the ***non-linear stochastic signal model*** (see Fig. 19.1). According to

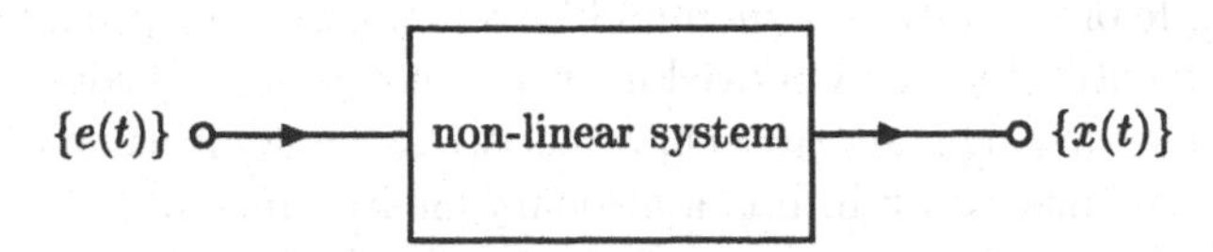

FIGURE 19.1. The non-linear stochastic signal model

[21], this non-linear system can be defined by the difference equation

$$x(t) = e(t) + h(x(t-1), \ldots, x(t-P), e(t-1), \ldots, e(t-Q), \boldsymbol{\theta}(t)) \tag{19.1}$$

with some non-linear function $h(\cdot)$ and, assuming finite parameter models, the N-dimensional parameter vector

$$\boldsymbol{\theta}(t) = (\theta_1(t), \ldots, \theta_N(t)) \in \mathbf{R}^N \tag{19.2}$$

With a known topology of the stochastic signal model, according to a known non-linear function $h(\cdot)$ in Eq. (19.1), the statistical properties of the stochastic signal process $\{x(t)\}$ are embodied in the system parameters $\theta_i(t)$, $i = 1(1)N$, and the variance σ_e^2 of the so-called *innovation process* $\{e(t)\}$.

In the past, a number of non-linear stochastic signal models were developed by appropriately defining the non-linear function $h(\cdot)$ [21, 26, 27]. A classical non-linear stochastic signal model is the *bilinear* model BL(p,q,m,n) given by the difference equation [21, 26]

$$\begin{aligned} x(t) \;=\; e(t) \;&-\; \sum_{i=1}^{p} a_i(t)x(t-i) + \sum_{j=1}^{q} b_j(t)e(t-j) \\ &+\; \sum_{i=1}^{m}\sum_{j=1}^{n} c_{i,j}(t)x(t-i)e(t-j) \end{aligned} \tag{19.3}$$

with BL-parameters $a_i(t)$, $b_j(t)$ and $c_{i,j}(t) \in \mathbf{R}$. The parameter vector $\boldsymbol{\theta}(t)$ of the BL(p,q,m,n) model is given by

$$\boldsymbol{\theta}(t) = \big(a_1(t), \ldots, a_p(t), b_1(t), \ldots, b_q(t), c_{1,1}(t), \ldots, c_{m,n}(t)\big) \tag{19.4}$$

with $N = p + q + mn$ components. Though simple in its structure, under mild conditions the bilinear model BL(p,q,m,n) is able to approximate to an arbitrary degree of accuracy any (discrete time) Volterra series representation of a time series over a finite time interval [21].

To analyse a stochastic signal process $\{x(t)\}$ with the help of a bilinear stochastic signal model, the parameter vector $\boldsymbol{\theta}(t)$ has to be estimated from one realization $x(t)$. The next section describes how adaptive bilinear predictors as specific adaptive filters can be applied to this parameter estimating task.

19.3 Adaptive Non-Linear Predictors

In digital signal processing, adaptive filters can be applied to a variety of tasks, e.g., system identification, adaptive equalization of data channels, echo cancellation, and stochastic signal estimation [6]. When a priori information about the statistical characteristics of the processed signals is unknown, the adaptive filter offers the capability of extracting the desired information from the input data. In Definition 1, the adaptive filter with scalar input and output signals is defined (see Fig. 19.2).

Definition 1 (adaptive filter) Given the realizations $x(t)$ and $y(t)$ of the stochastic signal processes $\{x(t)\}$ and $\{y(t)\}$, respectively, an adaptive filter $\mathcal{F}$ is defined as a system with input signal $x(t)$ and output signal

$$\widehat{y}(t) \;=\; \mathcal{F}\{\, x(t), \widehat{e}(t)\,\} \;\approx\; y(t) \tag{19.5}$$

that approximates signal $y(t)$ according to an error criterion. The residual value is $\widehat{e}(t) = y(t) - \widehat{y}(t)$. ◇

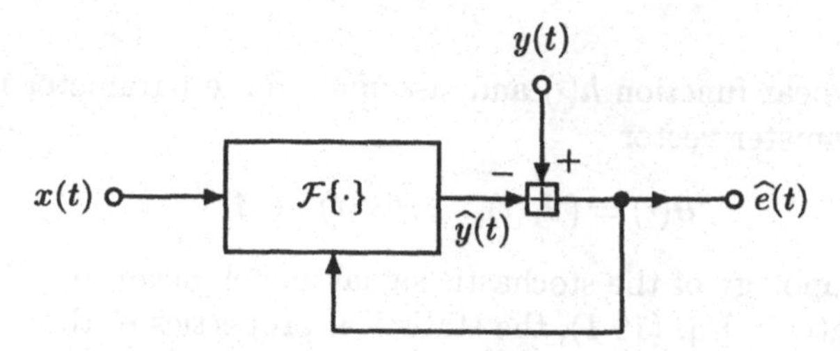

FIGURE 19.2. The adaptive filter

In general, a filter topology is defined by prescribing how $P + Q$ signal samples $x(t-1), \ldots, x(t-P)$ and $\widehat{e}(t-1), \ldots, \widehat{e}(t-Q)$ are combined.

To analyse a stochastic signal process $\{x(t)\}$ generated by the non-linear difference equation in Eq. (19.1), the *prediction error filter* with the signal flow graph in Fig. 19.3 can be used. The output of the prediction error filter is the *residual*

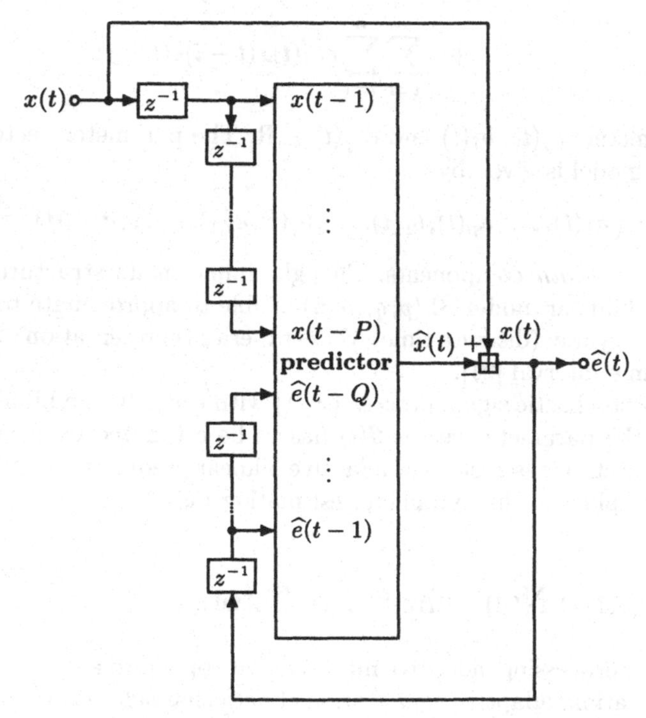

FIGURE 19.3. The prediction error filter (z^{-1} denotes the unit time delay)

$$\widehat{e}(t) = x(t) - \widehat{x}(t) \tag{19.6}$$

as the difference between the given signal value $x(t)$ and the predicted value obtained as the output signal $\widehat{x}(t)$ of a *non-linear predictor* defined by

$$\begin{aligned} \widehat{x}(t) \; = \; & h(\, x(t-1), \ldots, x(t-P), \\ & \widehat{e}(t-1), \ldots, \widehat{e}(t-Q), \widehat{\boldsymbol{\theta}}(t)\,) \end{aligned} \tag{19.7}$$

Since the innovation process $\{e(t)\}$ is not observable, the residual $\widehat{e}(t)$ is used in Eq. (19.7).

The non-linear predictor becomes adaptive by estimating the parameter vector

$$\widehat{\boldsymbol{\theta}}(t) = (\,\widehat{\theta}_1(t), \ldots, \widehat{\theta}_N(t)\,) \in \mathbf{R}^N \tag{19.8}$$

from one realization $x(t)$ of the stochastic signal process $\{x(t)\}$ with respect to the error criterion in Definition 2.

Definition 2 (error criterion) According to the method of *recursive least squares*, the error criterion requires minimizing

$$E(t) = \sum_{\tau=0}^{t} \beta^{\tau} \cdot \widehat{e}^2(t-\tau) \overset{\widehat{\boldsymbol{\theta}}(t)}{\longrightarrow} \text{minimum} \tag{19.9}$$

with weighting factor $0 < \beta \lesssim 1$. ◇

The calculation of $E(t)$ can be carried out recursively with the help of

$$E(t) = \beta \cdot E(t-1) + \widehat{e}^2(t) \tag{19.10}$$

For bilinear $\mathrm{BL}(p,q,m,n)$ stochastic signal models the corresponding adaptive bilinear predictor $\mathcal{F}$ is defined by the difference equation [9, 13, 14, 15, 17, 19, 20]

$$\begin{aligned} \widehat{x}(t) &= -\sum_{i=1}^{p} \widehat{a}_i(t)x(t-i) + \sum_{j=1}^{q} \widehat{b}_j(t)\widehat{e}(t-j) \\ &+ \sum_{i=1}^{m}\sum_{j=1}^{n} \widehat{c}_{i,j}(t)x(t-i)\widehat{e}(t-j) \end{aligned} \tag{19.11}$$

with $P = \max\{p, m\}$ and $Q = \max\{q, n\}$. The estimated parameter vector $\widehat{\boldsymbol{\theta}}(t)$ at time t according to Eq. (19.4) is given by

$$\widehat{\boldsymbol{\theta}}(t) = (\,\widehat{a}_1(t), \ldots, \widehat{a}_p(t), \widehat{b}_1(t), \ldots, \widehat{b}_q(t), \widehat{c}_{1,1}(t), \ldots, \widehat{c}_{m,n}(t)\,) \tag{19.12}$$

with $N = p + q + mn$ components.

In the next section, the application of the (μ, λ)-ES to the online adaptation of bilinear predictors is detailed.

19.4 (μ, λ)-Evolution Strategy with Self-Adaptation of Strategy Parameters

In *evolutionary computation* the *Evolution Strategy* (ES) is a robust optimization technique, especially for numerical parameter optimization problems [22, 23, 24]. In this paper the *multi-membered (μ, λ)-ES with self-adaptation of strategy parameters* is considered [1]. This specific *evolutionary algorithm* works on a *population*

$$\pi_\mu(t) := (\,\iota_1(t), \ldots, \iota_\mu(t)\,) \in I^\mu \tag{19.13}$$

consisting of μ *individuals* $\iota_k(t)$, $k = 1(1)\mu$. In the standard (μ, λ)-ES, each individual is given by $\iota_k(t) := (\,\widehat{\boldsymbol{\theta}}_k(t), \boldsymbol{\sigma}_k(t)\,)$. To obtain a computationally efficient algorithm that can work in non-stationary environments, $\iota_k(t)$ is augmented, leading to the quadruple

$$\iota_k(t) := (\,\widehat{\boldsymbol{\theta}}_k(t), \boldsymbol{\sigma}_k(t), \boldsymbol{\varepsilon}_k(t), \eta_k(t)\,) \in I \tag{19.14}$$

The corresponding ***object parameter vector***

$$\widehat{\boldsymbol{\theta}}_k(t) := (\,\widehat{\theta}_{k,1}(t), \ldots, \widehat{\theta}_{k,N}(t)\,) \in \mathbf{R}^N \tag{19.15}$$

represents one parameter vector of the adaptive bilinear predictor $\mathcal{F}$. Each individual quadruple comprises the ***strategy parameter vector***

$$\boldsymbol{\sigma}_k(t) := (\,\sigma_{k,1}(t), \ldots, \sigma_{k,N}(t)\,) \in \mathbf{R}_+^N \tag{19.16}$$

that is used in the mutation operator defined below to generate new individuals. Additionally, the ***memory component***

$$\boldsymbol{\varepsilon}_k(t) := (\,\widehat{e}_k(t-1), \ldots, \widehat{e}_k(t-Q)\,) \in \mathbf{R}^Q \tag{19.17}$$

is incorporated in each individual $\iota_k(t)$. This memory component $\boldsymbol{\varepsilon}_k(t)$ is a novel feature introduced here to the standard (μ, λ)-ES. Its objective is to allow for the recursive calculation of the error criterion as detailed below. The ***fitness*** assigned to individual $\iota_k(t)$ in the last time step $(t-1)$ is denoted by

$$\eta_k(t) = E_k(t-1) \in \mathbf{R}_+ \tag{19.18}$$

with the ***recursive least squares*** error criterion $E_k(t-1)$. Thus, $\iota_k(t) \in I$ with $I := \mathbf{R}^N \times \mathbf{R}_+^N \times \mathbf{R}^Q \times \mathbf{R}_+$. With these definitions, the structure of the modified multi-membered (μ, λ)-ES with self-adaptation of strategy parameters is given in Fig. 19.4.

To create the population $\pi_\mu(t+1)$, the operators ***recombination*** $\mathcal{R}$, ***mutation*** $\mathcal{M}$, and ***selection*** $\mathcal{S}$ are applied to population $\pi_\mu(t)$ [1]. In the first step, the offspring population $\pi'_\lambda(t) \in I^\lambda$ consisting of $\lambda > \mu$ offspring $\iota'_\ell(t)$, $\ell = 1(1)\lambda$, is created by the recombination operator $\mathcal{R}$. Using ***discrete sexual recombination*** of object parameters [1], two parents $\iota_{k_1(\ell)}(t)$ and $\iota_{k_2(\ell)}(t)$ are drawn uniformly at random from $\pi_\mu(t)$. The object parameter vector $\widehat{\boldsymbol{\theta}}'_\ell(t)$ of offspring $\iota'_\ell(t)$ is built by selecting each component $\widehat{\theta}'_{\ell,i}(t)$, $i = 1(1)N$, with equal probability from one of the two parental vectors $\widehat{\boldsymbol{\theta}}_{k_1(\ell)}(t)$ and $\widehat{\boldsymbol{\theta}}_{k_2(\ell)}(t)$.

$$\widehat{\theta}'_{\ell,i}(t) = \begin{cases} \widehat{\theta}_{k_1(\ell),i}(t) & \text{with probability } \frac{1}{2} \\ \widehat{\theta}_{k_2(\ell),i}(t) & \text{with probability } \frac{1}{2} \end{cases} \tag{19.19}$$

According to ***generalized intermediate recombination*** of strategy parameters [1], the vector $\boldsymbol{\sigma}'_\ell(t)$ is calculated with the help of the arithmetical mean

$$\sigma'_{\ell,i}(t) = \frac{1}{\mu} \sum_{k=1}^{\mu} \sigma_{k,i}(t) \tag{19.20}$$

Then the mutation operator $\mathcal{M}$ is applied to $\pi'_\lambda(t)$, leading to population $\pi''_\lambda(t) \in I^\lambda$. Each offspring $\iota'_\ell(t)$ is mutated with respect to the strategy parameters

$$\sigma''_{\ell,i}(t) = \sigma'_{\ell,i}(t) \cdot \mathrm{e}^{\alpha \cdot z^\sigma_{\ell,i}(t)} \tag{19.21}$$

Notation

$\pi_\mu(t) := (\iota_1(t), \ldots, \iota_\mu(t)) \in I^\mu$
$\pi_\lambda^{\prime,\prime\prime}(t) := (\iota_1^{\prime,\prime\prime}(t), \ldots, \iota_\lambda^{\prime,\prime\prime}(t)) \in I^\lambda$
$\iota_k(t) := (\widehat{\boldsymbol{\theta}}_k(t), \boldsymbol{\sigma}_k(t), \boldsymbol{\varepsilon}_k(t), \eta_k(t)) \in I$
$\widehat{\boldsymbol{\theta}}_k(t) := (\widehat{\theta}_{k,1}(t), \ldots, \widehat{\theta}_{k,N}(t)) \in \mathbf{R}^N$
$\boldsymbol{\sigma}_k(t) := (\sigma_{k,1}(t), \ldots, \sigma_{k,N}(t)) \in \mathbf{R}_+^N$
$\boldsymbol{\varepsilon}_k(t) := (\widehat{e}_k(t-1), \ldots, \widehat{e}_k(t-Q)) \in \mathbf{R}^Q$
$\eta_k(t) := E_k(t-1) \in \mathbf{R}_+$
$I := \mathbf{R}^N \times \mathbf{R}_+^N \times \mathbf{R}^Q \times \mathbf{R}_+$

Algorithm

$t := 0;$
initialize $\pi_\mu(0)$;
while end of adaptation $\neq$ true **do**
 recombine $\pi'_\lambda(t) := \mathcal{R}\{\pi_\mu(t)\}$;
 mutate $\pi''_\lambda(t) := \mathcal{M}\{\pi'_\lambda(t)\}$;
 evaluate $\pi''_\lambda(t)$;
 select $\pi_\mu(t+1) := \mathcal{S}\{\pi''_\lambda(t)\}$;
 $t := t+1;$
done

FIGURE 19.4. (μ, λ)-ES with self-adaptation of strategy parameters

with $\alpha \in \mathbf{R}$ and the Gaussian, independent, identically $N(0,1)$-distributed random numbers $z_{\ell,i}^\sigma(t)$ [1]. Using these mutated strategy parameters, the object parameter vector $\widehat{\boldsymbol{\theta}}_\ell''(t)$ is obtained via

$$\widehat{\theta}_{\ell,i}''(t) = \widehat{\theta}_{\ell,i}'(t) + \sigma_{\ell,i}''(t) \cdot z_{\ell,i}^\theta(t) \tag{19.22}$$

with Gaussian, independent, identically $N(0,1)$-distributed random numbers $z_{\ell,i}^\theta(t)$ [1].

In the last step, the *selection* operator $\mathcal{S}$ is applied to the offspring population $\pi''_\lambda(t)$, implementing the Darwinian principle of the *survival of the fittest*. To this end, each offspring $\iota''_\ell(t)$ has to be evaluated. In former *evolutionary computation* approaches the sum of squared errors

$$\sum_{\tau=0}^{L-1} \beta^\tau \ \left[\widehat{e}_\ell''(t-\tau)\right]^2 \tag{19.23}$$

was calculated for each offspring $\iota''_\ell(t)$ [2, 3, 11, 13, 14, 15, 16, 19, 28]. To reduce the computational complexity of the (μ, λ)-ES as an online adaptation procedure, this relationship is replaced by a recursive difference equation [12, 17, 18], taking into account Eq. (19.10). This is where the proposed memory component steps in. First of all, the memory component $\boldsymbol{\varepsilon}_{k_1(\ell)}(t)$ and the fitness $\eta_{k_1(\ell)}(t) = E_{k_1(\ell)}(t-1)$ of the first parent $\iota_{k_1(\ell)}(t)$ are transferred to offspring $\iota'_\ell(t)$ and $\iota''_\ell(t)$.

$$\boldsymbol{\varepsilon}_\ell''(t) = \boldsymbol{\varepsilon}_\ell'(t) = \boldsymbol{\varepsilon}_{k_1(\ell)}(t) \tag{19.24}$$

$$\eta_\ell''(t) = \eta_\ell'(t) = \eta_{k_1(\ell)}(t) \tag{19.25}$$

With the help of the object parameter vector $\widehat{\boldsymbol{\theta}}''_\ell(t)$ and the memory component $\boldsymbol{\varepsilon}''_\ell(t)$ the predicted value

$$\widehat{x}''_\ell(t) = h(x(t-1),\ldots,x(t-P), \widehat{e}''_\ell(t-1),\ldots,\widehat{e}''_\ell(t-Q),\widehat{\boldsymbol{\theta}}''_\ell(t)) \tag{19.26}$$

and the corresponding residual

$$\widehat{e}''_\ell(t) = x(t) - \widehat{x}''_\ell(t) \tag{19.27}$$

are calculated. Using $\eta''_\ell(t) = E''_\ell(t-1) = E_{k_1(\ell)}(t-1)$ leads to the *recursive least squares* error criterion (see Eq. (19.10))

$$\begin{aligned} E''_\ell(t) &= \beta\cdot\eta''_\ell(t) + [\widehat{e}''_\ell(t)]^2 \\ &= \beta\cdot E''_\ell(t-1) + [\widehat{e}''_\ell(t)]^2 \end{aligned} \tag{19.28}$$

The offspring quadruples $\iota''_\ell(t)$, $\ell = 1(1)\lambda$, are ranked according to $E''_\ell(t)$, and the best μ offspring are selected by the selection operator $\mathcal{S}$ as parents $\iota_k(t+1)$, $k = 1(1)\mu$, for the next time step $t+1$, i.e., assuming

$$E''_{\ell_1}(t) \le E''_{\ell_2}(t) \le \cdots \le E''_{\ell_\lambda}(t) \tag{19.29}$$

it follows that

$$\widehat{\boldsymbol{\theta}}_k(t+1) = \widehat{\boldsymbol{\theta}}''_{\ell_k}(t) \tag{19.30}$$

$$\boldsymbol{\sigma}_k(t+1) = \boldsymbol{\sigma}''_{\ell_k}(t) \tag{19.31}$$

The memory component $\boldsymbol{\varepsilon}_k(t+1)$ is updated

$$\begin{aligned} \boldsymbol{\varepsilon}_k(t+1) &= (\widehat{e}_k(t),\ldots,\widehat{e}_k(t-Q+1)) \\ &= (\widehat{e}''_{\ell_k}(t),\ldots,\widehat{e}''_{\ell_k}(t-Q+1)) \end{aligned} \tag{19.32}$$

and the fitness of $\iota_k(t+1)$ is obtained by

$$\eta_k(t+1) = E_k(t) = E''_{\ell_k}(t) \tag{19.33}$$

The exogenous parameters that must be specified for the multi-membered (μ,λ)-ES with self-adaptation of strategy parameters are μ, λ, and α. To produce a single estimate $\widehat{\boldsymbol{\theta}}(t)$ for the adaptive predictor's parameter vector at time t, the object parameter vector of the best offspring individual $\eta''_{\ell_1}(t)$ with index ℓ_1 is chosen, i.e.,

$$\widehat{\boldsymbol{\theta}}(t) = \widehat{\boldsymbol{\theta}}''_{\ell_1}(t) \tag{19.34}$$

19.5 Experimental Results

A multi-membered (μ,λ)-ES with self-adaptation of strategy parameters is applied to estimating bilinear stochastic signal models. The settings of the exogenous parameters of the (μ,λ)-ES are given in Table 19.1.

Number of parents	μ	10
Number of offspring	λ	50
Adaptation factor	α	0.01

TABLE 19.1. The exogenous parameters for the (μ, λ)-ES

The individual object and strategy parameters of $\pi_\mu(0)$ are initialized for $i = 1(1)N$ and $k = 1(1)\mu$ according to $\theta_{k,i}(0) = 0$ and $\sigma_{k,i}(0) = 0.005$, respectively. For the error criterion, the weighting factor $\beta = 0.95$ is used. To obtain statistically significant results, 50 simulation runs are carried out for each experiment and the averages $\overline{\widehat{\theta}}_i(t)$, $i = 1(1)N$, are calculated.

In the first experiment, the prediction of a non-stationary bilinear BL(3,0,1,1) stochastic signal process with BL-parameters

$$a_1(t) = -0.3, \quad a_2(t) = -0.6, \quad a_3(t) = 0.6$$

and

$$c_{1,1}(t) = \begin{cases} -0.1, & 0 \le t < 5000 \\ 0.1, & 5000 \le t \le 10000 \end{cases}$$

is considered in the time interval $0 \le t \le 10000$. Fig. 19.5 shows the averages $\overline{\widehat{\theta}}_i(t)$ of the estimated parameters $\widehat{\theta}_i(t)$ (solid lines) and the true parameters $\theta_i(t)$ (dotted lines). It is apparent that the (μ, λ)-ES can quickly approximate the true parameters and to track the time-varying parameter $c_{1,1}(t)$.

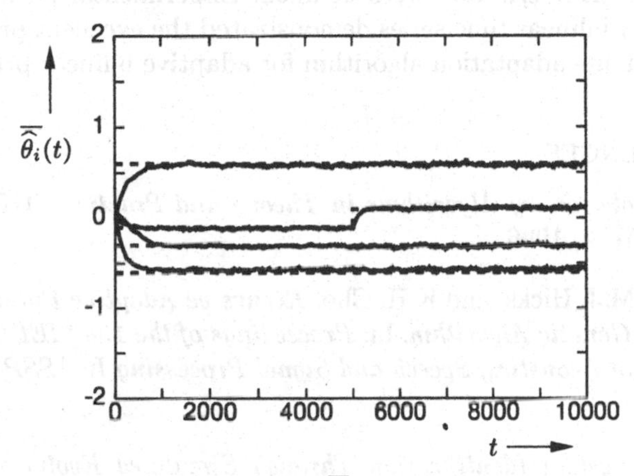

FIGURE 19.5. Averages $\overline{\widehat{\theta}}_i(t)$ for BL(3,0,1,1) model

In the second experiment, the non-stationary bilinear BL(2,0,2,1) stochastic signal process defined by the BL-parameters

$$a_1(t) = \begin{cases} -0.7, & 0 \le t < 5000 \\ -1.2, & 5000 \le t \le 10000 \end{cases}$$

and

$$a_2(t) = 0.9, \quad c_{1,1}(t) = -0.2, \quad c_{2,1}(t) = 0.3$$

is analysed in the time interval $0 \le t \le 10000$. As shown in Fig. 19.6 by the averages $\overline{\widehat{\theta}}_i(t)$, the (μ, λ)-ES again can estimate the true parameters $\theta_i(t)$ with good accuracy and is able to track the parameter $a_1(t)$.

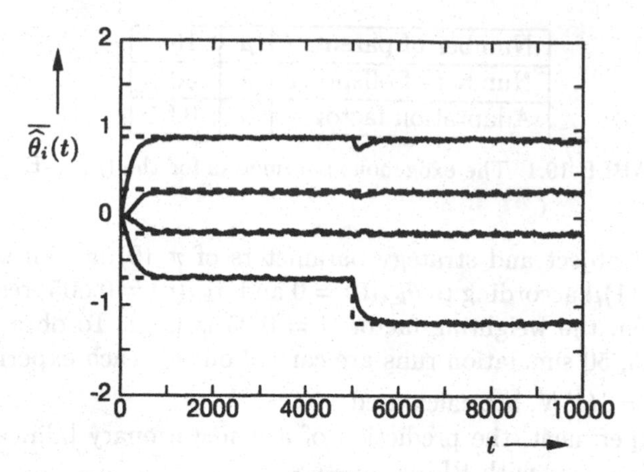

FIGURE 19.6. Averages $\overline{\widehat{\theta}_i}(t)$ for BL(2,0,2,1) model

19.6 Conclusion

The application of a multi-membered (μ, λ)-ES with self-adaptation of strategy parameters to the online adaptation of bilinear predictors was presented. Special emphasis was given to the tracking capabilities in non-stationary environments and to the development of a robust and computationally efficient adaptation algorithm. The modifications of the standard (μ, λ)-ES with respect to the recursive calculation of the error criterion were detailed. Experimental results for predicting non-stationary bilinear time series demonstrated the excellent properties of the (μ, λ)-ES as an online adaptation algorithm for adaptive bilinear predictors.

19.7 References

[1] T. Bäck. *Evolutionary Algorithms in Theory and Practice.* Oxford University Press, New York, 1996.

[2] D.M. Etter, M.J. Hicks, and K.H. Cho. *Recursive Adaptive Filter Design Using an Adaptive Genetic Algorithm.* In: *Proceedings of the 1982 IEEE International Conference on Acoustics, Speech and Signal Processing ICASSP'82.* 2:635–638, 1982.

[3] D.B. Fogel. *System Identification Through Simulated Evolution: A Machine Learning Approach to Modeling.* Ginn Press, Needham Heights, 1991.

[4] D.B. Fogel. *Evolutionary Computation: Toward a New Philosophy of Machine Intelligence.* IEEE Press, New York, 1995.

[5] L.J. Fogel, A.J. Owens, and M.J. Walsh. *Artificial Intelligence Through Simulated Evolution.* John Wiley & Sons, New York, 1966.

[6] S. Haykin. *Adaptive Filter Theory.* Prentice-Hall, Englewood Cliffs, N.J., 1986.

[7] J.H. Holland. *Adaptation in Natural and Artificial Systems: An Introductory Analysis with Applications to Biology, Control, and Artificial Intelligence.* First MIT Press Edition, Cambridge, 1992.

[8] J.R. Koza. *Genetic Programming: On the Programming of Computers by Means of Natural Selection.* The MIT Press, Cambridge, 1992.

[9] J. Lee and V.J. Mathews. *Adaptive Bilinear Predictors.* In: *Proceedings of the 1994 IEEE International Conference on Acoustics, Speech and Signal Processing ICASSP'94.* Adelaide, 3:489–492, 1994.

[10] V.J. Mathews. *Adaptive Polynomial Filters.* In: *IEEE Signal Processing Magazine.* 8(3):10–26, 1991.

[11] J.R. McDonnell and D. Waagen. *Evolving Recurrent Perceptrons for Time-Series Modeling.* In: *IEEE Transactions on Neural Networks: Special Issue on Evolutionary Computation.* 5(1):24–38, 1994.

[12] A. Neubauer. *Linear Signal Estimation Using Genetic Algorithms.* In: *Systems Analysis Modelling Simulation.* Gordon and Breach Science Publishers, Amsterdam, 18/19:349–352, 1995.

[13] A. Neubauer. *Real-Coded Genetic Algorithms for Bilinear Signal Estimation.* In: D. Schipanski (Hrsg.): *Tagungsband des 40. Internationalen Wissenschaftlichen Kolloquiums.* Ilmenau, Band 1, 347–352, 1995.

[14] A. Neubauer. *Non-Linear Adaptive Filters Based on Genetic Algorithms with Applications to Digital Signal Processing.* In: *Proceedings of the 1995 IEEE International Conference on Evolutionary Computation ICEC'95.* Perth, 2:527–532, 1995.

[15] A. Neubauer. *Genetic Algorithms for Non-Linear Adaptive Filters in Digital Signal Processing.* In: K.M. George, J.H. Carroll, D. Oppenheim., and J. Hightower. (Eds.) *Proceedings of the 1996 ACM Symposium on Applied Computing SAC'96.* Philadelphia, 519–522, 1996.

[16] A. Neubauer. *A Comparative Study of Evolutionary Algorithms for On-Line Parameter Tracking.* In: H.-M. Voigt, W. Ebeling, I. Rechenberg, and H.-P. Schwefel. (Eds.) *Parallel Problem Solving from Nature PPSN IV.* Springer-Verlag, Berlin, 624–633, 1996.

[17] A. Neubauer. *Prediction of Nonlinear and Nonstationary Time-Series Using Self-Adaptive Evolution Strategies with Individual Memory.* In: Bäck, Th. (Ed.): *Proceedings of the Seventh International Conference on Genetic Algorithms ICGA'97.* Morgan Kaufmann Publishers, San Francisco, 727–734, 1997.

[18] A. Neubauer. *On-Line System Identification Using The Modified Genetic Algorithm.* In: *Proceedings of the Fifth Congress on Intelligent Techniques and Soft Computing EUFIT'97.* Aachen, 2:764–768, 1997.

[19] A. Neubauer. *Genetic Algorithms in Automatic Fire Detection Technology.* In: *IEE Proceedings of the 2nd International Conference on Genetic Algorithms in Engineering Systems: Innovations and Applications GALESIA'97.* Glasgow, 180–185, 1997.

[20] A. Neubauer. *Adaptive Filter auf der Basis genetischer Algorithmen.* VDI-Verlag, Düsseldorf, 1997.

[21] M.B. Priestley. *Non-linear and Non-stationary Time Series Analysis.* 1st Paperback Edition, Academic Press, London, 1991.

[22] I. Rechenberg. *Evolutionsstrategie '94.* Frommann-Holzboog, Stuttgart, 1994.

[23] H.-P. Schwefel. *Numerical Optimization of Computer Models.* John Wiley & Sons, Chichester, 1981.

[24] H.-P. Schwefel. *Evolution and Optimum Seeking.* John Wiley & Sons, New York, 1995.

[25] P. Strobach. *Linear Prediction: A Mathematical Basis for Adaptive Systems.* Springer-Verlag, Berlin, 1990.

[26] T. Subba Rao and M.M. Gabr. *An Introduction to Bispectral Analysis and Bilinear Time Series Models.* Springer-Verlag, Berlin, 1984.

[27] H. Tong. *Non-linear Time Series: A Dynamical System Approach.* Oxford University Press, New York, 1990.

[28] M.S. White and S.J. Flockton. *Genetic Algorithms for Digital Signal Processing.* In: T.C. Fogarty. (Ed.): *Evolutionary Computing.* Springer-Verlag, Berlin, 291–303, 1994.

20

Non-Linear Dynamic Modelling with Neural Networks

Jose C. Principe[1]
Ludong Wang[2]
Jyh-Ming Kuo[3]

ABSTRACT
This paper discusses the use of artificial neural networks (ANNs) for dynamic modelling of time series. We briefly present the theoretical basis for the modelling as a prediction of a vector time series in reconstructed space, and address the important role of the delay operator to implement Takens' embedding theorem. Two types of dynamic models with vastly different topologies will be reviewed: the global models, and the local models.
Global models can be built from multilayer perceptrons extended with memory structures. In order to train and test dynamic models, we argue that iterated prediction is more appropriate to capture the dynamics, because it imposes more constraints during learning than single step prediction. We show how this method can be implemented by a recurrent ANN trained with trajectory learning. Local modelling partitions the phase space and each model specializes in the local dynamics. Each model can be rather simple and here linear models will be used. A modified Kohonen network is developed to first cluster and organize the trajectories in state space. We show that the weights of the Kohonen layer can be used directly to construct the local models. Experimental results corroborate the proposed methods.

20.1 Introduction

The search for a model of an experimental time series has been an important problem in science. Extracting a model from a time series using the prediction framework can be framed as a function approximation problem, since the model system tries to find the mapping that creates the next sample of the time series which is given as the desired response. In function approximation we attempt to approximate a complicated function *f(x)* by another function $\tilde{f}(x)$ composed of simpler functions,

$$\tilde{f}(x) = \sum_{i=1}^{N} w_i \varphi_i(x) \tag{20.1}$$

where w_i are scalar coefficients, $\varphi_i(x)$ are the basis functions and N is the dimension of the projection space. The error between *f(x)* and the approximation $\tilde{f}(x)$ should

[1]Computational NeuroEngineering Laboratory, University of Florida, Gainesville, Florida, U.S.A., E-mail: principe@cnel.ufl.edu
[2]Hughes Network Systems, Germantown, MD 20876, U.S.A.
[3]Kaohsiung Polytechnic Institute, Taiwan

be as small as possible in some norm. From a function approximation point of view there are three basic issues that must be addressed: what are the basis $\varphi(x)$, how to select the coefficients w_i and how to select the number of basis N.
For a long time the linear model was almost exclusively used to describe the system that produced the time series, which means that the basis are a linear combination of the past samples or/and of the previous model outputs [2]. Recently, non-linear adaptive models have been developed as alternatives to the linear model. This means that the basis functions in Eq. (20.1) are non-linear related to the input signal x.
In this paper we will describe and compare two non-linear choices for the basis implemented by artificial neural networks: global basis created by a multilayer perceptron (MLP) and local basis created by a Kohonen self-organizing feature map (SOFM). The task is dynamic modelling which is the identification a non-linear dynamical system with chaotic dynamics. This problem is an excellent test ground to address the accuracy of the modelling and of the training algorithms. Artificial neural networks have been proposed for this task [13] since they are universal approximators [17] and the coefficients w_i can be computed directly from the time series, yielding directly a solution that is ready to be delivered. It is too early to say which methodology is preferable [22], but it is important to extend to the non-linear case the extensive work carried out in linear time series analysis.

20.2 Principles of Dynamic Modelling

Figure 20.1 shows the block diagram that forms the basis for linear modelling. H(z) is a linear predictor for the time series, and after convergence its output is white noise. If the time series x(n) is produced by a linear unknown model U(z) with

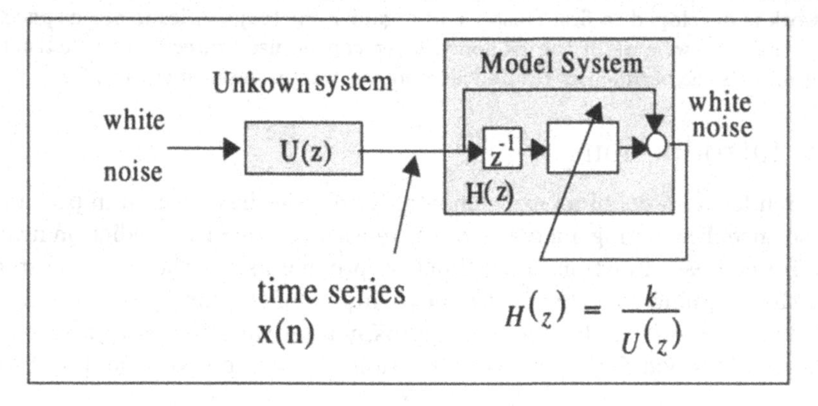

FIGURE 20.1. The linear model for time series prediction

constant coefficients, the linear model H(z) can be used to identify U(z), since the output of H(z) is white. There are two things to note in this methodology. First, there is an inverse relation between the model system H(z) and the unknown system. Secondly, prediction is the tool to identify the parameters of the system that produced the time series.
The goal in this paper is to substitute the linear model H(z) by a non-linear dynamical system with adaptable parameters. When one wants to extend the linear

methodology to non-linear systems, the relation between the model system and the unknown system has to be stated in different terms since the non-linear model system can no longer be described by a transfer function. The work of Takens [20], Casdagli [3] and many others have shown that an equivalent methodology exists for non-linear modelling, which is defined as the identification of the mapping $\Phi : R^d \rightarrow R^d$ that defines the unknown dynamical system(Figure 20.2). Figure 20.2

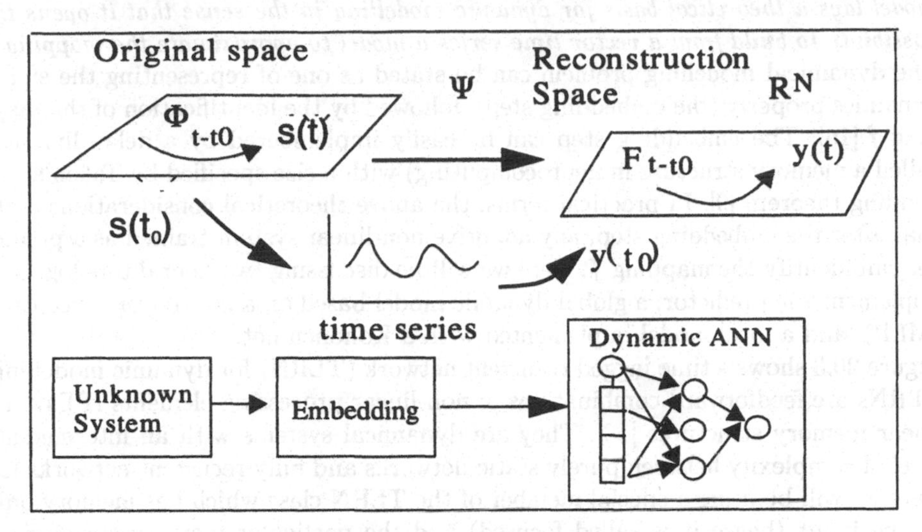

FIGURE 20.2 Non-linear modelling

shows the important steps required in non-linear modelling. Note that the unknown system is assumed autonomous (i.e. without an external input), unlike the linear case, which means that the complexity in the time series is solely due to model dynamics. First, the time series has to be transformed into a trajectory by using one of the embedding techniques [18] (the most common is a time delay embedding [20]. If the dimension N of the reconstruction space is larger than $2D_e + 1(D_e$ dimension of the original system), then the map Ψ is diffeomorphic and the flow F will preserve the dynamic invariants of the original system Φ. The dimension of the dynamical system can be estimated from the time series using for instance the ***correlation dimension algorithm*** [7]. Second, prediction of the next point in state space identifies the non-linear mapping F that governs the evolution of the trajectory in reconstruction space. In fact, according to Takens' embedding theorem, there exists a map $F : R^N \rightarrow R^N$ that transforms the current reconstructed state $x(t)$ to the next state $x(n + \tau)$, where τ is the normalized delay. For simplicity we will set $\tau = 1$, which means

$$x(n+1) = F(x(n)) \tag{20.2}$$

or

$$\begin{bmatrix} x(n+1) \\ \cdots \\ x(n-N+2) \end{bmatrix} = F(\begin{bmatrix} x(n) \\ \cdots \\ x(n-N+1) \end{bmatrix})$$

Note that Eq. (20.2) specifies a multiple input, multiple output system F built from several(non-linear) filters and a predictor. The predictive mapping is the center piece of modelling since once determined, F can be obtained from the predictive mapping by simple matrix operations. The predictive mapping $f:\ R^N \to R$ can be expressed as

$$x(n+1) = f(\mathbf{x}(n)) \tag{20.3}$$

where $\mathbf{x}(n) = [x(n)\ x(n-1)...\ x(n-N+1)]^T$. Eq. (20.3) defines a deterministic non-linear autoregressive (NAR) model of the signal. *The existence of this predictive model lays a theoretical basis for dynamic modelling in the sense that it opens the possibility to build from a vector time series a model to approximate the mapping f.* The dynamical modelling problem can be stated as one of representing the signal dynamics properly (the embedding step), followed by the identification of the mapping f [16]. The embedding step can be easily implemented by a delay line(also called a memory structure in neurocomputing) with a size specified by Taken's embedding theorem [9]. In practical terms, the above theoretical considerations state that after the embedding step, any adaptive non-linear system trained as a predictor can identify the mapping f. Here we will be discussing two neural topologies to implement the predictor: a global dynamic model based on a multilayer perceptron (MLP), and a local model implemented with a Kohonen net.
Figure 20.3 shows a time lagged recurrent network (TLRN) for dynamic modelling. TLRNs are feedforward combinations of non-linear processing elements (PEs) and linear memory structures [16]. They are dynamical systems with an intermediate level of complexity between purely static networks and fully recurrent networks [5]. Here we will be using a special member of the TLRN class which has memory only at the input (hence it is called focused) and the particular memory structure is made up of ideal delay operators (a tap delay line). In neurocomputing this topology is called a time delay neural network (TDNN) [12].
The memory structure at the input is a representation layer called the embedding layer because it effectively implements the embedding to reconstruct the state trajectory. The non-linear mapping achieved with the MLP is static, which assumes that the dynamical system parameters do not change over time. Using functional analysis, Sandberg [17] proved that a memory layer followed by a MLP is able to approximate any functional mapping as the ones we are addressing here. Dynamic neural networks are global models of the dynamics in the sense that the entire

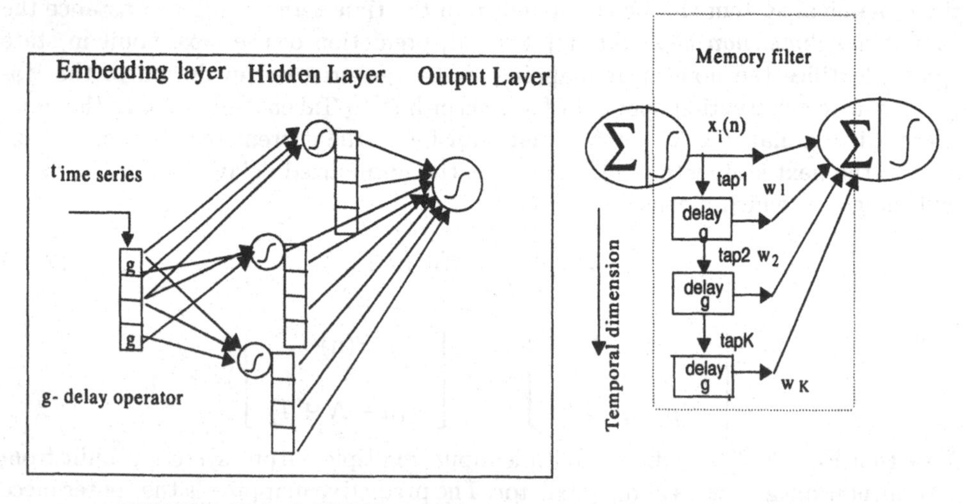

FIGURE 20.3. A TLRN architecture and detail of the memory PE

state space is being modeled by a single system. Due to the complexity of the trajectories, non-linear mappers with many degrees of freedom are normally required. Radial Basis Function(RBFs) networks have also been used for global modelling with good results [1] [8]. Although the basis of RBFs are local(Gaussians) the network is trained with data from the entire reconstruction space, so the model is global.
An alternative methodology is to partition the state space in local regions and model individually the local dynamics in each local region. Each local model can be potentially simpler, in fact a linear model. The problem is how to guarantee smoothness at the boundaries among models. Crutchfield [4] has shown that dynamic modelling fails if this condition is not imposed.
We will utilize a Kohonen map to represent the trajectories in reconstructed space. The Kohonen map is trained globally ensuring smoothness at the neighborhood boundaries, and it is more efficient than RBFs. Our work shows that the dynamics can be modeled by a set of local models, each directly fitted to the quantized state space obtained from the Kohonen self-organizing feature map (SOFM). The method has the advantages of compact state space representation of the original time series, simple state selection (winner take all), and state locality. The latter feature is ensured by the neighborhood preserving property which is the direct consequence of Kohonen's SOFM training. Figure 20.4 shows the neural topology for local dynamic modelling. As in the previous case we have an embedding layer built from a delay line to map the time series into state space. But now the embedding layer is followed by the Kohonen SOFM and the linear layer that extracts the linear models from neighborhoods of the Kohonen map. It is instructive to compare this topology with Eq. (20.1). According to this equation, the outputs of the PEs in the Kohonen map represent the local basis for function approximation.

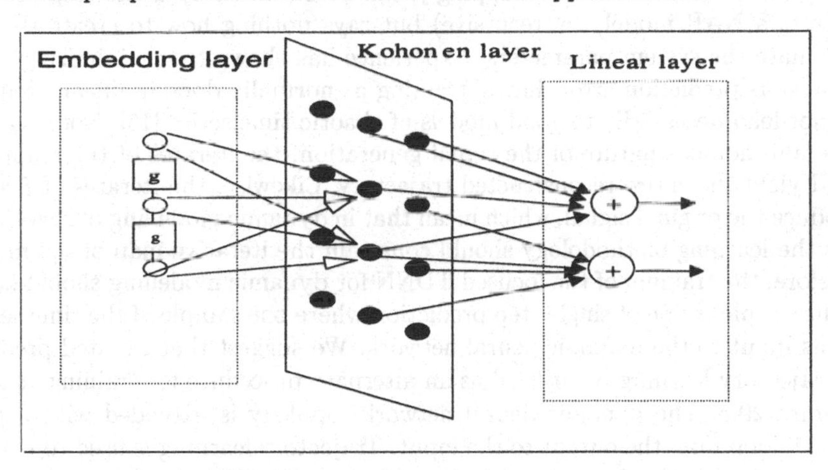

FIGURE 20.4. Structure of Kohonen SOFM for dynamic modelling

20.3 Neural Topologies for Global Modelling

We will treat first the case of dynamic neural networks with a memory layer built from ideal delay operators, which has been called TDNN in the literature. The view of the input delay line in the focused TDNN as an embedding layer implies that

its size can be determined using Takens' embedding theorem [20]. Takens' theorem specifies the size of the reconstruction space to avoid crossing trajectories, but leaves the normalized delay operator τ undetermined. However when the delay line is implemented it assumes a normalized delay of $\tau = 1$ time step. There are two ways to compensate for this: either a delay line of size $N = D_e\tau$ is used, or a delay line of size $N = D_e$ built from multiple delay elements $(z^{-\tau})$ is chosen. Both strategies span the same memory depth $(D_e\tau)$ and we prefer the latter since it reduces the number of network weights in the first layer.
Takens' theorem is a pessimistic bound for the dimension of the reconstruction space, and more refined methods to estimate the required memory depth for the embedding exist [14]. In the case of the focused TDNN we propose to estimate the product $D_e\tau$ directly from the correlation integral map as the dimension of the reconstruction space where the saturation in the correlation integral curves starts to appear because at this value of embedding dimension the attractor is fully enclosed [11]. Experimentally this method produce the same results as Mead's method, and does not require the knowledge of the dynamical equation.
The selection of the MLP topology is a nontrivial step. We recommend that weight decay be utilized during training to produce a compact model that generalizes well. Due to the universal mapping capabilities of the 1 hidden layer MLP this is the recommended topology. The output PE should be linear to conform with Eq. (20.1). Notice also that according to Eq. (20.1), the outputs of the hidden layer PEs are effectively providing the basis for function approximation.
One of the fundamental issues to derive accurate dynamic models from time series is the strategy to train the neural network. A comprehensive theory on how to best train the non-linear predictor for dynamic modelling only now is emerging [9]. Eq. (20.3) states that the mapping f can be obtained by a one step predictor (which is a NAR model, i.e. recursive) but says nothing how to create the error to estimate the system parameters. Experience has shown that minimizing the instantaneous prediction error during training as normally done in linear prediction does not lead necessarily to good models of chaotic time series [15]. Note that, due to the autonomous nature of the signal generation, the iterates of the mapping F should yield the entire reconstructed trajectory. Likewise, the iterates of f should reproduce the original signal, which mean that in dynamic modelling of chaotic time series the learning methodology should constrain the iterative map of the model. Therefore, the training of the focused TDNN for dynamic modelling should not follow the simple recipe of single step prediction where one sample of the time series is used as input to the dynamic neural network. We suggest that iterated prediction and trajectory learning be utilized as an alternate procedure for training as shown in Figure 20.5. The dynamic neural network topology is extended with a global feedback loop from the output to the input. Trajectory learning is used to train the network using the backpropagation through time (BPTT) algorithm or real time recurrent learning (RTRL) [10]. The input samples are progressively substituted by the output of the network during training by controlling the switch during training. In the beginning of training, most of the samples for the trajectory come from the input, but towards the end of training the input to the network comes from its output. The use of the input in the beginning of training facilitates training, but once the weights are at reasonable values the output of the system should substitute the input for more accurate modelling. This scheme is similar to teacher forcing but is implemented differently and does not produce model bias.

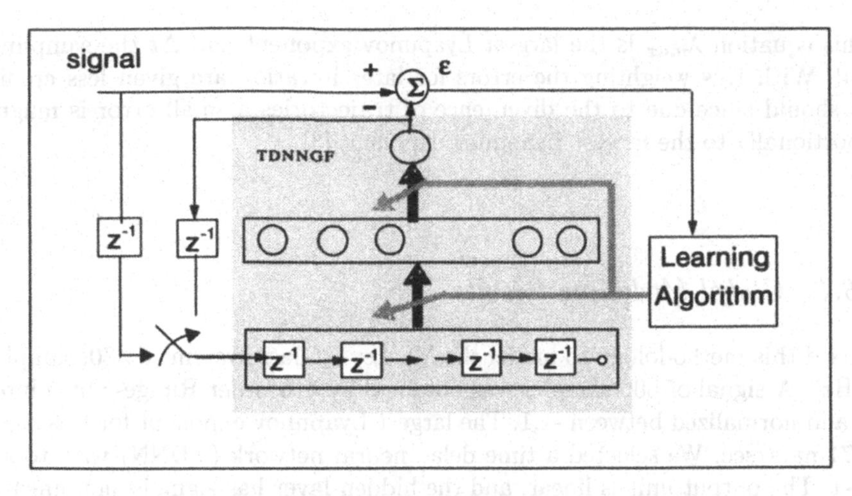

FIGURE 20.5. TDNN with global feedback (TDNNGF)

With this training methodology called iterated prediction, the dynamic model is effectively trained as an autonomous system, exactly as the assumption of Figure 20.2 for the generation of the time series. We found out that this training method constrains the iterates of the model as required for dynamic modelling, producing a neural network that works as a "chaotic" signal generator. In [9] we suggest that this training method is a kind of regularization, but instead of imposing smoothness constraints in the map, it uses the structure of the time series to regularize the weights.

Iterated prediction minimizes a cost function given by

$$E = \sum_{i=2m+1}^{K} dist(x(i+1) - \tilde{x}(i+1)) \tag{20.4}$$

where dist(.) is the L2 norm, K is the number of prediction steps (length of the trajectory) and $\tilde{x}(i+1)$ may be considered as an estimate of the map

$$\tilde{x}(i+1) = \tilde{f}(\tilde{x}(i-2D_e), ..., \tilde{x}(i)) \tag{20.5}$$

and

$$\tilde{x}(i) = \begin{cases} x(i) & 1 \le i \le 2D_e + 1 \\ \tilde{f}(x(i-2m-1), ..., x(i-1)) & i > 2D_e + 1 \end{cases} \tag{20.6}$$

The procedure described above must be repeated for several different segments of the time series. For each new training segment, $2D_e + 1$ samples of the original time series are used to seed the predictor. To ease the training we suggest that successive training sequences of length K overlap by q samples (q<K). For chaotic time series we also suggest that the error be weighted according to the largest Lyapunov exponent. Hence the cost function becomes

$$E = \sum_{i=0}^{r} \sum_{j=2m+1}^{K} h(i) \cdot dist(x(i+jq+1) - \tilde{x}(i+jq+1)) \tag{20.7}$$

where r is the number of training sequences, and

$$h(i) = (e^{\lambda_{max} \Delta t})^{-(i-2m-1)} \tag{20.8}$$

In this equation λ_{max} is the largest Lyapunov exponent and Δt the sampling interval. With this weighting the errors for later iteration are given less credit, as they should since due to the divergence of trajectories a small error is magnified proportionally to the largest Lyapunov exponent [3].

20.3.1 Global Modelling Results

We used this methodology to model the Mackey—Glass system (d=30, sampled at 1/6 Hz). A signal of 500 samples was obtained by 4th order Runge-Kutta integration and normalized between -1,1. The largest Lyapunov exponent for this signal is 0.0071 nats/sec. We selected a time delay neural network (TDNN) with topology 8-14-1. The output unit is linear, and the hidden layer has sigmoid non-linearities. The number of taps in the delay line was set at 8 using the tight bound for the embedding.
We trained a one-step predictor and the multistep predictor with the methodology developed in this paper to compare results. The single step predictor was trained with static backpropagation with no momentum and step size of 0.001. Trained was stopped after 500 iterations. The final MSE was 0.000288. After training, the predictor was seeded with the first 8 points of the time series and iterated for 3,000 times. Figure 20.6a shows the corresponding output. Notice that the waveform produced by the model is much more regular that the Mackey-Glass signal, showing that some fine detail of the attractor has not been captured.
Next we trained the same TDNN with a global feedback loop (TDNNGF). We displaced each training sequence by 3 samples (q=3 in Eq. (20.7)). BPTT was used to train the TDNNGF for 500 iterations over the same signal. The final MSE was 0.000648, higher than for the TDNN case. We could think that the resulting predictor was worse. The TDNNGF predictor was initialized with the same 8 samples of the time series and iterated for 3,000 times. Figure 20.6b shows the resulting waveform. It "looks" much closer to the original Mackey-Glass time series. We computed the average prediction error as a function of iteration for both predictors and also the theoretical rate of divergence of trajectories assuming an initial error ε_0 (Casdagli conjecture, which is the square of Eq. (20.8)) [3]. As can be seen in Figure 20.7 the TDNNGF is much closer to the theoretical limit, which means a better model. We also computed the correlation dimension [7] and the largest Lyapunov exponent [25] estimated from the generated time series, and the values obtained from TDNNGF are closer to the original time series (Table 20.1).

	Correlation Dimension	Largest Lyapunov
MG-30	2.70	0.0071 nats/sec
TDNN	1.60	0.0063 nats/sec
TDNNGF	2.65	0.0074 nats/sec

TABLE 20.1.

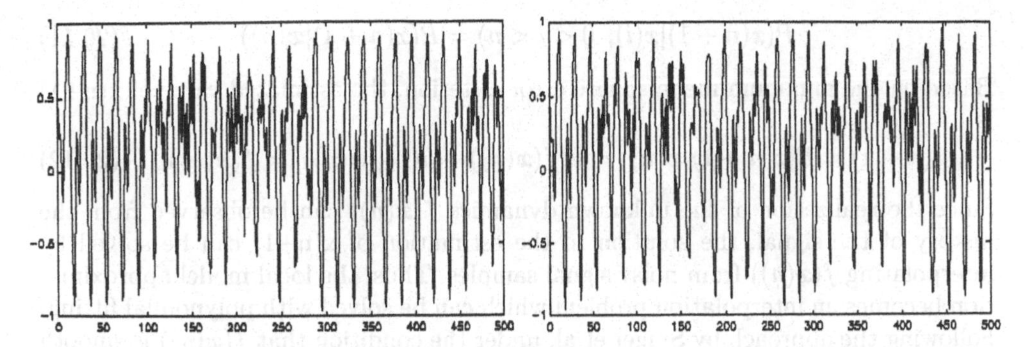

FIGURE 20.6. Generated time series with the TDNN (a) and with TDNNGF (b)

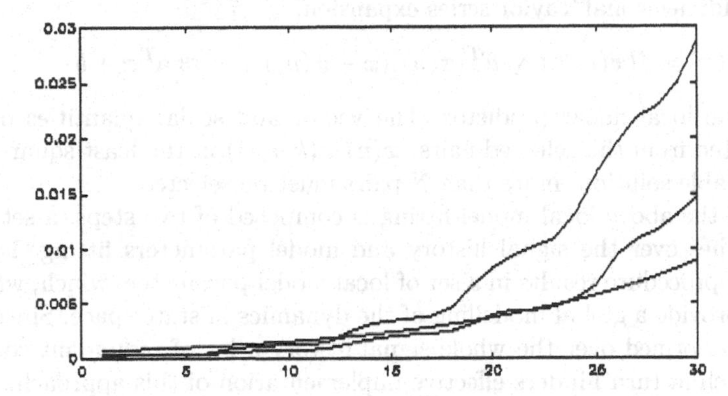

FIGURE 20.7. Comparison of normalized Prediction error versus iteration. Top: TDNN error, middle: TDNNGF error, bottom: Casdagli conjecture

20.4 Neural Topologies for Local Modelling

Different local prediction approaches were tested by Farmer and Sidorowich [6], and Casdagli [3] for a variety of low-dimensional chaotic systems. In this work, we pursue the locally linear approach, but instead of using the data samples directly, the local fitting is done over the output of the SOFM neural field.
As we saw in Eq. (20.2) the central problem of dynamic modelling is to find the predictive mapping F from R^N to R^N. An instant predictor at x(n) can be constructed based on its nearby neighbors, provided no crossing occurs among the manifolds. When the dynamic evolution F is approximated by a locally linear function F_x

$$x(n+1) = F_x(x(n)) = A_x x(n) + b_x \tag{20.9}$$

the matrix A_x and vector b_x can be solved by least squares fitting to all observations $(x(k), x(k+1))$ with $k < n$ such that x(k) is in the neighborhood of x(n).
From the point of view of signal processing, Singer, et al. [19] derived the locally linear prediction as an AR model generalization. In discrete time, a non-linear process can be described by a Nth order difference equations of the form

$$x(n+1) = f(x(n)) + u(n) \tag{20.10}$$

where $x(n) = [x(n), x(n-1), ..., x(n-N-1)]^T$, $f(x)$ represents the non-linear map from R^N to R^1, and *u(n)* is the white noise innovation term. Due to the statistical Markov structure of the non-linear dynamic model, we have

$$P(x(n+1)|x(i),\ 0<i<n) = P(x(n+1)|\boldsymbol{x}(n)) \tag{20.11}$$

Based on the minimum mean square error criterion, the estimated value of x(n+1) is

$$\hat{x}(n+1) = E[x(n+1)|\boldsymbol{x}(n)] = E[f(\boldsymbol{x}(n)) + u(n)|(\boldsymbol{x}(n))] = f(\boldsymbol{x}(n)) \tag{20.12}$$

Since the realization of the unknown dynamics $f(\boldsymbol{x}(n))$ can be observed from the history of the signal, the solution to the estimation of x(n+1) can be solved by interpolating $f(\boldsymbol{x}(n))$ from noisy signal samples. Thus, the local model approximation becomes an interpolation problem which can be solved with polynomial fitting. Following the approach by Singer et al, under the condition that $f(\boldsymbol{x}(n))$ is smooth enough in the vicinity of $\boldsymbol{x}(n), f(\boldsymbol{x}(n))$ can be approximated by the first few terms of its multidimensional Taylor series expansion,

$$\tilde{f}(\boldsymbol{x}) = f(\boldsymbol{x}(n)) + \nabla F^T(\boldsymbol{x}(n))(\boldsymbol{x} - \boldsymbol{x}(n)) + ... \approx \mathbf{a}^T\boldsymbol{x} + b \tag{20.13}$$

which is the local linear predictor. The vector and scalar quantities of $\mathbf{a}$ and b are estimated from the selected pairs $(\boldsymbol{x}(n), x(k+1))$ in the least square sense. To obtain a stable solution, more than N pairs must be selected.
In general, the above local model fitting is composed of two steps: a set of nearby state searches over the signal history and model parameters fitting. For a given signal, this procedure results in a set of local model parameters which, when pieced together, provide a global modelling of the dynamics in state space. Since the state search is performed over the whole signal history a lot of redundant computation results which in turn hinders effective implementation of this approach.

20.4.1 Localized Signal Representation with SOFM Modelling

Instead of direct sample selection from the signal history, we propose to alleviate these problems by the use of a Kohonen self-organizing feature map neural network. Kohonen [10] developed the self-organizing feature map (SOFM) procedure which is capable of transforming the input signal of arbitrary dimension into a lower (one or two) dimensional discrete space preserving topological neighborhoods. Let $\mathbf{\Phi}, \mathbf{X}, \mathbf{A}$ denote the SOFM mapping, input sample space and the discrete output space respectively. Since the feature map $\mathbf{\Phi}$ provides a faithful topologically organized output of the input vectors $\boldsymbol{x} \in \mathbf{X}$, the local model fitting can then be performed over the compact codebook domain A of the Kohonen output space.The proposed non-linear modelling scenario follows three steps:

- Reconstruction of the state space from the input signal (embedding).
- Train the SOFM neural field.
- Estimation of the locally linear predictors.

* *Embedding.* Following the approach by Takens, a sequence of $N = D_e\tau$ dimensional state vectors augmented by 1, $[\boldsymbol{x}(n)^T, x(n+\tau)]^T$ is created from the given training time series, where $\boldsymbol{x}(n) = [x(n-(N-1)\tau), x(n-(N-2)\tau), ..., x(n)]^T$ and τ is the appropriate time delay.
* *Training the neural field.* This step is accomplished via the Kohonen learning process. With each vector-scalar pair $[\boldsymbol{x}(n)^T, x(n+\tau)]^T$ presented as the input to the network, the learning process of Kohonen feature mapping algorithm adaptively discretizes the continuous input space $\mathbf{X} \subset \mathbf{R}^{N+1}$ into a set of K disjoint cells $\mathbf{A}$ to construct the mapping $\mathbf{\Phi} : \mathbf{X} \to \mathbf{A}$, according to

$$w_i(j+1) = w_i(j) + \eta(j)\Lambda(r_{i^o}, r_i)(x - w_i(j)) \qquad \Lambda(r_{i^o}, r_i) = exp[\frac{\| r_{i^o} - r_i \|^2}{2\sigma(j)}] \tag{20.14}$$

$$\eta(j) = \frac{1}{a_\eta + b_\eta j} \qquad \sigma(j) = \frac{1}{c_\sigma + d_\sigma j} \tag{20.15}$$

where i^o is the winning PE, $\Lambda(r_{i^o}, r_i)$ and η are the adaptive neighborhood function and learning rate and j is the iteration index. After learning, a neural field representation $\mathbf{A}$ of the input space $\mathbf{X}$ via the constructed mapping relationship $\mathbf{\Phi}$ is formed in terms of a set of disjoint units topologically organized in the output space.

* *Estimation of the locally linear predictors.* For each PE , $u_i \in \mathbf{a}$, its local linear predictor in terms of $[\mathbf{a}_i^T, b_i]$ is estimated based on $\alpha_i \subset \mathbf{A}$, which is a set of L PEs in the neighborhood of u_i including u_i itself. Each PE u_i, has a corresponding weight vector $[\boldsymbol{w}_i^T, w_{i(N+1)}]^T \in \mathbf{R}^{N+1}$, where $\boldsymbol{w}_i^T = [w_{i(1)}, w_{i(2)}, ..., w_{i(N)}]$. The local prediction model $[\mathbf{a}_i^T, b_i]$ is fitted in the least-square sense to the set of weights of the PEs in α_i, i.e.

$$w_{i(N+1)} = b + \mathbf{a}^T \boldsymbol{w}_i \tag{20.16}$$

To ensure a stable solution of the above equations, α_i must have $L > N+1$ elements. Thereafter each output PE $u_i \subset \mathbf{A}$, is associated with a unique linearly local model function $\tilde{f}(\mathbf{a}_i(.), b_i(.))$ in terms of the vector-scalar parameter pair $[\mathbf{a}_i^T, b_i]$. The global dynamics of the time series can be described by the set of all the constructed local models pieced together. For an input state vector $\boldsymbol{x}(n) = [x(n-N+1), x(n-N+2), ..., x(n)]^T$, the matched prototype element $u_{i^o} \in \mathbf{A}$ is found based on the SOFM competition among all elements in $\mathbf{A}$. The predicted value $\tilde{x}(n+1)$ is obtained by evaluating $\tilde{f}(\mathbf{a}(.), b(.))$ at $\boldsymbol{x}(n) = [x(n-N+1), x(n-N+2), ..., x(n)]^T$

$$\tilde{x}(n+1) = \tilde{f}(\mathbf{a}(u_{io}), b(u_{io}), \boldsymbol{x}(n)) = b(u_{io}) + \mathbf{a}^T(u_{io})\boldsymbol{x}(n) \tag{20.17}$$

In a similar manner, a K-step prediction x(n+K) based on x(n) can also be obtained by iterative prediction, i.e. feeding the output back to the input,

$$\tilde{x}(n+K) = \tilde{f}_K(\tilde{f}_{K-1}(...\tilde{f}_1(\mathbf{a}(u_{io}), b(u_{io}), \boldsymbol{x}(n)))) \tag{20.18}$$

where $\tilde{f}_j = \tilde{f}(a(u_{i^o}), b(u_{i^o}), x(n))$ is the prediction function at step j.That is, the first prediction generates a new state, which is used to find the new local model function. Evaluation of the new local model function at the new state produces in turn a new prediction until the final K-step prediction is reached.

20.4.2 Local Modelling Results

A non-linear time series from the Mackey-Glass system (delay of 30 as before) is modeled with the proposed scenario. A total of 2500 PEs, arranged on a 50 X 50 square lattice, constitute the SOFM output space. The dimension of the weight vectors $w_i(n)$ was chosen as 8, so the dimension of the state input during the training process is 9 (N+1). The learning rate and evolution of neighborhood function in Eq. (20.14) and Eq. (20.15) were used with $a_\eta = 1, b_\eta = 10^{-3}, a_\sigma = 1/30, b_\sigma = 0.6\text{x}10^{-4}$. A 10,000 samples Mackey-Glass time series is generated with d=30 and f_s=1/6 Hz. The Kohonen SOFM network is trained with this segment of time series for five epochs (50,000 samples). After training the weight vectors are frozen for local model estimation. A typical post-training output trajectory showing the winning PE on the neural field for 500 consecutive input samples is as shown in Figure 20.8. Note that the sequential winning PEs over the neural field form a trajectory which is a projection of the Mackey-Glass attractor.

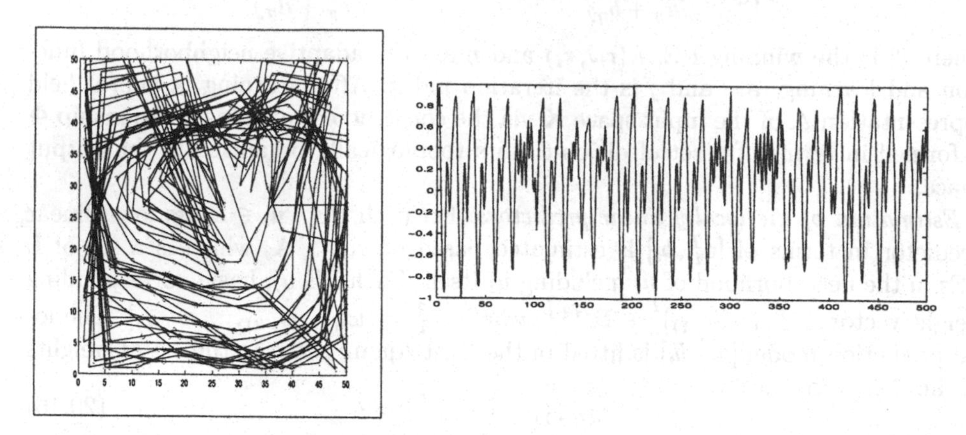

FIGURE 20.8. The output trajectory and the autonomous generation of the M-G time series

As shown in section 1.4, to ensure a stable solution in the least square sense, the subset α_i must contain at least N+1 neighbors for stable model estimation. We take 21 neurons surrounding the neuron u_i in the output space as its neighborhood subset $\alpha_i \subset \mathbf{A}$ to estimate the corresponding local linear prediction function $\tilde{f}_i = \tilde{f}(\bullet)$. Another different 5,000 sample Mackey-Glass time series is taken to test the prediction performance of the estimated local prediction model set. The testing is performed with iterated prediction ranging from 1 up to 20 samples ahead. Figure 20.9 shows the result for 20 steps ahead.The mean squared error normalized by the variance of the original signal is shown in Figure 20.10a, and it starts at 0.06 and increases to 0.15 for 20 step ahead prediction.

If the averaged Euclidean distance between weight vectors of two neighboring PEs is taken as the resolution of the neural field **A**, it is obvious that the larger its dimension the finer the resolution, which in turn provides more accurate local model estimation. With this notion, three SOFM networks with different lattice dimension (50 x 50, 60 x 60, 70 x 70) are compared in terms of the normalized MSE for 20 prediction steps and the result is consistent with the above observation as shown in Figure 20.10b.

Finally, Figure 20.8 shows an autonomously generated 500 point segment of the model output. This signal clearly shows that the dynamics of the system that produced the time series have been captured, and the trained model behaves just like a Mackey-Glass signal generator.

Table 20.2 presents the results of the estimation of the correlation dimension and largest Lyapunov exponent with the Kohonen self-organizing feature map. We can see that the SOFM produces a signal with very similar dynamics as the original time series. We believe that these results can be improved if the training is modified to reflect the complexity of fitting task, instead of simply training the SOFM to the input space density as is done in conventional Kohonen training.

The proposed SOFM dynamic model is also tested using the Lorenz signal. With

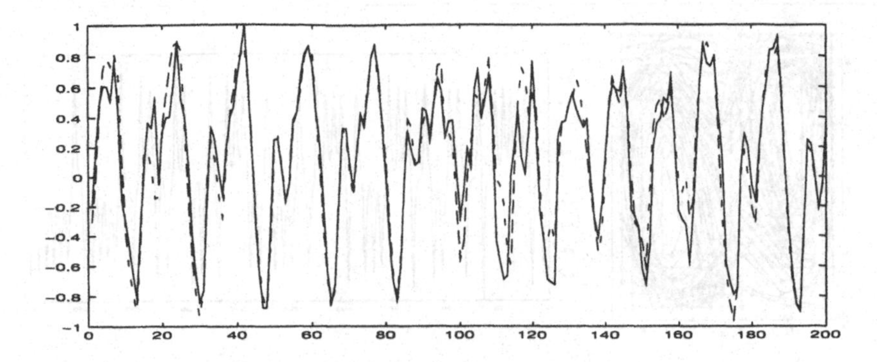

FIGURE 20.9. 20-step Predicted Sequence (Solid) vs. Original Signal (Dashed)

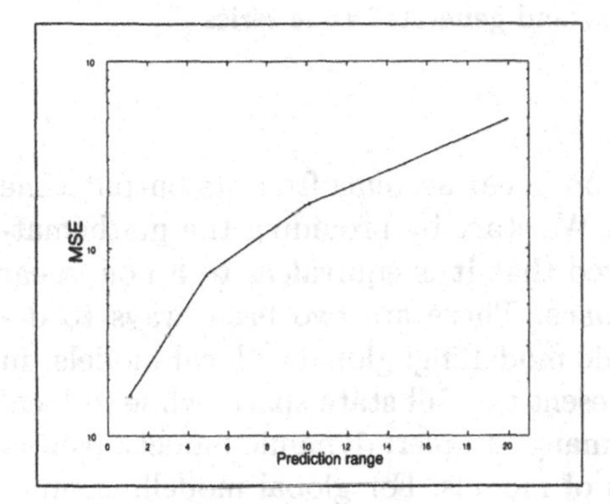

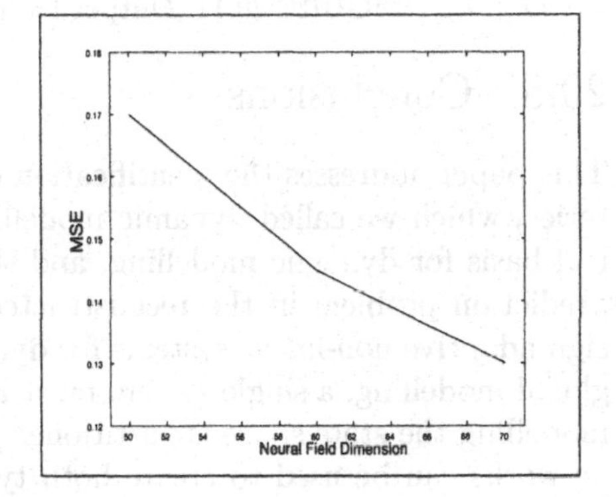

FIGURE 20.10. MSE vs. Prediction Interval (a) and MSE vs. Network size (b)

the selection $\alpha = 10, b = 8/3, r = 28$ the Lorenz system exhibits chaotic dynamics with large Lyapunov exponent. The Lorenz signal is sampled at 10 Hz. The SOFM system used here is composed of a 22x22 neural field. The embedding dimension is chosen as 4, and so the dimension of the state input during the training process is 5. The parameters for the learning rate and evolution of the neighborhood function are chosen as $a_\eta = 1, b_\eta = 1/500, c_\sigma = 1/8, d_\sigma = 1/4000$. The training of SOFM is performed over 3000 sample time series for 150 epoch. To test the whole structure, the whole system is iterated as an *autonomous system.*Figure 20.11 shows an output trajectory plot of a segment of the output and the corresponding autonomous prediction.

The output of autonomous prediction is analyzed with respects to the dynamic invariants, correlation dimension and largest Lyapunov exponent. These quantities estimated from the autonomous prediction and the original time series are highly consistent, which demonstrates that the underlying chaotic dynamics have been successfully captured by our system [21].

	Correlation Dimension	Largest Lyapunov
MG-30	2.70	0.0071 nats/sec
local SOFT	2.60	0.0059 nats/sec

TABLE 20.2.

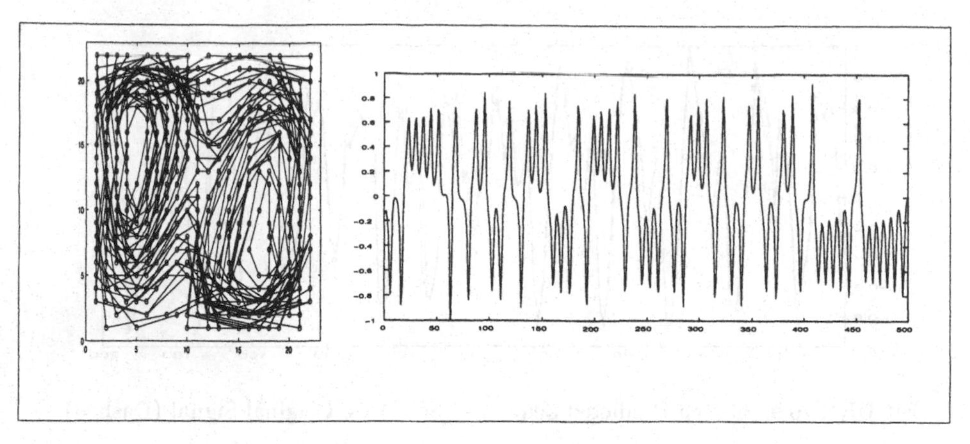

FIGURE 20.11. Output trajectory and generated time series

20.5 Conclusions

This paper addresses the identification of non-linear systems from its output time series, which we called dynamic modelling. We start by providing the mathematical basis for dynamic modelling, and showed that it is equivalent to a non-linear prediction problem in the reconstructed space. There are two basic ways to design adaptive non-linear systems for dynamic modelling: global or local models. In global modelling, a single system must represent the full state space, while in local modelling the state space is partitioned by many, simpler, dynamic models. Neural networks can be used to create both types of models. For global modelling, multilayer perceptrons (MLPs) extended with short term memory structures (called dynamic neural networks) are an obvious choice. We showed that the role of memory in the first layer of the MLP is effectively equivalent to the embedding step, and so one can provide some guidelines based on dynamics to set its size.

For local dynamic modelling, neural networks can also be utilized, but now the topology is very different. A Kohonen network is first used to represent (and cluster) the dynamics in state space. Each PE of the Kohonen layer will represent a partition of the state space. From its weights and of its neighbors a local linear model was derived. We called the approach local SOFM modelling.

One important issue is to compare the two methodologies for dynamic modelling and find out which is more appropriate for modelling signals produced by real world dynamical systems. Table 20.3 summarizes the comparison of the correlation dimension and largest Lyapunov exponent estimates produced by the global (TDNN and TDNNGF) models and the local SOFM model we have been discussing. We also show the logarithm of the normalized one step prediction error for all the models, including a 20 tap linear combiner (FIR) trained with the LMS algorithm.

This Table shows that the behavior of the linear and non-linear models is rather different. The FIR is not able to generate a chaotic time series in the autonomous mode as we would expect (the autonomous response of a linear system converges to a point attractor or to a limit cycle). However, it displays a good one step prediction error, in fact better than any of the other non-linear models (the non-linear models were trained for dynamic modelling, i.e. with the output error criterion discussed above, not for one step prediction). This result shows that short term prediction and dynamic modelling are really two different operations.

Predictor	Corr. Dim	Largest Lyapunov	log(e)
MG30	2.70	0.0071	NA
TDNNGF	2.65	0.0074	-1.38
TDNN	1.60	0.0063	-1.54
local SOFM	2.60	0.0059	-1.29
FIR	NA	NA	-1.68

TABLE 20.3.

Table 20.3 also shows that within the non-linear models, the local SOFM is comparable to the global models in terms of dynamic invariants, although it provides a larger normalized prediction one step prediction error. So we can say that the performance of global and local models is comparable in "synthetic time series", as the Mackey-Glass or Lorenz systems. Recently we successfully modeled a real world signal (the laser time series) with the local SOFM model, but we failed when global models were tried. It is too early to give up on the global models, but it seems that the local models handle better the intrinsic noise of real world signals, and with sufficiently large Kohonen layers, they also are able to represent arbitrary complex trajectories. Much more work is required to create a methodology to train and compare dynamic models for real world signals.

Acknowledgments: This work was partially supported by NSF grants No. ECS-9510715, ONR No. 1494-94-1-0858, NIH/NINDS R01NS31451-05A1 and NSF IRI-9526049.

20.6 References

[1] D. Broomhead, D. Lowe. *Multivariable Functional Interpolation and Adaptive Networks.* Complex System 2, pp. 321-355, 1988.

[2] G.E. Box and G.M. Jenkins. *Time Series Analysis, Forecasting and Control.* Holden Day, San Francisco, 1970.

[3] M. Casdagli. *Nonlinear Prediction of Chaotic Time Series.* Physica D 35, pp. 335-356, 1989.

[4] P. Crutchfield, B. McNamara. *Equations of Motion from a Data Series.* Complex System 1, pp. 417, 1987.

[5] B. de Vries and J.C. Principe. *The Gamma Model - a New Neural Model for Temporal Processing.* Neural Networks, vol. 5, no. 4 pp. 565-576, 1992.

[6] D. Farmer, J. Sidorowich. *Predicting Chaotic Time Series.* Phy. Rev. Let., vol. 59, no. 8, pp. 845-849, 1987.

[7] P. Grassberger, I. Procaccia. *Characterization of Strange Attractors.* Phy. Rev. Let., vol. 50, no. 5, pp. 346-349, 1983.

[8] S. Haykin. *Neural Networks Expand DSP Horizons.* IEEE SP Magazine, vol. 13, no. 2, pp. 24-29, 1996.

[9] S. Haykin and J.C. Principe. *Dynamic Modelling of Chaotic Time Series.* IEEE SP Magazine in press, 1998.

[10] T. Kohonen. *The Self-organizing Map.* proc. IEEE, vol. 78, no. 9, pp. 1-13, 1990.

[11] J.M. Kuo. *Nonlinear Dynamic Modeling with Artificial Neural Networks.* Ph.D. Dissertation, University of Florida, 1993.

[12] K. Lang, A Waibel, and G Hinton. *A Time Delay Neural Network Architecture for Isolated Word Recognition.* Neural Networks, vol. 3, no. 1, pp. 23-44, 1990.

[13] R. Lapedes and R. Farber. *Nonlinear Signal Processing Using Neural Network: Prediction and System Modeling.* Technical Report LA-UR87-2662, Los Aamos National Laboratory, Los Alamos, New Mexico, 1987.

[14] W.C. Mead, R.D. Jones, Y.C. Lee, C.W. Barnes, G.W. Flake, L.A. Lee,and M.K. O'rourke *Prediction of Chaotic Time Series Using CNLS-NET - Example: the Mackey-Glass Equation.* Technical Report: LA-UR-91-720, Los Aamos National Laboratory, Los Alamos, New Mexico, 1991.

[15] J.C. Principe, A. Rathie and J.M. Kuo. *Prediction of Chaotic Time Series with Neural Networks and the Issue of Dynamic Modeling.* International Journal of Biburcation and Chaos, vol. 2, no. 4, pp. 989-996, 1992.

[16] J.C. Principe, and J-M. Kuo. *Dynamic Modeling of Chaotic Time Series with Neural Networks.* Proc. Neural Inf. Proc. Sys., NIPS 7, pp. 311-318, 1995.

[17] I.W. Sandberg, and L. Xu *Uniform Approximation and Gamma Networks.* Neural Networks, vol. 10, pp. 781-784, 1997.

[18] T. Sauer, J.A. Yorke and M. Casdagli. *'Embedology,".* Journal of Statistical Physics, vol. 65, Nos. 3/4, pp. 579-616, 1991.

[19] C. Singer, G. Wornell, and A. Oppenheim. *Codebook Prediction: A Nonlinear Signal Modeling Paradigm.* IEEE ASSP, vol. V, pp. 325-329, 1992.

[20] F. Takens. *On the Numerical Determination of the Dimension of an Attractor.* In D. Rand and L.S. Young, editors, "Dynamical systems and Turbulence", Warwick 1980 Lecture Notes in Mathematics, vol. 898, pp. 366-381, Springer-Verlag, 1981.

[21] L. Wang. *Local Dynamic Modeling with self-organizing feature maps.* Ph.D. Dissertation, University of Florida, 1996.

[22] A.S. Weigend, B.A. Huberman and D.E. Rumelhart. *Predicting the Future: a Connectionist Approach.* International Journal of Neural System, vol. 1, pp. 193-209, 1990.

[23] P. Werbos. *Backpropagation through time: what it does and how to do it.* Proc. IEEE, vol. 78, no. 10, pp. 1550-1560, 1990.

[24] R.J. Williams and D. Zipser. *Experimental Analysis of the Real-time Recurrent Learning Algorithm.* Connection Science, vol. 1, no. 1, pp. 87-111, 1989.

[25] A. Wolf, J.B. Swift, H.L. Swinney and J.A. Vastano. *Determining Lyapunov Exponents from a Time Series.* Physica 4D, pp. 285-317, 1985.

21

Non-Linear Adaptive Prediction of Speech with a Pipelined Recurrent Neural Network and Advanced Learning Algorithms

Danilo Mandic[1]
Jens Baltersee[2]
Jonathon Chambers[1]

ABSTRACT
New learning algorithms for an adaptive non-linear forward predictor which is based on a Pipelined Recurrent Neural Network (PRNN) are presented. A computationally efficient Gradient Descent (GD) algorithm, as well as a novel Extended Recursive Least Squares (ERLS) algorithm are tested on the predictor. Simulation studies, based on three speech signals, which have been made public and are available on the World Wide Web (WWW), show that the non-linear predictor does not perform satisfactorily when the previously proposed gradient descent algorithm was used. The steepest descent algorithm is shown to yield a poor performance in terms of the prediction error gain, whereas consistently improved results are obtained using the ERLS algorithm. The merit of the non-linear predictor structure is confirmed by yielding approximately 2 dB higher prediction gain than only a linear structure predictor, which uses the conventional Recursive Least Squares (RLS) algorithm.

21.1 Introduction

Many signals are generated from an inherently non-linear physical mechanism and have statistically non-stationary properties, a classic example of which is speech. Linear structure adaptive filters are suitable for non-stationary characteristics of such signals, but they do not account for non-linearity, and associated higher order statistics [1]. Adaptive techniques which recognize the non- linear nature of the signal should therefore outperform traditional linear adaptive filtering techniques [2, 3].
The classic approach to time series prediction is to carry out an analysis of the time series data, which includes modelling, identification of the model and model

[1]Signal Processing Section, Department of Electrical Engineering, Imperial College of Science, Technology and Medicine, Exhibition Road, London, SW7 2BT, United Kingdom,
E-mails: {d.mandic, j.chambers}@ic.ac.uk
[2]Integrated Systems for Signal Processing, Aachen University of Technology, Templergraben 55, D-52056 Aachen, Germany, E-mail: balterse@ert.rwth-aachen.de

parameter estimation phases [4]. The design may be iterated by measuring the closeness of the model to the real data. This can be a long process, often involving the derivation, implementation and refinement of a number of models before one with appropriate characteristics is found.
In particular, the most difficult systems to predict are:

- those with non-stationary dynamics, where the underlying behaviour varies with time, typical example of which is speech,
- those which deal with physical data which are subject to noise and experimentation error,
- those which deal with short time series, providing few data points on which to conduct the analysis.

In all these situations, traditional techniques are severely limited and alternative techniques must be found [5, 6, 7, 8].
On the other hand, neural networks are powerful when applied to problems whose solutions require knowledge which is difficult to specify, but for which there is an abundance of examples [9, 10, 11]. As time series prediction is performed entirely by inference of future behaviour from examples of past behaviour, it is a suitable application for a neural network predictor. The neural network approach to time series prediction is non-parametric in the sense that it does not need to know any information regarding the process that generates the signal. For instance, the order and parameters of an Auto-Regressive (AR) or Auto-Regressive Moving Average (ARMA) process are not needed in order to carry out the prediction. This task is carried out by a process of learning from examples presented to the network and changing network weights in response to the output error. The training of the network is effected by minimising an error function using one of a number of possible non-linear optimisation methods [5, 12, 8]. *The training process itself can be very slow.*
In 1995, Haykin and Li [5] presented a novel, computationally efficient non-linear predictor based on a PRNN. The learning algorithm which was used by Haykin and Li for the PRNN was a gradient descent algorithm. This paper presents a new algorithm for the non-linear predictor, a novel Extended Recursive Least Squares (ERLS) learning algorithm. Three speech signals, available from the author's WWW homepage [13], were used to test the non-linear predictor trained with the new algorithm.

21.2 Non-Linear Models and Neural Networks

A general class of linear models used for forecasting purposes is the class of ARMA (p,q) models given by [4, 14]

$$x_t = \sum_{i=1}^{p} \phi_i x_{t-i} + \sum_{j=1}^{q} \theta_j e_{t-j} + e_t \tag{21.1}$$

where, it is assumed that

$$E(e_t | x_{t-1}, x_{t-2}, \cdots, x_1) = 0 \tag{21.2}$$

This condition is satisfied, for example, when the e_t are zero mean, Independent and Identically Distributed (IID) random variables, and independent of past x_t's. For the ARMA(p,q) model (21.1), the optimal predictor

$$\hat{x}_t = E(x_t | x_{t-1}, x_{t-2}, \cdots, x_1) \tag{21.3}$$

is given by

$$\hat{x}_t = \phi_1 x_{t-1} + \cdots + \phi_p x_{t-p} + \theta_1 \hat{e}_{t-1} + \cdots + \theta_q \hat{e}_{k-q} \tag{21.4}$$

where

$$\hat{e}_{t-j} = x_{t-j} - \hat{x}_{t-j}, \ j = 1, 2, \cdots, q \tag{21.5}$$

The natural generalisation of the linear ARMA(p,q) model to the non-linear case is given by

$$x_t = h(x_{t-1}, x_{t-2}, \cdots, x_{t-p}, e_{t-1}, \cdots, e_{t-q}) + e_t \tag{21.6}$$

where h is an unknown smooth function[3]. It is assumed that

$$E(e_t | x_{t-1}, x_{t-2}, \cdots) = 0 \tag{21.7}$$

and that the variance of e_t is σ^2. This model is called a Non-linear Auto-Regressive Moving Average (NARMA) model. In this case, the conditional mean predictor based on the infinite past of observations is

$$\hat{x}_t = E[h(x_{t-1}, \cdots, x_{t-p}, e_{t-1}, \cdots, e_{t-q}) | x_{t-1}, x_{t-2}, \cdots] \tag{21.8}$$

Suppose that the NARMA model is invertible in the sense that there exists a function g such that

$$x_t = g(x_{t-1}, x_{t-2}, \cdots) + e_t \tag{21.9}$$

Then, given the infinite past of observations $x_{t-1}, x_{t-2}, \cdots$, one can, in principle, use the above equation to compute e_{t-j} as a function f of past values of x

$$e_{t-j} = f(x_{t-j}, x_{t-j-1}, \cdots), \ j = 1, \cdots, q \tag{21.10}$$

In this case the conditional mean estimate is

$$\hat{x}_t = h(x_{t-1}, \cdots, x_{t-p}, e_{t-1}, \cdots, e_{t-q}) \tag{21.11}$$

where e_{t-j} is specified by (21.10).
Since in practice only a finite observation record is available, computation of (21.10) and (21.11) is not possible. However, by analogy with the recursive computation of the predictor $\hat{x}_t$ for the linear ARMA process, it seems reasonable to approximate the conditional mean predictor (21.11) by the recursive algorithm

$$\hat{x}_t = h(x_{t-1}, x_{t-2}, \cdots, x_{t-p}, \hat{e}_{t-1}, \hat{e}_{t-2}, \cdots, \hat{e}_{t-q}) \tag{21.12}$$

$$\hat{e}_j = x_j - \hat{x}_j, \ j = t-1, \cdots, t-q \tag{21.13}$$

---

[3]In fact, we wish to find some function f, which is continuos and at least twice- differentiable ($\in C^2$), because the values of network weights **w** are adjusted according to the first derivative of f, and we need a smooth learning trajectory in the error performance surface in order to "surf" along it looking for the steepest direction towards some minimum of the surface. In practice, we almost always use functions which are infinitely differentiable ($\in C^\infty$), such as *sigmoid* or *hyperbolic tangent*.

The approximate conditional mean predictor model (21.12), (21.13), can be approximated by the following NARMA(p,q) recurrent network model

$$\hat{x}_t = \sum_{i=1}^{I} W_i f(\sum_{j=1}^{p} w_{ij} x_{t-j} + \sum_{j=1}^{q} w'_{ij}(x_{t-j} - \hat{x}_{t-j}) + \theta_i) \tag{21.14}$$

The parameters W_i, w_{ij} and w'_{ij} can be estimated using the LS algorithm, e.g. by choosing the above parameters to minimise $\sum(x_t - \hat{x}_t)^2$. From Equation (21.14), it can be seen that the NARMA(p,q) model consists of a non-linear part, that is the set of non-linear functions $f(\cdot)$, whose arguments are the weighted sums of the past signal values and past prediction errors, and a linear combiner with weights $\{W_i \mid i = 1, \cdots, I\}$, which produces the predicted value $\hat{x}$ as a linear combination of the weighted values of the non-linear functions $f(\cdot)$.

21.3 The Non-Linear Adaptive Predictor

The non-linear predictor, proposed by Haykin and Li [5] is based on the idea of first linearizing the input signal, and then feeding the resulting data into a linear predictor to produce an one-step forward prediction of the original signal. This combination of non-linear and linear filtering should be able to extract both non-linear and linear information contained in an input signal. Therefore, it is expected that a non-linear predictor will outperform a linear predictor when applied to signals generated by some non-linear mechanism. The two step prediction process described above is shown in Fig. 21.1. The first operation, while performing non-linear pre-

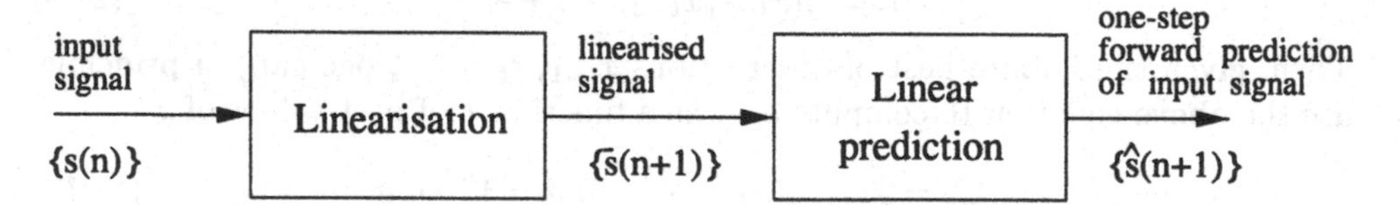

FIGURE 21.1. Two-step prediction process of the non-linear predictor

diction as proposed here, achieves a kind of mapping from the input signal space represented by $\{s(n)\}$ onto an intermediate space represented by $\{\bar{s}(n+1)\}$, in order to accomplish the linearization of the input signal. The linearization procedure itself, is performed by the PRNN. The linear prediction operation is performed by an adaptive Finite Impulse Response (FIR) filter trained by an appropriate commonly used algorithm.

21.3.1 The Pipelined Recurrent Neural Network (PRNN)

The PRNN is a modular neural network, and consists of a certain number M of Recurrent Neural Networks (RNNs) as its modules, with each module consisting of N neurons. The structure of a single RNN is shown in Fig. 21.2. The RNN consists of three layers:

- input layer
- processing layer
- output layer

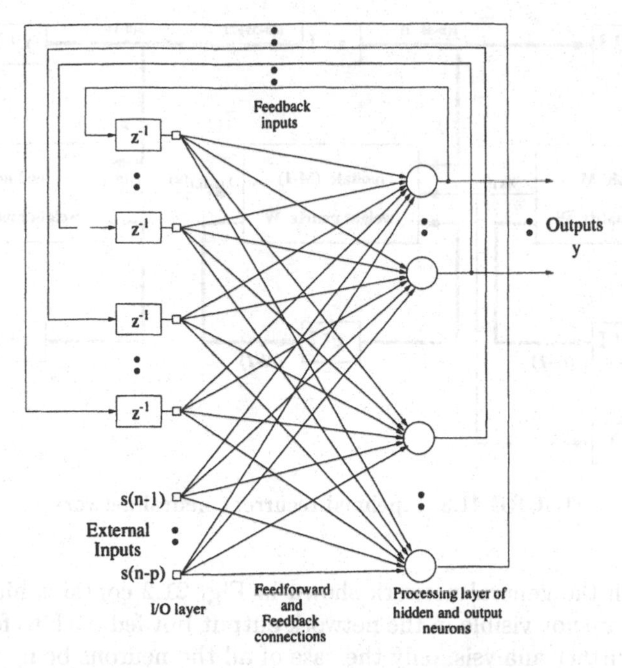

FIGURE 21.2. Single recurrent neural network

For each neuron k, $k \in [1, N]$, the elements u_i, $i \in [1, p+F+1]$, of the input vector to a neuron $\mathbf{u}$ (21.18), are weighted, then summed to produce an internal activation function of a neuron v (21.17), which is finally fed through a non-linear activation function Φ (21.15), to form the output of the k-th neuron y_k (21.16). The function Φ itself, is typically a monotonically increasing sigmoid logistic function, whose amplitude lies in the interval $[0, 1)$, and is given by

$$\Phi(v) = \frac{1}{1 + exp(-bv)} \tag{21.15}$$

For the kth neuron, its weights form a $(p+F+1) \times 1$ dimensional weight vector $\mathbf{w}_k^T = [w_{k,1}, \cdots, w_{k,p+F+1}]$, where p is the number of external inputs and F is the number of the feedback connections, one remaining element of the weight vector $\mathbf{w}$ being for the bias input weight. The feedback connections represent the delayed output signals of the RNN. In the case of the network shown in Fig. 21.2, we have N=F. Such a network is called a Fully Connected Recurrent Neural Network (FCRNN). For more details about recurrent neural networks, refer to the landmark paper by Williams and Zipser [15]. The following equations describe the FCRNN

$$\begin{aligned}
y_k(n) &= \Phi(v_k(n)), \ k \in [1, N] && (21.16)\\
v_k(n) &= \sum_{l=1}^{p+N+1} w_{k,l}(n) u_l(n) && (21.17)\\
\mathbf{u}_i^T(n) &= [s(n-1), \cdots, s(n-p), 1, && (21.18)\\
& \quad y_1(n-1), y_2(n-1), \cdots, y_N(n-1)]
\end{aligned}$$

where the $(p+N+1) \times 1$ dimensional vector $\mathbf{u}$ comprises both the external and feedback inputs to a neuron, with vector $\mathbf{u}$ having "unity" for the constant bias

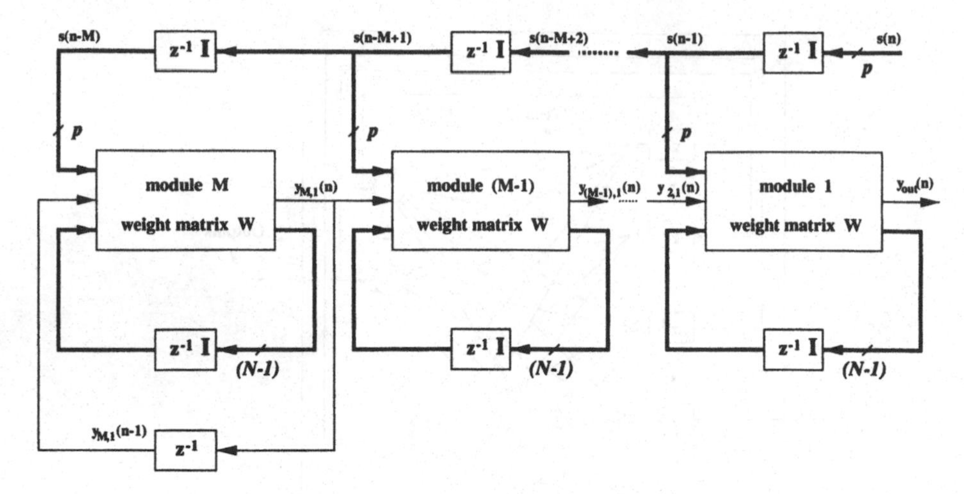

FIGURE 21.3. Pipelined recurrent neural network

input. Although the general network shown in Fig. 21.2 contains hidden neurons, whose outputs are not visible in the network output but fed back to form the input vector, in the further analysis, only the case of all the neurons being visible will be considered.

In the PRNN configuration, the M architectural modules, which are FCRNNs, are connected as shown in Fig. 21.3. The uppermost module M of the PRNN is simply a FCRNN, whereas in modules $\{M-1,\cdots,1\}$, the only difference is that the feedback signal of the uppermost neuron output within the module m, denoted by $y_{m,1}$, $m \in [1, M-1]$, is replaced with the appropriate output signal $y_{m+1,1}$, $m \in [2, M]$ from its left neighbour module $m+1$. The $(p \times 1)$ dimensional external signal vector $\mathbf{s}^T(n) = [s(n-1),\cdots,s(n-p)]$ is delayed by m time steps $(z^{-m}\mathbf{I})$ before feeding the module m, where z^{-m}, $m \in [1, M]$ denotes the m-step time delay operator, and $\mathbf{I}$ is the $(p \times p)$ dimensional identity matrix. The weight vectors $\mathbf{w}_k$ of each neuron k, are embodied in an $(p+N+1) \times N$ dimensional weight matrix $\mathbf{W} = [\mathbf{w}_1,\cdots,\mathbf{w}_N]$, with N being the number of neurons in each module. All the modules operate using the same weight matrix $\mathbf{W}$. The overall output signal of the PRNN is $y_{out}(n) = y_{1,1}(n)$, e.g. the output of the first neuron of the first module. A full mathematical description of the PRNN is given in the following equations

$$y_{i,k}(n) = \Phi(v_{i,k}(n)) \tag{21.19}$$

$$v_{i,k}(n) = \sum_{l=1}^{p+N+1} w_{k,l}(n) u_{i,l}(n) \tag{21.20}$$

$$\begin{aligned} \mathbf{u}_i^T(n) = {} & [s(n-i),\cdots,s(n-i-p+1),1, \\ & y_{i+1,1}(n), y_{i,2}(n-1),\cdots,y_{i,N}(n-1)] \\ & \text{for } 1 \le i \le M-1 \end{aligned} \tag{21.21}$$

$$\begin{aligned} \mathbf{u}_M^T(n) = {} & [s(n-M),\cdots,s(n-M-p+1),1, \\ & y_{M,1}(n-1), y_{M,2}(n-1),\cdots,y_{M,N}(n-1)] \\ & \text{for } i = M \end{aligned} \tag{21.22}$$

Given the input vectors $\mathbf{u}_i(n)$ for each module i, $i \in [1, M]$ at the time instant n, the outputs of all the neurons in the network can be calculated using the equations given above.
At the time step n, for each module i, $i \in [1, M]$, the one-step forward prediction error $e_i(n)$ associated to a module, is then defined as a difference between the desired response of that module $s(n-i+1)$, which is actually the next sample of the input speech signal, and the actual output of the i-th module $y_{i,1}(n)$, of the PRNN, i.e.

$$e_i(n) = s(n-i+1) - y_{i,1}(n) \tag{21.23}$$

Since the PRNN consists of M modules, a total of M forward prediction error signals are calculated. The goal is to minimize some measure of the error in the entire PRNN, termed a *cost function*, which is a weighted sum of all the error signals from individual modules. In such a performance criterion, a *forgetting factor* λ, $\lambda \in (0, 1]$, is introduced which determines the weighting of the individual modules. Thus, the overall cost function of the PRNN becomes

$$E(n) = \sum_{i=1}^{M} \lambda^{i-1} e_i^2(n) \tag{21.24}$$

where $e_i(n)$ is defined in Equation (21.23).
Since the predictor operates on the non-stationary input data, a learning algorithm has to be chosen which, at each time step, calculates the weight correction factor $\Delta\mathbf{W}$ in order to update the weight matrix $\mathbf{W}$. Hence, the updated weight matrix at time-step $(n+1)$ can be calculated as

$$\mathbf{W}(n+1) = \mathbf{W}(n) + \Delta\mathbf{W}(n) \tag{21.25}$$

Recall that $\mathbf{W}(n)$ is an $(N+p+1) \times N$ matrix consisting of N columns which represent weight vectors $\mathbf{w}_t$, for each of $t = 1, \cdots, N$ neurons within a module; $(N+p+1)$ being for the number of inputs to every neuron, where N denotes the number of feedback inputs, p denotes the number of external input signals, and one more input is for the constant bias input which is included in the weight matrix.
The merit of a pipelined recurrent neural network as compared to a single fully connected recurrent neural network is that its computational complexity is reduced for the same total number of neurons. Let a FCRNN contain a total of N neurons; if M FCRNN constitute M modules of the PRNN, then the total number of neurons in the PRNN is $M \times N$. Having in mind that the computational complexity of a FCRNN trained with the gradient descent algorithm increases with $\mathcal{O}(N^4)$ [15], than the PRNN approach reduces the computational complexity of an entire network containing $M \times N$ neurons to a mere $\mathcal{O}(M \times N^4)$ [5]. Another advantage of the PRNN over a FCRNN is its increased capability of tracking non-linearity, and therefore the underlying Higher Order Statistics (HOS) of the probability density function (pdf) of speech, owing to cascading of M modules containing FCRNNs as their architectural components.
The PRNN, as presented by Haykin and Li [5], employs a learning algorithm which is based on the gradient descent algorithm, developed first by Williams and Zipser [15, 16]. That algorithm is particularly suitable for the prediction of non-stationary signals (e.g. speech), because it enables continuous adaptation of the network's

weights while the network is running. A continuous adaptation refers to the updating of the weight matrix of a module $\mathbf{W}$, by adding a correction term $\Delta\mathbf{W}$ to the previous value of the weight matrix $\mathbf{W}$ at every discrete time instant. Note that this is not the case for most other learning algorithms, which commonly leave the weights fixed after an initial training period and are thus not suitable for prediction of non-stationary signals.

21.3.2 The Non-Linear Subsection

As already mentioned, the first operation while performing non- linear prediction as proposed here, is to perform a kind of mapping from the input signal space represented by $\{s(n)\}$ onto an intermediate space represented by $\{\bar{s}(n+1)\}$, as shown in Fig. 21.1, in order to accomplish the linearization of the input signal. The linearization procedure itself, is composed of the three following subtasks:

- *Prediction* - Compute the one-step forward non-linear prediction errors of the PRNN at the time instant n, using the procedure described above and Equation (21.23).

- *Weight updating* - A learning algorithm uses the suitably chosen overall cost function $E(n)$ (21.24) in order to calculate the weight matrix correction factor $\Delta\mathbf{W}$ which updates the weight matrix $\mathbf{W}$, as shown in Equation (21.25).

- *Filtering* - The linearized version $\{\bar{s}(n+1)\}$ of the input signal $\{s(n)\}$ is computed, using Equations (21.19) to (21.22). The updated input signal $\mathbf{u}_i(n+1)$ to every module i, $1 \leq i \leq M$ is formed by substituting the external signal input (speech) $\mathbf{s}_i(n) = [s(n-i), \cdots, s(n-i-p+1)]$ with the updated external signal input $\mathbf{s}_i(n+1) = [s(n-i+1), \cdots, s(n-i-p+2)]$. In the non-linear predictor proposed in [5], the feedback values contained in the overall input signal $\mathbf{u}_i(n)$ to the module i at the time instant n were not updated. Thus, the feedback values at the time instant $(n+1)$ consist of the filtered outputs of the appropriate neurons at the preceding time instant n.

21.3.3 The Linear Subsection

The output signal of the non-linear subsection $\bar{s}(n+1)$ (Fig. 21.1) is fed into the linear subsection in order to accomplish the one-step forward prediction process. The linear subsection itself, is a common linear adaptive filter which uses either a Least Mean Square (LMS) or a Recursive Least Squares (RLS) algorithm to update the filter weights [3]. The non-linear activation function Φ, used in the PRNN, with its amplitude which lies in the interval $(0, 1]$ cannot produce a zero-mean output. Therefore, a constant bias input should be introduced among the adaptive linear filter inputs so that the adaptation algorithms perform well on the nonzero mean input signals. That fact was neglected in Haykin and Li's paper [5]. With the bias input included, the length of the filter in the linear subsection becomes 12, which is a standard predictor length for telephonic quality speech, incremented by one for the bias input.

21.4 The Gradient Descent Learning Algorithm for the Non-Linear Predictor

The computation of the weights during the training aims at finding a system whose operation is optimal with respect to some performance criterion which may be either qualitative, e.g. (subjective) quality of speech reconstruction, or quantitative, e.g. maximising some measure of the signal to noise ratio. The goal is to define a positive *training function* such that a decrease of that function through modifications of the weights of the network leads to an improvement of the performance of the system [17, 5, 12, 8, 16]. In the case of non-adaptive training, the training function is defined as a function of all the data of the training set. The minimum of such a *cost function* corresponds to the optimal performance of the system.
In the case of adaptive training, it is impossible, in most instances, to define a time-independent cost function whose minimisation would lead to a system that is optimal with respect to the performance criterion. Therefore, the training function is time-dependent. The modification of the coefficients is computed continually from the gradient of the training function. The latter involves the data pertaining to a time window of a finite length, which shifts in time (sliding window), while the weights are updated at each sampling time.
According to the above analysis, the performance criterion function might be defined as

$$J = \sum_{\{s\}} ||e||^2 \tag{21.26}$$

where summation is carried out over a subset of given input data. While, strictly speaking the adjustment of the parameters should be carried out by determining the gradient of J in parameter space, the procedure commonly followed is to adjust it at every time instant based on the value of the forward prediction error at that time instant and a small step size η. If θ_j represents a typical parameter, then $\partial e/\partial \theta_j$ has to be determined in order to compute the gradient ΔJ. If the weights of the network are considered as the elements of a parameter vector Θ, then the learning process involves the determination of the vector Θ^* which optimizes a performance function J based on the output error. Using the backpropagation algorithm, the gradient of the performance function with respect to Θ is computed as $\nabla_\Theta J$ and Θ is adjusted along the negative gradient as

$$\Theta = \Theta_{nom} - \eta \nabla_\Theta J|_{\Theta=\Theta_{nom}} \tag{21.27}$$

where the step size η, is a suitably chosen constant, and Θ_{nom} denotes the nominal value of Θ at which the gradient is computed[4]. It is the concept that is used in

---

[4]Let $f(\Theta) = f(\theta_1, \theta_2, \cdots, \theta_n)$ be a scalar function of n variables $\theta_1, \theta_2, \cdots, \theta_n$. The gradient of $f(\Theta)$ with respect to the vector Θ is defined as the row vector

$$\nabla_\Theta f = \left[\frac{\partial f}{\partial \theta_1}, \frac{\partial f}{\partial \theta_2}, \cdots, \frac{\partial f}{partial\theta_n}\right]$$

The value of the gradient depends upon the point $\Theta_{nom} \in \mathcal{R}^n$ at which f_Θ is evaluated. If the operating point is changed from Θ_{nom} to $\Theta_{nom} + \Delta\Theta$, where $\Delta\Theta = \eta f_\Theta^T$, $\eta \ll 1$ is a positive constant, it follows that

$$f(\Theta_{nom} + \Delta\Theta) \geq f(\Theta_{nom})$$

since

$$f(\Theta_{nom} + \Delta\Theta) \approx f(\Theta_{nom}) + \eta \nabla_\Theta f \, \nabla_\Theta^T f$$

all gradient methods for the optimization of some performance index in static and dynamical systems. Strictly speaking, vector Θ is no longer a constant parameter vector, but a function of time, so the concept of a partial derivative has to be replaced by the concept of functional derivative. However, if η is sufficiently small, it can be assumed that the concept of partial derivative (gradient in parameter space) can still be applied.
In Fig. 21.4 [18], where $\mathbf{y} = \mathbf{N}[\mathbf{u} + z^{-1}\mathbf{I}\mathbf{y}]$, we have a neural network connected in feedback with $\mathbf{I}(z) = z^{-1}\mathbf{I}$, which is an alternative representation of the network shown in Fig. 21.2. The aim is to determine the derivatives $\bar{\partial} y_i(k)/\bar{\partial}\theta_j$, for $i =$

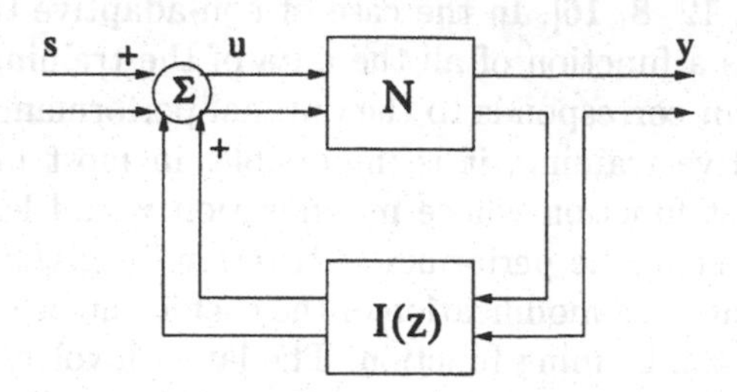

FIGURE 21.4. Alternative representation of the RNN

$1, 2, \cdots, m$ and all $k \geq 0$. In this case, $\bar{\partial} y_i(k)/\bar{\partial}\theta_j$ is the solution of a difference equation, e.g. it is affected by its past values [18, 19]

$$\frac{\bar{\partial}\mathbf{y}}{\bar{\partial}\theta_j} = \frac{\partial \mathbf{N}[\mathbf{u}]}{\partial \mathbf{u}} \mathbf{I}(z) \frac{\bar{\partial}\mathbf{y}}{\bar{\partial}\theta_j} + \frac{\partial \mathbf{N}[\mathbf{u}]}{\partial \theta_j} \tag{21.28}$$

In Eq. (21.28), $\bar{\partial}\mathbf{y}/\bar{\partial}\theta_j$ is a vector and $\partial\mathbf{N}[\mathbf{u}]/\partial\mathbf{u}$, and $\partial\mathbf{N}[\mathbf{u}]/\partial\theta_j$ are respectively the Jacobian matrix and a vector, which are evaluated around the nominal trajectory. Hence, it represents a linearized difference equation in the variables $\bar{\partial}\mathbf{y}/\bar{\partial}\theta_j$. Since, the partial derivatives of $\mathbf{N}$ can be computed at every instant of time, the desired partial derivatives can be generated as the output of a dynamical system shown in Fig. 21.5. The bar notation is used to distinguish between the partial derivatives. Following the above ideas, a gradient descent algorithm for the PRNN

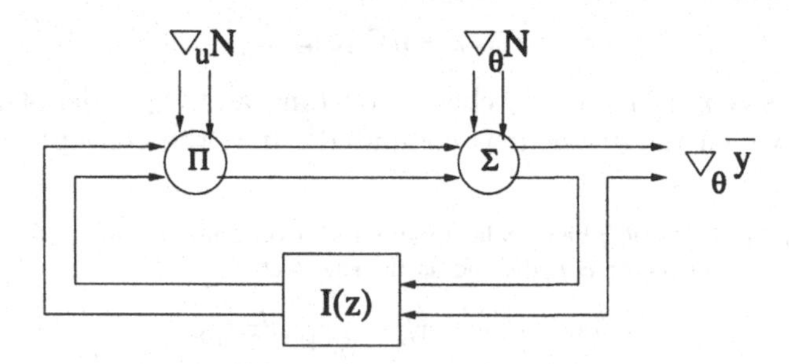

FIGURE 21.5. Generation of the gradient

can be derived. The aim is to calculate the corrections to the weight matrix $\Delta\mathbf{W}(n)$ according to the negative of the gradient of $E(n)$ with respect to the elements of the weight matrix $\mathbf{W}$. Hence, the correction for the lth weight of neuron k at the time instant n is derived as follows

$$\begin{aligned}\Delta w_{k,l}(n) &= -\eta\frac{\partial}{\partial w_{k,l}(n)}\left(\sum_{i=1}^{M}\lambda^{i-1}e_i^2(n)\right)\\ &= -2\eta\sum_{i=1}^{M}\lambda^{i-1}e_i(n)\frac{\partial e_i(n)}{\partial w_{k,l}(n)}\end{aligned} \tag{21.29}$$

Since the external signal vector $\mathbf{s}$ does not depend on the elements of $\mathbf{W}$, the error gradient becomes

$$\frac{\partial e_i(n)}{\partial w_{k,l}(n)} = -\frac{\partial y_{i,1}(n)}{\partial w_{k,l}(n)} \tag{21.30}$$

Using the chain rule, this can be rewritten as

$$\frac{\partial y_{i,1}(n)}{\partial w_{k,l}(n)} = \dot{\Phi}(v_{i,1}(n))\frac{\partial v_{i,1}(n)}{\partial w_{k,l}(n)} \tag{21.31}$$

Now, inserting (21.19)-(21.22) into (21.31) yields

$$\begin{aligned}&\dot{\Phi}(v_{i,1}(n))\frac{\partial v_{i,1}(n)}{\partial w_{k,l}(n)} =\\ &= \dot{\Phi}(v_{i,1}(n))\left(\sum_{\alpha=1}^{p+N+1}\left(\frac{\partial w_{1,\alpha}(n)}{\partial w_{k,l}(n)}u_{i,\alpha}(n) + \frac{\partial u_{i,\alpha}(n)}{\partial w_{k,l}(n)}w_{1,\alpha}(n)\right)\right)\end{aligned} \tag{21.32}$$

This term is zero except for $k = 1$ and $l = \alpha$, and the only elements of the input vector $\mathbf{u}$ that depend on the elements of $\mathbf{W}$ are the feedback values. Therefore, the Equation (21.32) can be simplified to

$$\begin{aligned}&\dot{\Phi}(v_{i,1}(n))\frac{\partial v_{i,1}(n)}{\partial w_{k,l}(n)} =\\ &= \dot{\Phi}(v_{i,1}(n))\left(\sum_{\alpha=1}^{N}\frac{\partial y_{i,\alpha}(n-1)}{\partial w_{k,l}(n)}w_{1,\alpha+p+1}(n) + \delta_{kl}u_{i,l}(n)\right)\end{aligned} \tag{21.33}$$

where

$$\delta_{kl} = \begin{cases}1, & k = l\\ 0, & k \neq l\end{cases} \tag{21.34}$$

21.5 A Novel Learning Algorithm for the PRNN Based on the Extended Recursive Least Squares (ERLS) Algorithm

The novel Extended Recursive Least Squares (ERLS) algorithm, which is presented here, is based upon the idea of the Extended Kalman Filter (EKF)[2, 3]. The cost function of the PRNN becomes

$$\epsilon(n) = \sum_{k=1}^{n}\xi^{n-k}E(k) \tag{21.35}$$

which is to be minimized with respect to the elements of the weight matrix $\mathbf{W}$. The newly introduced constant $\xi \in (0,1]$ represents a *forgetting factor* so that the resulting learning algorithm becomes suitable for the prediction task of non-stationary signals. The ERLS algorithm is used to solve the non-linear minimization problem of (21.35). In order to derive the ERLS algorithm, the vector state - vector observation Kalman filter equations are considered [3]

$$\mathbf{w}(n) = \mathbf{a}(\mathbf{w}(n-1)) + \mathbf{u}(n) \quad (21.36)$$

$$\mathbf{x}(n) = \mathbf{h}(\mathbf{w}(n)) + \mathbf{v}(n) \quad (21.37)$$

where $\mathbf{w}(n)$ becomes the $N(N+p+1) \times 1$ weight vector, $\mathbf{x}(n)$ is the $M \times 1$ observation (signal) vector, $\mathbf{u}(n)$ is a white Gaussian noise vector, $\mathbf{u} \sim \mathcal{N}(\mathbf{0}, \mathbf{Q})$, and $\mathbf{v}(n)$ is observation noise, WGN vector $\mathbf{v} \sim \mathcal{N}(\mathbf{0}, \mathbf{C})$[5]. Furthermore, we have the non-linear mapping functions

$$\mathbf{a} : \mathcal{R}^{N(N+p+1)} \rightarrow \mathcal{R}^{N(N+p+1)} \quad (21.38)$$

and

$$\mathbf{h} : \mathcal{R}^{N(N+p+1)} \rightarrow \mathcal{R}^{M} \quad (21.39)$$

which, respectively map the space spanned over the weighting vector $\mathbf{w}$ onto the same space, and the weight vector space onto the "output of the PRNN"-dimensional space. For prediction of speech, however, the function $\mathbf{a}(\cdot)$ is unknown, so that the state Equation (21.36) may be approximated by the random walk model [5]

$$\mathbf{w}(n) = \mathbf{w}(n-1) + \mathbf{u}(n) \quad (21.40)$$

Speech is a highly non-stationary signal, and therefore, the error performance surface of the PRNN processing speech changes its shape randomly. As for the EKF, the non-linear mapping function $\mathbf{h}(\cdot)$ is linearized using the first-order Taylor expansion around the estimate of $\mathbf{w}(n)$, based on the previous data, i.e. $\hat{\mathbf{w}}(n|n-1)$, which yields

$$\mathbf{h}(\mathbf{w}(n)) \approx \mathbf{h}(\hat{\mathbf{w}}(n|n-1) + \nabla \mathbf{h}^T[\mathbf{w}(n) - \hat{\mathbf{w}}(n|n-1)] \quad (21.41)$$

where the gradient of $\mathbf{h}(\cdot)$ can be written as

$$\nabla \mathbf{h}^T = \frac{\partial \mathbf{h}(\hat{\mathbf{w}}(n|n-1))}{\partial \hat{\mathbf{w}}(n|n-1)} = \mathbf{H}(n) \quad (21.42)$$

so that observation equation becomes

$$\mathbf{x}(n) = \mathbf{H}(n)\mathbf{w}(n) + \mathbf{v}(n) + [\mathbf{h}(\hat{\mathbf{w}}(n|n-1)) - \mathbf{H}(n)\hat{\mathbf{w}}(n|n-1)] \quad (21.43)$$

Moreover, the correlation matrix of the process state noise vector $\mathbf{u}(n)$ equals a scaled version of the minimum mean square error (MSE) matrix of the EKF [5, 3]

$$\mathbf{Q}(n) = E\{\mathbf{u}(n)\mathbf{u}^T(n)\} = (\xi^{-1} - 1)\,\mathbf{M}(n) \quad (21.44)$$

[5] Although the observation noise vector $\mathbf{v}$ is to be, generally speaking, described by its covariance matrix, $\mathbf{C}$, we will assume that that matrix is diagonal, e.g. that the observation noise $\mathbf{v}$ is satisfactorily described by its variance vector, together with its mean value

where ξ is the exponential forgetting factor of (21.35). Using (21.36),(21.43),(21.44), and the definition of the EKF in [3], the final equations of the ERLS algorithm become

$$\mathbf{K}(n) = \xi^{-1}\mathbf{M}(n-1)\mathbf{H}^T\left[\mathbf{C}(n)+\xi^{-1}\mathbf{H}(n)\mathbf{M}(n-1)\mathbf{H}^T(n)\right]^{-1} \tag{21.45}$$

$$\hat{\mathbf{w}}(n) = \hat{\mathbf{w}}(n-1)+\mathbf{K}(n)\left[\mathbf{x}(n)-\mathbf{h}(\hat{\mathbf{w}}(n-1))\right] \tag{21.46}$$

$$\mathbf{M}(n) = \xi^{-1}\left[\mathbf{I}-\mathbf{K}(n)\mathbf{H}(n)\right]\mathbf{M}(n-1) \tag{21.47}$$

For the PRNN, the $(M\times 1)$ dimensional vector $\mathbf{x}(n)$ becomes

$$\mathbf{x}^T(n) = [s(n), s(n-1), \cdots, s(n-M+1)] \tag{21.48}$$

which is the input speech signal itself. Furthermore, the $(M\times 1)$ dimensional vector, $\mathbf{h}(n)=\mathbf{h}(\mathbf{w}(n))$[6] becomes

$$\mathbf{h}^T(n) = [y_{1,1}(n), y_{2,1}(n), \cdots, y_{M,1}(n)] \tag{21.49}$$

Now, since by (21.36), (21.40)

$$\hat{\mathbf{w}}(n|n-1) = \hat{\mathbf{w}}(n-1|n-1) = \hat{\mathbf{w}}(n-1)$$

the gradient matrix $\mathbf{H}=\nabla\mathbf{h}$ becomes

$$\mathbf{H}(n) = \frac{\partial\mathbf{h}(\hat{\mathbf{w}}(n-1))}{\partial\hat{\mathbf{w}}(n-1)} \tag{21.50}$$

the elements of which are available from [5]

$$i=M \Rightarrow \frac{\partial y_{M,j}(n)}{\partial w_{k,l}(n)} \approx \dot{\Phi}(v_{M,j}(n))\left[\sum_{\alpha=1}^{N}\frac{\partial y_{M,\alpha}(n-1)}{\partial w_{k,l}(n-1)}w_{j,\alpha+p+1}(n)+\delta_{kj}u_{M,l}(n)\right] \tag{21.51}$$

$$i\neq M \Rightarrow \frac{\partial y_{i,j}(n)}{\partial w_{k,l}(n)} \approx \dot{\Phi}(v_{i,j}(n))\left[\frac{\partial y_{i+1,j}(n)}{\partial w_{k,l}(n)}w_{j,p+2}(n) + \sum_{\alpha=2}^{N}\frac{\partial y_{i,\alpha}(n-1)}{\partial w_{k,l}(n-1)}w_{j,\alpha+p+1}(n)+\delta_{kj}u_{i,l}(n)\right] \tag{21.52}$$

where δ_{kj} is given by (21.34), so that the derivation of the ERLS algorithm for the PRNN is now complete. Note that the covariance matrix of observation noise was set to $c\mathbf{I}$ during the simulations, $c\in\mathcal{R}$.

21.6 Experimental Results

The performance of the non-linear forward predictor was assessed via simulation. Three different input signals, denoted by *s1*, *s2* and *s3* were used to test the non-

[6] Actually, it can be found as $\mathbf{h}(n)=\mathbf{h}(\mathbf{w}(\mathbf{u}(n)),\mathbf{u}(n))$ or even more strictly speaking $\mathbf{h}(n)=\mathbf{h}(\mathbf{w}(\mathbf{u}(n\mid n-1,\cdots,0)),\mathbf{u}(n\mid n-1,\cdots,0))$

linear predictor. All the signals are available from [13]. Signal *s2* was identical to that used in [5]. The content of the speech signals used in simulations was:

- *s1*: speech sample "Oak is strong and ...", length 10000, sampled at 8kHz
- *s2*: speech sample "When recording audio data ...", length 10000, sampled at 8kHz
- *s3*: speech sample "I'll be trying to win ...", length 10000, sampled at 11kHz

The amplitudes of the signals were adjusted to lie in the range of the function Φ, i.e. $\in (0,1]$. The measure that was used to assess the performance of the predictors was the forward prediction gain R_p given by

$$R_p \triangleq 10\, log_{10} \left(\frac{\hat{\sigma}_s^2}{\hat{\sigma}_e^2} \right) dB \tag{21.53}$$

where $\hat{\sigma}_s^2$ denotes the estimated variance of the speech signal $\{s(n)\}$, whereas $\hat{\sigma}_e^2$ denotes the estimated variance of the forward prediction error signal $\{e(n)\}$. This approach to the definition of prediction gain is different from the one used in [5], which used the mean squared values of the signal and error instead of appropriate variance estimates. The usage of variance estimates is preferable, though, because the DC term contained in the mean squared values leads to biased results.
Various configurations of the non-linear predictor were tested and summarized below:

- PRNN with GD training algorithm in the non-linear subsection, and LMS in the linear subsection
- PRNN with GD training algorithm in the non-linear subsection, and RLS in the linear subsection
- PRNN with ERLS training algorithm in the non-linear subsection, and RLS in the linear subsection

21.6.1 The Initialisation of the Algorithms

In all the experiments performed, the RLS and LMS algorithms in the linear subsection had a constant bias input. The initialization of the weights $\mathbf{W}$ was achieved via epochwise training as is commonplace for neural networks with fixed weights. An initial weight matrix was chosen randomly. The first L samples of the input signal $\mathbf{s}$ were chosen as an input to the PRNN. The L samples were used for L weight update $\Delta\mathbf{W}$ calculations. Those L updates were summed to form an epoch weight update $\Delta\mathbf{W}_{epoch}$. Then, $\Delta\mathbf{W}_{epoch}$ was used instead of $\Delta\mathbf{W}$ to update $\mathbf{W}$. The whole procedure was then termed *an epoch*. The following Table 21.1 comprises all the relevant parameters used in simulations which determine the non-linear predictor. The forgetting factors and step-sizes for the algorithms were individually chosen for each input signal, in order to be as close as possible to the optimal predictor performance. The abbreviations concerning the above parameters are:

- Ω_P - Forgetting factor of the ERLS algorithm for the PRNN

Parameter	*GD algorithm*	*ERLS algorithm*
Length of the training sequence L *during initialisation*	300	300
No. of epochs during initialisation	200	5
No. of modules M *in the PRNN*	5	3
No. of neurons per module N	2	1
Slope b *in the activation function* Φ	1	2.75
System order p	4	8

TABLE 21.1. Parameters of the non-linear predictor initialisation

- μ_P - Step size of the GD algorithm for the PRNN
- Ω_L - Forgetting factor of the RLS algorithm (linear subsection)
- μ_L - Step size of the LMS algorithm (linear subsection)
- Ω - Forgetting factor when only the linear RLS based predictor is applied
- μ - Step size when only the linear LMS based predictor is applied

The values of the parameters themselves, are shown in Table 21.5.

21.6.2 The Experiments on Speech Signals

To confirm the improved performance of the newly presented ERLS algorithm over the existing GD-based learning algorithms, three experiments were undertaken. The simulations compare the performance of the non-linear predictor employing the two different learning algorithms, namely GD and ERLS, to only the RLS and LMS trained adaptive linear forward predictor. The simulation results are presented in Tables 21.2 to 21.4, which contain respectively results for the same experiment performed on the signals *s1* to *s3*. As the measure of performance the appropriate forward prediction gain was used. The appropriate forward prediction gains were denoted as follows:

non-linear prediction using	R_{PN}	R_{PRLS}	$R_{PN} - R_{PRLS}$
SG+LMS	10.25	12.70	-2.45
SG+RLS	13.01	12.70	0.31
LRLS+RLS, N=1	14.73	12.70	2.03
LRLS+RLS, N=2	14.77	12.70	2.07
	R_{PN}	R_{PLMS}	$R_{PN} - R_{PLMS}$
SG+LMS	10.25	9.24	1.01
SG+RLS	13.01	9.24	3.77
LRLS+RLS, N=1	14.73	9.24	5.49
LRLS+RLS, N=2	14.77	9.24	5.53

TABLE 21.2. Comparison of the non-linear predictor with RLS and LMS linear predictors using signal $s1$

- R_{PN} denotes the forward prediction gain when using only the non-linear predictor (PRNN)
- R_{PRLS} denotes the forward prediction gain when using the non-linear predictor with the RLS trained filter in the linear subsection
- R_{PLMS} denotes the forward prediction gain when using the non-linear predictor with the LMS trained filter in the linear subsection

non-linear prediction using	R_{PN}	R_{PRLS}	$R_{PN} - R_{PRLS}$
SG+LMS	9.49	11.55	-2.06
SG+RLS	11.80	11.55	0.25
LRLS+RLS, N=1	13.59	11.55	2.04
LRLS+RLS, N=2	13.40	11.55	1.58
	R_{PN}	R_{PLMS}	$R_{PN} - R_{PLMS}$
SG+LMS	9.49	8.06	1.43
SG+RLS	11.80	8.06	3.74
LRLS+RLS, N=1	13.59	8.06	5.53
LRLS+RLS, N=2	13.40	8.06	5.34

TABLE 21.3. Comparison of the non-linear predictor with RLS and LMS linear predictors using signal *s2*

Inspecting Tables 21.2-21.4, it can be seen that the advantage in the forward prediction gain of the non-linear predictor over the linear LMS-trained predictor ($R_{PN} - R_{PLMS}$) was 1.01*dB* for the speech signal *s1*, 1.43*dB* for the signal *s2*, and 0.99*dB* for the signal *s3* (row 6 in Tables 21.2- 21.4). The results confirm the use of the GD– trained non-linear predictor, since it achieves better performance compared to the linear LMS–trained adaptive predictor.
Furthermore, the GD–trained non-linear predictor has been shown to perform worse than only the RLS–trained linear predictor, for all the speech signals. The linear predictor had advantage in forward prediction gain ($R_{PN} - R_{PRLS}$) of 2.45*dB* over the non- linear predictor for signal *s1*, 2.06*dB* for signal *s2*, and 1.89*dB* for signal *s3* (row 1 in Tables 21.2-21.4). Since the RLS algorithm is computationally less demanding than the GD algorithm applied to the non-linear predictor, this means that the PRNN-based non-linear forward predictor doesn't perform satisfactorily when using the GD algorithm in the non-linear subsection of the PRNN, and the LMS algorithm in the linear subsection.
The previous experiment was then repeated, with the only difference being the RLS algorithm in the linear subsection of the PRNN instead of the LMS algorithm. In this case, the appropriate forward prediction gains achieved with the non-linear predictor were slightly higher than those achieved with only the RLS adaptive linear filter. The advantage in the forward prediction gain ($R_{PN} - R_{PRLS}$) for the three speech signals *s1*, *s2*, and *s3*, was respectively 0.31*dB*, 0.25*dB*, and 0.05*dB* (row 2 in Tables 21.2-21.4). In this case, the nonsubstantial improvements were achieved, despite the great increase in computational complexity compared to the sole RLS algorithm.
In the third experiment, using the same speech signals, the novel ERLS was used

non-linear prediction using	R_{PN}	R_{PRLS}	$R_{PN} - R_{PRLS}$
SG+LMS	7.3	9.19	-1.89
SG+RLS	9.24	9.19	0.05
LRLS+RLS, N=1	10.90	9.19	1.61
LRLS+RLS, N=2	9.85	9.19	1.11
	R_{PN}	R_{PLMS}	$R_{PN} - R_{PLMS}$
SG+LMS	7.30	6.31	0.99
SG+RLS	9.24	6.31	2.93
LRLS+RLS, N=1	10.80	6.31	4.49
LRLS+RLS, N=2	9.85	6.31	3.54

TABLE 21.4. Comparison of the non-linear predictor with RLS and LMS linear predictors using signal $s3$

non-linear predictor using	*signal* s1	*signal* s2	*signal* s3
SG+LMS	$\mu_P = 0.001$ $\mu_L = .18$	$\mu_P = 0.0001$ $\mu_L = .22$	$\mu_P = 0.0001$ $\mu_L = .18$
SG+RLS	$\mu_P = .001$ $\Omega_L = .999$	$\mu_P = .0001$ $\Omega_L = .9988$	$\mu_P = .0001$ $\Omega_L = .9988$
LRLS+RLS *N=1*	$\Omega_P = .993$ $\Omega_L = .999$	$\Omega_P = .993$ $\Omega_L = .9988$	$\Omega_P = .994$ $\Omega_L = .999$
LRLS+RLS *N=2*	$\Omega_P = .997$ $\Omega_L = .999$	$\Omega_P = .995$ $\Omega_L = .9988$	$\Omega_P = .9992$ $\Omega_L = .999$
LMS	$\mu = .22$	$\mu = .22$	$\mu = .14$
RLS	$\Omega = .999$	$\Omega = .9987$	$\Omega = .999$

TABLE 21.5. Parameter settings of the non-linear predictor for signals $s1$, $s2$, and $s3$

in the non-linear subsection, and the RLS algorithm in the linear subsection of the PRNN. The best forward prediction gains were achieved with only one neuron per PRNN module. This being the case, the non-linear predictor achieves the advantage in the forward prediction gain ($R_{PN} - R_{PRLS}$) of approximately $2dB$ for the signals $s1$ and $s2$, and $1.61dB$ for the signal $s3$ (row 3 in Tables 21.2- 21.4). Only for signal $s2$, did the performance slightly increase using two neurons per module, while for the remaining signals, the performance deteriorated (row 4 in Tables 21.2-21.4). That is due to "overdetermination" of the underlying NARMA model. This shows that usage of the ERLS algorithm is thoroughly supported, despite its demanding computational complexity.

21.7 Conclusions

New learning algorithms for an adaptive non-linear forward predictor based on the Pipelined Recurrent Neural Network (PRNN) have been presented. A Gradient Descent (GD) algorithm has been developed for such an architecture, and demonstrated that, in the PRNN structure, it does not outperform a conventional adaptive forward predictor based on a Recursive Least Squares (RLS) trained FIR adaptive filter, whatsoever.

An Extended Recursive Least Squares (ERLS) training algorithm for the PRNN has therefore been developed. The algorithm has been shown to yield consistently improved performance. Especially for highly nonstationary time–varying signals, such as speech, the ERLS training algorithm obtains significantly higher forward prediction gain values, than the GD algorithm, providing that the underlying linearization of the PRNN structure was appropriate. With the ERLS learning algorithm, the non-linear predictor achieves advantages of approximately 2dB in the forward prediction gain over the linear RLS algorithm. This algorithm works best with only one neuron in each of the three PRNN modules and is thus computationally less complex than the non-linear predictor presented in [5], which used two neurons and five modules. The benefit in the total number of neurons was due to the strength of the novel ERLS algorithm.
The above results demonstrate the potential of the combined linear & non-linear prediction schemes in prediction applications.

Acknowledgments: The authors would like to thank Dr. Li for providing the speech signal *s2*, that was used in simulations.

21.8 REFERENCES

[1] J. Shynk. Adaptive IIR filtering. *IEEE ASSP Magazine*, 6(2):4–21, 1989.

[2] S. Haykin. *Adaptive Filter Theory (Second Edition).* Prentice Hall, 1989.

[3] S.M. Kay. *Fundamentals of Statistical Signal Processing: Estimation Theory.* Prentice Hall, 1989.

[4] J. Makhoul. Linear Prediction: A Tutorial Overview. *Proceedings of the IEEE*, 63(4):561–580, 1975.

[5] S. Haykin and L. Li. Non-linear Adaptive Prediction of Non-stationary Signals. *IEEE Transactions on Signal Processing*, 43(2):526–535, 1995.

[6] L. Li and S. Haykin. A Cascaded Neural Networks for Real-Time Nonlinear Adaptive Filtering. *Proceedings of the IEEE International Conference on Neural Networks, (ICNN'93)*, San Francisco, USA, 2:857–862, 1993.

[7] M. Niranjan and V. Kadirkamanathan. A Nonlinear Model for Time Series Prediction and Signal Interpolation. *Proceedings of the IEEE International Conference on Acoustics, Speech and Signal Processing, (ICASSP-91)*, Toronto, Canada, 3:1713–1716, 1991.

[8] Y. Bengio. *Neural Networks for Speech and Sequence Recognition.* International Thomson Publishing, 1995.

[9] R.M. Dillon and C.N. Manikopoulos. Neural Net Nonlinear Prediction for Speech Data. *Electronics Letters*, 27(10):824–826, 1991.

[10] B. Townshend. Nonlinear Prediction of Speech. *Proceedings of the IEEE International Conference on Acoustics, Speech and Signal Processing, (ICASSP-91)*, Toronto, Canada, 1:425–428, 1991.

[11] C.R. Gent and C.P. Sheppard. Predicting time series by a fully connected neural network trained by back propagation. *Computing and Control Engineering Journal*, 109- -112, 1992.

[12] S.Z. Qin, H.T. Su, and T.J. Mc-Avoy. Comparison of Four Neural Net Learning Methods for Dynamic System Identification. *IEEE Transactions on Neural Networks*, 3(2):122- -130, 1992.

[13] http://www.ert.rwth-aachen.de/Personen/balterse.html

[14] J.T. Connor, R.D. Martin, and L.E. Atlas. Recurrent Neural Networks and Robust Time Series Prediction. *IEEE Transactions on Neural Networks*, 5(2):240- -254, 1994.

[15] R. Williams and D. Zipser. A Learning Algorithm for Continually Running Fully Recurrent Neural Networks. *Neural Computation*, 1:270–280, 1989.

[16] S. Haykin. *Neural Networks - A Comprehensive Foundation.* Prentice Hall, 1994.

[17] O. Nerrand, P. Roussel-Ragot, D. Urbani, L. Personnaz, and G. Dreyfus. Training Recurrent Neural Networks: Why and How? An Illustration in Dynamical Process Modelling. *IEEE Transactions on Neural Networks*, 5(2):178- -184, 1994.

[18] K.S. Narendra and K. Parthasarathy. Identification and Control of Dynamical Systems Using Neural Networks. *IEEE Transactions on Neural Networks*, 1(1):4- 27, 1990.

[19] K.S. Narendra and K. Parthasarathy. Gradient Methods for the Optimization of Dynamical Systems Containing Neural Networks. *IEEE Transactions on Neural Networks*, 2(2):252- 262, 1991.

22

Neural and Fuzzy Logic Video Rate Prediction for MPEG Buffer Control

Yoo-Sok Saw[1]
Peter M. Grant[2]
John M. Hannah[2]
Bernard Mulgrew[2]

ABSTRACT
Data rate management of compressed digital video has been a technically challenging task since it is vitally important in various audio-visual telecommunication services to achieve an effective video data rate (video rate) control scheme. It has a large influence on video quality and traffic congestion in B-ISDN networks. Up to date, this issue has been treated mainly from the teletraffic control point of view, i.e. by modelling congestion control via network protocols. Relatively less attention has been focused on video rate management in the source coding side. In this chapter we consider that it is more efficient and less costly to control video rate at the video source than handling network congestion (or overloading) due to an extremely large quantity of incoming variable bit rate (VBR) video traffic. Thus this chapter investigates effective rate control algorithms for video encoders. Considering the non-stationary nature of video rate derived from scene variations (i.e. the wide band nature of digital video), we adopted two nonlinear approaches; radial basis function (RBF) estimation using a neural network-based approach and fuzzy logic control as a nonlinear feedback control. The RBF network scheme is primarily discussed and then the fuzzy logic-based scheme is compared to it. The performance is evaluated using the criterion how effectively video rate is maintained within a specified range or at a value while achieving satisfactory video quality.

22.1 Introduction

Video rate control involves regulating video data rate in conjunction with video quality when video is encoded by statistically based compression schemes such as MPEG (Moving Picture Experts Group) standards. There is a clear tradeoff between data rate of compressed video and resulting quality. When the available bandwidth for compressed video is limited, whether in a transmission channel or storage medium, video rate and quality need to be controlled depending on the medium availability. In this process, a critical goal is how to regulate the data rate

---

[1]**Centre for Communications Research, Dept. of Electrical and Electronic Engineering, Univ. of Bristol, Bristol, BS8 1UB, United Kingdom, E-mail: Yoo-Sok.Saw@bristol.ac.uk**
[2]**Signals and Systems Group, Department of Electrical Engineering, Univ. of Edinburgh, Edinburgh, EH9 3JL, United Kingdom E-mail: Peter.Grant@ee.ed.ac.uk**

and also to maintain video quality and communication connectivity at the same time.

22.1.1 Background of Video Rate Control

An effective rate control algorithm becomes more demanding when the video contains rapid motion or frequent scene changes mixed with less scene variations. This wide band nature of digital video is a major cause of abrupt variations in video rate. The video rate variation can be controlled in many different ways. A good rate control algorithm should keep the video rate as stable as possible within a desired range by controlling the number of coded bits per unit time, with the smallest possible quality degradation in PSNR (peak signal-to-noise ratio). In this chapter attention is focused mainly on feed-forward schemes which can exploit scene change information in advance. Video with rapid scene variation, Figs. 22.1 and 22.2, cannot easily be accommodated in purely causal ways such as by prediction. Hence, it is necessary to introduce a priori knowledge (i.e. scene change features to represent variation in visual information) to improve the rate control performance of the MPEG video encoder.

The ISO13818-2 MPEG2 standard [14] and its evaluation model TM5 [13] specify the baseline of rate control for the MPEG2 video encoder interfaced with the video buffering verifier (VBV). This scheme is based on feedback control for constant bit rate (CBR) channels. It specifies the operation of the VBV buffer for constant bit rate channels, providing desired buffer occupancies and quantisation scale values as critical parameters. However, it describes global system aspects rather than the details of the rate control process itself. Although the MPEG2 evaluation model, TM5 specifies details of its rate control process, the scheme is based on a non-parametric prediction and is known to be inappropriate for video with large scene changes. TM5 is included as a performance reference for comparison purposes.

Digital video has short-term correlation in the temporal direction between adjacent frames as well as the spatial correlation within a picture. It is feasible to exploit these correlations when video rate control is performed [7]. This is well suited to the case of video with small motion or an aggregated video bit stream from multiple video sources, which is transmitted through a high capacity channel. Conventional approaches generally assume that the compressed video has high correlation so that linear predictive methods can be used effectively. However, for realistic videos such as movies, sports and advertisements, there can be many exceptions where the correlation may be abnormally low. In this approach the correlation is used in conjunction with scene change features on the basis that large variation in video content co-exists with highly correlated video.

22.1.2 Feed-Forward Video Rate Control: A Neural Network Approach

Time series prediction is one of the established signal processing areas, which has traditionally been treated in linear analytic ways [35]. A basic assumption here is that the signal falls into the category of linearity and stationarity with second-order statistics, with particular emphasis on Gaussian statistics. However, nonlinear prediction and estimation techniques do not make these assumptions, since realistic signals often contain very different statistical properties. Like other nonlinear

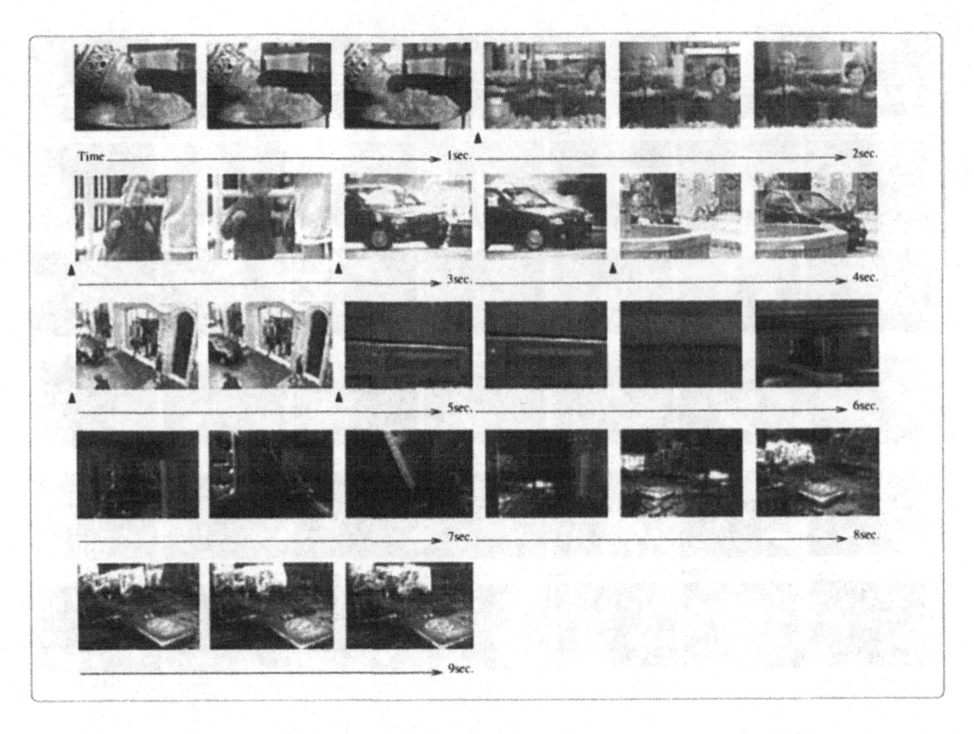

FIGURE 22.1. Selected frames from the "Adverts" sequence

predictors such as Voltera series [33] and neural networks based on multi-layer perceptrons [12], it has been shown that the RBF-network can universally approximate any function or time series [10, 25] on the basis of functional approximation theory [18]. This approximation capability can also be applied to the prediction of non-stationary video rate time series. Thus, one of novel architectures applied in this chapter is the nonlinear predictive video rate estimator using the RBF network. The RBF network is often classified as a special form of neural network [15] with a smaller number of hidden layers. It also has an advantage in the aspects of implementation, as it is a parallel structure [6, 1, 34]. The RBF network is known to have better estimation performance than linear predictors for non-stationary signals and has recently been used successfully in several engineering areas such as channel equalisation [4].

We employed a feed-forward predictive scheme in order to exploit short-term correlation of video and applied a nonlinear estimator (the RBF-network) as well as linear estimators, in order to improve the estimation performance. This approach starts from the widely accepted assumption that the video rate is a correlated time series [38, 24]. This implies that the video rate can be estimated by using predictive techniques. However, linear prediction performs effectively only when the video is highly correlated and contains small motion, i.e. no dramatic scene change occurs. A more effective rate control technique needs to be applied when video contains rapid motion or frequent scene changes. Although one can achieve a certain level of performance by using linear techniques which exploit the correlation of video, the performance can be further improved if nonlinear predictive technique is used. Thus a nonlinear feed-forward predictive scheme is used, in order to achieve further improvement. In this feed-forward scheme, a series of *scene change features*

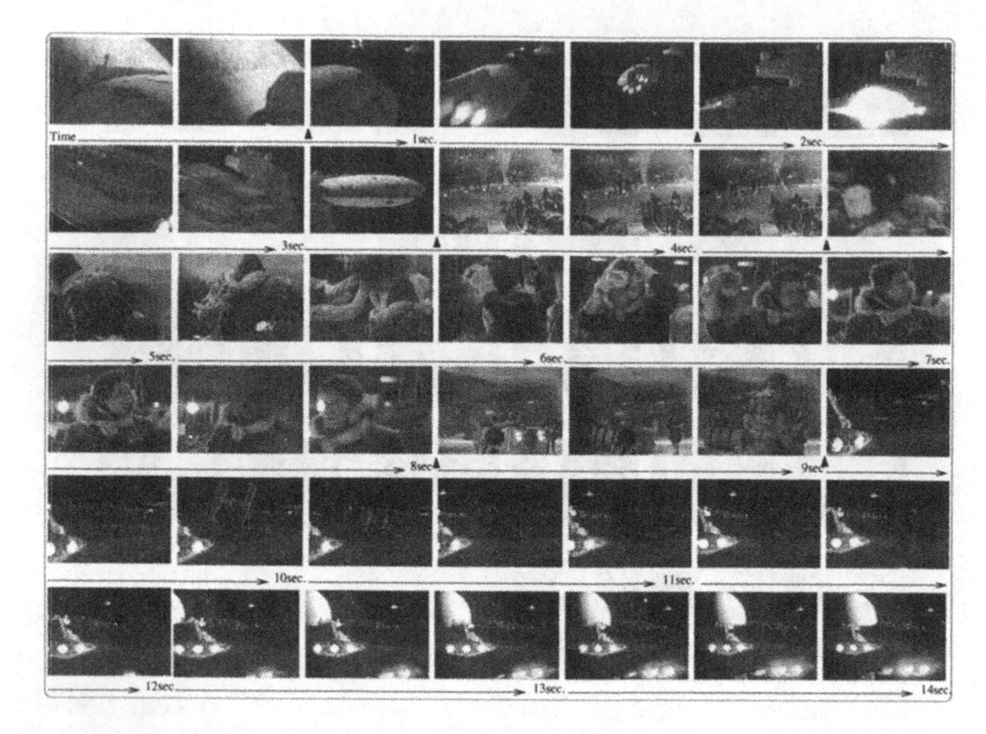

FIGURE 22.2. Selected frames from the "Starwars" sequence

are used as the input for video rate estimators. This is intended to improve the estimation performance by using a priori information rather than using previous video rate values. These features indicates how dramatically a scene has changed in comparison to the previous one. They represent the 1st and 2nd-order statistical measures: intra-frame variance, inter-frame variance plus picture type (I, P and B) values.

22.1.3 Feedback Video Rate Control: A Fuzzy Logic Approach

The buffer-based video rate control is fundamentally a feedback control in which the buffer occupancy is translated into a quantisation step size. As an improved feedback approach, we also investigated a fuzzy-logic based video rate control technique. It is considered that conventional fuzzy-logic based control (FLC) does not effectively control the two output variables (video data rate and video quality) which are mutually contradictory. The primary reason for this is that it is not easy to effectively project the control variables onto fuzzy rules due to the contradiction between these two variables on a rate-distortion theoretic basis. However, it is clear that the video quality should be considered as an equally important variable as well as the video rate since objectionable quality degradation should be prevented. We employed a FLC scheme which also takes the video quality into account by using feed-forward scaling factors whose inputs are scene change features. The performance of this scheme is comparable to the FLC scheme which does not take the quality into account.

22.1.4 Objectives and Organisation

The feed-forward predictive schemes will be first introduced, covering system identification, training and verification with the MPEG2 encoder. The fuzzy logic approach will then be discussed for comparison. The performance of the proposed schemes will also be compared with the MPEG2 TM5 and will be assessed in terms of buffer occupancy, video data rate (bits/frame) and peak signal-to-noise ratio (PSNR).

The main objective in this approach is to assess the practicality of the estimator-based rate control scheme, the performance of the conventional techniques and an emerging nonlinear technique based on radial basis function (RBF) networks, including the comparison with the fuzzy-logic based approach.

This chapter is organised as follows: Section 22.2 outlines the feed-forward video rate control techniques, including system identification using linear estimation. Section 22.3 discusses the nonlinear quantiser control technique. Section 22.4 deals with the feed-forward rate control scheme again, but focusing on a nonlinear rate estimator (the RBF network) including aspects of training and optimising the RBF network. In Section 22.5 the fuzzy logic approach is explained and compared with the RBF network. Section 22.6 presents simulation results and analysis. Finally, Section 22.7 concludes this chapter.

22.2 Feed-Forward Predictive Video Rate Estimation for MPEG

Two of the three scene change features used are framewise variances, $var\_org(k)$ and $var\_dif(k)$, which represent the variance of the input picture and the variance between the input picture and the previous picture, respectively, where k represents picture time index. The other scene change feature is picture type, $ptype(k)$, and a single value exists for the corresponding picture type, thus it forms a cyclic time series as k proceeds.

Two video sequences, "Starwars" and "Adverts", shown as a negative film-style representation in Figs. 22.1 and 22.2, are used in the evaluation process to give frequent scene changes and rapid motion video to the encoder. "Adverts" contains three television advertisements which exhibit rapid motion and dramatic scene changes. "Starwars" was digitised from its televised version (300 frames). The movie "Star Wars" is widely used in the field of video traffic modelling [23, 22]. It often exhibits dark background, slow motion, less colour change and many artificially synthesised scenes throughout the sequence. The sequence was captured from a part with relatively rapid motion and dramatic scene changes[3]. Dramatic scene changes are shown by attaching triangles between pictures. For display purposes one picture frame is taken out of every ten frames. Thus, the time interval is 33.3 ms, approximately. For example, in "Adverts" triangles indicate abrupt scene changes at frame numbers around 40, 70, 90, etc. The numerical representation of the scene change features are shown in Fig. 22.3 for the two video sequences on a logarithmic vertical scale

[3]These two video sequences used in simulations are colour pictures but for publication purposes black-and-white pictures are presented instead.

since the dynamic range of each input is considerably different from the others. The vertical scale represents variances and picture type values.

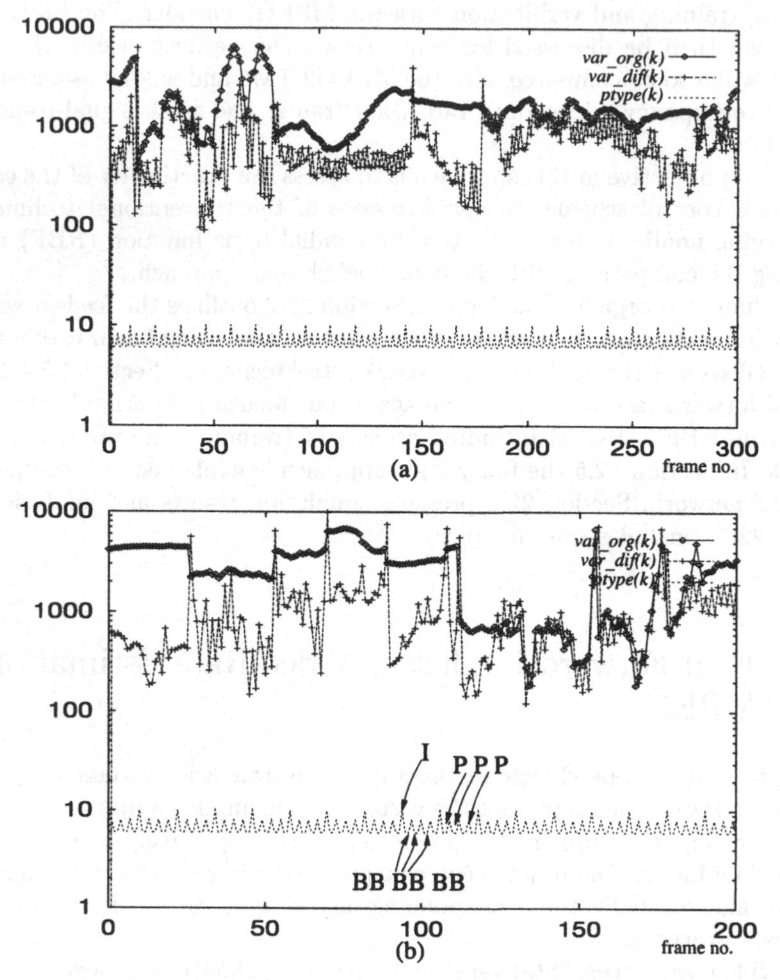

FIGURE 22.3. Three inputs, $var\_org(k)$, $var\_dif(k)$ and $ptype(k)$: (a) Starwars, (b) Adverts

The mechanism for controlling the occupancy can be enhanced by predicting the future occupancy which is assessed by a one-frame-ahead buffer capacity estimate before encoding the next picture. The quantiser then changes the step size in advance of the arrival of the picture data. The video rate estimator calculates the future occupancy by estimating a video rate value for an input picture using the previous and current scene change features. Fig. 22.4 shows the configuration of the generalised feed-forward video rate estimation scheme which comprises three main functions: scene change calculator (SCC), rate estimator (RE) and quantiser control (QC) based on nonlinear functional surfaces. Although any linear estimator can be used as a video rate estimator, two estimators will be investigated to highlight the difference between the linear and nonlinear estimators: a linear estimator trained with recursive least squares (RLS) algorithm and a nonlinear radial basis function (RBF) network.

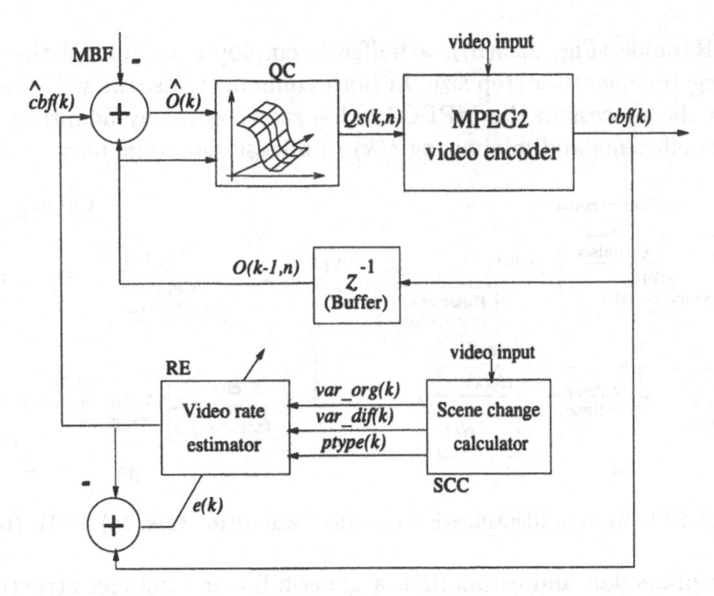

FIGURE 22.4. The structure of the feed-forward network for MPEG rate control

Considering the non-stationary nature of the inputs, the recursive linear estimator is used instead of the non-recursive least square predictor since the weights should be updated depending on changes in the input statistics. The scene change calculator operates in advance of actual encoding to estimate the framewise video data rate, $cbf(k)$. It outputs scene change features by calculating the variances and also passes picture type information, $ptype(k)$, for an input picture on to the rate estimator. $ptype(k)$ gives vital information on the video rate to the rate estimator since the MPEG2 video encoder processes picture frames in different coding modes according to the picture type repetition, i.e. IBBPBBP...[14].

The predicted video rate, $\widehat{cbf}(k)$, is added to the current occupancy, $O(k-1,n)$, to form the predicted occupancy, $\hat{O}(k)$. The nonlinear quantiser control finally outputs an appropriate quantisation scale value, $Qs(k,n)$. The MPEG2 video encoder accepts $Qs(k,n)$ as an input, and outputs $cbf(k)$. The transmission buffer is treated as a delay (z^{-1}) since it stores coded bits for a specific frame period. MBF is a constant determined by the current channel rate, which represents the mean bits allocated per frame.

22.2.1 System Identification via Linear Video Rate Estimation

Before applying the two video rate estimators to the MPEG2 video encoder, feasibility of the feed-forward approach is evaluated by conducting system identification simulations with linear estimation. The configuration of the system identification is shown in Fig. 22.5. In the VBR mode (Fig. 22.5(a)), as quantiser control is not necessary, as the buffer is not used. Therefore, the feedback path from the buffer is not connected to the quantiser. This implies that the configuration only takes into account the effect of the variation in the scene change on the estimated video rate, $\widehat{cbf}(k)$, since the quantisation scale is fixed. That is, the linear and nonlinear relationships are intended to be observed between scene change features and the predicted video rate without the influence from the buffering. On the other hand, in the CBR mode (Fig. 22.5(b)), a buffer is employed to control the video rate by adjusting the quantiser step size. In both configurations, the video rate estimator models the process of the MPEG2 video rate control by adaptively changing estimator coefficients and minimising $e(k)$ in a least square manner.

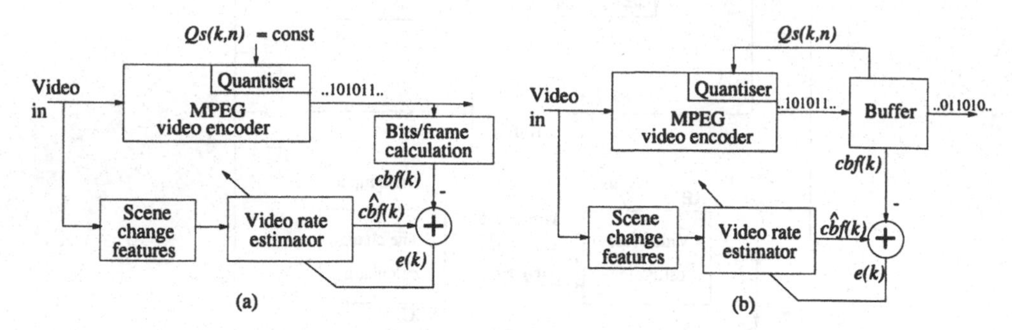

FIGURE 22.5. System identification of video rate estimators: (a) VBR, (b) CBR

In linear prediction and estimation, a generic linear combiner structure with a single input channel [8] is generally used:

$$\widehat{y}(k) = \sum_{i=0}^{N-1} h_i x(k-i) \tag{22.1}$$

where $\widehat{y}(k)$ is the predicted value of the actual output $y(k)$. $x(k-i)$ are the time lagged input data with N samples and the predictor coefficients of the linear combiner, h_i. As three scene change features are used as the input for the video rate estimator, the predicted video rate can now be written as:

$$\widehat{cbf}(k) = \sum_{i=0}^{N-1} h_{1i} var\_org(k-i) + \sum_{i=0}^{N-1} h_{2i} var\_dif(k-i) + \sum_{i=0}^{N-1} h_{3i} ptype(k-i) \tag{22.2}$$

where h_{1i}, h_{2i}, h_{3i} represent the estimator coefficients, respectively. The estimated number of coded bits per frame, $\widehat{cbf}(k)$, is the output of the linear combiner. We can evaluate the performance of the linear model by looking into how effectively the linear technique predicts the video rate. For training the estimator coefficients, both recursive and non-recursive methods are tested.

In the non-recursive least squares (LS) scheme predictor coefficients, $\mathbf{h}$, are obtained by applying matrix algebra in the linear equation, $\mathbf{Xh} = \mathbf{y}$ where $\mathbf{X}$ is the input matrix and $\mathbf{y}$ is the predicted output vector. The coefficient vector $\mathbf{h}$ is estimated by the singular value decomposition (SVD) technique [26, 11] using a finite number of input and output time series which are assumed to represent the overall linear property of the input. The predicted output, $\widehat{cbf}(k)$, is estimated from the $\mathbf{h}$ coefficients as shown in Fig. 22.6.

The recursive method, Fig. 22.7, is used to adapt the predictor coefficients, $h_i(n)$, depending on changes in the input. The recursive LS (RLS) algorithm is known to be appropriate for input whose statistical properties vary [11] since it can track slowly varying coefficients. The coefficients are updated when the error, $e(k)$, is calculated on a framewise basis. The input configuration is the same as the non-recursive case.

The objective of this evaluation is to examine the performance of linear estimators with given input and output data, rather than to evaluate the actual rate control

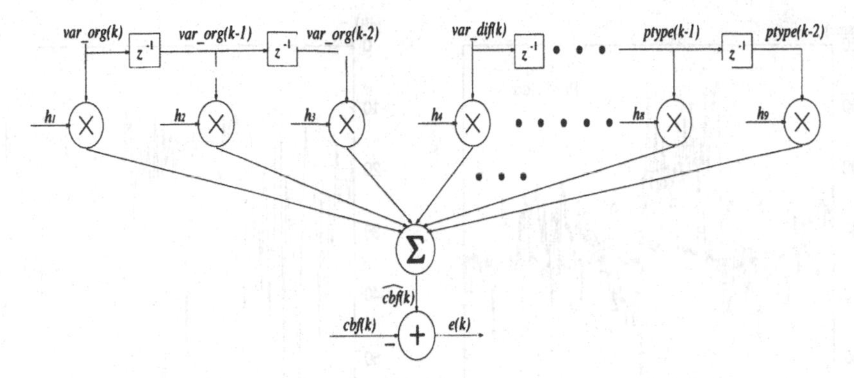

FIGURE 22.6. Linear estimator with least square training

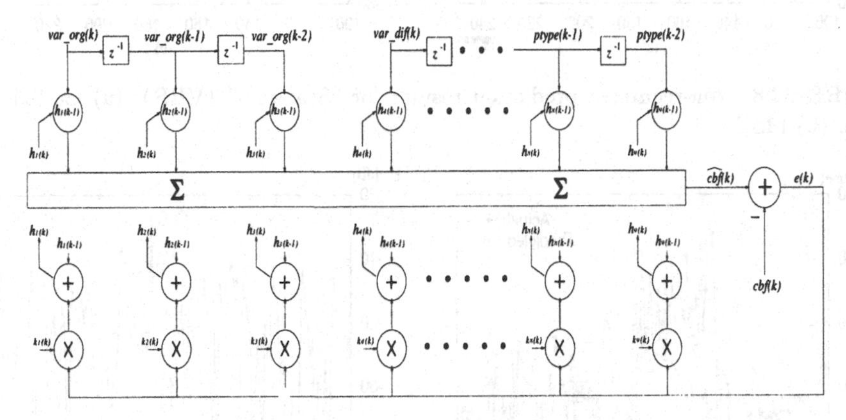

FIGURE 22.7. Recursive least square estimator

performance in a running MPEG2 encoder[4]. The input and output data are given by encoding the test video sequences at an effective data rate of 1280 kbits/s and in the variable bit rate mode. In the non-recursive technique (Fig. 22.6), 200 input samples in the mid range are taken for training among 300 available data samples. The rest of the data is used as a validating data set. The RLS forgetting factor α was chosen experimentally. A small value of α, say, smaller than 0.7, generally results in large variation in predicted signals since the 'memory' of the algorithm reduces. Thus, a large value 0.95 was consistently used in order to maintain the predicted signal with less noise-like variation. The relative performance is assessed by using the mean square error (MSE) in dB, which is given by:

$$\text{MSE} = 10 \log_{10} \frac{\textit{error power}}{\textit{signal power}} \tag{22.3}$$

where the *signal power* is the mean value of the squared sum of the actual output signal and the *error power* represents the error between the actual and the estimated signals. Figs. 22.8 and 22.9 show prediction results (i.e. the predicted signal and the actual output signal in bits per frame) for VBR coding, which are obtained by the LS estimator and the recursive estimator, respectively. Figs. 22.10 and 22.11 show the results for CBR coding at 1280 kbits/s for the "Starwars" video sequence. These results are summarised in Table 22.1.

[4]The latter case will be discussed in Section 22.6.

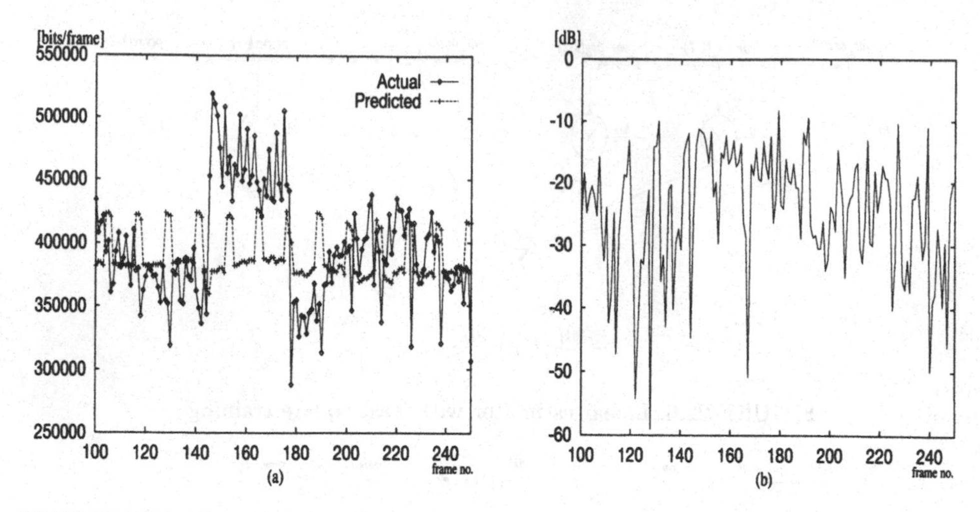

FIGURE 22.8. Non-recursive prediction results for "Starwars" (VBR): (a) Coded bits/frame, (b) MSE

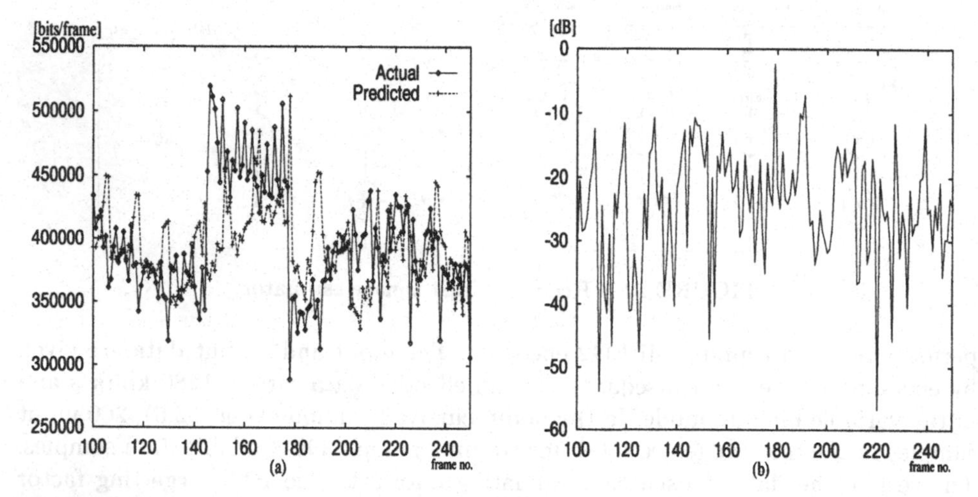

FIGURE 22.9. Recursive prediction results for "Starwars" (VBR): (a) Coded bits/frame, (b) MSE

The actual output signal in the CBR mode results (Figs. 22.10 and 22.11) does not show abrupt changes as appeared in the VBR results, and the resulting predicted signals show similar characteristics to the actual output. However, in the VBR case the actual output signal contains a dramatic increase and a decrease around the frame numbers 140 and 180, respectively, maintaining its mean value higher than other signal values outside the 140 to 180 frame number range. For this sudden change in the actual output signal, the recursive estimator appears to predict the actual output signal with more accuracy, as shown in Fig. 22.9(a) since it adaptively changes the predictor coefficients. The recursive estimator appears to have better performance than the LS estimator for the VBR case transmission with the MSE -23.40 dB versus -23.94 dB ("Starwars"). On the other hand, for the 1280 kbits/s case, where the video rate is controlled by the buffer, the MSE performance of the LS estimator is better than the recursive estimator, i.e. -24.25 dB versus

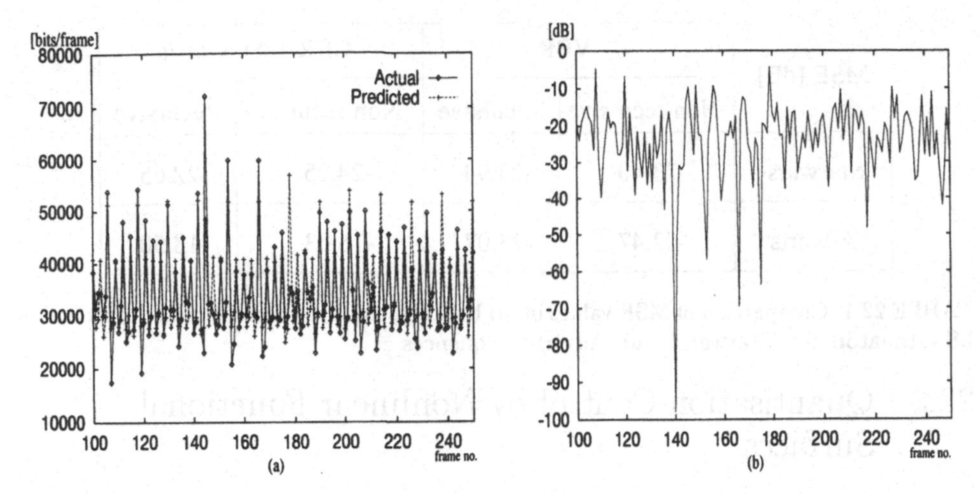

FIGURE 22.10. Non-recursive prediction results for "Starwars" at 1280 kbit/s: (a) Coded bits/frame, (b) MSE

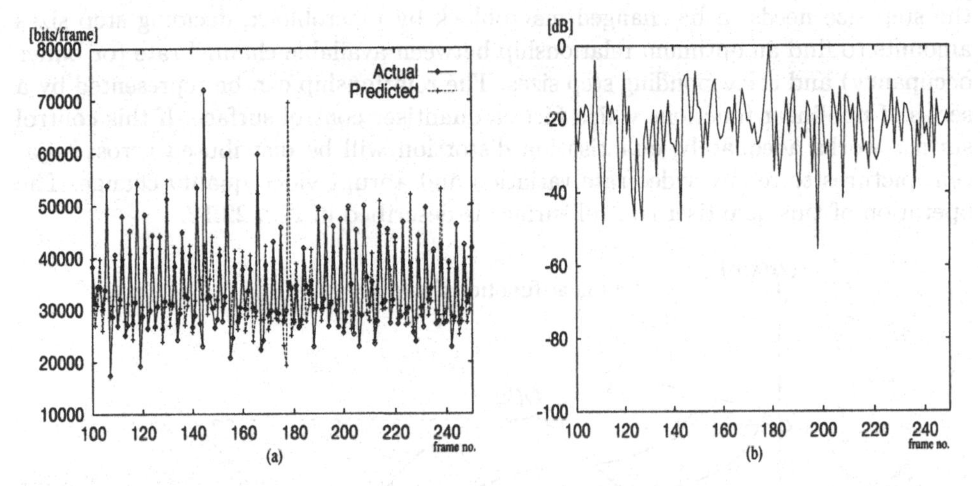

FIGURE 22.11. Recursive prediction results for "Starwars" at 1280 kbit/s: (a) Coded bits/frame, (b) MSE

-22.65 dB ("Starwars"). This may be interpreted as follows: the recursive estimator is better able to track the short-term, dramatic changes in the time series than the LS estimator while the latter works well with a time series with less dramatic changes as is the case for 1280 kbits/s. This outcome from the CBR mode simulations is the result of the rate control rather than a cause of degradation in the prediction performance. Therefore, these prediction results do not imply that the inputs are necessarily suitable for the VBR mode. On the contrary, the variance inputs, $var\_org(k)$ and $var\_dif(k)$, may be more suitable for the CBR mode as they are employed to improve the rate control performance by being used for estimating video rate within the framework of buffer control.

Although these simulation results do not guarantee the same performance for all types of video, this feed-forward approach can be justified in the sense that it is able to effectively estimate video rate using a priori information. The MSE values shown in Table 22.1 also exemplify the performance of the approach.

MSE [dB]	VBR		CBR 1280 kbits/s	
	Non-recursive	Recursive	Non-recursive	Recursive
Starwars	-23.40	-23.94	-24.25	-22.65
Adverts	-22.47	-23.07	-16.03	-18.50

TABLE 22.1. Comparison of MSE values in dB between the non-recursive and the recursive LS estimators for "Starwars" and "Adverts" sequences

22.3 Quantisation Control by Nonlinear Functional Surfaces

In the feed-forward rate control scheme, estimating video rate signifies that the quantiser step size is appropriately selected before a picture is encoded. Since the step size needs to be changed macroblock by macroblock, deciding step sizes amounts to find an optimum relationship between available channel rate (or buffer occupancy) and corresponding step sizes. The relationship can be represented by a series of nonlinear functions which form a quantiser control surface. If this control surface works adequately, quantisation distortion will be distributed across adjacent pictures to reduce video rate variation and abrupt video quality change. The operation of this quantiser control surface is described in Fig. 22.12.

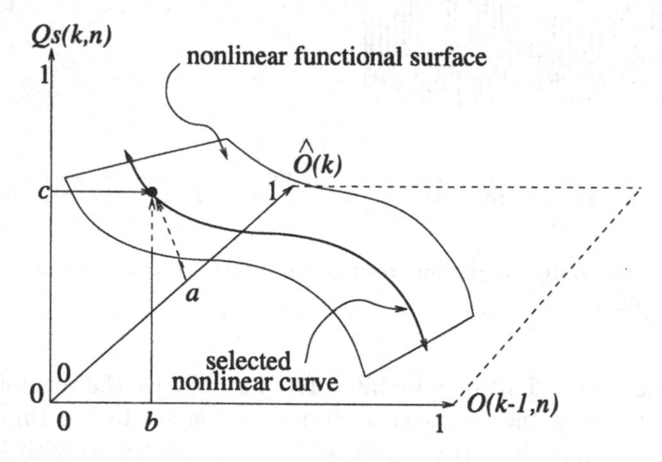

FIGURE 22.12. Quantiser control based on the nonlinear function surface

Let $O(k-1,n)$, $\widehat{O}(k)$ and $Qs(k,n)$ be current occupancy, predictive occupancy and quantisation scale, respectively. n is the macro block index which ranges from 0 to 329 for the 352 × 240 MPEG SIF image format. Thus, they signify thadualocc.epst the occupancies and the quantisation scale value vary on a macro-block-by-macro-block basis. $Qs(k,n)$ is calculated with the following equations (see Fig. 22.4):

$$\begin{aligned} Qs(k,n) &= f(O(k-1,n), \widehat{O}(k)) \\ \widehat{O}(k,n) &= O(k-1,n) + \widehat{cbf}(k) - \text{MBF} \end{aligned} \tag{22.4}$$

where the quantisation scale $Qs(k,n)$ (*c*) is determined by both $O(k-1,n)$ (*b*) and $\widehat{O}(k)$ (*a*) which is the sum of the current occupancy and the predicted video rate $\widehat{cbf}(k)$. The function $f()$ is a nonlinear functional surface to adaptively map the occupancies to a quantisation scale.

Two different control surfaces were used; sigmoidal and unimodal, as shown in Fig. 22.13. The way the two surfaces work is the same as described in Fig. 22.12. The lower the channel rate used, the more nonlinear the surface becomes, such as Fig. 22.13(b) and (d). If a dramatic change in the occupancy is predicted, then the control curve is made more nonlinear, otherwise, it selects a less nonlinear curve. The sigmoidal surface is formed by changing the curvature of a sigmoidal function. The unimodal surface comprises a combination of an exponential part and a logarithmic part.

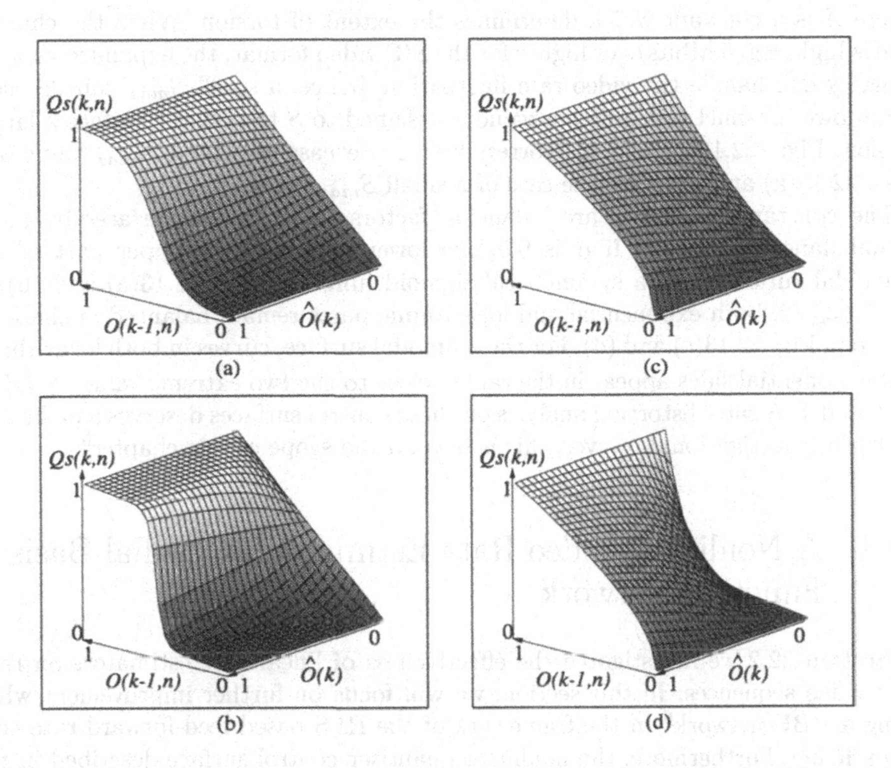

FIGURE 22.13. Quantisation surfaces: (a) sigmoidal (smaller torsion), (b) sigmoidal (larger torsion), (c) unimodal (smaller torsion), (d) unimodal (larger torsion)

The sigmoidal surface, SIGM, is given as follows:

$$
\begin{aligned}
\text{SIGM: } f(O(k-1,n),\widehat{O}(k)) &= \alpha\left(\frac{1}{\alpha}O(k-1,n)\right)^{(S\widehat{O}(k)+1)} \\
&\quad \times \mathit{trunc}(1+\alpha-O(k-1,n)) \\
&\quad + \left(1-(1-\alpha)\left(\frac{1}{1-\alpha}(1-O(k-1,n))\right)\right)^{(S\widehat{O}(k)+1)} \\
&\quad \times \mathit{trunc}\left(\frac{O(k-1,n)}{\alpha}\right)
\end{aligned} \tag{22.5}
$$

The quantiser step size in the unimodal surface, UNIM, is determined by the following equation which maps both occupancies to the step size:

$$\text{UNIM: } f(O(k-1,n), \widehat{O}(k)) = O(k-1,n)^{C/(S\widehat{O}(k)+1)} \quad (22.6)$$

In Equations (22.5) and (22.6), S is a steepness factor which represents the extent of torsion, and *trunc* stands for truncation function to truncate or hard limit an input to 1 or 0 depending on its value. The torsion factor S_{max}, which is the maximum value of S, varies with channel rates. There exists an inversely proportional relationship between the torsion factor S_{max} and the channel rate, expressed as follows:

$$S_{max} = \frac{A}{c\_rate} \quad (22.7)$$

where A is a constant which determines the extent of torsion. When the channel rate is high, e.g. 5 Mbits/s or higher for the SIF video format, the expanded channel capacity can handle the video rate fluctuation, hence, a small S_{max} can be used. For a lower channel rate a higher value is assigned to S to give the surface a larger torsion. Figs. 22.13(b) and (d) correspond to the case of a large S_{max} value and Figs. 22.13(a) and (c) show the case of a small S_{max} value.

The constants α and C are balancing factors to make the surfaces balanced or unbalanced in shape. If α is 0.5, the lower part and the upper part of the sigmoidal surface form a symmetrical sigmoid function, Fig. 22.13(a) and (b). If $C = S_{max}/2$, both exponential and logarithmic parts remain balanced in shape as shown in Fig. 22.13(c) and (d). For the unimodal surface, curves in both logarithmic and exponential sides appear in the ranges close to the two extreme values of $\widehat{O}(k)$, i.e. 0 and 1. A rate-distortion analysis on these control surfaces deserves a dedicated in-depth investigation, however, this is beyond the scope of this chapter[5].

22.4 A Nonlinear Video Rate Estimator: A Radial Basis Function Network

In Section 22.2 we investigated the effectiveness of linear rate estimators on realistic video sequences. In this section, we will focus on further improvement when using a RBF network[6] in the framework of the RLS-based feed-forward rate control scheme. Furthermore, the nonlinear quantiser control surface described in the previous section will also be adopted in the MPEG2 video encoder.

22.4.1 Configuration of the RBF-Network Estimator

The RBF-network rate estimator is viewed as a nonlinear (universal) functional approximator [10, 25], which approximates the video rate signal, real-valued function of the scene change feature vector $\mathbf{x}$. From the equations described in [10],

$$cbf(\mathbf{x}') = \langle \mathbf{x}', \mathbf{w} \rangle + \mathrm{b} \quad (22.8)$$

[5] Authors completed the investigation and submitted a full paper on this [28].

[6] A full paper on this technique will also be published in [30].

where **x'** and **w** are the processed version of the input vector **x** and a vector of linear weights, respectively, and b is a scalar bias. The term <**x'**, **w**> represents inner product. The input vector **x** can be transformed into **x'** by any continuous non-constant function. Thus, if this function is replaced with a radial basis function, e.g. Gaussian function, the resulting network becomes the radial basis function network which includes the Gaussian hidden layer.

The RBF network consists of centres with the radial basis function and linear weights given:

$$\widehat{cbf}(\mathbf{x}) = \sum_{i=1}^{N} w_i \phi(||\mathbf{x} - \mathbf{x}_i||)$$

$$\phi(||\mathbf{x} - \mathbf{x}_i||) = \exp(-\frac{||\mathbf{x} - \mathbf{x}_i||^2}{2\sigma^2}) \tag{22.9}$$

where $\widehat{cbf}(\mathbf{x})$ is the output of the RBF network, w_i is the linear weight. $\mathbf{x}_i$ represents the ith selected centre to represent a large number of input vectors in terms of Euclidean distance. $\phi()$ is the Gaussian function which outputs RBF layer values determined by Euclidean distance between the input vector **x** and the centre $\mathbf{x}_i$. The RBF network, shown in Fig. 22.14, may have as many centres as required by selecting input vectors, and it calculates the contribution of each input using the Gaussian function. For network efficiency, however, the RBF centres are usually selected by the orthogonal least square (OLS) algorithm [5] or by clustering algorithms [12]. The OLS algorithm selects representative RBF centres when supervised learning is used. However, in the case of the MPEG2 encoder, supervised learning cannot be used effectively, since statistical properties of the network input are non-stationary. Supervised learning is known to be effective to time series prediction where the signal is stationary or the statistical properties are known. Thus, unsupervised learning needs to be used in the RBF-network-based MPEG encoder. The k-means clustering algorithm is used for the unsupervised learning, as for RBF-network-based channel equalisation [3, 2] applications.

The centres are updated as follows [4]:

$$\mathbf{x}_i(k) = \mathbf{x}_i(k-1) + g_c(cbf(k) - \mathbf{x}_i(k-1)) \tag{22.10}$$

where the constant g_c controls the learning rate. This clustering technique updates centres in reference to a single previous centre since the time index for the previous centre is set to $k-1$. More previous centres can be taken into account to update the next centre. However, the value $k-1$ is particularly useful when centres are changing rapidly due to dramatic scene changes for a short period of time. Thus, the $k-1$ time index for centres is consistently used for all the RBF-network simulations.

22.4.2 Training and Optimising the RBF Video Rate Estimator

In consideration of the configuration of the MPEG2 video encoder, the number of RBF centres is set to 9. As shown in Fig. 22.14, the input consists of 3 scene change features, each of which has 3 taps. The number of time delays is equal to the number of B pictures between P pictures. It is assumed that the cyclic repetition of the video rate is determined by the number of B pictures, and the correlation of the video rate varies with an interval of two pictures. Different numbers of centres - up

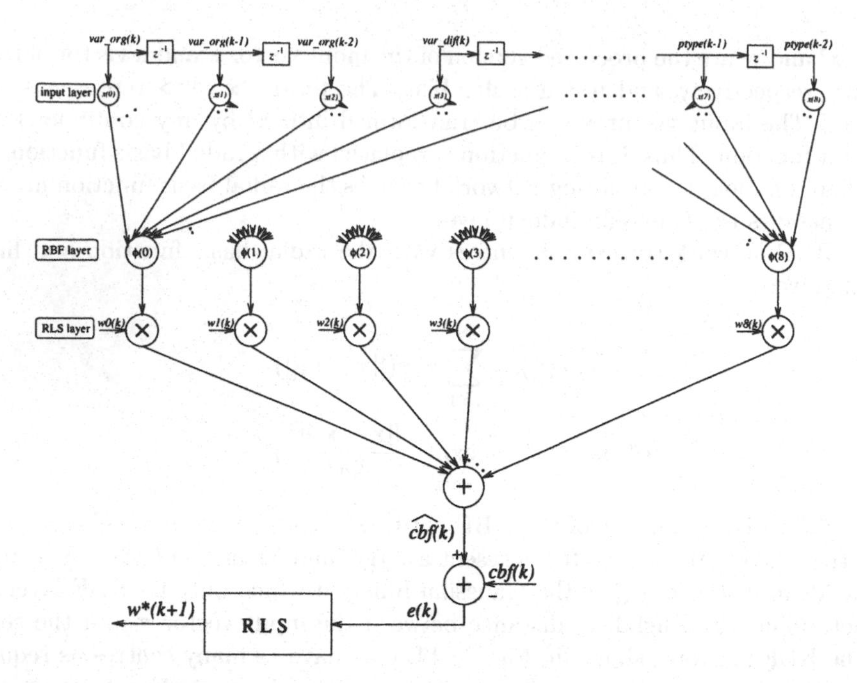

FIGURE 22.14. RBF predictor with 3 inputs and 9 taps

to 50 - were also simulated and their performance was compared in normalised mean square error (NMSE), as shown in Fig. 22.15 and Table 22.2. We used realistic video sequences including two other sequences, "Cascaded" and "JFK"[7], to see the effect of the RBF network estimator. For all four video sequences tested, the 6 to 9 centres appeared to exhibit the same performance as cases of the larger number of centres. It is considered that the number of RBF centres may be set in conjunction with the number of the B pictures. This implies that the system complexity of a RBF-network-based rate control can be dramatically reduced by selecting an appropriate number of centres from video encoding parameters.

	MSE[dB]	*var_org(k)*	*var_dif(k)*
Cascaded	-24.49	1224.8	579.6
Starwars	-22.44	1671.2	713.4
Adverts	-20.67	2888.5	1143.9
JFK	-15.33	3329.9	1914.7

TABLE 22.2. Mean NMSE and mean variances of video sequences

[7]"Cascaded" is composed of three standard sequences ("Miss America", "Football" and "Susie"), each of which has 90 pictures. "JFK" is a movie sequence which contains many dramatic scene changes.

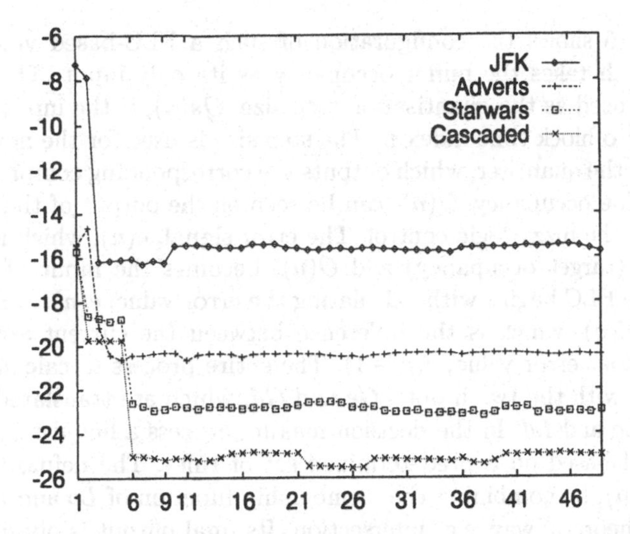

FIGURE 22.15. Variation in NMSE profile with the number of centres

22.5 Fuzzy-Logic Based Video Rate Control: A Feedback Approach

In this section, another nonlinear video rate control scheme is described: fuzzy logic control. This scheme is fundamentally a feedback method to control a target quantity. A further performance improvement is obtained by using a feed-forward method to adjust the fuzzy logic parameters.

22.5.1 A Basic FLC Model for Video Rate Control

Fuzzy logic and fuzzy set theory has been extensively used, particularly, for industrial and commercial control applications [32, 9] since its concept [37] was first published. Fuzzy logic control is known to be effective in conveying the meaning of linguistic variables to systems in control [19]. Although an analogy can be found in other industrial applications, video rate control is not just a matter of controlling the level of liquid reservoir which is treated as a typical FLC application. The occupancy of the video buffer has direct influence on video quality while the liquid level is not related to the output fluid quality in the same manner as is video rate and quality. In order to improve the FLC performance for video rate control, we introduced adaptive scaling factors which vary, depending on the scene change features. First, we examined the conventional FLC where the number of fuzzy variables is one, and its performance was evaluated in various settings of the fuzzy control parameters. As an adaptive algorithm, a FLC scheme with feed-forward scaling factors was designed and its performance was compared to the conventional FLC.

Recently, a few leading researchers have applied the FLC to video sequence coding standards. The techniques aim to improve the video rate control performance for JPEG [36, 31] and H.261 [20] by adaptively controlling the quantiser and the buffer occupancy. The fuzzy logic-based control techniques used in these researches have the same technical base in that they appeared to follow a series of common processes; fuzzification, decision making and defuzzification.

Figure 22.16 shows the configuration of such a FLC-based video rate control (FLC-R) which takes the buffer occupancy as its only input. The control input, Go, which is used as the quantisation step size, $Qs(n)$, is the input to the encoder with the macro block time index n. The step size is used for the next macro block, $mb(n+1)$, by the quantiser, which outputs the corresponding compressed bit stream $Cv(n+1)$. The occupancy, $O(n)$, can be seen as the output of the encoder and is controlled by the fuzzy logic control. The error signal, $e(n)$, which is the difference between O_T (target occupancy) and $O(n)$, becomes the input of the FLC. The process of the FLC begins with calculating the error value, $e(n)$, and its differential error value, $d(n)$, which is the difference between the current error value, $e(n)$, and the previous error value, $e(n-1)$. The entire process to calculate the output, Go, proceeds with the two inputs, Ge and Gd, which are translated into linguistic expressions Le and Ld. In the decision making process a linguistic judgement, Lo, is determined based on a predetermined set of rules. The defuzzification process calculates, $o(n)$, by combining the membership function of Lo and those of Le and Ld in a set theoretic way, e.g. intersection. Its final output is obtained by a series of arithmetic operations, e.g. the centre of gravity method [15]. All scaling factors (ge, gd and go) are constants which are generally tuned by expert knowledge.

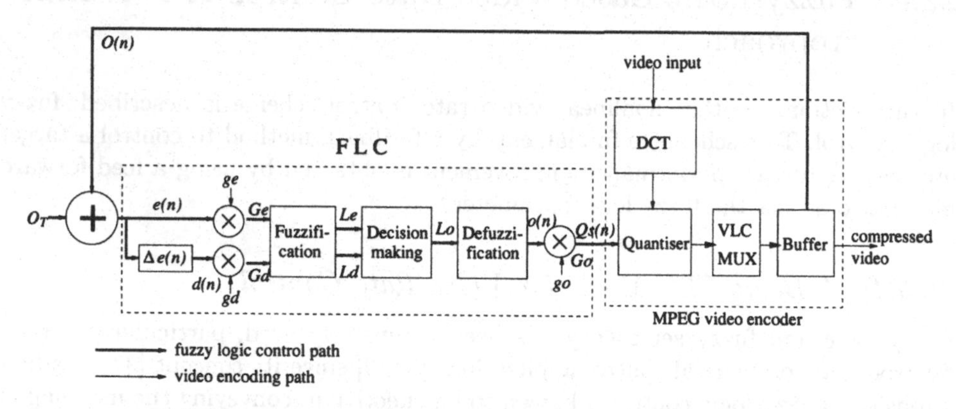

FIGURE 22.16. Configuration of the FLC-based rate control

The first step of designing FLC is initiated by transforming a series of expert knowledge into a set of rules comprising linguistic expressions [21] so that a decision for the output value can be made. A fuzzy set is expressed in a non-numerical form carrying linguistic meanings, e.g. very big, big, small or very small. Three fuzzy logic variables are listed in Table 22.3, each of which consists of 7 fuzzy sets.

A triangular membership function, Fig. 22.17(a), which is mapped on to a normalised range from -.5 to .5, is used for two inputs (Ge and Gd) and the output (Go) under the assumption that all the control variables have similar dynamic property associated with the membership functions. The complete representation of the rules can be given either in **IF ... THEN** statements or in a tabular form which is usually called fuzzy associative memory (FAM) [17]. The FAM representation, Fig. 22.17(b), is known to be more efficient in handling a complicated organisation of the rules [17]. The organisation of FAM follows a common method which locates fuzzy variables in the 45-degree diagonal. Defuzzification, which converts Ge and Gd to an output value $o(n)$, is performed using the rules representing Lo and the membership function values derived from input values (Ge and Gd). Each selected membership value is used to defuzzify the output into $o(n)$ using the centre of gravity method or one of its simplified versions (Mamdani's model or Larsen's product operation rule) [16, 15]

Fuzzy logic variables					
Occupancy (FVO)		Differential occ. (FVD)		Quantisation scale (FVQ)	
FL	full	PB	posituve big	HG	huge
CF	close to full	PM	positive medium	LG	large
HH	higher than HF	PS	positive small	LM	larger than MD
HF	half full	ZE	zero	MD	medium
LH	lower than HF	NS	negative small	SM	smaller than MD
CE	close to empty	NM	negative medium	SL	small
ET	empty	NB	negative big	TN	tiny

TABLE 22.3. Fuzzy logic variables for fuzzy control input and resulting output

The scaling factors can be tuned depending on the dynamic ranges of corresponding inputs, $e(n)$ and $d(n)$, and output, $o(n)$. The bigger values they take, the quicker response can be achieved. In video rate control go is fixed at 1.0 since the actual output Qs is multiplied by 31 to adjust it to the legal range of MPEG2 quantisation scale. A small change of go can cause wide fluctuation in the quantisation scale. ge, gd can be set to specific values suitable to accommodate the dynamic ranges of $e(n)$ and $d(n)$. Alternatively, the scaling factors can be adaptively controlled by a supervisory or adaptive control function.

The membership function can take different shapes, different inter-rule spacing, etc. It is well known that using non-triangular shapes does not provide substantial difference in the performance [19], hence, the triangular shape is used here. The formation can be asymmetrical since the positive section of the $e(n)$ value can have different significance from the negative section. In video rate control, however, both sections here are assumed to have unbiased linguistic interpretation. Thus, the membership functions shown in Fig. 22.17(a) are symmetrical with respect to the centre value 0.

The 3-dimensional representation of FAM, Fig. 22.17(c), reflects the dynamic property of an organisation of rules and membership functions. Figure 22.17(d) shows the resulting control surface for the FAM configuration of Fig. 22.17(b) and (c) with the two input variables, $d(n)$ and $e(n)$.

22.5.2 FLC with Adaptive Scaling Factors

Scene change features provide vital information about the resulting number of bits for an incoming picture in advance of encoding it. Here, they are incorporated with the non-adaptive scaling factors (ge, gd and go) of FLC-R, as shown in Fig. 22.18, in order to adaptively scale inputs of the fuzzification process, $e(n)$ and $d(n)$. This configuration forms the adaptive FLC scheme (FLC-FS). The three previous scene change features are supplied to the scaling factor calculation block (Mapping function) which generates time varying scaling factor values, $ge(k)$ and $gd(k)$. The

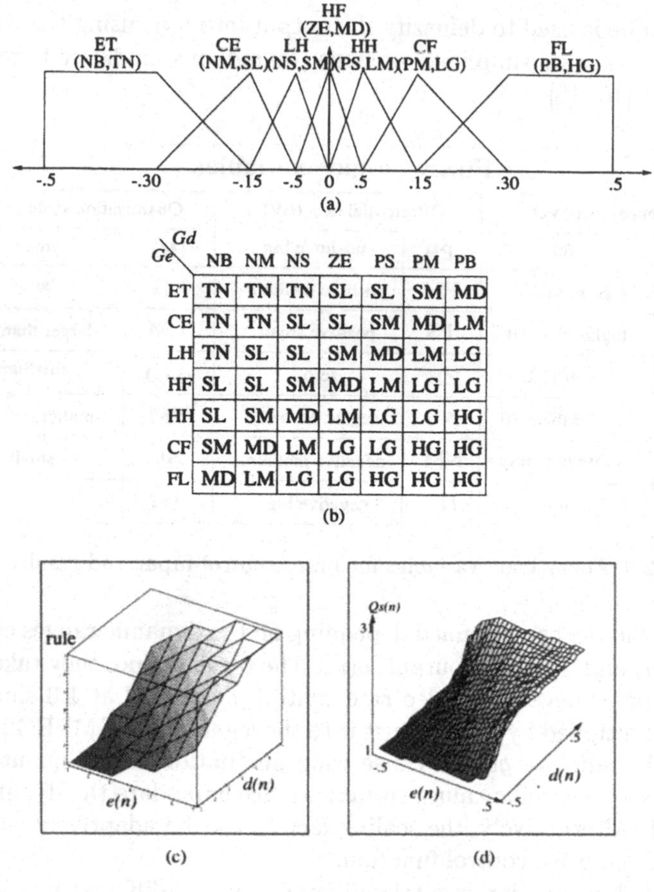

Ge \ Gd	NB	NM	NS	ZE	PS	PM	PB
ET	TN	TN	TN	SL	SL	SM	MD
CE	TN	TN	SL	SL	SM	MD	LM
LH	TN	SL	SL	SM	MD	LM	LG
HF	SL	SL	SM	MD	LM	LG	LG
HH	SL	SM	MD	LM	LG	LG	HG
CF	SM	MD	LM	LG	LG	HG	HG
FL	MD	LM	LG	LG	HG	HG	HG

FIGURE 22.17. Fuzzy logic-based parameters for video rate control. (a) membership function, (b) FAM, (c) 3-dimensional representation of the FAM, (d) the resulting control surface

remainder of the FLC-FS operation is the same as FLC-R depicted in Fig. 22.16. In the mapping function the equation maps the scene change features to the scale factors as follows:

$$ge(k) = gd(k) = \frac{\log_{10} var\_org(k)}{\log_{10} var\_dif(k)} \times ptype(k) \qquad (22.11)$$

In this scheme, the scaling factors vary depending on the scene changes features so that the video rate can be controlled effectively for the next video image. This enhances the performance of the FLC-based rate control scheme. The performance will be evaluated in the following section.

22.6 Simulation Studies

The performance of the feed-forward rate estimators is first evaluated and FLC schemes will be compared with rate estimators on the test bed of the MPEG2 software encoder. The new configuration of the MPEG2 encoder was verified with

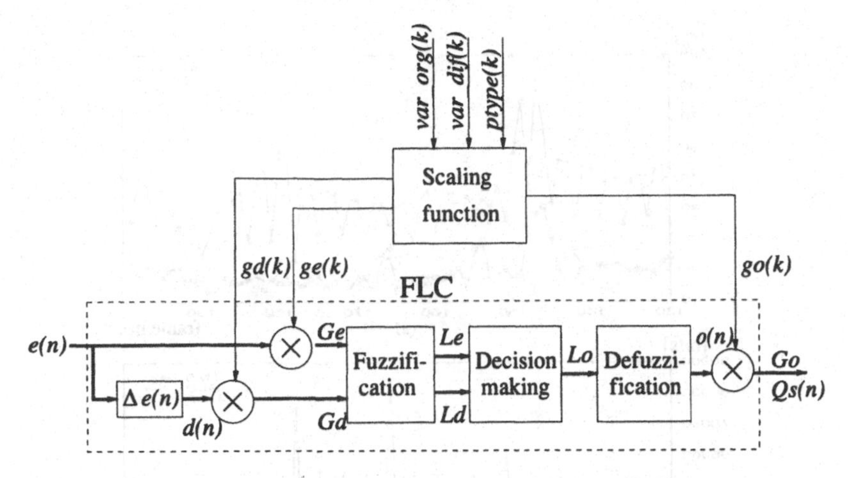

FIGURE 22.18. Configuration of the FLC-FS

a variety of different settings of encoding parameters. However, we only present representative simulation results for the "Starwars" sequence.

The MPEG2 video encoder based on TM5 was set to operate at a channel rate of 1024 kbits/s and a frame rate at 30 frames/s. It has a buffer with the size of twice the MBF. This size of buffer is specified for low delay mode codings in bi-directional communications. The unimodal function was used as the quantiser control surface since it possesses better performance for video rate control [27]. For the feed-forward rate control scheme, three techniques are evaluated: TM5 the performance reference and the two rate estimator-based schemes, RLS and RBF.

Figure 22.19 shows the performance of the three rate control schemes for the critical parts of the "Starwars" sequences. Table 22.4 shows the mean and standard deviation for each of the performance measure over the whole sequence. NFVR in the middle column stands for normalised fluctuation of the video rate which represents variation of $cbf()$. It is expressed in the following equation:

$$\mathrm{NFVR} = \frac{\sigma}{1+\sigma}, \qquad \sigma^2 = E\left[\left(\frac{cbf(k)}{\mathrm{MBF}} - 1\right)^2\right] \tag{22.12}$$

where $\frac{cbf(k)}{\mathrm{MBF}}$ represents instantaneous fluctuation.

TM5 exhibits the worst occupancy performance, often reaching buffer overflow on some sequences such as "JFK", with more rapid movement [29]. The schemes based on predictors, RLS and RBF, show better performance than TM5 in both bit rate and PSNR in Fig. 22.19. RBF appeared to be capable of maintaining the occupancy lower with a smaller standard deviation in comparison to RLS. RBF also exhibits a similar PSNR value to RLS. This result implies that the nonlinear rate estimator, RBF-network, works more effectively for non-stationary video with many scene changes and rapid motion without further degradation in video quality.

The basic FLC model (FLC-R), described in Section 22.6, possesses a considerable flexibility to change the fuzzy control parameters. Scaling factors were assessed by applying 8 different values ranging from 2 to 16 increasing by 2. The scaling factor for the output $go(k)$ was set to 1. The target occupancy was set to 30%, in order to observe the effect of using a lower value than 50%. Figure 22.20 shows encoding results for different scaling factor values. For simplicity, only two extreme cases are

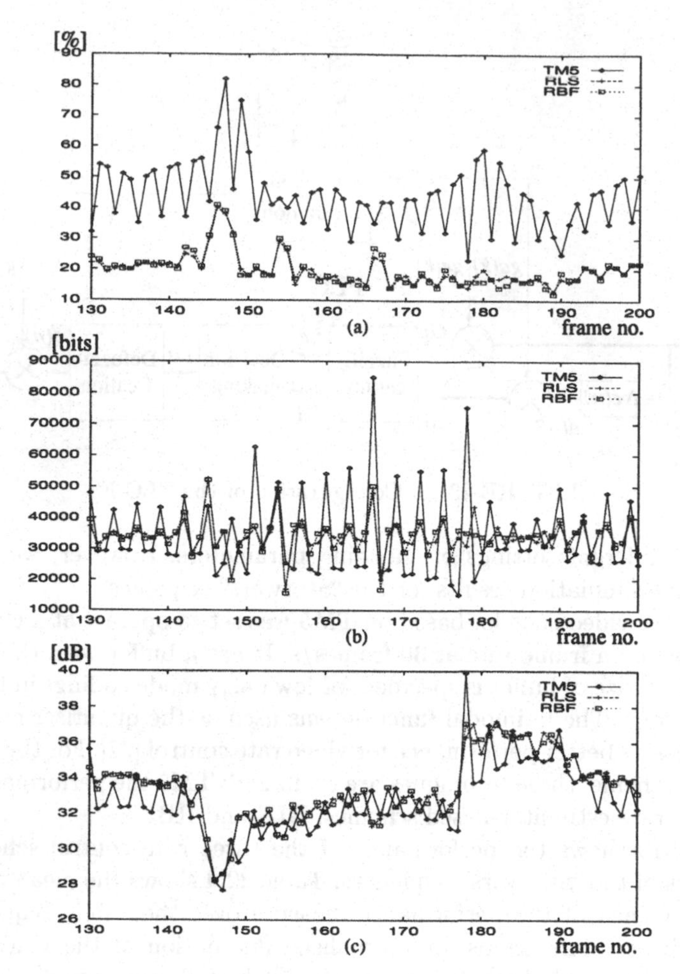

FIGURE 22.19. Performance of feed-forward rate control algorithms ("Starwars"): (a) occupancy, (b) coded bits/frame, (c) PSNR

displayed in Fig. 22.20(b) and (c). Bigger values of $ge(k)$ and $gd(k)$ (e.g. 16) exhibit far better control capability over the occupancy and bits/frame. However, PSNR is noticeably lower than the case of smaller scaling factor values since, in FLC-R, the quality is not taken into account in the control process. Pictures, which entail a large number of bits, will be given much stronger control action. When they are I or P pictures, the distortion caused by rate control will affect the next coming pictures. For this reason, as shown in Fig. 22.20(c), PSNR remains low when scaling factors are large. The performance difference can be found easily around the frame number 140 to 180.

The FLC model assisted by the feed-forward scaling factors (FLC-FS) was assessed with respect to FLC-R. Fig. 22.21 shows a critical part of the encoded results. Table 22.4 summarises the performance for the entire video sequence (300 frames). While FLC-R is superior in controlling the video rate or the occupancy to FLC-FS, the latter shows wider variations since its scaling factors change depending on scene change. Hence, profiles of the occupancy and video rate may vary dramatically depending on the scene change. The video rate profile exhibits similarly changing patterns.

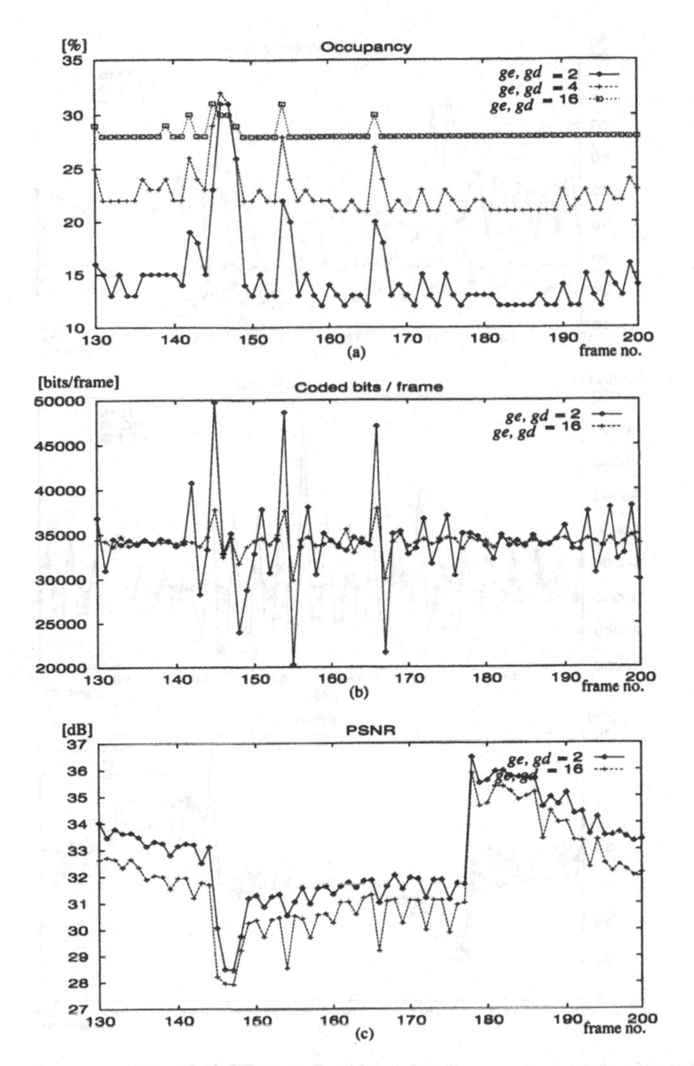

FIGURE 22.20. Performance of FLC-R depending on scaling factors ("Starwars")

FLC-R consists of fuzzy control rules based only on the occupancy-related rules. Therefore, it appeared to be powerful in controlling the occupancy. However, it does not take account of the quality, and it shows lower figures in PSNR. In the simulation, the scaling factors are set to 8 to see the performance of controlling the occupancy. For FLC-FS, they are allowed to change within a 1-to-8 range depending on the scene change features according to Eq. (22.11). In FLC-FS, the scaling factors $ge(k)$ and $gd(k)$ scale up the error signal, $e(n)$, Fig. 22.18, to adaptively change the actual input of $e(n)$. The scaling factors are generally smaller than 8, and the resulting performance of the occupancy control appear to be inferior to FLC-R. Accordingly, FLC-FS shows more fluctuating profiles. However, the feed-forward scaling factors ($ge(k)$ and $gd(k)$) improve the video quality by making the most of the occupancy margin below 30%. Therefore, it is concluded that FLC-FS performs better in terms of PSNR achieved as well as the occupancy.

Comparing the FLC-FS scheme with the RBF estimator-based scheme, the performance shows no noticeable difference as shown in Table 22.4. Thus, comparison

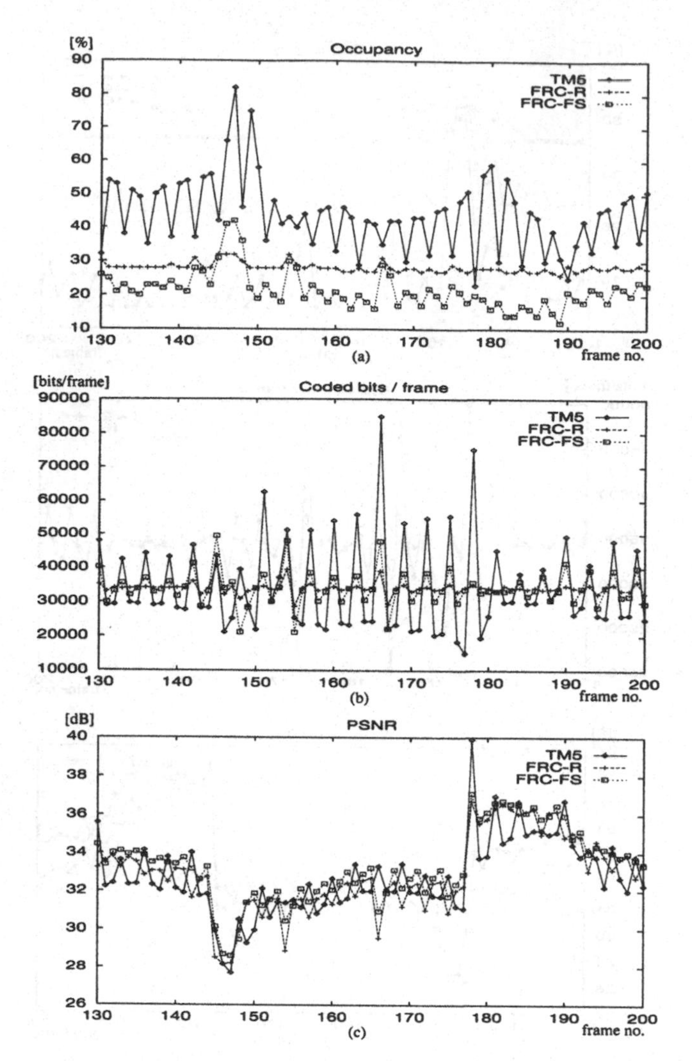

FIGURE 22.21. Performance profiles for FLC-R and FLC-FS ("Starwars") (a) occupancy, (b) coded bits/frame, (c) PSNR.

between these two schemes should be considered from a different point of view: performance for particular scene change types, implementational complexity, etc.

22.7 Conclusions

Improvements on video rate control have been achieved using nonlinear approaches: RBF-network and fuzzy logic control. These two approaches have different structures and properties, thus, performance comparison is not straight forward. However, a comparison can be conducted from the user's point of view. The fuzzy logic approach is considered to be computationally simpler and less complicated, which is critical to implementation. However, the simulation results show that the RBF-network scheme does exhibit a lower mean buffer occupancy than the fuzzy logic-base control schemes. This feature is advantageous for low-delay applications

Starwars	Occupancy(%)		Coded bits / frame		PSNR (dB)	
	mean(max.)	std.dev.	NFVR	std.dev.	mean	std.dev.
TM5	41 (75)	10.8	0.285	13704	33.27	2.69
RLS	18 (38)	4.5	0.127	4519	33.89	2.54
FLC-R	27 (32)	1.5	0.086	1321	33.40	2.59
FLC-SC	19 (42)	5.3	0.146	5058	33.86	2.55
RBF	12 (34)	4.3	0.124	4288	33.87	2.54

TABLE 22.4. Performance comparison between FLC-R and FLC-FS ("Starwars")

since shorter delay can be achieved by maintaining low buffer occupancy. The RBF-network scheme also appears to better exploit the occupancy margin by allowing more occupancy variation than the fuzzy logic-based control. Furthermore, the parameters of the RBF-network estimator examined in this chapter were not selected by an analytical approach. The number of centres, which determines the complexity, was only determined by an experimental method. More intelligent selection of the centres, i.e. the training problem, is another important issue to be investigated. Hence, it is likely that further improvement can be achieved from the RBF-network-based scheme by enhancing the training process, tuning the parameters and introducing more adaptivity. An in-depth research on these matters is a topic for future research. However, a conclusion can be drawn in respect of system complexity. In the RBF-network-based scheme, supplementary processing is required on the estimated video rate information to derive a corresponding quantisation step size to achieve the predictive rate control function. On the other hand, the fuzzy logic-based control can accomplish the equivalent task by using a set of rules, a pair of control variables and adaptive scaling factors. This configuration is generally simpler than the approach for predictive scheme. Hence, as the performance is very similar, the fuzzy logic approach is more promising in terms of system complexity than the RBF-network-based scheme.

Acknowledgments: Yoo-Sok Saw acknowledges the support of *The British Council* and *LG Information and Communications Ltd., Korea.*

22.8 References

[1] O.T.-C. Chen, B.J. Sheu, and W.-C. Fang. Image compression using self-organisation networks. *IEEE Trans. Circuits and Systems for Video Technology*, 4(5):480–489, 1994.

[2] S. Chen. Non-linear time series modelling and prediction using Gaussian RBF networks with enhanced clustering and RLS learning. *Electronics Letters*, 31(2):117–118, 1995.

[3] S. Chen, B. Mulgrew, and P.M. Grant. A clustering technique for digital communications channel equalisation using radial basis function networks. *IEEE Trans. Neural Networks*, 4(4):570–579, 1993.

[4] S. Chen, B. Mulgrew, and S. McLaughlin. Adaptive Bayesian equaliser with decision feedback. *IEEE Trans. Signal Processing*, 43(5):1937–1946, 1995.

[5] E.S. Chng, S. Chen, and B. Mulgrew. Efficient computational schemes for the orthogonal least square algorithm. *IEEE Trans. Signal Processing*, 43(1):373–376, 1995.

[6] R.D. Dony and S. Haykin. Neural network approaches to image compression. *IEEE Proceedings*, 83(2):288–303, 1995.

[7] V.S. Frost and B. Melamed. Traffic modeling for telecommunications networks. *IEEE Communications Magazine*, 32(3):70–81, 1994.

[8] P.M. Grant, C.F.N. Cowan, B. Mulgrew, and J.H. Dripps. *Analogue and Digital Signal Processing and Coding.* Chartwell-Bratt, 1989.

[9] M.M. Gupta. *Approximate reasoning in expert systems.* Elsevier, 1985.

[10] E.J. Hartman. Layered neural networks with Gaussian hidden units as universal approximations. *Neural computation*, 2(2):210–215, 1990.

[11] S. Haykin. *Adaptive Filter Theory.* Prentice-Hall, 1991.

[12] S. Haykin. *Neural networks: a comprehensive foundation.* Macmillan, 1994.

[13] ISO-IEC/JTC1/SC29/WG11. Test Model 1, 1st ed., 1992.

[14] ISO/IEC JTC1/SC29/WG11. ISO/IEC DIS 13818-2: Information Technology - Generic Coding of Moving Pictures and Associated Audio Information: Video, 1994.

[15] J.-S.R. Jang and C.-T. Sun. Neuro-fuzzy modeling and control. *IEEE Proceedings*, 83(3):378–406, 1995.

[16] P.J. King and E.H. Mamdani. The application of fuzzy control systems to industrial processes. *Automatica*, 13:235–242, 1977.

[17] B. Kosko. *Neural networks and fuzzy systems: a dynamical systems approach to machine intelligence.* Prentice-Hall, 1992.

[18] E. Kreyszig. *Introductory functional analysis with applications.* John, Wiley & Sons, 1978.

[19] C.C. Lee. Fuzzy logic control systems: Fuzzy logic controller - part I. *IEEE Trans. Sys. Man and Cybern.*, 20(2):404–418, 1990.

[20] A. Leone, A. Bellini, and R. Guerrieri. An H.261-compatible fuzzy-controlled coder for videophone sequences. In *3rd Int. Conf. on Fuzzy Systems*, vol. 1, pp. 244–248. IEEE, Orlando, Florida, 1994.

[21] R.J. Marks. Fuzzy models - what are they, and why? *IEEE Technology updates series: Fuzzy logic technology and applications*, 1:3–7, 1994.

[22] B. Melamed and D. Raychaudhuri. TES-based traffic modeling for performance evaluation of integrated networks. In *IEEE INFOCOM '92*, vol. 1, pp. 75–84. IEEE, 1992.

[23] B. Melamed, D. Raychaudhuri, B. Sengupta, and J. Zdepski. TES-based video source modeling for performance evaluation of integrated networks. *IEEE Trans. Commun.*, 42(10):2773–2777, 1994.

[24] M. Nomura and T. Fujii. Basic characteristics of variable bit rate video coding in ATM environment. *IEEE Journal on SAC*, 7:752–760, 1989.

[25] J. Park and I.W. Sandberg. Universal approximation using radial-basis-function networks. *Neural computation*, 3(2):246–257, 1991.

[26] W.H. Press and B.P. Flannery. *Numerical Recipes in C*. Cambridge University Press, 1988.

[27] Y.-S. Saw, P.M. Grant, and J.M. Hannah. Feed-forward buffering and rate control based on scene change parameters for MPEG video coder. In *EUSIPCO'96*, vol. II, pp. 727–730. EURASIP, 1996.

[28] Y.-S. Saw, P.M. Grant, and J.M. Hannah. Rate-distortion analysis of nonlinear quantisers for MPEG video coders: sigmoidal and unimodal quantiser control functions. *IEE Proc. Vision, Image and Signal Processing*, submitted.

[29] Y.-S. Saw, P.M. Grant, J.M. Hannah, and B. Mulgrew. Nonlinear predictive rate control for constant bit rate mpeg video coders. In *ICASSP'97*. IEEE, 1997.

[30] Y.-S. Saw, P.M. Grant, J.M. Hannah, and B. Mulgrew. Video Rate Control using a Radial Basis Function Estimator for Constant Bit Rate MPEG Coders. *EURASIP Signal Processing Image Communication.*

[31] S.H. Supangkat and K. Murakami. Quantity control for JPEG image data compression using fuzzy logic algorithm. *IEEE Trans. Consumer Electronics*, 41(1):42–48, 1995.

[32] H. Takagi. Application of neural networks and fuzzy logic to consumer products. *IEEE Technology updates series: Fuzzy logic technology and applications*, 1:8–12, 1994.

[33] H. Tong. *Non-linear time series A dynamical system approach.* Oxford University Press, 1990.

[34] N.P. Walker, S.J. Eglen, and B.A. Lawrence. Image compression using neural networks. *GEC Journal of Research*, 11(2):66–75, 1994.

[35] A.S. Weigend and N.A. Gershenfeld. The future of time series: Learning and understanding. In *Time Series Prediction: Forecasting The Future and Understanding The Past*, vol. 15, pp. 1–70, 1992. Proceedings of the NATO Advanced Research Workshop on Comparative Time Series Analysis.

[36] C.J. Wu and A.H. Sung. Application of fuzzy controller to JPEG. *IEE Electronics Letters*, 30(17):1375–1376, 1994.

[37] L.A. Zadeh. Fuzzy sets. *Information and control*, 8:338–353, 1965.

[38] J. Zdepski and D. Raychaudhuri. Statistically based buffer control policies for constant rate transmission of compressed digital video. *IEEE Trans. Commun.*, 39:947–957, 1991.

23

A Motion Compensation Algorithm Based on Non-Linear Geometric Transformations and Quadtree Decomposition

Andrea Cavallaro[1]
Stefano Marsi[1]
Giovanni L. Sicuranza[1]

ABSTRACT
In this paper we present an efficient very low bit-rate coder for CIF image sequences at 10 frames/s. The proposed algorithm performs motion compensation using a second-order geometric transformation and a quadtree structure. High compression ratios have been reached adapting the features of the transformation and the block size to the motion. A linear transformation is used when there is translational motion, whereas a non-linear transformation is applied when a block has to be warped. In addition the block size is increased in homogeneous zones and reduced in detailed moving zones. Simulation results show that the proposed coding method improves the predicted image quality and reduces the blocking artifacts with respect to traditional hybrid block-based techniques.

23.1 Introduction

Very low bit-rate image sequence coding is very important for video transmission and storage application. There are many practical applications of this kind of coder, ranging from video transmission by narrow-band channels, such as conventional telephone lines or mobile telephone networks, to multimedia applications used in databases and electronic mail. Very low bit-rate video coding requires very high compression ratios that can be obtained by removing spatial and temporal redundancy contained in digital image sequences. Many different approaches, such as pixel based [1], motion compensated pixel based [2], object-based [3, 4], model based [5, 6, 7], and fractal based [8] techniques, have been presented in the literature in order to exploit these statistical redundancies. A very effective technique is the so called *hybrid coding*, characterised by a combination of predictive coding and transform coding. Temporal redundancy is reduced using a motion-compensated prediction of the current frame from the previous reconstructed one, whereas spa-

[1]University of Trieste - DEEI, Via A. Valerio 10, 34127 Trieste, Italy,
E-mails:{marsi, sicuranza}@univ.trieste.it

tial redundancy left in the prediction error is reduced by a transform coder.

Typical motion compensation algorithms can be divided into four main groups: gradient [9], pel recursive [10], block matching [11], and frequency-domain techniques. Their goal is to reduce temporal redundancy by estimating the ***optical flow***, i.e., the spatiotemporal variation of intensity that is due not only to the motion of the objects in the scene, but also to the global motion (camera motion) and to the variation of the illumination. Among the motion compensation techniques the block matching algorithm has been adopted for video coding standards, such as MPEG and H.261/3, thanks to its simplicity for computation and to the small number of information required to describe the motion. Nevertheless the conventional block matching techniques have important limits: the definition of unrealistic motion fields and the inability to maintain acceptable quality at a very low bit-rate. In fact the image is divided into non overlapping square blocks of fixed dimensions and to each one is assigned a motion vector that describes the estimated motion of the block between two consecutive frames. The use of blocks and the use of a simple transformation approximates the motion by translational displacements. This approach limits the amount of motion information but reduces every kind of motion to the translational one. These considerations explain that the prediction error could be significant.

Consequently for most video coders there are two important drawbacks for motion estimation: the fixed block size and the use of a unrealistic transformation to find the best matched block in the previous frame. The former brings artificial discontinuities in the motion field and the latter causes an unrealistic compensation of the true motion in the scene.

The solution of this problem should be the adaptation of block shape and dimensions to the characteristic of the objects, but this approach forces an unacceptable increase of information for very low bit-rate coding. Quite a good approximation can be achieved by modifying only block dimensions, using a bigger block size for background and for big objects and smaller for particulars. Besides, it is possible to improve the approximation of the movement using a more realistic motion estimation that allows not only the modelling of translation but also modelling of other kind of motion such as rotation, shearing, warping, and uneven stretching. In this way we obviously have to increase the amount of information to represent the prediction, and a tradeoff between motion information and prediction error is required.

Algorithms based on more realistic motion estimation have been presented in [12, 13] with an affine transformation, in [14] with a bilinear transformation, and in [15] with a second order geometric transformation. The first transformation allows modelling every kind of linear movement, such as translation, rotation, zooming and, shearing. It is not possible in this case to follow movements that change the shape of objects, such as warping and uneven stretching. These are important kinds of motion that bring significant prediction errors. Therefore it would be interesting to find an effective way to represent them. The former can be modelled by a bilinear transformation that can map a quadrilateral of any shape from the previous frame to the current block. In this way, however, it is not possible to warp the edges of a block, and for this reason this transformation cannot compensate for a general complex motion. Good capability in estimating movements is given by the transformation presented in [15] that can also warp the edges of a block to stretch it unevenly. The problem connected with such a transformation is high computational cost. To overcome this problem, we present a fast algorithm to calculate the non-linear geometric transformation and a multi-resolution structure that selects this transformation only in the regions of the image that really need to be warped.

The section is organised as follows. The non-linear geometric transformation and its employment are presented in Subsection 23.2. In Subsection 23.3 the quadtree structure is introduced in which the motion compensation algorithm is inserted. Subsection 23.4 describes how the proposed very low bit- rate coder selects and codes the prediction error. Simulation results are presented in Subsection 23.5. The performances of the proposed algorithm are evaluated using CIF format images at 10 frames per second. Finally, Subsection 23.6 draws the conclusions.

23.2 Geometric Transformations

In the block-matching technique the same displacement vector is assigned to all the pixels of each block and therefore the quality of the estimation is conditioned by the block size. Large blocks may include pixels whose real displacement is very different from that represented by the calculated displacement vector. The use of small blocks generates a high number of motion vectors. Therefore transformations capable of specifying a different motion vector for each pixel within a block have been presented in the literature [12, 14, 15]. It has been shown that more complex motion models can provide a significant improvement in the accuracy of the prediction error in the new image using large size blocks. The choice of the transformation depends on the kind of motion we want to compensate for with motion estimation. The best tradeoff between computational complexity and estimation properties has to be found. Let us write a generic second-order geometric transformation as

$$\begin{aligned} x &= d_1u^2 + d_2u + d_3uv + d_4v^2 + d_5v + d_6 \\ y &= d_7u^2 + d_8u + d_9uv + d_{10}v^2 + d_{11}v + d_{12} \end{aligned} \tag{23.1}$$

where (u, v) and (x, y) are the spatial coordinates, respectively, of the current and of the previous frame. If we use more parameters, the prediction improves but the transmission cost grows. On the other hand, a complex method allows a better description of the motion in a sequence and thus it can efficiently represent the motion of the image. Our aim is to overcome the problem of computational complexity related to the use of non-linear transformations using a fast technique for parameter calculation and for their application. To define the deformation properties of (23.1), the parameters d_i have to be fixed, through the knowledge of the mapping of a number of pixels (*control points*) related to the number of the parameters. A transformation characterised by $2k$ parameters needs at least k control points to be defined.

A complete processing for parameter calculation should perform a global optimisation within each block based on the following stages:

- displacement of *one* control point,
- parameters calculation,
- application of the transformation,
- calculation of the prediction error.

These steps are repeated for all the positions of the control points within the search window. The parameters that lead to the minimum prediction error are selected as the block parameters. We suggest a different method for the evaluation of the parameters based on a local, instead of global, optimisation. The steps of this process are as follows:

- determination of the new position of the control points by block matching
- parameters calculation.

Using this sub-optimal method we have to carefully choose the number and the position of the control points. Experimentally we found that the best tradeoff between computational complexity and prediction accuracy is given by the symmetrical disposition of nine points: the four vertices, the midpoints of each side, and the centre of the block (Fig. 23.1). Now we have to define the way to calculate the parameters of (23.1).

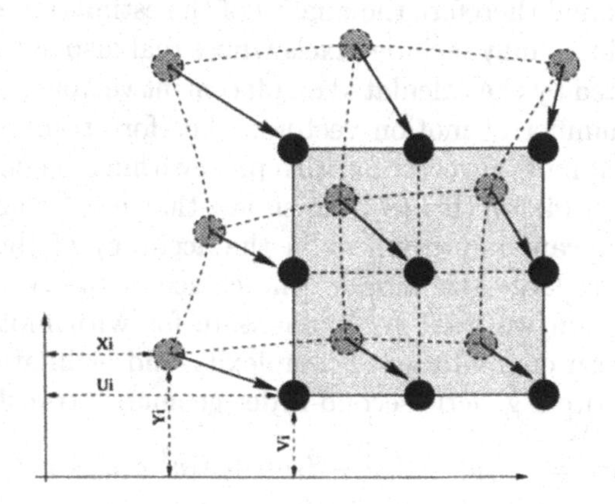

FIGURE 23.1. Location of the control points in the block and the uneven stretching effect

$$
\begin{aligned}
\mathbf{d}' &= (d_1, d_2, \ldots, d_6)^T \\
\mathbf{d}'' &= (d_7, d_8, \ldots, d_{12})^T
\end{aligned}
\tag{23.2}
$$

Let (u_i, v_i) and (x_i, y_i) be the generic control points in the current and in the previous frame, respectively. Let their coordinates be collected in the vectors

$$
\begin{aligned}
\mathbf{x} &= (x_1, x_2, \ldots, x_9)^T \\
\mathbf{y} &= (y_1, y_2, \ldots, y_9)^T
\end{aligned}
\tag{23.3}
$$

The matrix

$$
\mathbf{A} = \begin{pmatrix}
u_1^2 & u_1 & u_1 v_1 & v_1^2 & v_1 & 1 \\
u_2^2 & u_2 & u_2 v_2 & v_2^2 & v_2 & 1 \\
\vdots & & \cdots & & & \vdots \\
u_9^2 & u_9 & u_9 v_9 & v_9^2 & v_9 & 1
\end{pmatrix}
\tag{23.4}
$$

made up of the values of the coordinates of the control points in the current image, is a rectangular matrix. Therefore, the solutions of the equations

$$\begin{aligned} \mathbf{x} &= \mathbf{A}\mathbf{d}' \\ \mathbf{y} &= \mathbf{A}\mathbf{d}'' \end{aligned} \tag{23.5}$$

are the vectors $\mathbf{d}'$ and $\mathbf{d}''$ calculated using the generalised inverse of matrix A:

$$\begin{aligned} \mathbf{d}' &= (\mathbf{A}^T\mathbf{A})^{-1}\mathbf{A}^T\mathbf{x} \\ \mathbf{d}'' &= (\mathbf{A}^T\mathbf{A})^{-1}\mathbf{A}^T\mathbf{y} \end{aligned} \tag{23.6}$$

The existence of $(\mathbf{A}^T\mathbf{A})^{-1}$ is very easy to verify. All of the rows of matrix $\mathbf{A}$ are linearly independents thanks to the particular coordinates of the control points in the actual frame. So matrix $(\mathbf{A}^T\mathbf{A})$ could not be singular.

Once we have the parameters, we can apply the transformation using an efficient method to reduce the computational complexity of the non-linear geometric transformation: the *line scanning technique* [14]. Referring to 23.1 and writing a geometric transformation as

$$\begin{aligned} x &= F(u,v) \\ y &= G(u,v) \end{aligned} \tag{23.7}$$

we can calculate the mapping of the pixels with coordinates $(u, v+1)$ and $(u+1, v)$ starting from the knowledge of the mapping of (u, v):

$$\begin{aligned} F(u,v+1) &= F(u,v) + [d_4 + d_5 + d_3u + 2d_4v] \\ G(u,v+1) &= G(u,v) + [d_{10} + d_{11} + d_9u + 2d_{10}v] \end{aligned} \tag{23.8}$$

for row pixels, and

$$\begin{aligned} F(u+1,v) &= F(u,v) + [d_1 + d_2 + d_3v + 2d_1u] \\ G(u+1,v) &= G(u,v) + [d_7 + d_8 + d_9u + 2d_7u] \end{aligned} \tag{23.9}$$

for column pixels of each warped block. In this way it is possible to greatly reduce the number of operations required to perform the non-linear transformation.

23.3 Quadtree

The classical video and image compression algorithms are based on the division of the image into square blocks of fixed size and on the independent coding of each block. The choice of the block size constitutes one of the problems of this kind of algorithm. Large-size blocks allow a higher compression degree but with a degradation of subjective quality of the coded image in highly detailed moving regions. On the other hand, small blocks allow better adaptation but with reduced compression efficiency.

All of the problems raised by the division of images into blocks of fixed sizes suggest the segmentation of the image into blocks of variable size. Therefore images will be divided into large or small size blocks according to the complexity of the region to be coded. Furthermore, in the moving areas we are proposing an algorithm that allows us to use not only variable size blocks but also different kinds of transformations to estimate the motion.

The algorithm has been designed to obtain a coder with a high degree of adaptation and an implementation complexity as low as possible. Adaptability is a very important feature in very low bit-rate coding, as it allows efficient management of the limited number of available bits. When the coder is adapted to the spatial and temporal variations of the image features, it is possible to improve the quality of the coded images for a constant channel bit-rate over traditional methods. Quadtree decomposition allows natural division of the image with careful treatment of highly detailed areas. This is achieved by the proposed method with two sizes of blocks and a three-stage motion compensation algorithm. By comparison with traditional hybrid block-based techniques, like H.261 - H.263, the increase of information due to the definition of the quadtree structure is compensated for by the use of larger blocks.

The proposed method works on two resolution levels: 32×32 and 16×16 size blocks (Fig. 23.2). The peak signal-to-noise ratio (PSNR) of the luminance component is used as a distortion measure (Table 23.5). The first stage at each level deals with the stationarity of the block. If the block is not stationary, the algorithm looks for the best matched block in the previous frame using the traditional two-parameter transformation. At the first stage the block-matching technique cannot give a satisfactory prediction, the algorithm decomposes the 32×32 block into its 16×16 quadrants and repeats the process on each subblock. If the translational compensation does not allow good prediction at the second level, then the block has to be warped with a non-linear transformation.

In Fig. 23.3 two consecutive frames of the test sequence "Miss America" and the corresponding quadtree decomposition are presented. The three tonalities correspond to the different kind of motion compensation chosen by the algorithm: no motion compensation (light gray), translational motion compensation (dark grey), and non-linear motion compensation (black). Between the two frames the speaker closes her eyes, a typical non-translation movement. As a consequence the algorithm chooses to warp the corresponding block so that it is possible to improve the prediction.

The ability of the proposed method to approximate warping an uneven stretching is displayed in Fig. 23.4. The particular is taken from the sequence "Miss America" and shows the different predictions obtained by linear and non-linear transformations when an eye closes. The former cannot give a good prediction at all, whereas the latter succeeds in providing at least a coarse approximation of the real motion. This improvement corresponds to a decrease of the prediction error.

The limit of the fixed subdivision can be partially overcome, without a significant increase of the computational complexity and of the overhead bits, using a further subdivision that does not need calculation of new motion vectors. If neither the warped block succeeds in being over the PSNR threshold, it is subdivided into its 8×8 quadrants, and then they are weighted with the corresponding quadrants of the translated block. Therefore each quadrant can be treated in a different way according to its kind of motion. Let $\mathbf{A}$ be the translated 16×16 block and $\mathbf{B}$ the

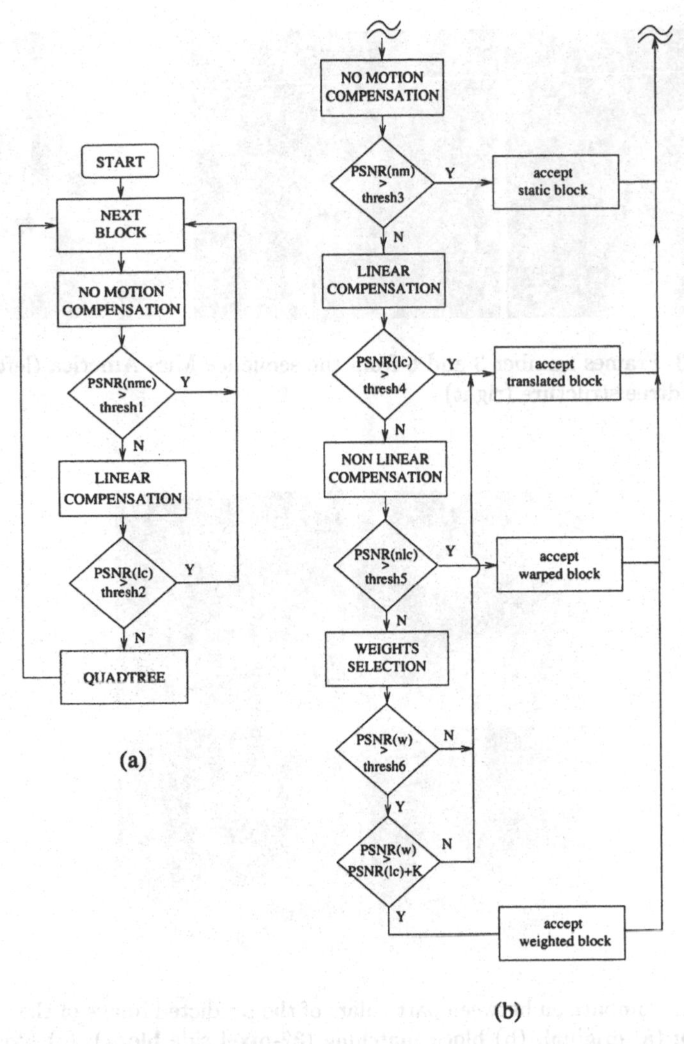

FIGURE 23.2. Block diagrams of the proposed algorithm: (a) 32 × 32 resolution level and (b) 16 × 16 resolution level, starting from linear compensation

warped one. We consider the corresponding quadrants $\mathbf{A}_i$ and $\mathbf{B}_i$, so that

$$\mathbf{A} = \begin{pmatrix} \mathbf{A}_1 & \mathbf{A}_2 \\ \mathbf{A}_3 & \mathbf{A}_4 \end{pmatrix} \quad , \quad \mathbf{B} = \begin{pmatrix} \mathbf{B}_1 & \mathbf{B}_2 \\ \mathbf{B}_3 & \mathbf{B}_4 \end{pmatrix}$$

and we weight them using the coefficients α_i and β_i with $\beta_i = 1 - \alpha_i$ and $\alpha_i \in [0, 1]$. To code the weights effectively, α_i is chosen with step 0.1. Therefore there is a fixed set of possible coefficients that reduces the number of bits required for transmission. The generic 8 × 8 weighted quadrant is given by

$$\mathbf{C}_i = \alpha_i \cdot \mathbf{A}_i + \beta_i \cdot \mathbf{B}_i$$

and therefore

$$\mathbf{C} = \begin{pmatrix} \mathbf{C}_1 & \mathbf{C}_2 \\ \mathbf{C}_3 & \mathbf{C}_4 \end{pmatrix}$$

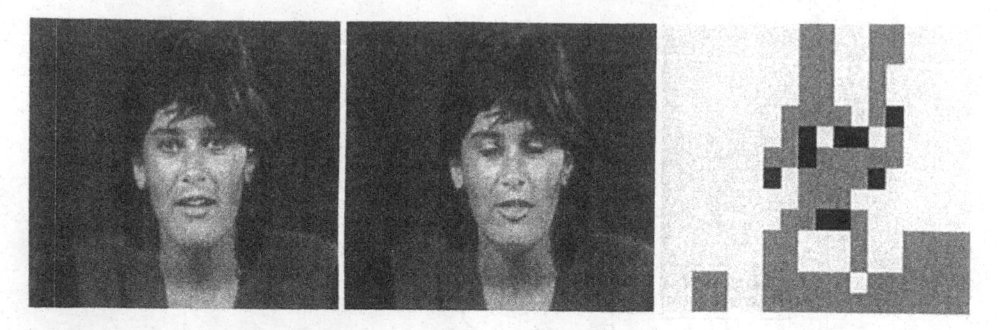

FIGURE 23.3. Frames number 3 and 6 from the sequence Miss America (left) and corresponding quadtree structure (right)

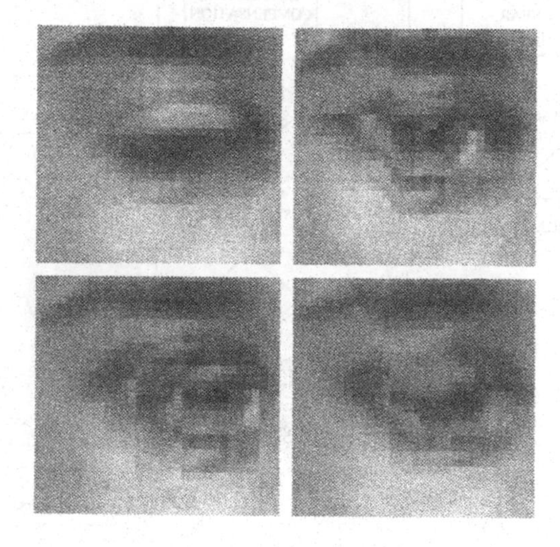

FIGURE 23.4. Comparison between particulars of the predicted image of the test sequence Miss America: (a) original, (b) block matching (32-pixel side block), (c) block matching (16-pixel side block) and (d) proposed method (32- and 16-pixel side block)

will be the 16×16 generic weighted prediction. The method will choose the block $\mathbf{C}$ made up of the quadrants $\mathbf{C}_i$ that led to the best PSNR. Finally, to consider the overhead related to the weighting process, the block $\mathbf{C}$ is selected by the quadtree structure only if the resultant compensation is significantly better than block matching (block $\mathbf{A}$).

This strategy together with the non-linear motion compensation allows us to improve the prediction as shown in Fig. 23.5.

Once a block is motion-compensated, the following information has to be transmitted to the decoder: the type of the motion compensation technique applied and the block size, the displacement vectors (for linear transformation) or the non-linear transformation parameters and, finally, the weights. For the motion compensation technique and the block size, the same code word is used. According to the statistics on simulation results, a Huffman table is used for each information, except for the parameters of the nonlinear tranformation that are quantised and then transmitted.

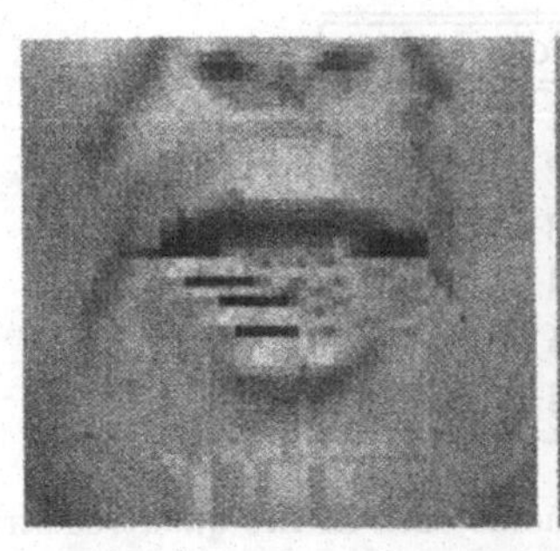
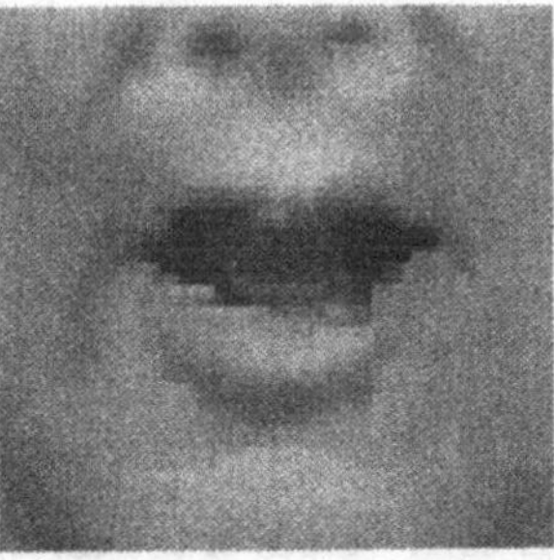

FIGURE 23.5. Comparison between the particulars of the predicted image of the test sequence Miss America. Left: linear method; right: proposed method

23.4 Prediction Error

In addition to the quadtree information, it is necessary to transmit the prediction error according to the residual bandwidth.

The difference between the predicted frame and the current one is the prediction error that should be transmitted. According to the actual coding conditions of very low bit-rate, there is often only a small amount of bits left for the transmission of this information, so we cannot generally provide the decoder with the whole prediction error. Therefore we have to choose only a subset of regions in which to code it.

The strategy that the proposed method carries out is based on the individuation of the most important error areas. It neglects the isolated and the smaller errors and then uses the conventional two-dimensional DCT transform, transmitting only the low-frequency coefficients. The algorithm finds the pixels of the image whose error is higher than a fixed threshold at first and then performs as many 8×8 DCT blocks as possible, according to the residual number of bits, starting from the blocks characterised by the largest errors.

23.5 Simulation Results

The coder has been used for simulations with the luminance component of several sequences on channels at very low bit-rate. The original sequences have a spatial resolution of 288×352 (CIF) and a temporal resolution of 30 images per second. Input sequences have been temporally downsampled at 10 Hz. The system has been set up to work at a constant bit-rate.

In Fig. 23.6 the last reconstructed frames of a typical test sequence coded at 16 Kb/s are presented, and experimental results are shown in Fig. 23.7. These plots describe the peak signal-to-noise ratio as a function of the frame number of the original sequence. The results show that the proposed method based on non-linear transformations and quadtree decomposition outperforms a similar method (labelled reference) in which these techniques have been neglected (as in H.261 - H.263) and the related data have been substituted by the transmission of the prediction errors in a larger area. There is a significant improvement in the subjective image quality that can be appreciated in the detailed areas, such as the eyes and the mouth of the speakers. This improvement is especially due to the reduction of the blocking artifacts and to the more realistic prediction of non-translational movement obtained with the proposed method.

Threshold	Value
Thresh1	33 dB
Thresh2	32 dB
Thresh3	32 dB
Thresh4	31 dB
Thresh5	31 dB
Thresh6	27 dB

TABLE 23.1. Values of the thresholds used in the quadtree structure

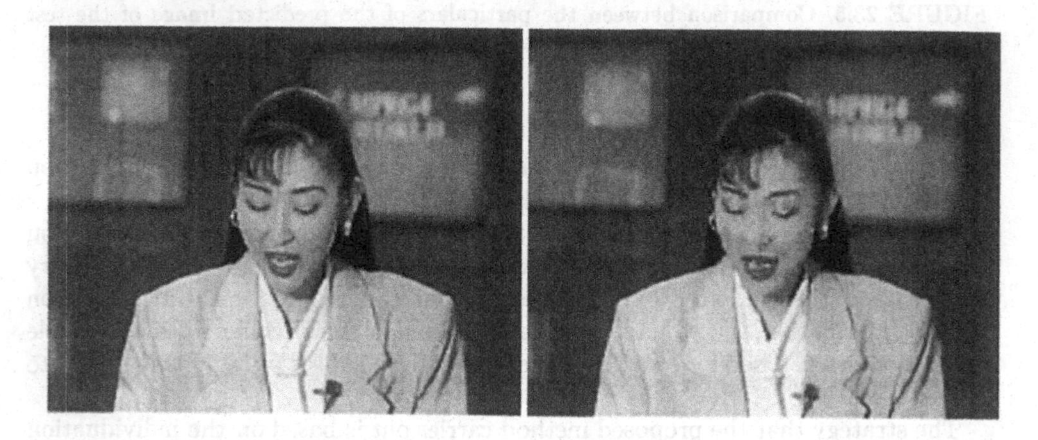

FIGURE 23.6. Last frames of the test sequences. Left: original frames; right: reconstructed frames

23.6 Conclusions

A motion compensation method that improves the quality of predicted images in a hybrid coding scheme has been proposed. This technique works on two resolution levels and uses a linear transformation, where motion is purely translational, and an efficient application of a non-linear transformation in detailed moving areas. The result obtained with the warping process, if required, is weighted with that obtained with the linear transformation. Thanks to this solution, it is possible to achieve a remarkable reduction of the prediction errors and blocking artifacts with respect to typical linear prediction-based methods.

Large block sizes let us reduce the total number of bits, and the fast approach for non-linear parameter calculation and the line scanning technique allow us to reduce computational complexity. Experimental results have shown that the use of non-linear geometric transformations on the areas of the images characterised by complex motion permits the quality improvement of the reconstructed frames of a very low bit-rate coder.

Acknowledgments: This work was partially supported by "MURST" and "ESPRIT-LTR 20229 Noblesse" projects.

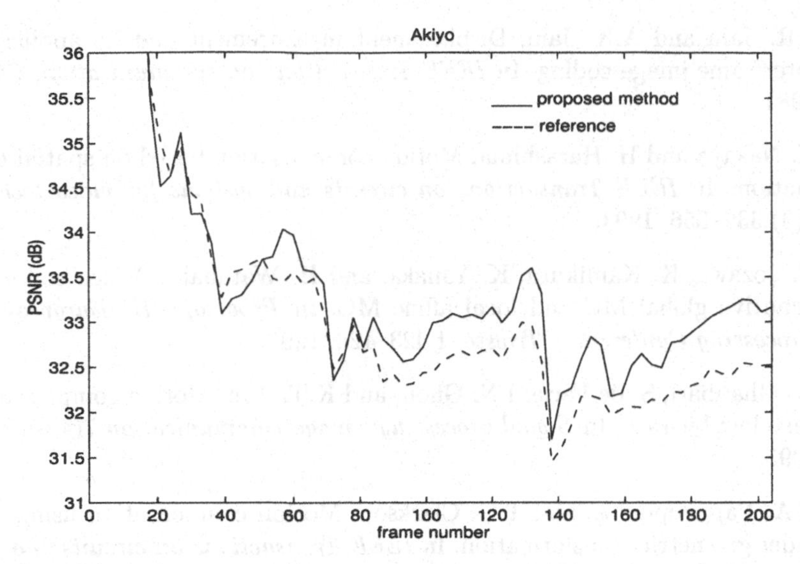

FIGURE 23.7. PSNR of the decoded sequence CIF. Akiyo at 16 kb/s

23.7 REFERENCES

[1] A.K. Jain. Image data compression: a review. In *Proceedings of the IEEE*, 69(3):349–389, 1981.

[2] A.N. Netravali and J.D. Robbins. Picture coding: a review. In *Proceedings of the IEEE*, 68(3):366–406, 1980.

[3] M. Kunt, A. Ikonomopoulos, and M. Kocher. Second generation image-coding techniques, In *Proceedings of the IEEE*, 73(4):549–574, 1985.

[4] H.G. Musmann, M. Hotter, and J. Ostermann. Object oriented analysis-syntesis coding of moving images. In *Signal Process.: Image Commun.*, 1(2):117–138, 1989.

[5] D.E. Pearson. Developments in model-based video coding. In *Proceedings of the IEEE*, 83(6):892–906, 1995.

[6] K. Aizawa and T.S. Huang. Model-based image coding advanced video coding techniques for very low bit-rate applications. In *Proceedings of the IEEE*, 83(2):259–271, 1995.

[7] M.F. Chowdhury, A.F. Clark, A.C. Downtown, E. Morimatsu, and D.E. Pearson. A switched model-based coder for video signals. In *IEEE Transactions on circuits and sistems for video technology*, 4(3):216–227, 1994.

[8] A.E. Jacquin. Fractal Image Coding: a review. In *Proceedings of the IEEE*, 81(10):1451–1465, 1993.

[9] H. Nagel. Displacement vectors derived from second order intensity variations in image sequences. In *Computer Graphics and Image Process*, 21:85–117, 1983.

[10] D.R. Walker and K.R. Rao. Improved pel-recursive motion compensation. In *IEEE Trans. Commun.*, COM-32:1128–1134, 1984.

[11] J.R. Jain and A.K. Jain. Displacement measurement and its application in interframe image coding. In *IEEE Transactions on communication*, COM-29, 1981.

[12] Y. Nakaya and H. Harashima. Motion compensation based on spatial tranformations In *IEEE Transactions on circuits and systems for video technology*, 4(3):339–356, 1994.

[13] H. Jozawa, K. Kamikura, K. Yanaka, and H. Watanabe. Video coding using adaptive global MC and local affine MC. In *Proc. of VIII European Signal Processing Conference, Trieste,* 1:423–426, 1996.

[14] M. Ghanbari, S. de Faria, I.N. Ghon, and K.T. Tan. Motion compensation for very low bit-rate. In *Signal processing: image communication* 7(4-6):567–580, 1995.

[15] C.A. Papadopoulos, and T.G. Clarkson. Motion compensation using second-order geometric transformation. In *IEEE Transactions on circuits and sistems for video technology* 5(4):319–331, 1995.

[16] H.G. Musmann, P. Pirsch, and H.J. Grallert. Advances in picture coding. In *Proceedings of the IEEE* 73(4):523–548, 1985.

[17] M. Hoetter. Differential estimation of the global motion parameters zoom and pan, In *Signal Process* 16, 1989.

[18] F. Dufaux and F. Moscheni. Motion estimation techniques for digital TV: a review and a new contribution, In *Proceedings of the IEEE* 83(6):858–876, 1995.

[19] T. Ebrahimi, E. Reusens, and W. Li. New trends in very low bitrate video coding. In *Proceedings of the IEEE*, 83(6):877–891, 1995.

[20] Y. Huang, X. Zhuang, and C. Yang. Two block-based motion compensation methods for video coding. In *IEEE Transactions on circuits and systems for video technology*, 6(1):123–126, 1996.

[21] A. Netravali and A. Lippman. Digital television: a perspective. In *Proceedings of the IEEE*, 83(6):834–842, 1995.

[22] S. Okubo. Reference model methodology - a tool for the collaborative creation of video coding standards. In *Proceedings of the IEEE*, 83(2):139–150, 1995.

[23] R. Schafer and T. Sikora. Digital video coding standards and their role in video communications. In *Proceedings of the IEEE*, 83(6):907–924, 1995.

[24] S.F. Wu and J. Kittler. A differential method for simultaneous estimation of rotation, change of scale and rotation. In *Signal Processing: Image Communication*, 2(1):69–80, 1990.

[25] M.F. Zanuy and X.D. Reguant. Interblock redundancy reduction using quadtrees. In *Proc. of VIII European Signal Processing Conference, Trieste,* 1:340-343, 1996.

24

Smoothing Techniques for Prediction of Non-Linear Time Series

Svetlana Borovkova[1]
Robert Burton[2]
Herold Dehling[3]

ABSTRACT
We address the problem of non-linear modelling of time series and give a brief introduction to the method of state space reconstruction - embedding one-dimensional data into a higher-dimensional space. This method is based on the fundamental result in the area of chaotic time series, the Takens reconstruction theorem. Then we consider, among other non-linear methods of prediction, the kernel estimation of autoregression, and introduce a variation of this method. We apply it to an experimental time series and compare its performance with predictions by feed-forward neural networks as well as with fitting a local and a global linear autoregression.

24.1 Non-Linear Models and Reconstruction

Prediction of future observations is an important problem in the analysis of time series. Given a time series $\{Y_n\}_{n \in \mathbb{N}}$, the question is to find a predictor for Y_{n+s} as a function of a certain number of previous observations, i.e., we are looking for

$$\begin{aligned} \hat{Y}_{n+s} &= F(y_n, y_{n-1}, ..., y_{n-k+1}) \\ &:= \mathbf{E}(Y_{n+s} | Y_n = y_n, ..., Y_{n-k+1} = y_{n-k+1}) \end{aligned} \tag{24.1}$$

Beginning with Yule's invention in 1927 of the linear autoregression for the analysis of the sunspot data, linear models dominated time series analysis for the following half century. In traditional linear models driven by noise, such as AR and ARMA, a future observation is taken as a linear combination of a certain number of previous observations and random, mostly Gaussian, innovations. Thus, in this case the function F is restricted to a linear form. However, there are simple examples

---

[1]Shell Research and Technology Centre, Amsterdam, P.O.Box 38000, 1030 BN Amsterdam, The Netherlands, E-mail: Svetlana.A.Borovkova@opc.shell.com
[2]Oregon State University, Department of Mathematics, Kidder Hall 368, Corvallis, OR, 97331 U.S.A, E-mail: burton@math.orst.edu
[3]University of Groningen, Department of Mathematics, P.O.Box 800, 9700 AV Groningen, The Netherlands, E-mail: dehling@math.rug.nl

of time series, for instance, those related to chaotic dynamical systems, for which linear models are inadequate, because of the strong presence of non-linearities.

Suppose a time series $\{y_n\}_{n\in\mathbf{N}}$ is obtained by the following rule:

$$y_{n+1} = f(y_n) \tag{24.2}$$

for some $y_0 \in \mathbf{R}$, where $f : \mathbf{R} \longrightarrow \mathbf{R}$ is a non-linear function. In this case the time series satisfies the first-order autoregression model with non-linear autoregression function f. Even in this simple deterministic one-dimensional case, the function f can be such that the behaviour of the resulting time series is so complex that it appears random. Perhaps the most well-known example is the *logistic map* $f(x) = \lambda x(1-x)$, which is known to exhibit chaotic behaviour for certain parameter values, e.g., $\lambda = 4$.

Deterministic models of the type (24.2) can be written more generally as

$$X_{n+1} = G(X_n) \tag{24.3}$$

for some X_0, where $X_n \in \mathbf{R}^k, n \geq 0$, and $G : \mathbf{R}^k \longrightarrow \mathbf{R}^k$ is a vector-valued function. The difference equation (24.3) describes a discrete dynamical system in $\mathbf{R}^k$ with the evolutionary map G.

In reality, observations rarely evolve according to the model (24.3): usually there are some observational or measurement noise or other random disturbances present in the system. Moreover, our deterministic model will inevitably be inexact as far as modelling of real data is concerned. Therefore, it is more realistic to replace (24.3) by a model

$$X_{n+1} = G(X_n, e_{n+1}) \tag{24.4}$$

where $G : \mathbf{R}^{k+1} \longrightarrow \mathbf{R}^k$ and $\{e_n\}_{n\in\mathbf{N}}$ is a sequence of random variables such that e_n is independent of $\{X_i\}_{i<n}$. The model (24.4) is called a (discrete) *stochastic dynamical system.* The sequence of random vectors $\{e_n\}$ is called the *dynamic noise.* For convenience of analysis it is usually assumed that the dynamic noise is *additive*, so that (24.4) reduces to the model with additive noise

$$X_{n+1} = G(X_n) + e_{n+1} \tag{24.5}$$

where $G : \mathbf{R}^k \longrightarrow \mathbf{R}^k$. If we have the following vector representations:

$$\begin{aligned} X_n &= (y_n, y_{n-1}, ..., y_{n-k+1}) \\ G(X_n) &= (F(X_n), y_n, ..., y_{n-k+2}) \\ e_n &= (\epsilon_n, 0, ..., 0) \end{aligned} \tag{24.6}$$

then (24.5) implies that

$$y_{n+1} = F(y_n, y_{n-1}, ..., y_{n-k+1}) + \epsilon_{n+1} \tag{24.7}$$

and the function F is non-linear if G is. Relationship (24.7) defines a *non-linear autoregression model of order* k for the time series $\{y_n\}$. Conversely, the model (24.7) for the time series $\{y_n\}$ can be written as a stochastic dynamical system (24.5) by vectorising $\{y_n\}$ and using (24.6).

The question of separating non-linear time series from linear ones is, in general, quite complex and cannot be answered just by visual observation. However, there are ways (graphical as well as analytical) to distinguish non-linear time series and time series related to deterministic systems. A simple but rather useful tool in the analysis of non-linear time series is the so-called ***delay-1 map***: plotting y_{n+1} vs. y_n. The use of return maps for discriminating non-linear time series is closely related to the ***state space reconstruction***. This technique is based on a remarkable result in the area of chaotic time series, the ***Takens reconstruction theorem*** [9]. First we shall try to give a flavour of the reconstruction technique.

When observing delay maps for a time series, we study the behaviour of pairs of observations, i.e., vectors (y_{n-1}, y_n). In the case of a non-linear time series this can, in principle, provide a better understanding of the dynamics that generated this time series. Intuitively, one may expect that the behaviour of vectors (y_{n-1}, y_n) provides more insight into the underlying dynamics than the behaviour of the single observations y_n. Consecutively, for more complicated cases, more information about the dynamics can be provided by studying three-dimensional delay maps, i.e., vectors (y_{n-2}, y_{n-1}, y_n), or even k-dimensional vectors $(y_{n-k+1}, ..., y_n)$ for some higher k. This is the basic idea of reconstruction: to replace a time series $\{y_n\}$ by a series of vectors $\{(y_{n-k+1}, ..., y_n)\}$, which are called the ***reconstruction vectors***. The reconstruction vectors "live" in a higher-dimensional space $\mathbf{R}^k$, also called the ***embedding space***, and its dimension k is called the ***embedding dimension***. Note that both the space $\mathbf{R}^k$, where a stochastic dynamical system (24.5) evolves, and its dimension are naturally related to the embedding space and the embedding dimension.

We illustrate the reconstruction technique on the example of the Henon dynamical system. This system evolves in $\mathbf{R}^2$ according to the law

$$(x, y) \longrightarrow (1 - ax^2 + y,\ bx)$$

For the parameter values $a = 1.4$, $b = 0.3$, it is known to exhibit chaotic behaviour. In Fig. 24.1 a trajectory of this system in the original coordinates (x, y) is shown. The corresponding time series, which was taken as a time series of the first coordinate $\{x_n\}$, is shown in Fig. 24.2. In Fig. 24.3 we observe the delay-1 plot, i.e., the plot of reconstruction vectors $\{(x_n, x_{n+1})\}$ generated by the time series. The two-dimensional embedding space is given in this case by the delay coordinates (x_n, x_{n+1}). Comparing Fig 24.1 with Fig. 24.3, we see a very similar structure in the embedding space and in the original phase space.

Contrary to this example, in most realistic situations neither an original phase space nor a dynamical law is known. What is actually observed is a real-valued time

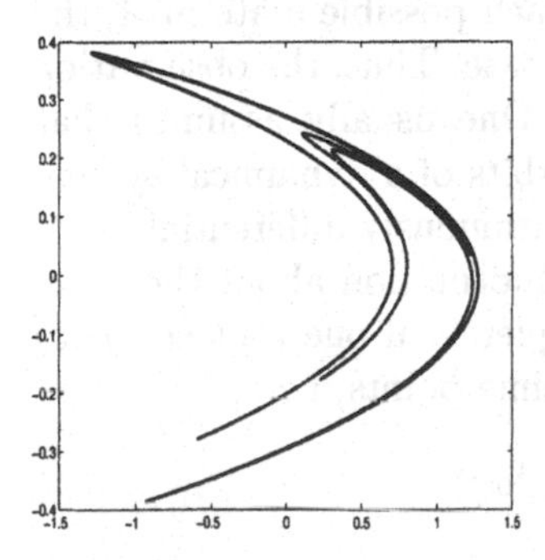

FIGURE 24.1. A trajectory of the Henon dynamical system

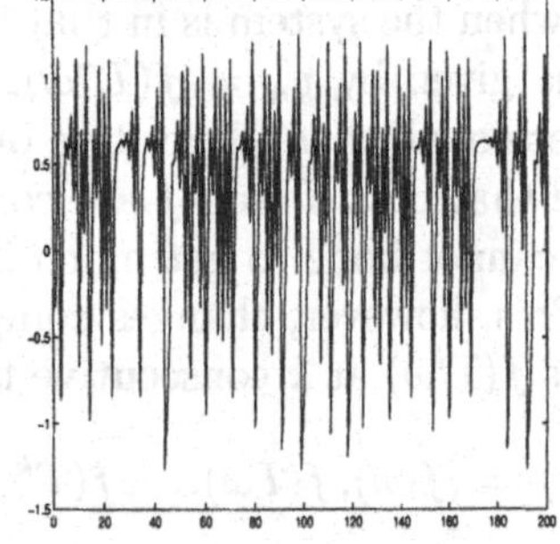

FIGURE 24.2. Time series of the Henon map

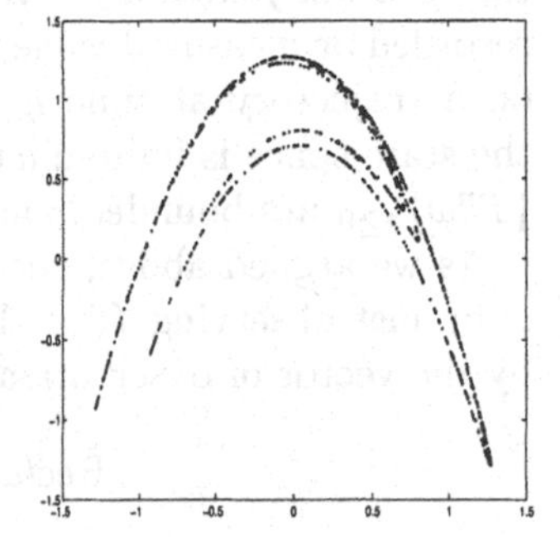

FIGURE 24.3. Delay-1 map

series of measurements made on some physical system. As an example, we consider a time series of pressure measurements made in a *fluidized bed*, which is a certain type of chemical reactor, where a bed of solids is brought into a fluid-like state by pressing gas from the bottom into the reactor. A part of this data set is shown in Fig. 24.4. This time series is believed to be a good example of a chaotic time series coming from a low-dimensional chaotic dynamical system with a low level of measurement noise (see, for example, [10]). Observing the delay map at lag 35 in Fig. 24.5 (i.e., the plot of y_{n+35} vs. y_n), one confirms this belief to a certain degree.

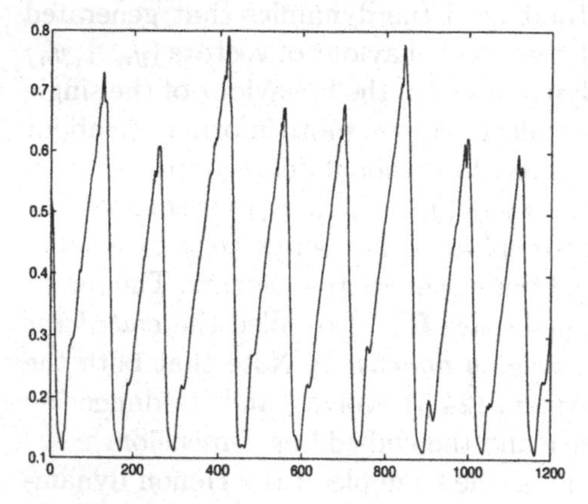

FIGURE 24.4. Pressure fluctuations in the fluidized bed

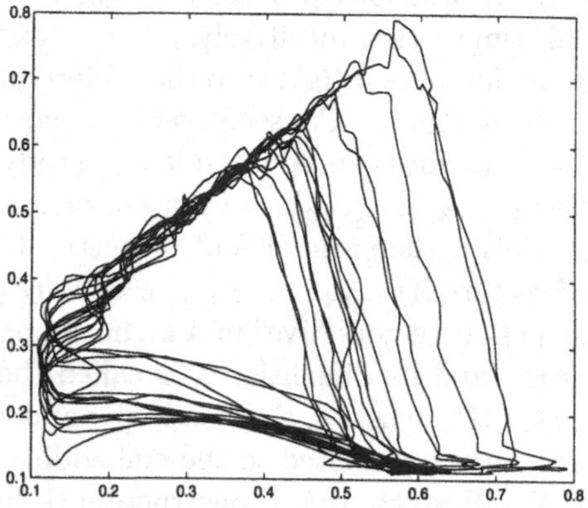

FIGURE 24.5. Delay map for the fluidized bed time series: X_n vs. X_{n+35}

If one tries to study such a system from an observed time series, then a fundamental problem arises because the physical system and the observed data are in different spaces. The importance of the reconstruction theorem is that it assures that a time series of only one scalar variable contains enough information to reconstruct the dynamics in the original multivariate phase space without any prior knowledge of it. Now we shall express this in a more formal way.

Let $(\mathcal{X}, T)$ be a *discrete time dynamical system*, i.e., a combination of a state space $\mathcal{X} \subseteq \mathbf{R}^d$ and a time evolution map $T : \mathcal{X} \longrightarrow \mathcal{X}$. For an initial state $\omega \in \mathcal{X}$, the iterations of T give rise to a trajectory $(\omega, T\omega, T^2\omega, ...) = \{T^n\omega\}_{n\in\mathbf{N}}$. The relationship between a dynamical system and a time series is drawn by means of the *read-out function* $f : \mathcal{X} \longrightarrow \mathbf{R}$, which assigns to each possible state in $\mathcal{X}$ the recorded or measured value when the system is in that state. Thus, the observation on a trajectory at time n is given by $y_n = f(T^n\omega)$. One usually assumes that the state space is finite-dimensional, that all positive orbits of a dynamical system $\{T^n\omega\}_{n\geq 0}$ are bounded, and that both T and f are continuously differentiable.

As we argued above, one cannot hope to get much information about the state ω by just observing $f(\omega)$. This, however, changes completely if one replaces $f(\omega)$ by the vector of observations $f(T^i\omega)$ at k consecutive time points, i.e.,

$$\text{Rec}(\omega) := (f(\omega), f(T\omega), ..., f(T^{k-1}\omega))$$

Then the Takens reconstruction theorem assures that in generic situations the mapping $\text{Rec} : \mathcal{X} \rightarrow \mathbf{R}^k$ defines an embedding, provided $k \geq 2\dim(\mathcal{X}) + 1$. Consequently we can obtain information about the state space and the dynamics of T by studying the process of reconstruction vectors

$$X_n = \mathrm{Rec}(T^n \omega),\ n \geq 0$$

For practical purposes, i.e., for reconstructing the state space from the observed real-valued time series $\{y_n\}_{n \in \mathbf{N}}$, we define the reconstruction vectors by

$$X_n = (y_n,\ y_{n+1},\ \ldots,\ y_{n+k-1})$$

for $k \geq 2\dim(\mathcal{X}) + 1$. Then the trajectory $\{T^n \omega\}_{n \in \mathbf{N}}$ in $\mathcal{X}$ is an image of the trajectory generated by the sequence of reconstruction vectors $\{X_n\}_{n \in \mathbf{N}}$ in $\mathbf{R}^k$ under the reconstruction map. In other words, if the embedding dimension k is high enough, then the trajectory generated by the series of reconstruction vectors in the embedding space replicates the evolution of the system in the phase space.

When it is not known whether the time series is produced by a deterministic dynamical system or whether there is significant noise present in the time series, then it is harder to interpret the results of the state space reconstruction. Also one should not forget that the choice of the embedding dimension k is essential. The problem of selecting a sufficiently large k can be difficult since the dimension of $\mathcal{X}$ is usually unknown, and hence the criteria $k > 2 \cdot \dim(\mathcal{X})$ of the Takens theorem cannot be applied directly. There are a number of methods for choosing k suggested in the literature, see, for example, [4, 3].

The reconstruction theorem also assures that the essential features of the original dynamics can be deduced from the behaviour of the reconstruction vectors. Characteristics defined by the sequence $\{X_n\}_{n \in \mathbf{N}}$, such as the invariant measure, the dimension, etc., do not depend on the read-out function f but describe intrinsic properties of the dynamics. This, in turn, gives a possibility for classification of time series according to characteristics of the underlying dynamics. Two characteristics often used for classification are the *correlation integral* and the *correlation dimension*. For a broad discussion on dimensions as well as their estimation procedures see [5]. Apart from dimension-based methods, there are many other methods of discriminating between deterministic and stochastic time series introduced in the literature, such as the BDS test [2], classification based on prediction [3], etc. For a good review on this subject, see [8].

24.2 Non-Linear Predictors

The Takens reconstruction theorem implies that time series coming from a chaotic dynamical system in general satisfy a non-linear autoregression model. Recall that the reconstruction vectors $X_1, X_2, \ldots \in \mathbf{R}^k$ are defined from the observed time series $\{y_n\}_{n \in \mathbf{N}}$ by

$$X_i = (y_i, y_{i+1}, \ldots, y_{i+k-1})$$

A consequence of the Takens reconstruction theorem is that the original dynamical system $(\mathcal{X}, T)$ together with the reconstruction map induce a dynamical system $(\mathrm{Rec}(\mathcal{X}), G)$, where $\mathrm{Rec}(\mathcal{X}) \subset \mathbf{R}^k$ and the map $G : \mathbf{R}^k \longrightarrow \mathbf{R}^k$ is conjugate to T. In terms of reconstruction vectors, we get

$$X_{i+1} = G(X_i) \tag{24.8}$$

Denoting the first coordinate of G by $F : \mathbf{R}^k \longrightarrow \mathbf{R}$, we obtain the following relation for the original time series

$$y_{i+k} = F(y_i, y_{i+1}, ..., y_{i+k-1}) \tag{24.9}$$

The original transformation T was assumed to be non-linear, so F is also a non-linear function. The relationship (24.9) implies that the time series $\{y_i\}_{i \in \mathbf{N}}$ satisfies the kth order non-linear autoregression model, where the autoregression function F is some unknown non-linear function not restricted to any parametric form. Now we also assume that there is measurement noise in the system, which for simplicity we assume to be additive. Thus, we observe

$$\tilde{y}_{i+k} = y_{i+k} + \epsilon_{i+k} = F(y_i, y_{i+1}, ..., y_{i+k-1}) + \epsilon_{i+k} \tag{24.10}$$

where ϵ_i are mean zero and finite variance errors.

In this case and in the case of other non-linear time series (not necessarily chaotic), one expects that non-linear methods of prediction have an advantage over traditional linear methods.

There are many non-linear methods of prediction introduced in the literature. Most of them are oriented toward understanding and predicting the behaviour of a time series by studying its regularities in the past. Among these methods the local methods of prediction are of particular importance. Their main idea is to capture the local dynamics of the time series, when the effects of non-linearities and the amplification of noise are not yet that strong. A method of prediction frequently mentioned in the literature on chaotic time series is the so-called (k, ϵ)-method (or local linear predictors). On the local scale we expect that if the non-linear function F is smooth enough, it can be successfully interpolated by a linear function. Then the method can be briefly described as follows: suppose that having observed k last values of the time series $\mathbf{x} := y'_1, ..., y'_k$, we want to predict y'_{k+1}. We collect all vectors of length k from the time series which are within distance ϵ from $\mathbf{x}$, as well as the corresponding observations 1 step after, and then fit a linear regression to the collected data. The pair (k, ϵ) is chosen by minimizing some measure of prediction error. This method turns out to be quite successful when the time series is indeed coming from a low-dimensional chaotic dynamical system and enough data are available.

We shall concentrate here on another method, known from nonparametric regression theory, namely kernel regression smoothing. This method also has a local character but is more flexible.

Let $\{y_n\}_{n \in \mathbf{N}}$ be an observed time series and define the delay vectors $X_i = (y_i, y_{i+1} ..., y_{i+k-1})$. Let $K : \mathbf{R} \longrightarrow \mathbf{R}$ be a *kernel function*, i.e. a continuous bounded symmetric and unimodal function which integrates to 1. Then the kernel estimator of the function F is given by

$$\hat{F}_h(\underline{x}) = \frac{\sum_{i=k}^{n-1} \mathbf{K}_h(\underline{x} - X_i)\, y_{i+k}}{\sum_{i=k}^{n-1} \mathbf{K}_h(\underline{x} - X_i)} \tag{24.11}$$

where $\mathbf{K}_h$ for $k > 1$ is defined via a kernel function K by

$$\mathbf{K}_h(\underline{x}) = \prod_{l=1}^{k} K(x_l/h)$$

where $\underline{x} = (x_1, ..., x_k)$. $\mathbf{K}_h$ is also called a *product kernel.* For $k = 1$, one takes $\mathbf{K}_h(y) = K(y/h)$. Here $h = h_n$ is a sequence of scaling parameters called the *bandwidth sequence.* In regression estimation the estimator (24.11) is usually referred to as the *Nadaraya-Watson estimator.*

The application of the kernel method to time series was studied by Collomb (1984), Delecroux (1987) (see [6] and the references therein), and it was shown (Delecroux (1987)) that this estimator is consistent if the time series is a stationary ergodic process and the bandwidth sequence converges to zero at some specified rate. (See also [1]).

In applications, a very important question is the bandwidth selection. Choosing a bandwidth too small leads to a higher variance of the kernel estimator, and choosing a bandwidth too large increases its bias. In practice we have to balance these two factors. For this consider some measure of accuracy of the kernel estimator, for example the average squared error:

$$ASE[\hat{F}_h] = \frac{1}{n}\sum_{i=k}^{n}[\hat{F}_h(X_i) - F(X_i)]^2$$

which is the sum of the variance and the squared bias components (in principle, we could take the mean squared error or the integrated squared error here, the average squared error was chosen as the actually observed error). A value of the bandwidth, which minimizes the ASE, is desirable. Such a value is selected by a data-driven approach called *cross-validation.* We estimate the ASE by the *cross-validation function*

$$CV(h) = \frac{1}{n}\sum_{i=k}^{n}[y_{i+k} - \hat{F}_{h,i}(X_i)]^2$$

where $\hat{F}_{h,i}(X_i)$ is the leave-i-out kernel estimate of $F(X_i)$:

$$\hat{F}_{h,i}(X_i) = \frac{\sum_{i\neq j}\mathbf{K}_h(X_i - X_j)\, y_{j+k}}{\sum_{i\neq j}\mathbf{K}_h(X_i - X_j)}$$

The cross-validation function $CV(h)$ is an asymptotically unbiased estimator of ASE with the advantage that it can be computed directly from the data. Then we choose the value of the bandwidth $\hat{h}_{opt}$ which minimizes $CV(h)$. Then the question is whether $\hat{h}_{opt}$ also (asymptotically) minimizes the ASE. Haerdle and Vieu [7] have shown that, if the sequence (X_i) is strongly mixing, and under some additional conditions on K, F, and the distribution of (X_i), the cross-validation procedure is asymptotically optimal, i.e., that the bandwidth chosen by means of cross-validation asymptotically minimizes the ASE.

One of the main disadvantages of the methods mentioned above is the fixed choice of the order k, which results into taking entirely into account the previous k observations and completely disregarding the rest. Moreover, in applications k is usually unknown and has to be estimated in some way (traditionally via linear models).

In the next section we suggest a variation of kernel autoregression smoothing which overcomes this disadvantage and is in general more flexible.

24.3 Variation of the Kernel Smoothing Method

One way to apply the kernel smoothing method for prediction in the case $k > 1$ involves product kernels. Another way is to define in some suitable way a distance $d(\underline{x}, \underline{y})$ between two k-dimensional vectors $\underline{x}, \underline{y}$ and define the kernel regression estimator via a one-dimensional kernel function as

$$\hat{F}_h(\underline{x}) = \frac{\sum_{i=1}^{n} K[d(\underline{x}, X_i)/h]\, y_{i+k}}{\sum_{i=1}^{n} K[d(\underline{x}, X_i)/h]} \tag{24.12}$$

The distance $d(\cdot, \cdot)$ can be taken as the Euclidean or the maximum distance. However, the choice of these distances would be purely formal and would not exploit the fact that the vectors X_i's are parts of a time series, i.e., sequences of observations ordered in time. Here we suggest a variation of the estimator (24.12) and choose such a distance between the vectors which takes the specifics of the time series setting into account.

We define a distance between two vectors $\underline{x} = (x_1, x_2, ..., x_k)$ and $\underline{y} = (y_1, y_2, ..., y_k)$ by

$$d(\underline{x}, \underline{y}) = \sum_{i=1}^{k} (x_i - y_i)^2 \gamma_i \tag{24.13}$$

where $\{\gamma_i\}_{i=1}^{k}$, $\gamma_i \in [0, 1]$, is a collection of weights.

In general, we will choose decreasing weights $\gamma_1 \geq \gamma_2 \geq ... \geq \gamma_k$, expressing the idea that the influence of past observations on the prediction should be discounted as the time lag grows. More specifically, we suggest to take

$$\gamma_i = \gamma^i \tag{24.14}$$

for some $\gamma \in (0, 1]$. One motivation for this is the exponential divergence of trajectories (whose starting points are close) in chaotic dynamical systems. However, this technique is appropriate for many time series, including those with stochastic properties. Thus, the more recent observations give more precise information about the present state of the system, and the information decreases exponentially as time passes.

Observations further in the past are less relevant for future predictions since they are most affected by noise. In chaotic dynamical systems this is expressed via the notion of the sensitive dependence on initial conditions, which is considered as the characteristic feature of chaos. The exponential growth of errors, which are always present in a real-life systems, makes the chaotic evolution self-independent of its own past. Quantitatively, this is described by the Lyapunov exponents of a chaotic map.

In outline, the Lyapunov exponent measures the mean exponential rate at which nearby orbits diverge with time. Suppose that $\{x_i\}_{i \in \mathbf{N}}$ is the orbit of a chaotic discrete dynamical system corresponding to the initial condition x_0. If we change slightly the position of the initial point: $x_0 \longrightarrow x_0' = x_0 + \Delta x_0$, the point at the time n will also be changed. Generally speaking, one expects that if Δx_0 is small, Δx_n is also small. But, due to the sensitive dependence on the initial condition, when n becomes large, the small initial distance grows exponentially fast: $\Delta x_n \sim \Delta x_0 exp(\lambda n)$, where the mean rate of divergence of the orbits λ is the Lyapunov exponent.

In dimension one a dynamical system has one Lyapunov exponent. It can be formally defined as

$$\lambda = \lambda(x) = \lim_{n \to \infty} \frac{1}{n} \log |(T^n)'_x|$$

where $(T^n)'_x$ is a derivative of T^n evaluated at x.

For a one-dimensional dynamical system, a positive Lyapunov exponent implies sensitive dependence on initial conditions, and so is an indication of chaotic behaviour. In dimensions higher than one the mean exponential rate with which nearby orbits diverge with time is measured by the highest Lyapunov exponent, and the existence of at least one positive Lyapunov exponent is evidence for sensitive dependence on initial conditions.

In the context of chaotic time series we are more interested in backward divergence of orbits, which is responsible for the decreasing influence of past observations on future ones. In that case the mean rate of divergence of backward orbits is measured by the highest Lyapunov exponent of the inverse map T^{-1} (provided T is invertible). It is given by the inverse of the lowest Lyapunov exponent of the original map T.

For our applications this situation can be illustrated by the following example. If $\underline{x} = (x_1, ..., x_k)$ and $\underline{y} = (y_1, ..., y_k)$ are segments of two orbits of an invertible chaotic dynamical system or two parts of the same orbit and if $\| x_k - y_k \| = \Delta$, then, because of the exponential divergence of backward trajectories, we have that $\| x_{k-1} - y_{k-1} \| = \Delta e^{\lambda}$, $\| x_{k-2} - y_{k-2} \| = \Delta e^{2\lambda}$, etc., where λ is the highest Lyapunov exponent of the inverse of the underlying chaotic map. The distance between $\underline{x}$ and $\underline{y}$ is given by

$$d(\underline{x}, \underline{y}) = \sum_{j=1}^{k} \| x_{k-j} - y_{k-j} \|^2 \gamma^j = \sum_{j=1}^{k} \Delta^2 e^{2j\lambda} \gamma^j \tag{24.15}$$

and it is completely determined by δ, whereas the factors $e^{j\lambda}$ indicate the natural divergence of trajectories with rate λ. This reasoning suggests taking $\gamma = e^{-2\lambda}$ to "discount" for this exponential divergence. Due to the state space reconstruction, the above reasoning also applies when $\underline{x} = (x_1, ..., x_k)$ and $\underline{y} = (y_1, ..., y_k)$ are not parts of an orbit but segments of a chaotic time series.

In practice we will choose γ in a different way by optimising the value of γ experimentally using the cross-validation technique.

We believe our method is more flexible than that of local linear predictors, as well as the other methods mentioned above. Rather than having sharp cutoff points (k, ϵ) in time we have a smoother way to assign the significance of past observations. The choice of the parameter γ in a way replaces the choice of the order of autoregression k. By taking the decreasing weights (24.14), the choice of k becomes less important because the dependence on the previous observations is not cut off at the number k but decreases smoothly with the decay of γ_i.

The influence of the parameter γ on our kernel estimate is comparable to the bandwidth influence. Thus, the choice of γ and of h determined the quality of our predictions. At the same time the choice of γ and of h is not independent: the bigger the bandwidth h, the higher the value of γ we should choose. Again we use a cross-validation algorithm. But since the parameters h and γ are bound together, the selection of their optimal values should be carried on simultaneously, i.e., we choose the ***optimal pair*** $(h,\gamma)_{opt}$ as

$$(h,\gamma)_{opt} = \text{argmin}\{CV(h,\gamma)\}$$

where $CV(h,\gamma)$ is a cross-validation function of the estimator (24.12).

The consistency of our estimator follows from standard results for the traditional kernel estimator (see [1, 6]). The double cross-validation procedure is asymptotically optimal. This may be seen using arguments similar to those found in [7].

24.4 Application to an Experimental Time Series

In this section we apply our kernel method to a time series of the pressure measurements in a fluidized bed. A part of this data set was shown in Fig. 24.4.

The prediction step for this time series was chosen 35 measurements ahead, and since the time series is obviously oversampled, we introduce a time delay $\tau = 35$ and look for a predictor of the form

$$\hat{y}_{n+35} = F(y_n, y_{n-\tau}, ..., y_{n-5\tau}) \tag{24.16}$$

We used a larger part of the time series as the basis for predictions and disjoint smaller parts as test sets. As a measure of quality of our predictions, we take the ***Average Absolute Error:***

$$AAE_N = \frac{1}{N}\sum_{j=1}^{N} \frac{|y_j^{(t)} - \hat{y}_j^{(t)}|}{R} \cdot 100\% \tag{24.17}$$

where R is the range of values of the time series, $y_1^{(t)}, ..., y_N^{(t)}$ is a test set, and $\hat{y}_1^{(t)}, ..., \hat{y}_N^{(t)}$ are the obtained predictions.

We took the Gaussian kernel function and geometrically decreasing weights $\gamma_i = \gamma^i$ $(i = 1, ..., 5)$ for computation of the distance (24.13). Selection of the bandwidth h and the parameter γ was done by simultaneous cross-validation. The cross-validation surface is shown in Fig. 24.6, the optimal pair of parameters being $(h,\gamma) = (0.03, 0.6)$.

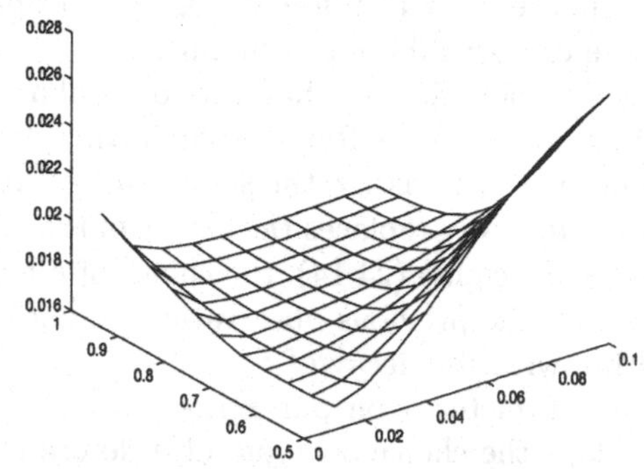

FIGURE 24.6. Cross-validation function $CV(h,\gamma)$

Fig. 24.7 shows two test sets of length 600 (solid line) together with predictions (dashed line). The value of the AAE is 5.75%. In general the quality of prediction is quite good, and the dashed curve is rather smooth, but, as we could expect, on highs and lows where the fluctuations are most noticeable we get slightly worse predictions than on intermediate parts of the time series.

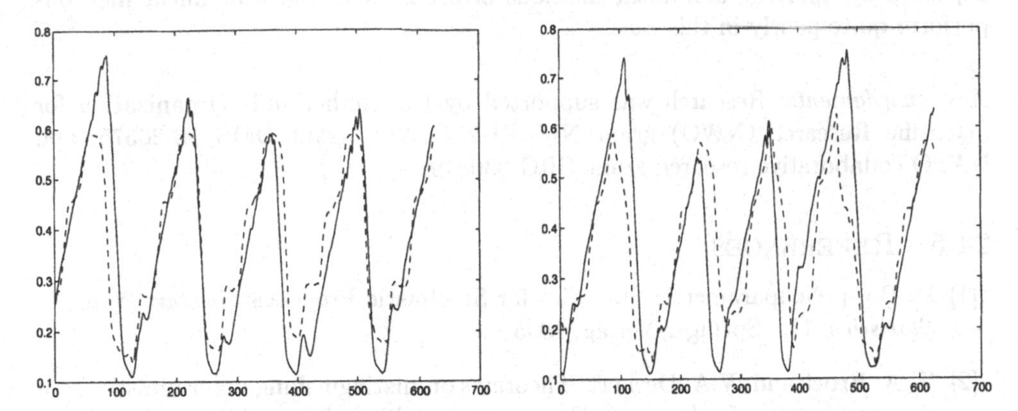

FIGURE 24.7. Fluidized bed time series with predictions

Next, we compare the results with the performance of a neural net. Looking again for a predictor of the form (24.16), we used the one-hidden-layer feedforward neural networks with five inputs, three hidden neurons and one output, which was trained to be the prediction for a value 35 measurements ahead. We used 10^5 training iterations performed on the training set of length 3000. The example of a test set together with neural net predictions is shown in Fig. 24.8. The AAE obtained is 8.7%. Comparison of Fig. 24.7 with Fig. 24.8 shows that the kernel method works better not only in terms of average error, but it has a definite advantage in predicting high and low values, where the neural net failed completely.

Finally, we apply local and global linear predictors to our data set. For the (k, ϵ)-method, we take into account ϵ-close vectors of k past values, sampled with time delay 35, and we base the choice of the pair (k, ϵ) on the minimization of the AAE. The optimal pair is $k = 3$, $\epsilon = 0.2$. A test set together with local linear predictions is shown in Fig. 24.9. The AAE in this case is 8.4%, but note that again the predictions of high and low values are of poor quality.

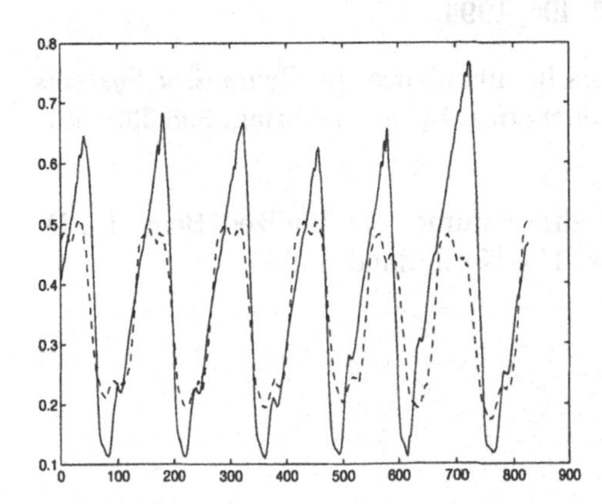

FIGURE 24.8. 5:3:1-Neural net predictions for fluidized bed data

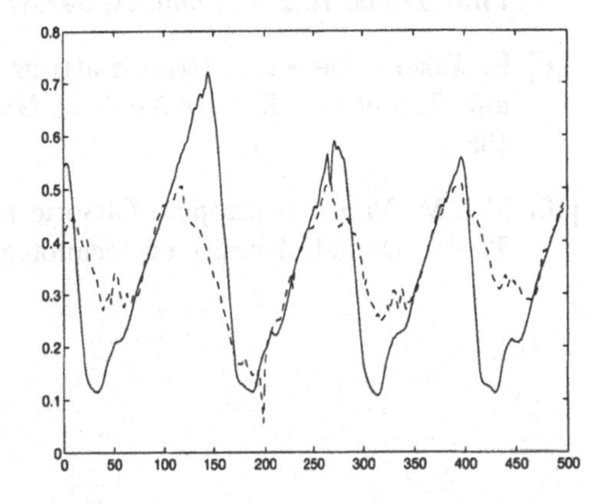

FIGURE 24.9. Local linear predictions for fluidized bed data

Fitting global linear autoregression from the past five values sampled with time delay $\tau = 35$ (as in (24.16), with F linear) gave a much higher error level of 15%. This shows that strong non-linear dependence in this data set is indeed best captured by applying non-linear methods of predictions and that linear methods perform quite poorly in this case.

Acknowledgments: Research was supported by the Netherlands Organization for Scientific Research (NWO) grant NLS 61-277, NSF grant DMS 96-26575, and NATO collaborative research grant CRG 930819.

24.5 References

[1] D. Bosq. Nonparametric Statistics for Stochastic Processes. *Lecture Notes in Statistics*, 110, Springer-Verlag, 1995.

[2] W.A. Brock and W.A. Dechert. Theorems on distinguishing deterministic from random systems. In: Dynamic Economic Modelling, *Proc. of the 3d Int. Symp. on Economic Theory and Econometrics*, Cambridge University Press, 1998.

[3] M. Casdagli. Chaos and Deterministic *vs.* Stochastic Non-linear Modelling. *J. R. Statist. Soc. B*, 54(2):303–328, 1991.

[4] B. Chen, H. Tong. Nonparametric function estimation in noisy chaos. In: *Developments in Time Series Analysis*, pp. 183–206, Chapman & Hall, London, 1993.

[5] C.D. Cutler. Some results on the behaviour and estimation of fractal dimension of distributions on attractors. *J. Stat. Phys.*, 62:651-708, 1991.

[6] L. Gyorfi, W. Haerdle, P. Sarda, and P. Vieu. Nonparametric Curve Estimation from Time Series. *Lecture Notes in Statistics*, Springer-Verlag, 60, 1989.

[7] W. Haerdle and P. Vieu. Kernel regression smoothing of time series. *J. Time Ser. Anal.*, 13(3):209–232, 1992.

[8] G. Sugihara. Nonlinear forecasting for the classification of natiral time series. *Phil. Trans. R. Soc. Lond. A*, 348:477–495, 1994.

[9] F. Takens. Detecting strange attractors in turbulence. In *Dynamical Systems and Turbulence. Lecture Notes in Mathematics*, Springer-Verlag, 898:336–381, 1981.

[10] M.L.M. Van der Stappen. Chaotic Hydrodynamics of Fluidized Beds. Ph.D. Thesis, Delft University of Technology, The Netherlands, 1996.

25

Non-Parametric Forecasting Techniques for Mixing Chaotic Time Series

D. Guégan[1]
L. Mercier[2]

ABSTRACT
We consider mixing chaotic time series and present two prediction methods for such systems. The methods we develop are the Nearest Neighbors method and the Radial Basis Functions method. We discuss the optimal prediction horizon according to the sampling time step. We also discuss a reliable method measuring the prediction error. We illustrate our results with simulations.

25.1 Introduction

In recent years, there has been a growing interest in search of evidence of nonlinear dynamical systems and chaotic processes. One aspect of these models is their tendency to display very complex, seemingly random behavior, even when simple cases are analyzed. Consequently, it seems difficult to predict the correct behavior of a chaotic system, even if we can reconstruct the equations that generate this behavior.

Searching for nonlinear dynamics is important. Indeed, if the dynamics are nonlinear, using linear models will not lead to good predictions because non-linear systems can display highly variable phenomena that cannot be accounted for by the usual linear systems. Should the system under study be chaotic, long term prediction would be impossible but short term prediction may be possible.

An important property of chaotic systems is the mixing property. The mixing property plays an important role in the study of chaotic time series. This property is often a prerequisite for the methods used to describe, analyse and make inference on such time series. In this paper we shall focus on this class of models. We precisely describe the different notions related to this property and explain what is the chaos which verifies this property.

Then we study the problem of prediction for these models using two methods whose main properties are known for simulated chaotic time series [14]. An im-

[1]ENSAE-CREST, Timbre J120, 3 av. P. Larousse 92245 Malakoff Cedex, France, E-mail: guegan@ensae.fr
[2]CREST, Laboratoire de statistique, Timbre J340, 3 av. P. Larousse 92245 Malakoff Cedex, France, E-mail: mercier@ensae.fr

portant feature of simulated chaotic systems is that there is some exogenous noise which perturbs the system. We analyse this problem, too. The problem of interaction between noise and chaos is rather difficult and complex. There are several kinds of noise and different ways to deal with them. We can do some prewhitening of the data, we can use methods robust to noise, or we can use a method that deals simultaneously with the noise and the dynamics of the system. What is done here is closer to the second approach. Our simulations have shown that, provided the noise level is low, prediction of a chaotic system is still possible in the short term.

The two methods used here are non-parametric. The first belonging to the more general class of local methods, is the Nearest Neighbors method. The second, belonging to semi-local methods, is the Radial Basis Functions method. A third class of non-parametric methods is a global method like a multi-layer neural network, but we do not use it here.

There are many works in the literature concerning prediction of chaotic systems. An interesting review paper has been written by Lillekjendlie, Kugiumtzis and Christophersen [20]. Concerning the local approach, readers can also consult papers [6, 8], among others. Further papers describe the semi-local approach [26, 27] and the global approach [19].

Another approach could be used here to make predictions. It is the analytical approach with the help of consistent estimates of the Lyapunov exponents, but the results of this method have to be improved consistently to be used on real data potentially contaminated with noise. That is why we do not consider this method here; see for example the works of Delecroix, Guégan and Léorat [2].

We consider a chaotic time series $X(t)$, where $t \in [0,T]$, which is generated by the following differential system:

$$\frac{dX}{dt} = \Phi(X) \tag{25.1}$$

starting from some initial condition $X(0)$ in $\Re^m$. This time series is sampled with a time step h giving $N+1$ points $X_0, \ldots, X_N$ with $X_i = X(i.h)$. We denote in the following φ_h the non-linear map which represents a discrete version of the system (1.1) and which verifies the following recursive scheme:

$$X_i = \varphi_h(X_{i-1})$$

for all i and all $X(0)$. Here φ_h is some nonlinear map from D to D, where D is a compact subset of $\Re^m$ and φ_h is measurable with respect to the Borel field $B(D)$. Following paper by Eckmann and Ruelle [5], we call chaos a dynamical system which is sensitive to initial conditions and is ergodic.

Our work deals essentially with *short term* predictions. When $X_0, \ldots, X_N$ are known, we can expect to forecast $X_{N+1}, \ldots, X_{N+H}$ where the prediction horizon H is small. The future of a nonlinear system in the chaotic phase, however, is not predictable for a *mean-term* time interval because forecasting errors grow exponentially fast. We can say that we make a *long-term* prediction when the prediction horizon is bigger than several pseudo-periods (when pseudo-periods can be defined). In that case, provided the historical data are long enough, we can compute the density of the prediction if we know the invariant measure. We do not consider this case in this paper.

The remainder of this paper is organized as follows. In Section 25.2 we discuss the mixing properties and the classification that we can do inside the broad class of chaotic systems. Section 25.3 presents the two forecasting methods we use. In Section 25.4 a comparison criterion is chosen. In Section 25.5 we show the influence of the sampling time step on predictions. We present some simulations. Section 25.6 discusses the problem of the noise which pollutes the data and concludes.

25.2 Chaos and Mixing

Recall that a stationary process X is said to be ergodic if

$$lim_{T\to\infty}\sum_{t=0}^{T}\int_0^T P(A\cap\varphi^t B)dt = P(A)P(B)$$

for any A, $B \in F_X$, where F_X is the σ-field generated by X. X is said to be mixing if

$$lim_{T\to\infty}\frac{1}{T}P(A\cap\varphi^t B) = P(A)P(B)$$

for any A, $B \in F_X$. X is said to be weakly mixing if the above limit holds with t restricted to a set of density one which may depend on A and B. Alternatively, X is weakly mixing if

$$lim_{T\to\infty}\frac{1}{T}\sum_{t=0}^{T}|P(A\cap\varphi^t B) - P(A)P(B)|dt = 0$$

for any A, $B \in F_X$, which shows that weak mixing is an intermediate property between ergodicity and mixing [24].

The notion of mixing is really important one in the study of chaotic systems. Indeed, computer calculations showed that there is in general a well-defined mean behavior. Time averages of most observables along single trajectories exist and do not depend on the specific trajectory. This suggests that many of these systems support very special measures for which some sort of generalised ergodic theorem is expected to hold. The knowledge of mixing properties for chaotic systems gives us a lot of possibilities for understanding their probabilistic behavior and for obtaining convergence for the different estimates that we can construct and sometimes also rates of convergence.

Mixing properties have been established for piecewise expanding maps of the unit interval. There are many examples of piecewise expanding maps of the unit interval where even explicit expressions for the ergodic measures are known; see the work of Lasota and Mackay [18] for a review of these maps. For some piecewise monotonic and C^1 maps, it has been shown that the mixing coefficients in the weak Bernouilli property decrease at an exponential rate. In particular, for the r-adic map, see [16, 30] and for more details, see [1, 17].

The extension of the theory into high dimensions faces serious difficulties. These are connected with the problem of finding the appropriate functional spaces for the Frobenius Perron operator (for a good introduction to this operator, see [18]), when its spectrum can be understood satisfactorily. This has been proved for a class of expanding maps of the unit cube in R^k by Mayer [22], and examples have been given in this article. A review concerning the results on functionals of such models can be found in [3].

Another class of chaotic systems denoted as Smale's Axiome A systems can be considered. They are systems characterized by the property that they partly expand and partly contract directions uniformly. The contraction is responsible for their very peculiar behavior. For these systems the existence of an asymptotic measure has been shown. Their support is often a complicated Cantor-like set called a strange attractor, see Bowen and Ruelle for complete development of the subject.

Even if the mixing property is established for certain classes of transformations, we can conjecture that it remains valid for a broader class of chaotic systems. We shall consider the class of mixing chaotic time series throughout this paper. Recent developments on this subject can be found in the thesis by Mercier [23].

25.3 Forecasting Methods

In this section, we present briefly the two prediction methods we use. We consider here a chaotic time series without noise. The role of noise will be discussed in Section 25.4.

For real data, the state vector $X(t)$ introduced in (25.1) and the true map Φ are unknown but we have a transformation (for instance a projection) of $X(t)$ denoted x_t. Here we assume that x_t is real. Using Taken's theorem [29], we embed this series in dimension m to have a diffeomorphism of the true series. Now we study the series $(x_t, x_{t-\tau}, \ldots, x_{t-\tau(m-1)})$ where m is called the embedding dimension and τ the lag. To simplify the notations, we assume that this vector is exactly $X(t)$, but what we really have is only a diffeomorphism of this series.

The choices of m and τ are difficult, and there is no standard way to estimate these values directly from the data. We also suppose that the observed series is sampled with a fixed time step h. Now the model under study can be written as

$$\begin{pmatrix} x_{N+h} \\ x_{N+h-\tau} \\ \vdots \\ x_{N+h-(m-1).\tau} \end{pmatrix} = \varphi_h \begin{pmatrix} x_N \\ x_{N-\tau} \\ \vdots \\ x_{N-(m-1).\tau} \end{pmatrix}$$

Our approach is new because the sampling time step h and the lag τ can take different values.

Now the problem is to estimate the map $\widehat{\varphi}_h$. This is done with the two following methods. From now on we denote

$$X_i = \begin{pmatrix} x_i \\ x_{i-\tau} \\ \vdots \\ x_{i-(m-1).\tau} \end{pmatrix}$$

25.3.1 Nearest Neighbours

This method is a local one. It is relatively easy to implement. If we want to use only one neighbor, we search among $X_0, \ldots, X_{N-1}$ for the point X_i which is the closest to X_N (it is this point that we want to approximate) in the sense of a certain norm (for example, the euclidean norm). Then we define

$$\widehat{\varphi}_h(X_N) = X_{i+1}$$

The future of X_i is used to forecast the future of X_N. It is possible to use more than one neighbor, to take more than one past value for each one, or to consider various weights. We use K neighbors $X_{i_1}, ..., X_{i_K}$ of X_N, a weighting function w and a norm $\| . \|$ of $\Re^m$. Then we get

$$\widehat{\varphi}_h(X_N) = \sum_{k=1}^{K} w(\|X_N - X_{i_k}\|)\ X_{i_k+1}$$

Here, when more than one neighbor is implied, we use the method chosen by Finkenstädt and Kuhbier [8]. That is, we assume that the weight of each neighbor is inversely proportional to the exponential of its distance to X_T.

25.3.2 *Radial Basis Functions*

With this semilocal method, we approach φ_h with a set of radial basis functions. This method is often related to neural networks, but its expression is a lot simpler. We construct $\widehat{\varphi}_h$ in the following way :

$$\widehat{\varphi}_h(X_N) = \sum_{c=1}^{C} \lambda_c w_c(\|X_N - Y_c\|)$$

The functions w_c are radial functions like a Gaussian or a multiquadric function. The choice of the function is arbitrary as for the nearest neighbors weights before. Here we choose inverse multiquadric functions for computational efficiency:

$$w_c(r) = \frac{1}{\sqrt{r^2 + r_c}}$$

where r_c is the radius of $w_c(r)$. The points Y_c are the centers of the radial functions.

25.4 Criteria for Comparison

There are several methods for comparing the efficiency of the predictors. The most classical is the mean square error of the prediction for a given prediction time. We can also compare the coefficients of correlation between the real values and the predictors [28]. We can find other procedures in [4].

Actually, for chaotic systems, it seems reasonable to consider different criteria with respect to the prediction barrier. We can consider the previous methods but it also seems interesting to find out what is the limit of the "small term" for chaotic systems. For the systems we are studying, we can conjecture that the prediction will get worse as the horizon increases. We define the prediction horizon as the date when the prediction differs by more than some small distance d from the true value. We use this last criterion here. Note that this prediction horizon can also be called predictability time.

25.5 Influence of the Sampling Time Step

25.5.1 An Example: The Lorenz Attractor

We simulate here a well known chaotic system to show the role of the sampling time step for predictions.

The time series is generated by the following differential equations

$$\begin{aligned}\frac{dx}{dt} &= 16\,(y\,(t) - x\,(t)) \\ \frac{dy}{dt} &= x\,(t)\;(45.92 - z\,(t)) - y\,(t) \\ \frac{dz}{dt} &= x\,(t)\;y\,(t) - 4\,z\,(t)\end{aligned}$$

We choose a distance d equal to 5% of the diameter of the attractor. For values of N (number of points) between 100 and 10000 and h (sampling time step) between 0.0001 and 1, we forecast $X_{N+h}, X_{N+2.h}, \ldots$ and calculate the prediction horizon Hh for each forecast. For each pair (N, h), 100 series are simulated with various initial conditions. The following figure shows the prediction horizon plotted against the $\log(N)$ and $\log(h)$. We can observe the following:

1. For a fixed h, the prediction horizon Hh increases with the sample size N, but it seems to reach a maximum value.

2. For a fixed N, the prediction horizon Hh increases as the sampling time step h increases (more orbits are build), but after a threshold the prediction horizon Hh diminishes (the points are more and more random on the attractor). But the existence of a unique optimal time step h^* is not proven.

This figure was drawn using the Nearest Neighbors predictor but the Radial Basis Functions yield the same shape with a slightly better prediction horizon. The same phenomenon has been observed with other chaotic systems.

25.5.2 Optimal Sampling Time Step

The role of the sampling time step h seems to be very important in the case of chaotic time series. Should the same phenomenon be observed for financial intraday data, this would give a good basis for studying this kind of data.

We illustrate this phenomenon in the case of one nearest neighbour prediction. In this case, the short term error for chaotic time series is given by

$$\left\| X_{N+j} - \widehat{X}_{N+j} \right\| \approx \|X_N - X_{i_1}\|\; e^{\lambda.j.h}$$

with a first-order approximation for a j step ahead prediction. Here λ is the local Lyapunov exponent of the system in X_N. Under this approximation the predictability time Hh verifies $\|X_N - X_{i_1}\|\; e^{\lambda\, H\, h} = d_0$. Thus

$$Hh = \frac{1}{\lambda}\log(d_0) - \frac{1}{\lambda}\log(\|X_N - X_{i_1}\|)$$

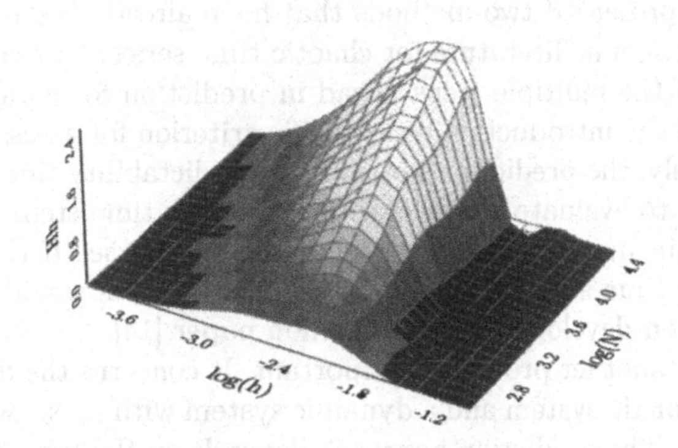

FIGURE 25.1. Average prediction horizon for the Lorenz system. h denotes the sampling time step, N is the number of points in the series, Hh is the prediction horizon (in seconds)

We are looking for the optimal time step h^* that maximizes Hh for a fixed amount of data N.

We denote $X(t_N)$ the intersection of the closest trajectory to X_N and the hyperplane orthogonal to the direction $\vec{\Phi}(X_N)$, γ_T the distance $\|X_N - X(t_N)\|$, m_l the local embedding dimension, and δ_h the distance between $X(t_N)$ and the closest observed point on this trajectory which is assumed to be X_{i_1}. γ_T is a function of T and scales as $T^{-\frac{1}{m_l-1}}$ with first-order approximation. We get

$$\gamma_T = \Gamma\, T^{-\frac{1}{m_l-1}}$$

The distance δ_h is on average equal to $\frac{h}{4}\left\|\vec{\Phi}(X_N)\right\|$. It follows that

$$\|X_N - X_{i_1}\| \approx \sqrt{\gamma_T^2 + \delta_h^2}$$

When we consider the derivative $\frac{\partial Hh}{\partial h}\Big|_N$ equal to zero, this implies that

$$\left.\frac{\partial \gamma_T}{\partial h}\right|_N \gamma_T + \left.\frac{\partial \delta_h}{\partial h}\right|_N \delta_h = 0$$

Thus,

$$-\frac{\Gamma^2}{m_l - 1} N^{-\frac{2}{m_l-1}} (h^*)^{-\frac{m_l+1}{m_l-1}} + \frac{\left\|\vec{\Phi}(X_N)\right\|^2}{16} h^* = 0$$

and finally

$$h^* = \left(\frac{4\,\Gamma}{\left\|\vec{\Phi}(X_N)\right\| \sqrt{m_l - 1}}\right)^{\frac{m_l-1}{m_l}} N^{-\frac{1}{m_l}}$$

The quantities Γ, λ and m_l depend on the local properties of the map. Rough approximations have been made here but they account fairly well for the observed horizon map and its maximum. For more details see [23].

25.6 The Role of Noise

In this paper we presented two methods that have already been tested in the economic and non-economic literature for chaotic time series. Our contribution develops in a new way the multiple steps ahead in prediction for highly nonlinear time series. It involved the introduction of a specific criterion for assessing the quality of predictions, namely, the prediction horizon (or predictability time) and it provided a suitable setting to evaluate the role of the sampling time step.

One of our main motivations in this research was to see if the existence of an optimal sampling time step for purely deterministic series would appear with real data. This has been developed in a companion paper [14].

With real data, another problem is important. It concerns the difference between a chaotic deterministic system and a dynamic system with noise, which is a stochastic model [10, 11]. The prediction approach depends on the type of system at hand [14, 15].

In the deterministic case the data $\tilde{X}_i$ that we observe corresponds to chaotic deterministic system with a measurement error that we can describe in the following way:

$$\tilde{X}_{i+1} = \phi_h (X_i, \epsilon_i)$$

A possible representation of this is $\tilde{X}_i = X_i + \epsilon_i$, and the underlying process is given by $X_i = \varphi_h (X_{i-1})$. Here, we predict X_{N+H}.

In the stochastic case, the data that we observe are generated by

$$X_{i+1} = \phi_h (X_i, \epsilon_i)$$

and the noise directly affects the underlying dynamics. In the latter case, we are close to what is called "noisy chaos" in the literature and which is in fact a stochastic system. The prediction concerns X_{N+H} which is completely different from the previous one. The latter can be computed in terms of conditional expectation, which is not the case for the former.

Here we suppose we are in the first case, that is, a chaotic deterministic system observed with a measurement error. We use two methods to estimate φ_h with the points $\tilde{X}_0, ..., \tilde{X}_N$ instead of the true points $X_0, ..., X_N$.

Acknowledgments: We are grateful to C. Dunis for giving us the data exchange rates used in this study.

25.7 References

[1] R. Bowen. Bernouilli maps of an interval. *Israel J. Math.*, 28:298–314, 1982.

[2] M. Delecroix, D. Guégan, and G. Léorat. Determining Lyapunov Exponents in Deterministic Dynamical Systems. *Computational Statistics* 12:93–107, 1997.

[3] M. Denker and G. Keller. Rigourous statistical procedures for data from dynamical systems. *Journ. of Stat. Physics*, 44:67–93, 1986.

[4] F.X. Diebold and R.S. Mariano. Comparing Predictive Accuracy. *Journal of Business & Economic Statistics*, 13 253–263, 1995.

[5] J.P. Eckmann and D. Ruelle. Ergodic theory of chaos and strange attractors. *Reviews of Modern Physics*, 57:617, 1985.

[6] J.D. Farmer and J.J. Sidorowich. Predicting chaotic time series. *Physical Review Letters* 59:845–848, 1987.

[7] J.D. Farmer and J.J. Sidorowich. Exploiting chaos to predict the future and reduce noise. *Evolution, Learning, and Cognition*, ed. Y. C. Lee, World Scientific, Singapore, pp. 277, 1988.

[8] B. Finkenstadt and P. Kuhbier. Forecasting nonlinear economic time series: A simple test to accompany the nearest neighbor approach. *Empirical Economics* 20:243–263, 1995.

[9] J. Geweke. Inference and forecasting for deterministic nonlinear time series observed with measurement error. Preprint, 1988.

[10] D. Guégan. How can noise be brought out in dynamical chaos. *Proceedings of the Workshop: Methods of Non Equilibrium Processes in Economics and Environment Sciences*, Matrafured (Hungary), 1996.

[11] D. Guégan. Stochastic versus deterministic systems. *Working Paper INSEE 9438*, 1994.

[12] D. Guégan and G. Léorat. What is the good identification theory for noisy chaos. *Working Paper INSEE 9619*, 1996.

[13] D. Guégan and L. Mercier. Rising and falling prediction in financial intra-day data. *Proceedings of the 3rd International Conference Sponsored by Chemical Bank and Imperial College: Forecasting financial markets*, 1996.

[14] D. Guégan and L. Mercier. Methods and Comparisons Using Simulations. To appear in *Signal Processing*, 1997.

[15] D. Guégan and L. Mercier. Predictions for financial data using chaotic methods. To appear in the *European Journal of Finance.*

[16] F. Hofbauer and G. Keller. Ergodic properties of invariant measures for piecewuse monotonic transformations. *Math. Z.*, 180:119–140, 1982.

[17] A. Lasota and J. Yorke. Asymptotic periodicity of the iterates of Markov operators. *Trans. Am. Math. Soc.*, 286:751–764, 1984.

[18] A. Lasota and Mackay. *Chaos, Fractals and Noise.* Springer-Verlag, 1987.

[19] B. Lebaron and A.S. Weigend. Evaluating neural network predictors by bootstrapping. Preprint, 1995.

[20] B. Lillekjendlie, D. Kugiumtzis, and N. Christophersen. Chaotic time series: System identification and prediction. Preprint, 1994.

[21] F. Lisi, O. Nicolis, and M. Sandri. Combining Singular-Spectrum Analysis and neural networks for time series forecasting. *Neural Processing Letters*, 2(4):6–10, 1995.

[22] D.H. Mayer. Approach to equilibrium for locally expanding maps in R^k. *Comm. Math. Phy.*, 95:1–5, 1984.

[23] L. Mercier. Séries temporelles chaotiques appliquées à la finance. *Thèse de doctorat Paris IX*, 1998.

[24] K. Petersen. *Ergodic Theory.* Cambridge University Press, Cambridge, 1983.

[25] J. Scheinkman and B. LeBaron. Nonlinear Dynamics and Stock Returns. *J. of Business*, 311–337, 1989.

[26] L.A. Smith. Identification and prediction of low dimensional dynamics. *Physica D*, 58:50–76, 1992.

[27] K. Stokbro and D.K. Umberger. Forecasting with weighed maps. *Nonlinear Modeling and Forecasting*, M. Casdagli and S. Eubank, eds., Addison-Wesley, Reading, MA, 1992.

[28] G. Sugihara and R.M. May. Nonlinear forecasting as a way of distinguishing chaos from measurement error in time series. *Nature*, 344:734–741, 1990.

[29] F. Takens. Detecting strange attractors. *Dynamical Systems and Turbulence*, Lecture Notes in Mathematics 898:366–381, 1981.

[30] S. Wong. A central limit theorem for piecewise monotonic mappings of the unit interval. *Ann. of Probab.*, 7:500–514, 1979.

Part IV

SPEECH AND BIOMEDICAL SIGNAL PROCESSING

26

Signal Analysis and Modelling for Speech Processing

Gernot Kubin[1]

ABSTRACT
This chapter presents a survey of standard and advanced methods for the analysis and modelling speech signals. First it introduces several speech processing functions as part of voice communication systems technology and proceeds to a brief description of human speech production. From this, a two-tier physical model of speech emerges which embraces the speech organ movements at the articulatory tier and the coupled aerodynamic flow and sound propagation at the aero-acoustic tier. Both of these *physical* tiers appear as separate components in most *computational* speech signal models. Their discussion addresses both the standard view of linear short-time stationarity and more advanced concepts from non-stationary processes (underspread processes, cyclostationarity) and non-linear systems (neural networks, non-linear oscillators).

26.1 Speech Processing for Voice Communication Technology

Among the many forms of human communication, spoken language is maybe the most natural and heavily used in our daily lives. Voice communications technology started out with speech transmission by analog telephony and with speech storage by the invention of the phonograph, both in the last century. Electronic speech synthesis became possible in the 1930's, and studies into automatic speech recognition followed later. However, only over the last decade, advances in digital signal processing (DSP) methods have teamed up with rapid increases in the computational power of DSP hardware and concomitant decreases in cost to move voice communications technology from scientific research into applied telecommunications. This process is being accelerated by an ever-increasing demand for high-quality, easy-to-use communications services where speech is expected to provide the key to a human-oriented interface.

Voice communication technology can be defined as the application of a variety of speech processing functions to telecommunication systems (in particular to telephony, multimedia systems, and personal information terminals for the global

[1]Institute of Communications and High-Frequency Engineering, Vienna University of Technology, Gusshausstrasse 25/389, A–1040 Vienna, Austria, E-mail: G.Kubin@ieee.org

information infrastructure). These functions include the coding, synthesis, recognition, and enhancement of speech as well as speaker recognition and spoken language recognition. For a comprehensive survey of the state-of-the-art, see the compilation in [24]. More specific reviews are provided by [19, 31, 26, 21, 35, 32, 54, 11, 56] and applications are discussed in [64, 65, 70].

Relationships among Speech Processing Functions

The block diagram in Figure 26.1 shows how two human users can communicate with each other or with a machine. Depending on the amount of processing, speech coding can operate at the waveform, parameter, or speech segment level, with a corresponding dramatic decrease in bit rate. If speech analysis is pushed further to speech recognition, a text representation of the speech input is obtained and can be either processed by a machine (e.g., in a dialog system) or automatically translated. The latter is the research objective for automatic translating telephony which should allow for a telephone conversation between speakers of different languages. To this end, the translated input has to be re-synthesized in the output language, a common task in text-to-speech systems which drive parametric control units and signal synthesizers from textual inputs.

Speech enhancement basically follows the processing path of speech coders at the parametric level or at the segment level (in hidden Markov model based systems); speaker recognition and spoken language recognition basically follow the processing path of speech recognition.

This discussion illustrates that most speech processing functions are realized with the same set of basic signal processing algorithms which directly reflect the modelling assumptions about the acoustic waveform generation processes and the linguistic processes at a symbolic level. These will be specified in the next section.

26.2 A Two-Tier Model of Speech

26.2.1 Model Structure

A model of human speech (the so-called "speech chain") should span a range of levels or tiers from the mental processes that govern *what we want to say* over the physiological mechanisms *how we speak* to a representation of the *audible speech utterance.* Leaving aside the higher processing levels that essentially operate on linguistic symbols[2], we have to focus on the two-tiered structure shown in Fig. 26.2.

The Articulatory Tier

This upper tier corresponds to the mechanics of speaking and models the kinematics/dynamics of speech organs or articulators (such as lips, tongue, or velum) which move under nervous control. These movements or *speech gestures* are organized in

[2]Of course, these levels are critical for speech understanding or natural language generation systems, but they barely touch the issues of speech signal processing per se.

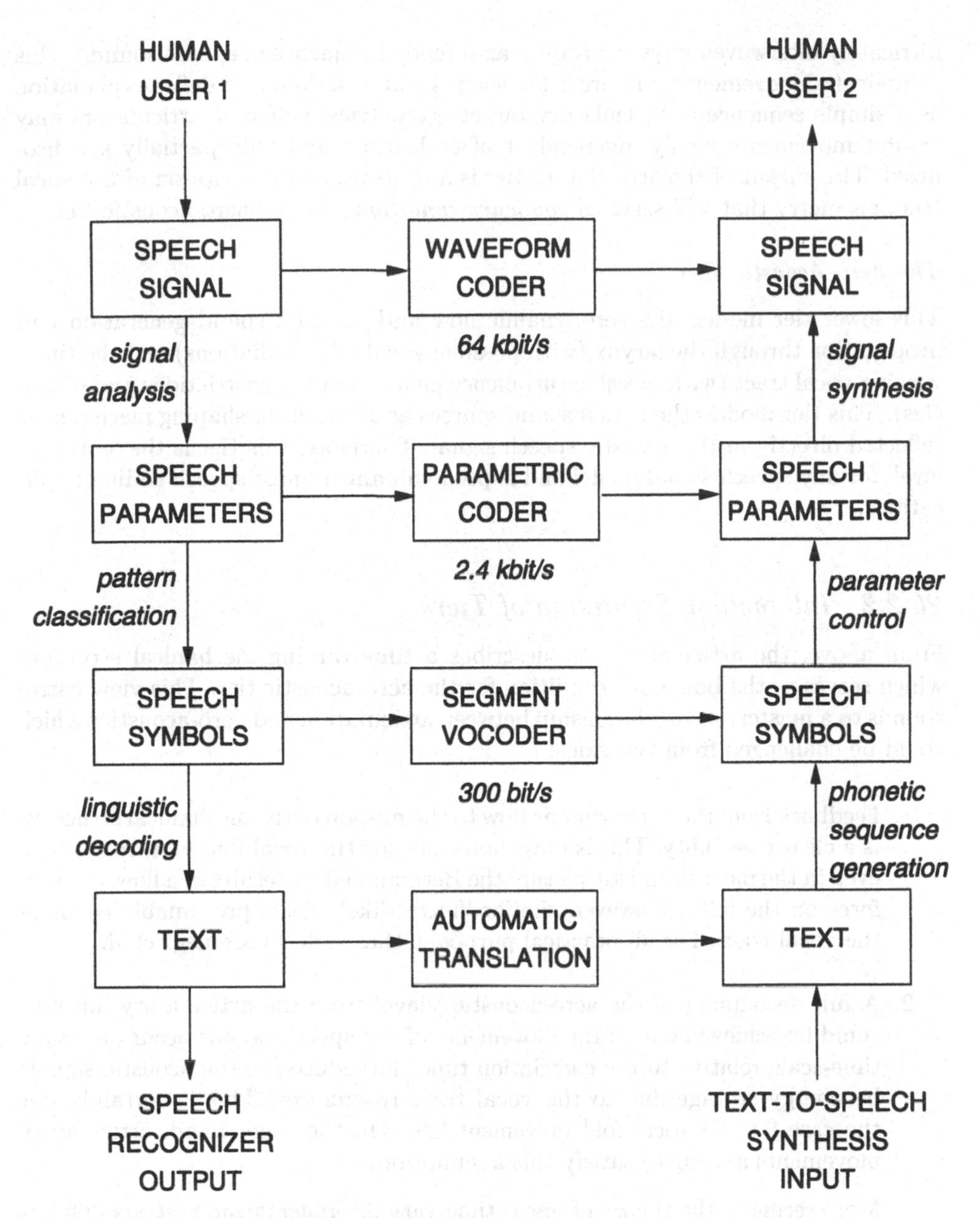

FIGURE 26.1. Speech processing functions and their mutual relationships

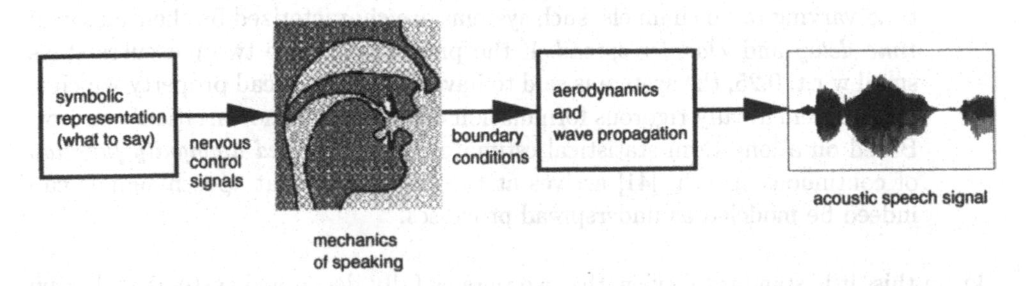

FIGURE 26.2. A two-tier physical model of human speech production

intricately interwoven ways to produce an intended sequence of speech sounds. This complicated movement structure is known as ***co-articulation*** and defies explanation as a simple sequence of articulatory target geometries. Different articulators may exhibit movements highly independent of each other and only partially synchronized. The output of the articulatory tier is a time-varying description of the vocal tract geometry that will serve as ***boundary conditions*** for the aero-acoustic tier.

The Aero-Acoustic Tier

This lower tier models the aerodynamic flow and acoustic sound generation and propagation through the larynx (with possible vocal fold oscillations) and the time-varying vocal tract (with possible turbulence generation at constrictions and obstacles). This tier models the actual sound sources and waveform shaping mechanisms reflected directly in the acoustic speech signal. Therefore, this tier is the 'entrance level' for any speech signal model and is predominant in most speech coding applications.

26.2.2 Interaction/Separation of Tiers

From above, the articulatory tier describes a time-varying mechanical structure which serves as the boundary condition for the aero-acoustic tier. This view corresponds to a master/slave relationship between articulation and aero-acoustics which could be challenged from two sides:

1. Feedback from the aerodynamic flow to the motion of the mechanical structure is a clear possibility. This is most notorious for the vocal fold vibration where even in the most simplistic picture the Bernoulli effect results in a flow-induced force on the folds. However, similar 'flutter-like' effects presumably occur in the vocal tract. For all practical purposes, these effects are neglected.

2. A full decoupling of the aero-acoustic 'slave' from the articulatory 'master' could be achieved only if the movements of the speech organs occur on a slow time-scale relative to the correlation times introduced in the acoustic signals by energy storage due to the vocal tract resonances. This is certainly not the case for the vocal fold movement [29]. On the other hand, articulatory movements appear to satisfy this assumption.

   More recently, the theory of linear time-varying ***underspread systems*** [39] has been applied to speech modelling [41]. With a terminology borrowed from time-varying radio channels, such systems are characterized by their maximal ***time delay*** and ***Doppler spread.*** If the product of these two measures stays small w.r.t. 0.25, the system is said to have the underspread property which is the mathematically rigorous formulation of the 'slow time-variation' concept. Based on a long-term statistical estimate of the ***expected ambiguity function*** of continuous speech, [41] arrives at the conclusion that speech signals can indeed be modeled as underspread processes.

From this, it is standard to view the two tiers as fully decoupled systems and, even more, to disregard any Doppler-like effects on the aero-acoustic tier.

26.3 The Articulatory Tier

26.3.1 Standard View: Frame-by-Frame Processing

The underspread characterization of speech provides some justification for the standard model of the time-variation found on the articulatory tier. In this model, the statistical properties of speech are presumed constant over short time windows or ***frames,*** typically some 20 to 30 ms long. This frame-by-frame model results in a staircase approximation of the temporal evolution of the signal statistics and model parameters even if adjacent analysis frames are allowed some overlap to minimize the effect of discontinuities. This approach is the basis for the ***nonparametric spectrogram analysis*** of speech which was established over fifty years ago [62], an example of which is given in Fig. 26.3. It is also the basis for several ***parametric*** speech models such as the ***linear predictive vocoder*** [3] which uses frame-by-frame or block-adaptive parameter estimation algorithms [51]. In both the parametric and non-parametric case, the choice of a ***window*** for the definition of analysis frames is mostly based on experience from spectral estimation for stationary processes, emphasizing the trade-offs between spectral resolution and window length and between spectral leakage and window shape.

26.3.2 Non-Stationary Aspects

Resampling and Filtering

A first attempt to improve the tracking of articulatory movements in speech signal models is to use frames with maximal overlap, i.e., to shift a frame one sample at a time. This allows an efficient recursive implementation of the signal processing operations and it increases the temporal resolution of the analysis results. However, it ***does not necessarily improve the tracking performance.*** For weighted recursive least-squares parameter estimation, it has been shown in [57, 58] that the weighting function operates as a (low-pass) filter on the 'true' parameter trajectories. Obviously, the thus filtered estimated trajectories can be subsampled to a certain extent without introducing aliasing distortions which brings us back to a representation with overlapping blocks. One can demonstrate that ***the windowed covariance method for linear prediction analysis [69] is exactly equivalent to a subsampled recursive least-squares algorithm using the window as its weighting function.*** In this context, the choice of the ***window*** is governed mostly by its ability to suppress spurious pitch-induced fluctuations in the parameter trajectories [44] as illustrated in the following.

Figure 26.4 shows the magnitude and the group delay of transfer functions associated with a rectangular window and a Hamming window, respectively. They have been normalized so that they behave identically for stationary data: bias-free estimates (unity gain at 0 Hz) and identical estimation variance (identical sum of squared window samples [61]).

Therefore, a Hamming window of 190 samples is equivalent to a rectangular window of 138 samples. At an 8 kHz sampling rate, the window lengths match the typical values around 20 ms as used for speech analysis. The rectangular window has a relatively slow decay toward higher frequencies whereas the Hamming window has over 40 dB attenuation for all frequencies above 80 Hz. From this,

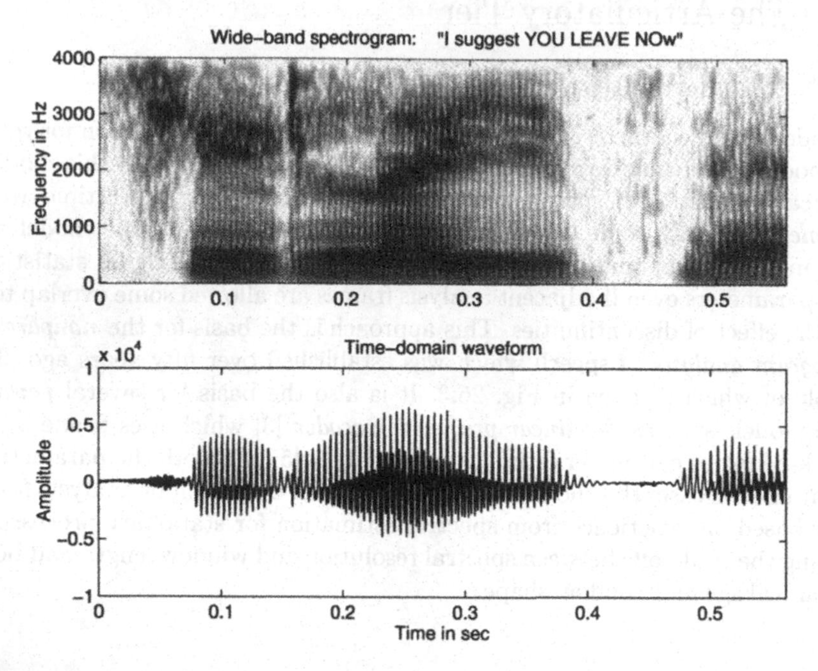

FIGURE 26.3. Wide-band spectrographic analysis of speech: The lower graph shows the time-domain waveform of the utterance *"... you leave now"* and the upper graph the spectrogram, i.e., a time-frequency distribution of the signal's energy where darker areas correspond to higher energy densities. The analysis is based on a short time-domain window which allows recognizing high-resolution temporal details, such as the periodic vertical striations due to the periodic excitation of voiced speech. On the other hand, this results in a wide frequency-domain smoothing of the spectrogram so that only the dominant resonance frequencies (the *formant frequencies)* are visible but not the harmonics of the fundamental frequency. Unvoiced speech gives rise to irregular, noise-like structures with less pronounced spectral peaks, see the beginning of the utterance

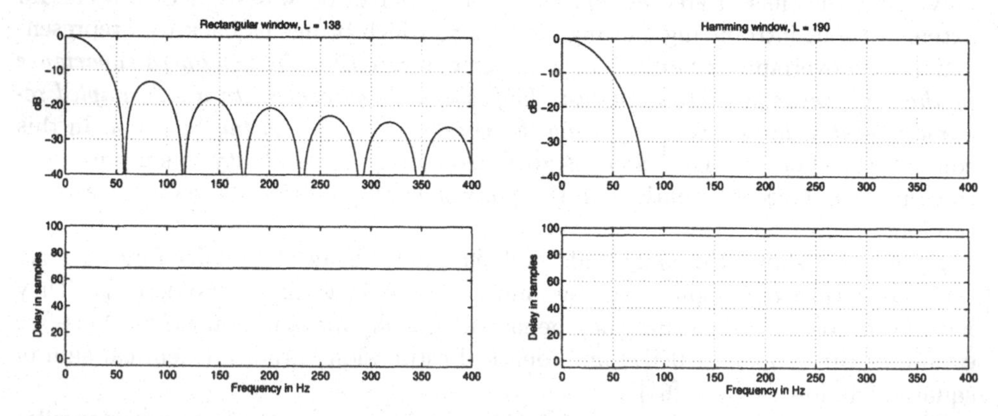

FIGURE 26.4. Frequency response magnitude and group delay of two 'noise-equivalent' windows used for parameter tracking with windowed least-squares methods: Rectangular window of 138 samples length (left) and Hamming window of 190 samples length (right). Both plots assume an 8 kHz sampling frequency

a Hamming-windowed covariance method is automatically robust w.r.t. the frame position relative to the pitch cycle. In such non-synchronized analyses, parameter estimates obtained from the rectangular-windowed covariance method will exhibit significant fluctuations with the fundamental frequency or, after subsampling to 40 or 50 Hz (block processing), at an aliased frequency [69].

In another context, filtering of frame-wise speech parameter estimates is used to get closer to the underlying kinematics of the speech gesture (e.g., first temporal differences in the so-called Delta-Cepstra [32]) or to improve the robustness of the estimates (RASTA processing [30]).

Gabor Frames and Hypermodels

In this subsection, we understand by ***frame representation*** a signal decomposition or expansion into sets of signals which are not necessarily orthogonal but which allow a numerically stable reconstruction of the original signal from its expansion coefficients [39]. The classical *Gabor representation* is specified by a *lattice* or *tiling* in the time-frequency plane and an associated *window*. Since speech signals are modeled as underspread processes, both the lattice and the window can be optimized to achieve maximal decorrelation of the expansion coefficients. This optimization problem has been solved approximately in [38, 41] and results in Gaussian-like windows on a hexagonal (non-separable) time-frequency lattice [40].

On the parametric side, [28, 34, 5, 20, 43, 49] and several others have shown that a priori hypermodels of the temporal evolution can improve non-stationary parameter estimation. However, again this improvement is achieved only if the non-stationarity satisfies the underspread condition, in which case their performance is similar to frame representations matched to the signal statistics. An application of a vector-autoregressive hypermodel to speaker identification has been discussed in [12].

Non-Deterministic Time Evolution: Hidden Markov Models

For completeness, we mention hidden Markov models as a non-deterministic approach to modelling the temporal evolution on the articulatory tier. Excellent surveys of this important model class are available elsewhere [63, 32]. We briefly note that, in its standard form, it is restricted to abrupt jumps between stationary states (just like the standard frame-by-frame model) but that refinements do exist that attempt to incorporate articulatory kinematics within model states [22, 27].

26.3.3 Non-Linear Aspects

Temporal Decomposition

In an attempt to approximate the overlapping and asynchronous nature of speech gestures, the temporal decomposition technique [2, 10, 73, 53] represents the multivariate parameter trajectories of speech by a linear combination of (overlapping) scalar interpolative functions weighted by vector-valued coefficients (so-called target vectors). Unlike signal expansions into orthogonal bases or into frames, this method includes the signal-adaptive optimization of the interpolative functions themselves and, therefore, is a highly non-linear signal model. Its practical application is difficult due to the added complexity of the very interpolative functions which may require a significant share of the bit rate in speech coding and due to the fact that co-articulation cannot be reduced to simple interpolation between parameter vector targets.

Time-Delay Neural Networks

Although a neural network by itself may not be well suited to represent a continuous stream of data as needed on the articulatory tier, it can still be used to capture some of the kinematics of the underlying speech gesture. To this end, a dynamic memory is included in the neural network, either by internal, delayed, recurrent branches (as in recurrent networks) or by forward/open-loop chains of delays which represent local temporal neighborhoods and which can be organized into a hierarchical cascade. The latter approach has been implemented with good success in the time-delay neural networks of [75]. For a further discussion, see [54].

Non-Linear Dynamics

The hypothesis that the temporal evolution on the articulatory tier is governed by non-linear dynamics receives further support from physiological modelling. The control loops for many human body movements (locomotion, human gait etc.) and other physiological processes form non-linear oscillators [50, 4, 66] and speech production movements are no exception to this [33, 72].

26.4 The Aero-Acoustic Tier

26.4.1 Standard View: Wold's Decomposition for Stationary Processes

In the standard frame-by-frame approach on the articulatory tier, speech is regarded as a stationary process within each frame. Stationary processes always allow Wold's decomposition into two additive components [59], i.e.,

1. a ***regular component*** represented as an uncorrelated noise filtered through a causal and stable linear filter (IIR in general) and
2. a ***singular component*** which consists essentially of a sum of sinusoids; this sum can be countable infinite, and the sinusoids need not be harmonically related.

Linear Predictive Vocoder

The simplest implementation of this decomposition is found in the linear predictive vocoder [3]. The simplifications are three-fold:

- A hard switch defines a complete speech frame as either regular or singular corresponding to the *voiced/unvoiced* feature of human speech.
- The singular component is restricted to harmonically related sinusoids such that it can be generated by passing a *periodic delta pulse train* through a causal and stable linear filter.
- The linear filter used in the representation of both the regular and the singular components is required to be *minimum phase* such that it has a stable and causal inverse.

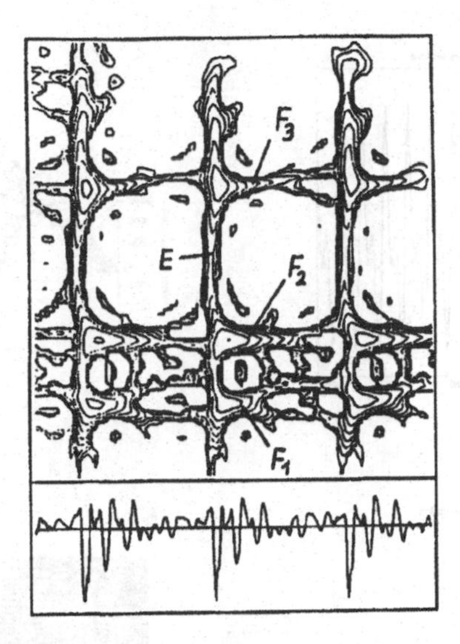

FIGURE 26.5. Wigner distribution analysis of three pitch periods of sustained vowel [a:], after [76]: Excitation impulse E, formants F_1, F_2, and F_3, horizontal time axis (total duration 30 ms), vertical frequency axis (total frequency range 4 kHz) for the contour plot in the upper graph, time-domain waveform in the lower graph

The last property guarantees that both the linear prediction analysis and synthesis filters are stable, a key requirement for coding applications. On the other hand, it destroys the original phase information in the singular component.

Sinusoidal Model

The sinusoidal model [52] follows a complementary approach toward simplification:

- The regular component is generated by a sum of sinusoids with randomized phases (following spectral representation theory; see [59]).
- No hard switch between regular and singular components is required. A frame can be modeled by an additive mixture of the two by controlling the phases of (groups of) sinusoids accordingly.
- As no filtering operations are involved, stability or minimum-phase properties are not an issue. Therefore, the original phases of the singular component's sinusoids can be maintained (if desired/affordable).

26.4.2 Non-Stationary Aspects

Careful analysis, e.g., by high-resolution time-frequency analysis with the smoothed pseudo-Wigner distribution [76] reveals that even sustained, voiced speech changes its spectral distribution rapidly over the duration of a short time frame (20 to 30 ms). With every new pitch cycle, there is a new broad-band excitation impulse that excites several formants recognized as narrow-band decaying signal components; see Fig. 26.5.

This pitch-synchronous variation of signal statistics [18] calls for the appropriate mathematical representation found in the *theory of cyclostationary processes* [25].

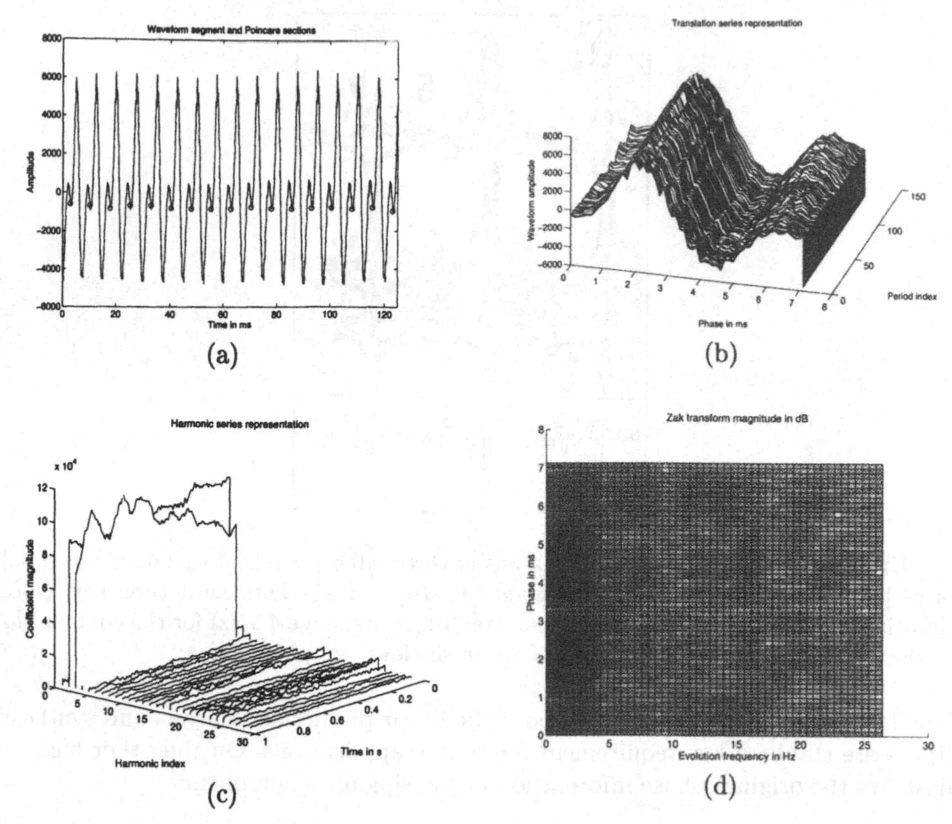

FIGURE 26.6. Cyclostationary analysis of sustained vowel [i:] (a) periodic waveform segmentation using Poincaré sections [47] (the circles indicate synchronization marks); (b) translation series representation or polyphase decomposition; (c) harmonic series representation; and (d) Zak transform [16]

Its application to speech has been discussed in [46] and leads to a unified picture for various pitch-synchronous speech processing methods (such as PSOLA [55]) with waveform interpolation [36] and decomposition [37] techniques used in low bitrate speech coding. An example for cyclostationary speech analysis is shown in Fig. 26.6.

The first graph (a) shows how the signal is segmented into (almost) periodic segments using a synchronization technique known as Poincaré sections [47]. These segments are stacked in a waterfall-like display, the *translations series representation* or polyphase decomposition (b). There, each signal sample is indexed by an integer period index and a continuous phase variable that runs from zero to the length of a pitch period. The period-to-period variability is clearly seen on top of a strong cyclic mean. The *harmonic series representation* (c) offers a dual description of the same data where the coefficients of the harmonic series become stationary stochastic processes. Only for strictly periodic data, these processes would degenerate to constants. The *Zak transform* [16] (d) is obtained from the translation series representation by a Fourier transform w.r.t. the discrete period index. It distributes the signal energy on a finite-support rectangle extending over all cycle phases and evolution frequencies (where darker areas stand for more energy just as in the spectrogram). In our sustained speech example, most of the energy is concentrated below 5 Hz, i.e., the period-to-period variability is described as a low-pass process.

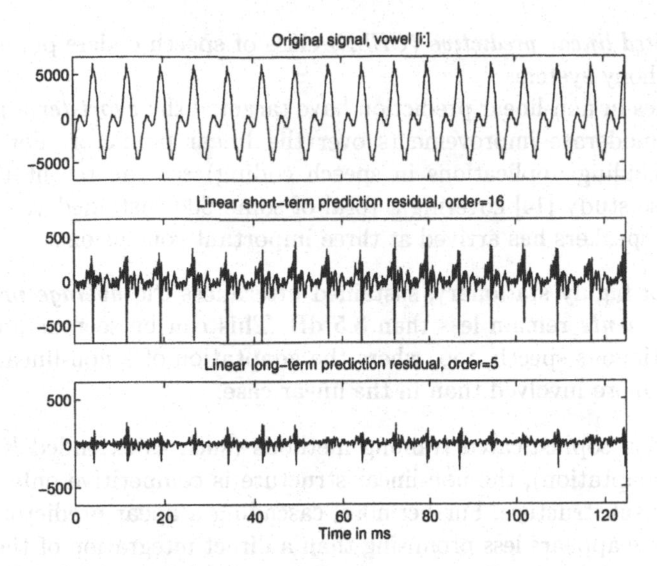

FIGURE 26.7. Linear prediction errors or residuals: Original (top), short-term prediction residual (middle), long-term prediction residual (bottom)

26.4.3 Non-Linear Aspects

Non-linearity is clearly involved in the ***physics*** of speech production. The vocal folds constitute a ***non-linear dynamic system*** capable of ***self-sustained oscillation***, and ***turbulence*** is a primary or secondary sound source in unvoiced and voiced speech, respectively. Non-linear ***computational models*** at the signal waveform level have appeared only in the past few years and can be classified into ***predictive signal models*** and ***oscillator models***.

Predictive Signal Models

Non-linear prediction should improve on existing linear prediction techniques for speech signals which are illustrated in Fig. 26.7. The original waveform (top diagram) is a segment of the sustained vowel [i:] sampled at 8 kHz. The prediction error or residual of the linear least-squares one-step ahead predictor of order 16 is shown as the middle waveform. Note from the different scaling of the ordinates that its power is much reduced w.r.t. the original waveform (***prediction gain***). The waveform shape resembles a periodic impulse train which is the crude approximation used in the conventional LPC vocoder [3]. More refined linear predictive techniques cascade this ***short-term predictor*** with a ***long-term predictor*** that attempts to extrapolate over a full pitch period. As this period is generically no integer multiple of the sampling frequency, a ***fractional*** prediction delay [42] is usually achieved with a long-term predictor order greater than one (in our example, it is set to five). The bottom graph in Fig. 26.7 shows the residual of this cascaded predictor structure which exhibits a further reduction of signal power and a reduced 'spikiness'. However, it is well known that a stable linear time-invariant system with a stable inverse (minimum-phase property) is ***not capable of modelling periodicity*** because its output signal is periodic if and only if its input is periodic. Therefore, even the long-term prediction residual for a segment of sustained voiced speech has some periodicity, i.e., some deterministic structure which needs to be transmitted as a waveform pattern in so-called ***linear predictive analysis-by-synthesis*** coders, see the *code-excited linear predictive (CELP)* class of speech coders popular in digital cellular telephony systems.

Most studies in non-linear prediction have targeted the *short-term predictor* and report only moderate improvements over the linear predictor. For an excellent overview (including applications in speech coding) see the recent thesis [71]. A large database study [14] covering a total of some 600 sustained vowel utterances from over 80 speakers has arrived at three important conclusions:

- Even for highly stationary, sustained utterances the *average prediction gain improvements* remain less than 3.5 dB. This can be seen as an upper limit for continuous speech, too, where the adaptation of a non-linear predictor is usually more involved than in the linear case.

- Even with sophisticated training methods (such as extended Kalman filter-based adaptation), the non-linear structure is competitive only if it includes a linear substructure. Furthermore, cascading a linear predictor with a non-linear one appears less promising than a direct integration of the linear component into the non-linear structure as realized by a *Radial-Basis Function network (RBF)* for gradual transition between *local linear models* (so-called RBF-AR models [74]).

- The improvement due to non-linear prediction is highest for those test utterances which suffer from the lowest linear prediction gain. Over the database, the linear prediction gain varies between 10 and 40 dB, but when it is only 10 dB, the non-linear predictor achieves about 15 dB. This shows that there is some promise of increasing the *robustness of predictive signal models* by extending them to non-linear structures.

Unlike short-term prediction, *non-linear long-term prediction* has appeared only in a few studies so far [71, 15]. The latter reports *significant* advantages of non-linear prediction over linear prediction as *any prediction delay between one sample and a few pitch periods can be allowed without compromising performance.* The linear long-term predictor essentially cannot predict any sample in the distant future unless it can copy it from its predictor memory. Usually, this memory spans only a few samples, restricting linear prediction to the neighborhood of integer multiples of the pitch period. Thus, if uniform performance over all possible delays is required, the linear predictor memory has to be increased to a full pitch period. Contrary to that, the non-linear predictor works with a short memory and relies on just three delayed signal samples (or taps) to achieve an even higher prediction gain than the long-memory linear predictor; see Fig. 26.8.

Non-linear oscillator model

Predictive signal models are an instance of *source-filter models* for speech [23]. Figure 26.9 compares this model class to the *non-linear oscillator model* for speech [48]. Both models require slow time variation of their parameters to cope with the processes on the articulatory tier, but the self-oscillating non-linear oscillator model circumvents the direct specification of a source signal (which eventually is transmitted as a waveform pattern in speech coding), and it addresses the observed cyclostationarity of speech on the aero-acoustic tier (on the basis of the synchronization properties of oscillators; see [47]). In the oscillator model, a slowly time-varying,

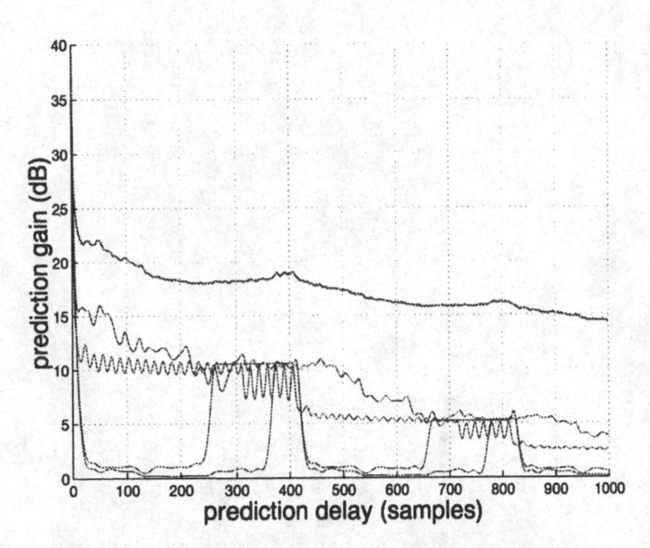

FIGURE 26.8. Linear vs. non-linear long-term prediction, after [15], prediction gain in dB as a function of prediction delay for a sustained vowel sampled at 48 kHz; curves from top to bottom: theoretical upper bound from mutual information analysis, non-linear three-tap predictor with RBF-AR structure, 144-tap linear predictor, 33-tap linear predictor, and one-tap linear predictor

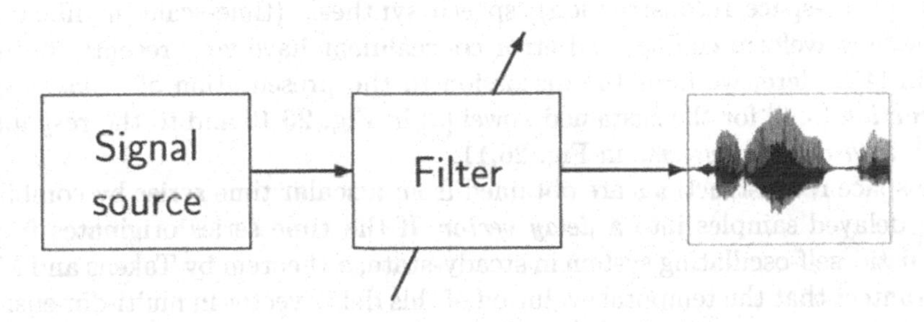

(a) Signal model with slowly time-varying filter

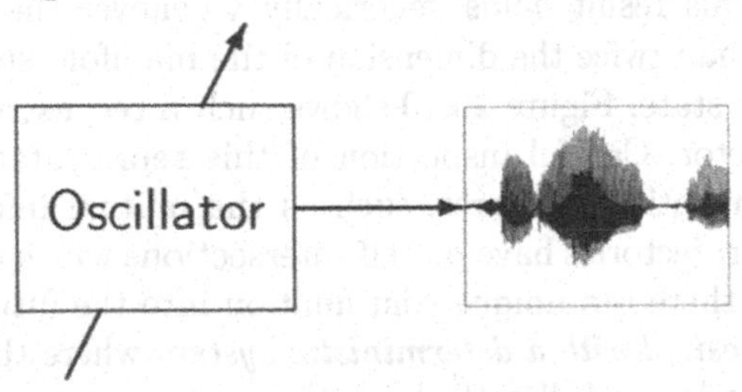

(b) Signal model with slowly time-varying oscillator

FIGURE 26.9. Source-filter versus oscillator model structures for speech

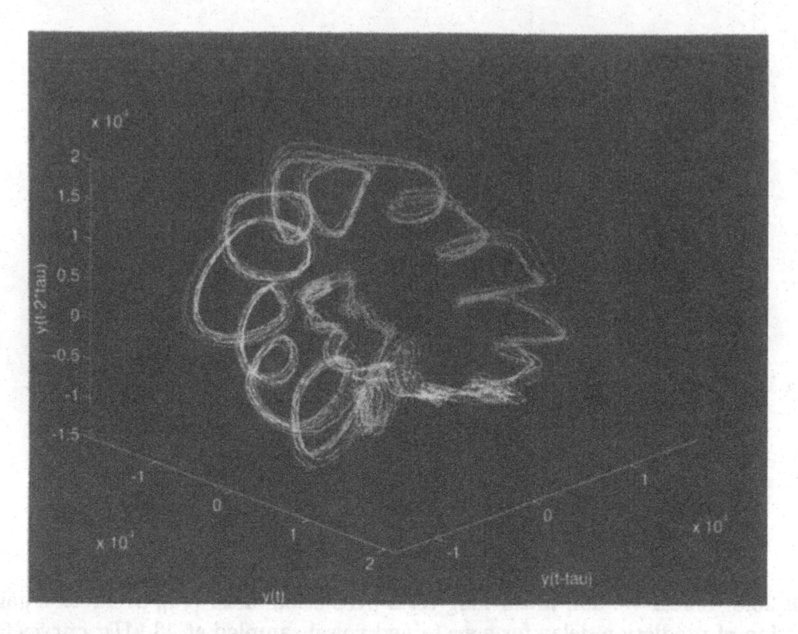

FIGURE 26.10. Phase-space reconstruction of vowel [i:], grey scale indicates depth of axionometric view

deterministic dynamic system is controlled in its non-linear feedback function such that the oscillator output closely matches the given continuous speech waveform. The signal processing algorithms needed to work with this model originate from chaos theory [60, 17, 1] and information theory, e.g., [9]. Its applications to speech analysis (phase-space reconstruction), speech synthesis (time-scale modification), synchronous waveform coding, and error concealment have very recently been reviewed in [45]. Here, we limit the discussion to the presentation of a ***phase-space reconstruction*** [8, 6] for the sustained vowel [i:] in Fig. 26.10 and to the ***resynthesis of speech with chaotic systems*** in Fig. 26.11.

Phase-space reconstructions are obtained from a scalar time series by combining multiple delayed samples into a ***delay vector***. If the time series originates from a deterministic, self-oscillating system in steady-state, a theorem by Takens and Mañé [67] guarantees that the temporal evolution of this delay vector in multi-dimensional space has a one-to-one relationship with the evolution of the state vector of the underlying system. This result holds generically whenever the dimension of the delay vector is more than twice the dimension of the manifold spanned by the true state vector in steady state. Figure 26.10 shows such a reconstruction for a three-dimensional delay vector. Careful inspection of this representation (by animated rotation and by quantitative measures, such as the mutual information function [7]), reveals that the trajectories have no self-intersections which shows that, at any state vector location, there is a unique continuation into the future. Therefore, ***the signal can be re-synthesized with a deterministic system*** where the ***non-linear state transition map*** can be learned directly from the phase space reconstruction. Any universal approximator may serve this purpose, and Radial Basis Function (RBF) neural networks are a good choice [13, 68]. The quality of the synthetic signal depends critically on the accuracy of the approximated non-linearity such that stable operation cannot be guaranteed unconditionally. However, if re-synthesis is

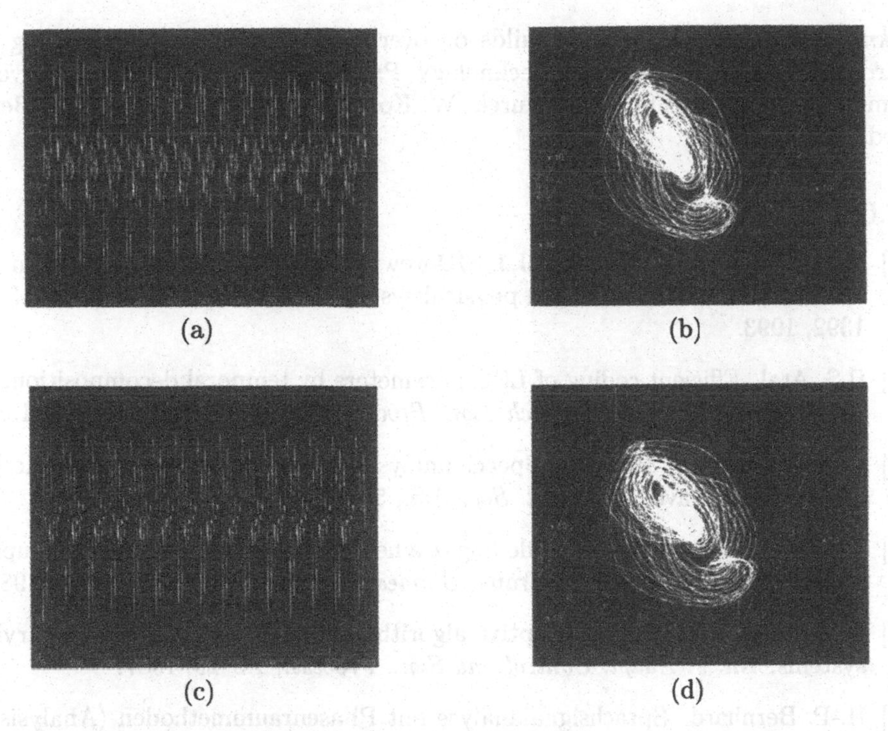

FIGURE 26.11. Resynthesis of sustained vowel [a:] with a recursive RBF network operating in the chaotic regime; (a) original time-domain waveform (125 ms); (b) original phase-space reconstruction; (c) synthetic time-domain waveform (125 ms selected from the 2 s elongated synthesis result); and (d) synthetic phase-space reconstruction

successful, it allows obtaining elongated versions of the speech segment used for training. Figure 26.11 illustrates such an experiment where 375 ms of the sustained vowel [a:] have been used to train an RBF network with 20 nodes and Gaussian activation functions whose inputs are delayed versions of its own scalar output. After training, its parameters are frozen and its steady-state oscillations are observed. Even after 2 seconds, it does not settle on a periodic or quasi-periodic attractor but continues to produce waveform patterns with an irregular appearance reminiscent of the variability seen in the training signal. As the system is time-invariant and has no internal noise sources or external driving inputs, the most plausible interpretation is that this RBF oscillator *operates in the chaotic regime*. When listening to the synthetic signal, its internal variability is perceived as highly natural - *chaos helps to add naturalness to synthetic speech.*

26.5 Conclusion

The standard speech signal model consists of a pulse/noise source and a linear filter where both change their characteristics on a frame-by-frame basis. Both on the articulatory and aero-acoustic tiers of speech models, significant extensions have been made in the past based on insights from non-stationary and non-linear signal processing. They are crucial in understanding the present state-of-the-art in speech processing and for the future advancement of the field.

Acknowledgments: This survey builds on over a decade of speech processing research at Vienna University of Technology. Particular thanks are due to several former doctoral students: W. Wokurek, W. Kozek, M. Birgmeier, and H.-P. Bernhard.

26.6 References

[1] H.D.I. Abarbanel, R. Brown, J.J. SIDorowich, and L.Sh. Tsimring. The analysis of observed chaotic data in physical systems. *Rev. Mod. Phys.*, 65(4):1331–1392, 1993.

[2] B.S. Atal. Efficient coding of LPC parameters by temporal decomposition. In *Proc. Int. Conf. Acoust. Speech Sign. Process.*, pp. 81–84, Boston, MA, 1983.

[3] B.S. Atal and S.L. Hanauer. Speech analysis and synthesis by linear prediction of the speech wave. *J. Acoust. Soc. Am.*, 50(2 (Part2)):637–655, 1971.

[4] J.S. Bay and H. Hemami. Modelling of a neural pattern generator with coupled nonlinear oscillators. *IEEE Trans. Biomed. Eng.*, BME-34(4):297–306, 1987.

[5] A. Benveniste. Design of adaptive algorithms for the tracking of time-varying systems. *Int. J. Adapt. Control and Sign. Process.*, 1:3–29, 1987.

[6] H.-P. Bernhard. Sprachsignalanalyse mit Phasenraummethoden (Analysis of speech signals with phase space methods, in German). In *Fortschritte der Akustik - DAGA'95*, pp. 1015–1018. Deutsche Gesellschaft für Akustik, Oldenburg, Germany, 1995.

[7] H.-P. Bernhard. The Mutual Information Function and its Application to Signal Processing. Ph.D. thesis, Vienna University of Technology, Vienna, Austria, 1997.

[8] H.-P. Bernhard and G. Kubin. Speech production and chaos. In *Proc. XIIth Int. Congr. Phonetic Sci.*, pp. 394–397, Aix-en-Provence, France, Aug. 1991.

[9] H.-P. Bernhard and G. Kubin. A fast mutual information calculation algorithm. In M.J.J. Holt et al., eds., *Signal Processing VII: Theories and Applications.* 1:50–53. Elsevier, Amsterdam, 1994.

[10] F. Bimbot et al. Temporal decomposition and acoustic-phonetic decoding of speech. In *Proc. Int. Conf. Acoust. Speech Sign. Process.*, pp. 315–318, New York, 1988.

[11] F. Bimbot, G. Chollet, and A. Paolini, eds. Special section on automatic speaker recognition, identification and verification. *Speech Commun.*, 17(1–2), 1995.

[12] F. Bimbot et al. Standard and target driven AR-vector models for speech analysis and speaker recognition. In *Proc. Int. Conf. Acoust. Speech Sign. Process.*, II-5–II-8. San Francisco, CA, 1992.

[13] M. Birgmeier. Kalman-Trained Neural Networks for Signal Processing Applications. Doctoral dissertation, Vienna University of Technology, Vienna, Austria, 1996.

[14] M. Birgmeier. Nonlinear prediction of speech signals using radial basis function networks. In *Proc. VIII Europ. Signal Process. Conf., EUSIPCO'96*, pp. 459–462, Trieste, Italy, 1996.

[15] M. Birgmeier, H.-P. Bernhard, and G. Kubin. Nonlinear long-term prediction of speech signals. In *Proc. Int. Conf. Acoust. Speech Sign. Process.*, pp. 1283–1286, Munich, Germany, 1997.

[16] H. Bölcskei and F. Halwatsch. Discrete Zak transforms, polyphase transforms, and applications. *IEEE Trans. Signal Process.*, 45(4), 1997.

[17] M. Casdagli et al. Nonlinear modelling of chaotic time series: theory and applications. In J.H. Kim and J. Stringer, eds., *Applied Chaos*, pp. 335–380. Wiley, New York, 1992.

[18] P.R. Cook. Noise and aperiodicity in the glottal source: a study of singer voices. In *Proc. XIIth Int. Congr. Phonetic Sci.*, 1:166–170, Aix-en-Provence, France, 1991.

[19] M. Cooke, S. Beet, and M. Crawford, eds.. *Visual Representations of Speech Signals.* Wiley, Chichester, England, 1993.

[20] A. De Lima Veiga and Y. Grenier. A multi-step excited model for speech parameter trajectories. In *Proc. Int. Conf. Acoust. Speech Sign. Process.*, pp. 67–70, New York, 1988.

[21] J.R.B. de Marca and M. Copperi, eds. Special issue on speech coding for telecommunications. *Europ. Trans. Telecomm.*, 5(5), 1994.

[22] L. Deng. A generalized hidden Markov model with state-conditioned trend functions of time for the speech signal. *Signal Process.*, 27:65–78, 1992.

[23] G. Fant. *Acoustic Theory of Speech Production*, 2nd ed. Mouton, The Hague (The Netherlands), 2nd ed., 1970.

[24] S. Furui and M.M. Sondhi, eds.. *Advances in Speech Signal Processing.* Marcel Dekker, New York, 1992.

[25] W.A. Gardner, ed.. *Cyclostationarity in Communications and Signal Processing.* IEEE Press, New York, 1994.

[26] A. Gersho. Advances in speech and audio coding. *Proc. IEEE*, 82(6):900–918, 1994.

[27] O. Ghitza and M.M. Sondhi. Hidden Markov models with templates as nonstationary states: Application to speech recognition. *Comp. Speech Lang.*, 2:101–119, 1993.

[28] Y. Grenier. Time-dependent ARMA modelling of nonstationary signals. *IEEE Trans. Acoust. Speech Signal Process.*, ASSP-31(4):899–911, 1983.

[29] G.C. Hegerl and H. Höge. Numerical simulation of the glottal flow by a model based on the compressible Navier-Stokes equations. In *Proc. Int. Conf. Acoust. Speech Sign. Process.*, pp. 477–480, Toronto, Ont, 1991.

[30] H. Hermansky and N. Morgan. Rasta processing of speech. *IEEE Trans. Speech Audio Process.*, 2(4):578–589, 1994.

[31] N. Jayant, J. Johnston, and R. Safranek. Signal compression based on models of human perception. *Proc. IEEE*, 81(10):1385–1422, 1993.

[32] B.H. Juang and L.R. Rabiner. *Fundamentals of Speech Recognition.* Prentice-Hall, Englewood Cliffs, NJ, 1994.

[33] J.A.S. Kelso et al. A qualitative dynamic analysis of reiterant speech production: Phase portraits, kinematics, and dynamic modelling. *J. Acoust. Soc. Am.*, 77(1):266–280, 1985.

[34] G. Kitagawa and W. Gersch. A smoothness prior time-varying AR coefficient modelling of nonstationary covariance time series. *IEEE Trans. Autom. Contr.*, AC-30(1):48–56, 1985.

[35] W.B. Kleijn and K.K. Paliwal, eds.. *Speech Coding and Synthesis.* Elsevier, Amsterdam, 1995.

[36] W. Bastiaan Kleijn and W. Granzow. Methods for waveform interpolation in speech coding. *Digital Signal Processing,* 1(4):215–230, 1991.

[37] W. Bastiaan Kleijn and J. Haagen. A speech coder based on decomposition of characteristic waveforms. In *Proc. Int. Conf. Acoust. Speech Sign. Process.*, pp. 508–511, Detroit, MI, May 1995.

[38] W. Kozek. Matched generalized Gabor expansion of nonstationary processes. In *Proc. IEEE Int. Conf. Signals, Systems, and Computers*, pp. 499–503, Pacific Grove, CA, Nov. 1993.

[39] W. Kozek. Matched Weyl–Heisenberg Expansions of Nonstationary Environments. Ph.D. thesis, Vienna University of Technology, Vienna, Austria, 1996.

[40] W. Kozek. Adaptation of Weyl-Heisenberg frames to underspread environments. In Hans G. Feichtinger and Thomas Strohmer, eds., *Gabor Analysis and Algorithms - Theory and Applications.* chap. 10. Birkhäuser, Boston, 1997.

[41] W. Kozek and H.G. Feichtinger. Time-frequency structured decorrelation of speech signals via nonseparable Gabor frames. In *Proc. Int. Conf. Acoust. Speech Sign. Process.*, Munich, Germany, Apr. 1997.

[42] P. Kroon and W.B. Kleijn. Linear-prediction based analysis-by-synthesis coding. In W. B. Kleijn and K. K. Paliwal, eds., *Speech Coding and Synthesis*, pp. 70–119. Elsevier, Amsterdam, The Netherlands, 1995.

[43] G. Kubin. Coefficient filtering - a common framework for the adaptation in time-varying environments. In D. Docampo and A.R. Figueras, eds., *Adaptive Algorithms: Applications and Non-Classical Schemes*, pp. 91–110, Vigo, Spain, 1991.

[44] G. Kubin. A mixed bag of tools for WI speech coding and beyond. AT&T Bell Laboratories, Murray Hill, NJ, 1995.

[45] G. Kubin. Nonlinear processing of speech. In W. B. Kleijn and K. K. Paliwal, eds., *Speech Coding and Synthesis*, pp. 557–610. Elsevier, Amsterdam, 1995.

[46] G. Kubin. Voice processing - beyond the linear model. In *PRORISC/IEEE Workshop on Circ., Systems, and Signal Process.*, pp. 393–400, Mierlo, The Netherlands, 1996.

[47] G. Kubin. Poincaré section techniques for speech. In *Proc. 1997 IEEE Workshop on Speech Coding for Telecomm.*, pp. 7–8, Pocono Manor, PA, 1997.

[48] G. Kubin and W.B. Kleijn. Time-scale modification of speech based on a nonlinear oscillator model. In *Proc. Int. Conf. Acoust. Speech Sign. Process.*, I–453–I–456, Adelaide, Australia, 1994.

[49] L. Lindbom. A Wiener Filtering Approach to the Design of Tracking Algorithms - With Applications in Mobile Radio Communications. Ph.D. thesis, Uppsala University, Uppsala, Sweden, 1995.

[50] M.C. Mackey and L. Glass. Oscillation and chaos in physiological control systems. *Science*, 197:287–289, 1977.

[51] J.D. Markel and A.H. Gray, Jr. *Linear Prediction of Speech.* Springer, Berlin, 1976.

[52] R.J. McAulay and T.F. Quatieri. Speech analysis/synthesis based on a sinusoidal representation. *IEEE Trans. Acoust. Speech Signal Process.*, ASSP-34(4):744–754, 1986.

[53] Claude Montacié et al. Cinematic techniques for speech processing: Temporal decomposition and mutivariate linear prediction. In *Proc. Int. Conf. Acoust. Speech Sign. Process.*, I–153–I–156, San Francisco, CA, 1992.

[54] N. Morgan and H. Bourlard. Continuous speech recognition. *IEEE Signal Process. Mag.*, 12(3):24–42, 1995.

[55] E. Moulines and F. Charpentier. Pitch-synchronous waveform processing techniques for text-to-speech synthesis using diphones. *Speech Commun.*, 9(5/6):453–467, 1990.

[56] Y.K. Muthusamy, E. Barnard, and R.A. Cole. Reviewing automatic language identification. *IEEE Signal Process. Mag.*, 11(4):33–41, 1994.

[57] M. Niedźwiecki. First-order tracking properties of weighted least squares estimators. *IEEE Trans. Autom. Contr.*, AC-33(1):94–96, 1988.

[58] M. Niedźwiecki. On tracking characteristics of weighted least squares estimators applied to nonstationary system identification. *IEEE Trans. Autom. Contr.*, AC-33(1):96–98, 1988.

[59] A. Papoulis. *Probability, Random Variables, and Stochastic Processes*, 2nd ed. McGraw-Hill Int., Tokyo, 2nd ed., 1984.

[60] T.S. Parker and L.O. Chua. Chaos: a tutorial for engineers. *Proc. IEEE*, 75(8):982–1008, 1987.

[61] B. Porat. Second-order equivalence of rectangular and exponential windows in least-squares estimation of Gaussian autoregressive processes. *IEEE Trans. Acoust. Speech Signal Process.*, ASSP-33(5):1209–1212, 1985.

[62] R.K. Potter, A.G. Kopp, and H.C. Green. *Visible Speech.* Van Nostrand, New York, 1947.

[63] L.R. Rabiner. A tutorial on hidden Markov models and selected applications in speech recognition. *Proc. IEEE*, 77(2):257–286, 1989.

[64] L.R. Rabiner. Applications of voice processing to telecommunications. *Proc. IEEE*, 82(2):199–228, 1994.

[65] D.B. Roe and S. Furui, eds. Special issue on interactive voice technology for telecommmunication application. *Speech Commun.*, 17(3–4), 1995.

[66] E.S. Saltzmann. Dynamics and coordinate systems in skilled sensorimotor activity. In *Status Report on Speech Research*, SR-115/16:1–15, Haskins Laboratories, New Haven, CT, 1993.

[67] T. Sauer, J.A. Yorke, and M. Casdagli. Embedology. *J. Stat. Phys.*, 65:579–616, 1991.

[68] T. Schlögl. Synthese von Sprachsignalen mit rückgekoppelten neuralen Netzen (Synthesis of speech signals with feedback neural networks, in German). INTHFT - student project report, Vienna University of Technology, Vienna, Austria, 1997.

[69] S. Singhal and B.S. Atal. Improving performance of multi-pulse LPC coders at low bit rates. In *Proc. Int. Conf. Acoust. Speech Sign. Process.*, 1.3.1–1.3.4, San Diego, CA, 1984.

[70] V. Steinbiss et al. Continuous speech dictation - From theory to practice. *Speech Commun.*, 17(1–2):19–38, 1995.

[71] J. Thyssen. Non-Linear Analysis, Prediction, and Coding of Speech. Ph.D. thesis, Technical University of Denmark, Lyngby, Denmark, 1995.

[72] R. Togneri, M.D. Alder, and Y. Attikiouzel. Dimensions and structure of the speech space. *IEE Proceedings-I*, 139(2):123–127, 1992.

[73] A.M.L. van Dijk–Kappers and S.M. Marcus. Temporal decomposition of speech. *Speech Commun.* 8:125–135, 1989.

[74] J.-M. Vesin. On Some Aspects of Non-Linear Signal Modelling and its Real World Applications. Ph.D. thesis, EPFL, Lausanne, Switzerland, 1992.

[75] A. Waibel et al. Phoneme recognition using time-delay neural networks. *IEEE Trans. Acoust. Speech Signal Process.*, 37:328–339, 1989.

[76] W. Wokurek, G. Kubin, and F. Hlawatsch. Wigner distribution - a new method for high-resolution time-frequency analysis of speech signals. In *Proc. XIth Int. Congress Phonetic Sciences*, pp. 44–47, Tallinn, Esthonia, 1987.

27

Modulation Spectrum in Speech Processing

Hynek Hermansky[1]

ABSTRACT
The work questions adequacy of the short-term spectral envelope as the dominant carrier of the phonetic identity of a given speech instant and suggests temporal dynamics of components of the spectral envelopes as more reliable means for carrying the linguistic context of the speech message. It summarizes some research which employs processing of the so called modulation spectrum of speech. In argues that dominant linguistic information bearing components of modulation spectrum are found in the vicinity of 4 Hz and demonstrates that suppressing slow and fast components of trajectories of spectral envelopes can be useful in alleviating effects of acoustic environment in processing of speech.

27.1 Short-Term Spectral Envelope in Speech Processing

Spectral analysis of sufficiently short segment of speech signal yields a short-term spectrum of speech, each spectral vector describing about 10 ms of the speech signal. The short-term analysis dominates speech processing for more than half century.

Spectral envelope of this short-term spectrum reflects resonance properties of the vocal tract. The resonant properties depend on the tract shape.

Parameters which characterize spectral envelope can be derived in a number of ways such as spectral integration, cepstral truncation, or all-pole modelling. Often, some combination of methods is used. Changes of vocal tract shape are reflected in temporal patterns of such speech parameters.

Most automatic speech recognition (ASR) systems use speech features based on the short-term spectral envelope of speech spectrum, implicitly accepting dominance of the spectral envelope as the prime carrier of linguistic information in speech. Current phoneme-based ASR typically utilizes a simple context-dependent phoneme models (sometimes confusingly called "triphones" implying a dependency on the left and right neighboring phoneme). In a typical large vocabulary stochastic ASR system, the identity of sub-word unit (typically a part of context-dependent phoneme) is derived by stochastic matching of short-term speech parameters. Each phoneme model has three different states, representing piece-wise approximation of the internal phoneme dynamics. The overall dynamics of speech is then captured by concatenating individual context dependent phoneme models.

---

[1]Oregon Graduate Institute of Science & Technology, Department of Electrical and Computer Engineering, P.O. Box 91000, Portland, OR 97291, E-mail: hynek@ece.ogi.edu

Within each state, the parameters are assumed to be stationary, identically distributed, and independent of the neighboring analysis frames. This assumption implies that there is no constraint on a particular temporal order of speech parameters within each state. Such piece-wise stationary model is rather crude approximation of the rich temporal dynamics of speech. The fact is that patterns of speech parameters vary gradually within each distinct segment of speech and their particular dynamics is an important cue to the identity of a given segment.

27.2 Motivation for The Current Work

The spectral envelope is well accepted as the prime carrier of the phonetic identity of the given speech instant. However, the spectral envelope can be very easily corrupted by common distortions such as linear filtering or additive environmental noise, which have only a minimal effect on human speech communication process.

We therefore wish to suggest that the prime role of correlations between individual components of the spectral envelopes may be to increase robustness of human speech communication by providing necessary redundancies in the signal. Since such correlations are easily affected by nonlinguistic factors, they should not be relied on as reliable carriers of the linguistic information.

We would like to suggest that the temporal dynamics of components of the spectral envelopes rather than the envelopes themselves may serve more reliable cues for communicating the linguistic context of the speech message. Information about this temporal dynamics is available in the so called modulation spectrum of speech.

Pushing the argument a little further, we speculate that the main reason for frequency-selective properties of human auditory system is not to derive frequency content of a given segment for phonetic classification but rather to provide means for optimal choice of high signal-to-noise (SNR) regions for deriving reliable sub-band based modulation spectrum by temporal analysis of the high SNR sub-bands of the signal. This part of the argument is not further elaborated on in the current work but it forms basis of our ongoing work on the multi-stream ASR of speech [12].

27.3 Modulation Spectrum of Speech

Spectral analysis of temporal trajectories of spectral envelopes of speech yields modulation spectrum of speech [15]. The concept of the modulation spectrum of speech is illustrated in Fig. 27.1. Dominant components of the modulation spectrum indicate dominant rate of change of the vocal tract shape. The modulation spectrum of continuous and uninterrupted speech is dominated by components between 2 Hz and 8 Hz, reflecting syllabic and phonetic temporal structure of speech [9]. It is interesting but not surprising that human auditory system is most sensitive to modulation frequencies around 4 Hz (see e.g. [12, 10] for the review of evidence available in the literature). Further, this frequency is also hypothesized to be dominant in cortical part of human brain [9]. The modulation spectrum is affected by noise and reverberation and a practical engineering technique for estimation of intelligibility of speech in enclosed spaces is build around this fact [15].

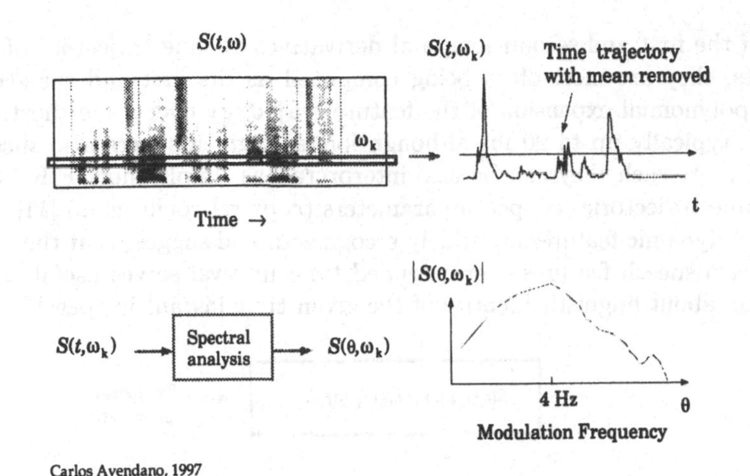

FIGURE 27.1. Modulation spectrum of speech can be obtained by spectral analysis of temporal trajectory of power spectral component of speech

27.3.1 Modulation Spectrum of Speech in Current ASR

In spite of well-known speech production phenomena of coarticulation, auditory perception phenomena of forward masking, and linguistic concept of syllable, all of which point to temporal phenomena over medium-term time interval of the order of several hundreds of ms, such medium-term temporal dynamics has not been extensively studied and utilized in speech processing. One of reasons may be that current ASR systems are not well suited for handling modulation aspects of the speech signal.

To obtain sufficient spectral resolution at low modulation frequencies, rather large time spans of speech (of the order of at least several hundreds ms) are required to compute the modulation spectrum of speech. Then, any acoustic event, even a very short one, will extend its effect over larger time span, introducing rather significant "sluggishness" of the processing. Such sluggishness may not present any significant problem in a single-utterance ASR (apart of some easily manageable problems with initialization of the processing). As a matter of fact, as discussed in [12], the single-utterance ASR may benefit from spreading of the effect of short acoustic events such as plosives into larger segments of the short-term features since this may improve chances of their detection in the subsequent pattern matching. However, it may present a challenge in current sub-word large vocabulary ASR since it makes the current instant more dependent on the phonetic environment which may span several neighboring phonemes. Our intuition is that use of such large vocabulary sub-word units need to be seriously re-examined in order to derive full benefits from modulation spectrum approaches. However, as discussed below, benefits of the modulation spectrum-based processing can be substantial and may well justify some changes in the current main-stream sub-word based ASR paradigm.

27.3.2 Dynamic Features, RASTA Processing, and Processing of Modulation Spectrum of Speech

So called dynamic cepstral features [8] represent one of first successful attempts to process the modulation spectrum of speech. Even though introduced as as estimates of the first and second temporal derivatives of time trajectory of cepstral coefficients, they are most often being computed as the first and the second orthogonal polynomial expansion of the feature trajectory over some short segment of speech (typically up to 90 ms although longer segments were also successfully used [8, 1]). As such they can be also interpreted as simple short FIR filters applied to time trajectories of speech parameters (cepstral coefficients) [11]. Benefits from use of dynamic features are widely recognized. and suggest that the dynamics of short-term speech features over extended time interval serves useful additional information about linguistic identity of the given time instant in speech.

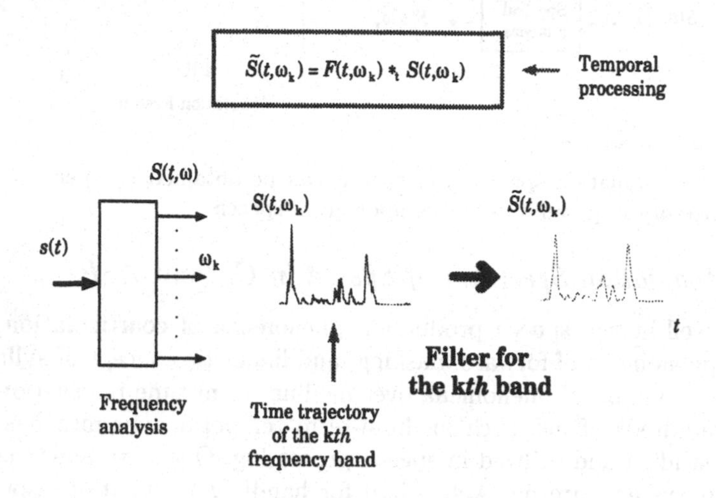

FIGURE 27.2. Principle of processing of modulation spectrum of speech

RASTA processing [11] generalizes dynamic features by proposing linear filtering sandwiched between two non-linearities. The rationale behind such processing is that the rate of change of target features is within certain limits (in the case of speech determined by physical constraints on the rate of change of a vocal tract). The components of the feature time trajectories which are outside such limits are due to noise and should be alleviated by the RASTA filtering. Since the filtering operation requires access to some temporal environment of the current data point, RASTA effectively utilize rather larger time-span of the signal (impulse response of the original RASTA filter [11] has effective length about 220 ms).

Over the time we came to realize that the success of dynamic features and RASTA processing points to the so far relatively untapped domain of processing of the modulation spectrum of speech and to new possibilities offered there [10, 9]. To modify the modulation spectrum of speech one needs to filter time trajectories of spectral energies (see Fig. 27.2). This is done in an ad hoc manner by both the computation of dynamic features (linear filtering of cepstrum can be always interpreted in terms of the linear filtering of logarithmic spectrum) and in RASTA.

Therefore, over past several years, we and others have been experimenting with speech processing techniques which employ and modify the modulation spectrum of speech. This experience has been described in a number of publications, some of which are referred in the text and is summarized below. This overview is not presented chronologically but it is ordered by the logical flow of our understanding of such processing.

27.4 Relative Importance of Various Components of Modulation Spectrum of Speech

27.4.1 Perceptual Evidence

Drullman et al. [6, 7] have been studying intelligibility of Dutch syllables and meaningful sentences with temporally modified spectral envelopes. Their experiments employed high-pass and low-pass filtering of time trajectories of power spectral envelopes derived by octave/ half-octave, and third-octave filtering. The care was taken not to modify the fine spectral structure of the speech signal. Their experiments indicate relatively low contributions of very low (below 1 Hz) and high (above 16 Hz) components of the modulation spectrum of speech to the intelligibility of speech.

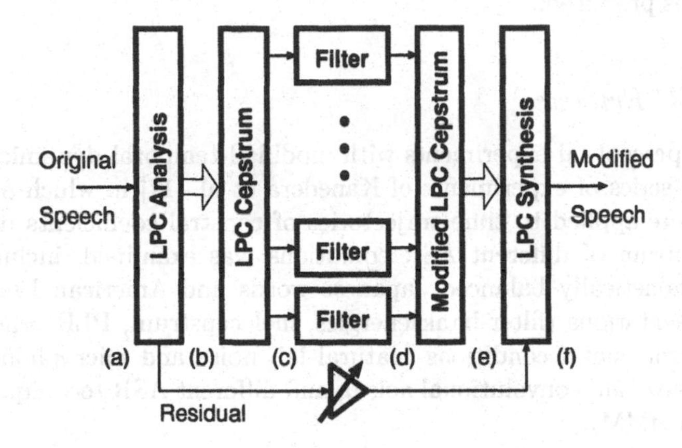

FIGURE 27.3. Experimental system for modifying dynamics of spectral envelopes of speech

Arai et al. [4] extended Drullman's experiments using band-pass filtering in addition to the low-pass and high-pass filtering. Their experiments used residual-excited LPC vocoder with band-pass filtered trajectories of LPC cepstral coefficients. The experimental system is illustrated in Fig. 27.3, results are summarized in Fig. 27.4.

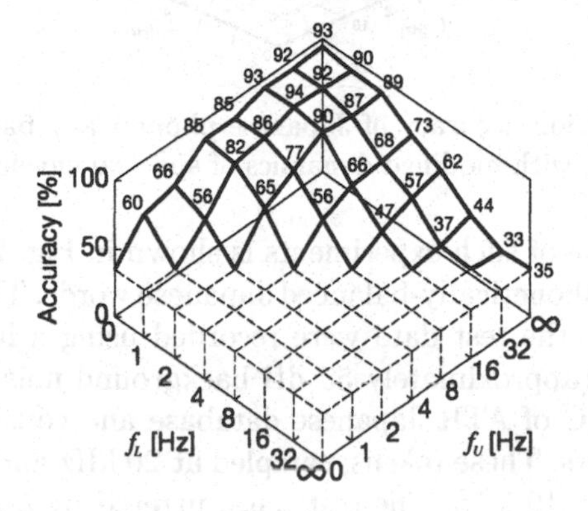

FIGURE 27.4. Intelligibility of Japanese syllables with modified dynamics of spectral envelopes

The Figure indicates upper cutoff frequency of the bandpass filter on the x-axis, the lower cutoff frequency on the z-axis and the resulting intelligibility of speech with artificially modified temporal dynamics of spectral envelopes on the y-axis. Thus, the intelligibility of the original (unmodified) speech is indicated at the beginning of the coordinates It is seen that as long as components of modulation spectrum which lay between 1 and 16 Hz are preserved, the intelligibility of Japanese nonsense syllables is not seriously affected. This result is important because it supports a hypothesis (utilized in RASTA processing) that no significant linguistic information is contained in very slow and very fast changes of spectral envelopes of speech (thus in the very slow and very fast changes of the vocal tract shapes). Even though the speech signal may contain slow and fast modulation spectrum components, such components likely carry non-relevant information such as information about identity of the speaker or about acoustic environment in which the speech was produced.

27.4.2 ASR Evidence

Extension of perceptual experiments with modified temporal dynamics of spectral envelopes are series of experiments of Kanedera et al. [17] in which narrow band-pass filters were applied to time trajectories of cepstral coefficients used in ASR. A whole spectrum of different ASR conditions was examined, including various databases (phonetically balanced Japanese words and American English digits), feature representations (filter-bank energies, mel cepstrum, PLP cepstrum), mismatched environmental conditions (natural lab noise and microphone mismatch, artificial additive and convolutional noise), and different ASR techniques (template matching and HMM).

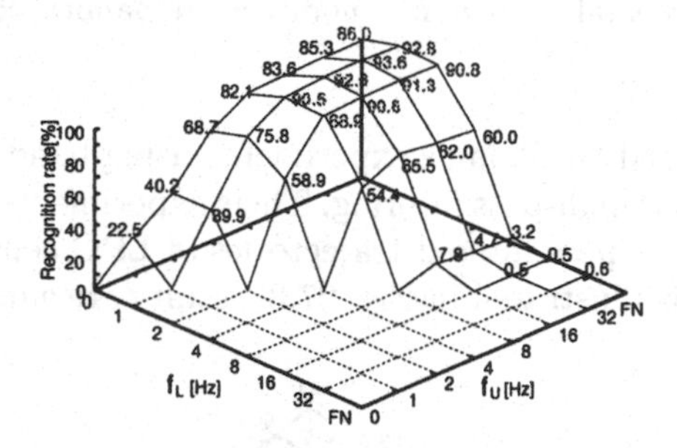

FIGURE 27.5. Recognition accuracy of Japanese phonetically balanced words produced in realistic environment with modified dynamics of spectral envelopes

Typical result of one of such experiments is shown in Fig. 27.5. This experiment used database of 216 phonetically-balanced Japanese words. The system was trained on clean speech while the test data were recorded using a hand-held microphone in a computer room (approximately 55 dB background noise). The training data were drawn from set C of ATR Japanese database and consisted of samples from 10 male native speakers. These tokens, sampled at 20 kHz and quantized to 16 bits, were down-sampled to 10 kHz. The test data, uttered by 5 male native speakers, were sampled at 11.025 kHz (also 16 bit quantization). Each utterance was passed through a FFT-based equal-Q 16-band filter bank. The time trajectories of such logarithmic energies were filtered with a 511-tap linear phase temporal FIR filter. As the main interest was mainly in the relative changes in ASR performance with changes of temporal filter shape rather than in highest obtainable scores, the ASR was done using a simple-but-efficient DTW-based recognizer.

The Figure follows the same conventions as the Fig. 27.4, The vertical axis shows the recognition accuracy using features with modified temporal dynamics of features, while the other axes indicate the lower cutoff frequency f_L and the upper cutoff frequency f_U of the band-pass filters. A similarity of this ASR result with result of the perceptual experiment from Fig. 27.4 is quite obvious. Components of modulation spectrum with frequencies higher than 16 Hz hardly contribute any relevant information for the ASR process. However, unlike in the perceptual experiment where the best performance was observed when all components of the modulation spectrum were preserved, for some band-pass conditions (indicated by bold letters) there is a noticeable improvement in the recognition accuracy comparing to the accuracy (86%) of the baseline (i.e., no filtering) system. This happens when the components with frequencies lower that 1 Hz are eliminated. This can be attributed to the fact that experiment was carried out under a environmental mismatch between training and test data. Since both convolutional distortions due to different microphone and the steady additive background noise present in the test data show mainly in the slowly-varying changes of spectral envelopes, eliminating the low frequency components of the modulation spectrum of speech improves robustness of processing in such mismatched environment, supporting rationale behind the original RASTA processing [11].

Clearly, the dominant contribution to the ASR process comes from components of the modulation spectrum covering frequencies between 2-8 Hz. This can be deduced from the fact that any elimination of components from this range results in a significant degradation of the ASR performance. This observation carries over the wide range of different ASR conditions studied by Kanedera et al [17].

27.5 Enhancement of Degraded Speech

Both perceptual and ASR experiments described above indicate that some components of modulation spectrum are more important for speech communication than others. Further, it appears that the linguistically dominant factors of the speech signal may occupy different parts of the modulation spectrum than do some non-linguistic factors such as steady additive noise. This suggests that a proper processing of modulation spectrum of speech may improve quality of noisy speech.

Indeed, some of our preliminary experiments with a simple ad hoc RASTA filtering of time trajectories of short-term power spectrum of speech (which eliminate the slowly and fast changing components of the trajectories) in the overlap-add analysis-resynthesis system yielded results comparable to the conventional spectral subtraction based technique for enhancement of noisy speech [11]. Further, Hirsch et al. [14] used high-pass filtering of time trajectories of power spectrum of speech for alleviating effects of additive noise in ASR.

To investigate possibilities of the modulation spectrum domain for enhancement of noisy speech, Avendano and his colleagues [13, 2]. worked with parallel recordings of clean and noisy speech to derive FIR filters which would optimally (in the RMS

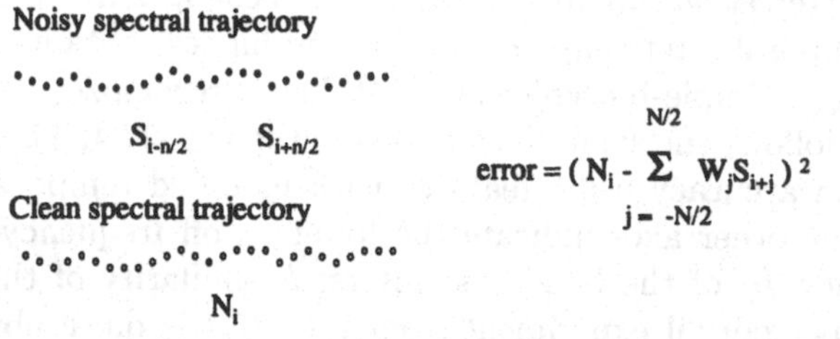

Noisy spectral trajectory

$S_{i-n/2}$ $S_{i+n/2}$

Clean spectral trajectory

N_i

$$\text{error} = \left(N_i - \sum_{j=-N/2}^{N/2} W_j S_{i+j} \right)^2$$

FIGURE 27.6. Deriving FIR RASTA filters by minimizing RMS error between clean and noisy speech

sense) map several points of the time trajectory of the amplitude compressed power spectrum of speech on the single point of the time trajectory of the clean speech. The frequency characteristics of the FIR filters derived in this way are illustrated in Fig. 27.7.

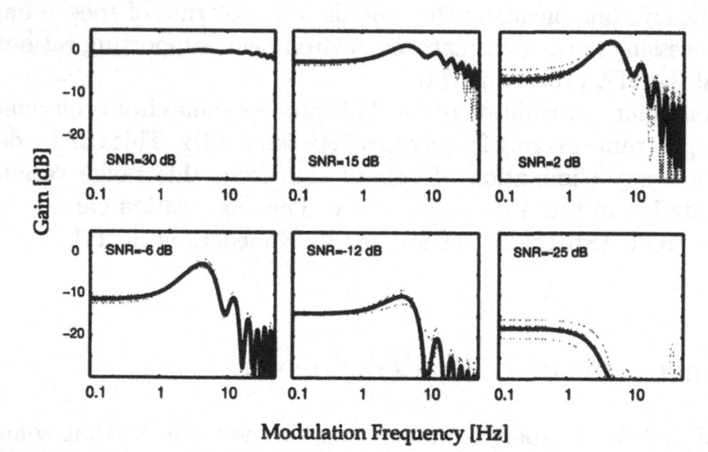

FIGURE 27.7. Temporal RASTA filters designed by minimizing RMS error between clean and noisy speech

The filters appear to be relatively independent of their respective carrier frequencies (dotted lines in the Figure indicate frequency characteristics of the filters from different frequency channels) but are dependent on the signal-to-noise ratio (SNR)in the respective channel (different panels in the figure correspond to different SNRs). Namely, for the high SNRs, the filters pass all components of the modulation spectrum and for the extremely poor SNRs they attenuate all components. However, for the moderate SNRs, the filters attempt to enhance the modulation spectrum components in the vicinity of 4 Hz while suppressing the components at extremely low and high modulation frequencies.

This result is important considering that the filters were derived using no a priori assumptions but the FIR architecture of the temporal filtering (20-30 point FIR filters spanning 200-300 ms of temporal trajectories were found optimal). It further supports dominance of modulation spectrum components in the vicinity of 2-8 Hz in speech communication.

27.6 Alleviation of Non-Linguistic Components in ASR

The work on RASTA-based speech enhancement yielded general methodology for data-driven design of temporal RASTA filters using data containing sources of undesirable variability. The one technique which appears to be promising is the technique based on linear discriminant analysis (LDA) [3, 12, 18].

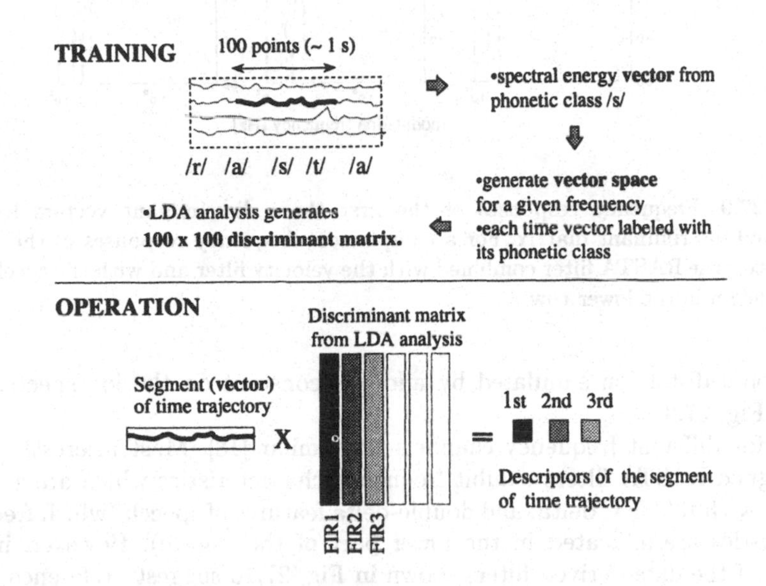

FIGURE 27.8. Data-driven design of FIR RASTA filters using LDA technique

The basic principle of such filter design is illustrated in Fig. 27.8. Rather long (about 1 s) segments of temporal trajectories of critical-band logarithmic spectrum are assigned to phonetic classes. (Either hand-labeled speech or labels resulting from forcefully aligned HMM models on the training data were used). All vectors which centers lay within boundaries of the given phoneme are assigned to the given class. Each class then contains 100 point (for 10 ms analysis step data) vectors which describe a temporal evolution of spectral energy inside and around the given phoneme.

The LDA [16, 5] yields a matrix of discriminant eigenvectors ordered by their eigenvalues. Each discriminant eigenvector represents an FIR filter to be applied to the original temporal trajectory. Each FIR filter then provides one particular view of the 1 s of speech around the current time instant. Since the corresponding eigenvalues die off rather rapidly, (typically, the first three eigenvectors account for more than 95 the first few FIR filters dominate the process.

Frequency responses of the first three FIR filters derived from about 20 minutes of hand-labeled Switchboard data, appended by the same data with the additional

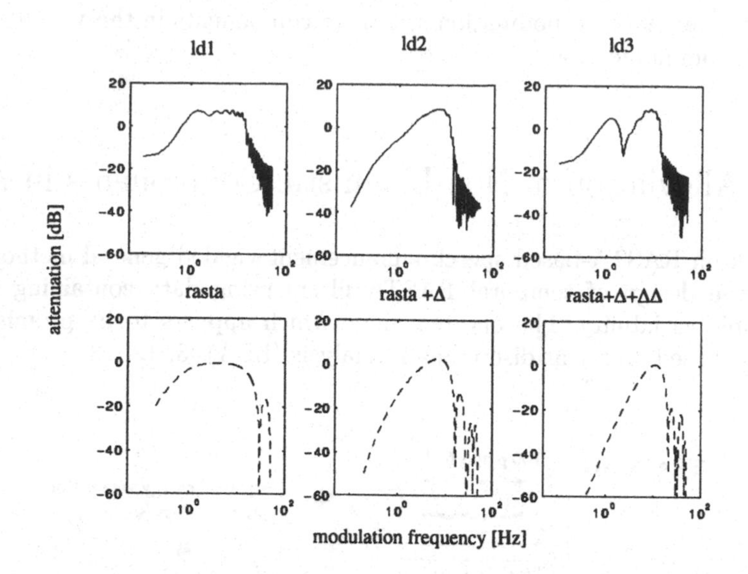

FIGURE 27.9. Frequency responses of the first three discriminant vectors from the LDA-derived discriminant matrix. For a comparison, frequency responses of the original RASTA filter, the RASTA filter combined with the velocity filter and with the acceleration filter are shown in the lower row

convolutional distortion simulated by additive constant on the log spectrum are shown in Fig. 27.9.

Filters for different frequency channels are similar [18]. Most interestingly, the data-designed RASTA filters exhibit frequency characteristics which are generally consistent with RASTA, delta, and double-delta features of speech (which frequency characteristics are indicated in the lower part of the Figure). However, impulse responses of the data-derived filters shown in Fig. 27.10 suggest preference for the zero-phase filters. Effective parts of the impulse responses appear to span at least 250 ms.

The general characteristics of the data-derived RASTA filters appear to be relatively independent of the particular database used for their design.

27.7 Conclusion

Based on our experience with temporal processing of speech features, we are convinced that better understanding and utilization of frequency-specific temporal dynamics in speech will lead to significant improvements of current speech technology. The purpose of this work is to review techniques which we have been experimenting with, to discuss some results obtained so far, and above all, to show new possibilities being opened by stepping out of the conventional short-term frame-based speech analysis techniques. We would like to advocate more systematic attempts for utilizing the medium-term temporal dynamics of speech features in hope for spurring more interest of speech research community in this relatively ignored area.

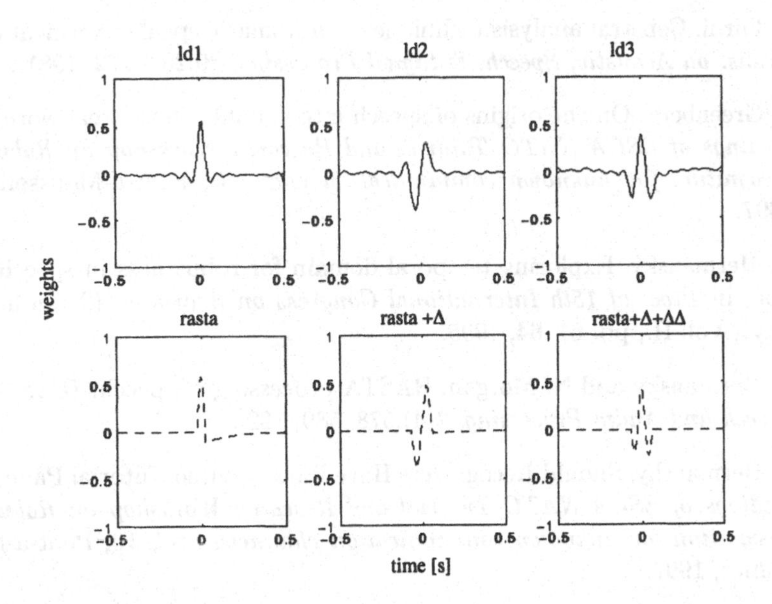

FIGURE 27.10. Impulse responses of the first three discriminant vectors from the LDA-derived discriminant matrix. For a comparison, impulse responses of the original RASTA filter, the RASTA filter combined with the velocity filter and with the acceleration filter are shown in the lower row

27.8 References

[1] T.H. Applebaum and B.A. Hanson. Regression features for recognition of speech in quiet and in noise. In *Proceedings of the 1991 IEEE International Conference on Acoustics, Speech and Signal Processing*, pp. 985–989, Toronto, Canada, 1991.

[2] C. Avendano and H. Hermansky. On the properties of temporal processing for speech in adverse environments, to appear in *Proceedings of 1997 Workshop on Applications of Signal Processing to Audio and Acoustics*, Mohonk Mountain House, New Paltz, New York, 1997.

[3] C. Avendano, S. van Vuuren, and H. Hermansky. Data based filter design for RASTA-like channel normalization in ASR. In *Proc. ICSLP-96*, Philadelphia, 1996.

[4] T. Arai, M. Pavel, H. Hermansky, and C. Avendano. Intelligibility of speech with filtered time trajectories of spectral envelopes. In *Proc. ICSLP-96, Philadelphia*, 1996.

[5] P. Brown. The Acoustic-Modeling Problem in Automatic Speech Recognition. Ph.D. Thesis, Carnegie Mellon University, 1987.

[6] R. Drullman, J.M. Festen, and R. Plomp. Effect of temporal envelope smearing on speech reception. *J. Acoust. Soc. Am.*, 95:1053–1064, 1994.

[7] R. Drullman, J.M. Festen, and R. Plomp. Effect of reducing slow temporal modulations on speech reception. *J. Acoust. Soc. Am.*, 95:2670–2680, 1994.

[8] S. Furui. Cepstral analysis technique for automatic speaker verification. *IEEE Trans. on Acoustic, Speech, & Signal Processing*, 29:254–272, 1981.

[9] S. Greenberg. On the origins of speech intelligibility in the real world. In *Proceedings of ESCA-NATO Tutorial and Research Workshop on Robust speech recognition for unknown communication channels*, Pont-a-Mousson, France, 1997.

[10] H. Hermansky. Exploring temporal domain for robustness in speech recognition. In *Proc. of 15th International Congress on Acoustics*, (Trondheim, Norway), Vol. II., pp. 61–64, 1995.

[11] H. Hermansky and N. Morgan. RASTA processing of speech. *IEEE Trans. on Speech and Audio Processing*, 2(4):578–589, 1994.

[12] H. Hermansky. Should Recognizers Have Ears? Invited Tutorial Paper, In *Proceedings of ESCA-NATO Tutorial and Research Workshop on Robust speech recognition for unknown communication channels*, pp.1–10, Pont-a-Mousson, France, 1997.

[13] H. Hermansky, E. Wan, and C. Avendano. Speech enhancement based on temporal processing. In *Proceedings of the 1995 IEEE International Conference on Acoustics, Speech and Signal Processing*, Detroit, Michigan, 1995.

[14] H.G. Hirsch, P. Meyer, and H. Ruehl. Improved speech recognition using highpass filtering of subband envelopes. In *Proc. Eurospeech '91*, Genova, Italy, 1991.

[15] T. Houtgast and H.J.M. Steeneken. A review of the MTF concept in room acoustics and its use for estimating speech intelligibility. *J. Acoust. Soc. Am.*, 95(3):1593–1602, 1994.

[16] M.J. Hunt. A Statistical Approach to Metrics for Word and Syllable Recognition. *J. Acoust. Soc. Am.*, *66*(S1), S35(A), 1979.

[17] N. Kanedera, T. Arai, H. Hermansky, and M. Pavel. On the importance of various modulation frequencies for speech recognition. In *Proceedings of EUROSPEECH 97*, Rhodos, Greece, 1997.

[18] S. van Vuuren and H. Hermansky. Data-driven design of RASTA-like filters. In *Proceedings of EUROSPEECH 97*, Rhodos, Greece, 1997.

[19] S. van Vuuren, T. Kamm, J. Luettin, and H. Hermansky. Presentation of the 1997 Summer Workshop on Innovative Techniques for Continuous Speech ASR, Johns Hopkins University, 1996, available on the http//www.clsp.jhu.edu.

28

Detection and Discrimination of Double Talk and Echo Path Changes in a Telephone Channel

Catharina Carlemalm[1]
Fredrik Gustafsson[2]

ABSTRACT
It is important to detect a change in the echo path quickly but not confuse it with double talk, since the echo canceler should react differently for these two phenomena. In this chapter, a sequential detection scheme is presented and, based on model assumptions, the maximum a posteriori probabilities of the occurrence of double talk and abrupt changes in the echo path are derived. By utilizing the fact that the most active part of a typical impulse response of a telephone channel is short and by using a low complexity time-delay estimation algorithm to find these most active taps and not estimating the rest of the taps, low computational complexity of the proposed detection method is achieved. The scheme is verified experimentally in computer simulations using a real speech signal and impulse responses created from measured impulse responses from real hybrids of a telephone channel.

28.1 Introduction

The objective of this chapter is to propose a new method for detecting and discriminating double talk and changes in the echo path in a telephone channel. This chapter is based on results presented in [1, 2].

One ultimate goal in telecommunications is to provide a perfect transmission between the users, i.e., the transmitted information must not be distorted. One severe problem is the echoes that occur during long-distance calls. Furthermore, when a telephone call is over, the same channel is available to a different call between two completely new end stations. Therefore, the echo path is quite variable and changes from call to call. Even during the progress of a call, there could be changes if, e.g., an additional telephone goes off-hook or an additional party is conferenced in. Today, in the new wireless communication systems where digital processing techniques are used, echo control has also become a prominent problem [3]. When speech is transmitted through the network, the audible echoes obstruct

[1]Division of Automatic Control, Department of Signals, Sensors and Systems, Royal Institute of Technology, SE-100 44 Stockholm, Sweden, E-mail: catharina.carlemalm@s3.kth.se.
[2]Division of Automatic Control, Department of Electrical Engineering, Linköping University, SE-581 83 Linköping, Sweden, E-mail: fredrik@isy.liu.se.

conversation and might be annoying. To eliminate or at least reduce sufficiently the echoes in the telephone network has been an extensively studied problem during the last three decades.

Commonly utilized methods for reducing echo in the telephone network are echo cancelers. The idea of the echo canceler is to use a finite impulse response (FIR) to approximate the transfer function of the echo path and let the filter coefficients be adjusted on the bases of the calculated prediction error [4]. Typically, the echo cancelers use an FIR filter and adapt it with the least mean squares (LMS) algorithm [5, 6]. This results in relatively low computational complexity, good stability, relatively good robustness against implementation errors and rapid suppression of echoes [7]. Attenuation of the echo by 30 dB is typically possible. Furthermore, if the input signal approximates white noise, initial convergence of the canceler is accomplished in one-half to one second. Unfortunately, this simple solution to the problem of echo control is not applicable because of the existence of double talk, i.e., both subscribers talk simultaneously. This phenomenon also gives rise to an abrupt increase of the prediction error. One must avoid making large corrections of the echo path during double talk in a doomed-to-failure attempt to cancel the echo. Thus, the adaptation rate should be decreased during double talk.

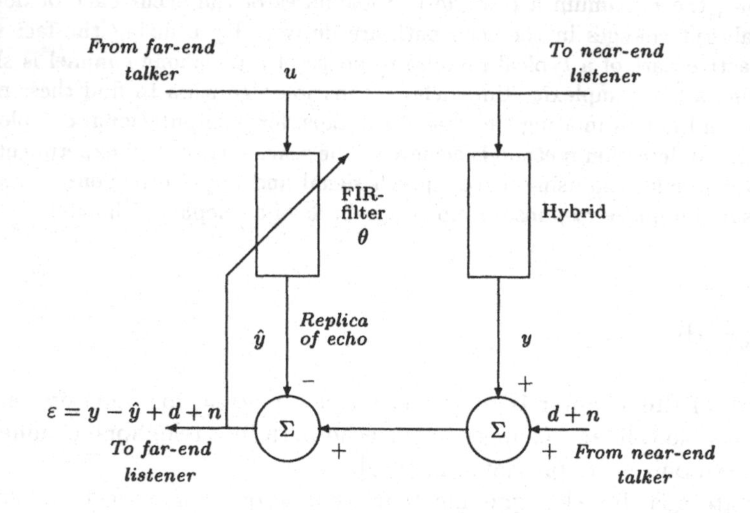

FIGURE 28.1. Schematic of the communication system

The schematic of a typical echo canceler is depicted in Fig. 28.1. The undesired echo is denoted y. The echo canceler generates the echo replica $\hat{y}$ by applying the *known* reference signal, u, to the FIR filter. The remaining echo energy is used to improve the replica. Thus, the idea of the echo canceler is to use an FIR filter to approximate the transfer function of the echo and let the filter coefficients be adjusted on the bases of the calculated prediction error ε. n denotes the background noise and d the possible double talk. When the listener is silent, i.e. $d = 0$, the adaptation of the filter should be based on the prediction error ε. During double talk, though, ε will increase drastically, and the adaptation rate should be reduced to prevent the filter parameters from diverging. Numerous double talk detectors based on more or less ad hoc principles have been suggested. In [8], the suggested detector alarms for double talk if $(d+n)$ is larger than half the largest magnitude of u over the preceding I seconds, where I is a fixed chosen number.

The conclusions are that merely measuring the prediction error will not discriminate between double talk and echo path changes and that the echo canceler must react differently whether double talk or a change in the echo path has occurred. Furthermore, the detection and discrimination algorithm must be fast to prevent the FIR filter in the echo canceler from being misadjusted.

The key contribution of this chapter is a novel algorithm for detecting and discriminating double talk and echo path changes in a telephone channel based on a maximum likelihood approach. Depending on whether double talk or an echo path change occurs, different parameters in the channel model change abruptly. Based on model assumptions, the maximum a posteriori (MAP) probabilities of the two phenomena are derived. A global channel model is compared with a local model over a sliding window, both estimated with the recursive least squares (RLS) algorithm. By utilizing the fact that the most active part of a typical impulse response of a telephone channel is short, i.e. about 4 ms long (i.e. 32 samples with the standard 8 kHz sampling frequency), and using a low-complexity time-delay estimation algorithm suggested in [9] to find these most active taps and not estimating the rest of the taps, low computational complexity in the proposed detection method is achieved. Thereafter the RLS method is used to achieve good estimates of these active taps.

The outline of this chapter is as follows. In Section 28.2, the problem is stated. Furthermore, the model and the notation used in the following sections are introduced, and assumptions are made. In Section 28.3, the proposed algorithm is derived. Next, in Section 28.4, some experimental results are presented when real speech signal and impulse responses created from measured impulse responses from real hybrids of a telephone channel are used. Finally, Section 28.5 gives some concluding remarks.

28.2 Problem Description

28.2.1 Signal Model

A finite impulse response (FIR) is generally used to describe a telephone channel. Actually, the autoregressive exogenous input (ARX) model is a better model. The motivation can partly be found, e.g., in [10]. By introducing poles in the transfer function, a more accurate description of long-impulse responses with "tails" consisting of taps close to zero is achieved.

Though, the detection and discrimination method suggested in this paper detects the *active* part of the impulse response. Therefore, the "tail" of the impulse response is not modeled. Thus, the following signal model is used to determine whether an abrupt change in the parameters has occurred at time t_0.

$$y(t) = \begin{cases} \varphi^T(t)\theta^0 + w^0(t), & \text{if } t \leq t_0 \\ \varphi^T(t)\theta^1 + w^1(t), & \text{if } t > t_0 \end{cases} \tag{28.1}$$

Here, the following notation has been used. $y(t)$ denotes the known output signal at time t, and $\varphi(t) = [u(t-n_a), u(t-n_a+1), \ldots, u(t-n_b)]^T$, where n_a and n_b denote the number of the first and last active parameters, respectively, $u(t)$ denotes the known input signal at time t, $\theta^j \in R^{n_b-n_a+1}$, and $w^j(t)$, which models the double talk signal added to the background noise, is a zero-mean, white Gaussian noise process with covariance $\Lambda(t)$.

28.2.2 Statement of Hypotheses

Consider the signal model (28.1). When a change in the echo path occurs, the parameter vector should change, i.e., $\theta^0 \neq \theta^1$. On the other hand, when double talk happens the variance of the noise should increase abruptly, i.e., $\text{Var}(w^0) \neq \text{Var}(w^1)$.

To sum up, the following hypotheses are used:

$$\begin{aligned} H_0: &\quad \theta^0 = \theta^1 \quad \text{and } \text{Var}(w^0) = \text{Var}(w^1) \\ H_1: &\quad \theta^0 \neq \theta^1 \quad \text{and } \text{Var}(w^0) = \text{Var}(w^1) \\ H_2: &\quad \theta^0 = \theta^1 \quad \text{and } \text{Var}(w^0) \neq \text{Var}(w^1) \end{aligned}$$

The situation H_1 occurs when an echo path change happens at time t_0, the situation H_2 when double talk appears at time t_0, and, finally, the null hypothesis when neither an echo path change nor double talk has occurred at time t_0.

A prior probability of each event can be assigned. Finally, it is assumed that $\text{Prob}(H_1 \cap H_2) = 0$.

28.2.3 Objectives

The objective of this chapter is to suggest a detection scheme based on a sliding window approach to detect double talk and echo path changes in a telephone channel, that is, given the data $y(1), \ldots, y(t)$, it is desired to perform the two steps described below.

1. *Channel identification:* Use a method of low complexity to find the active taps of the communication channel. Apply a sliding window approach to compute parameter estimates of the communication channel using the recursive least squares (RLS) algorithm. Then, fit a model M_0 to the data $y(1), y(2), \ldots, y(t-L)$ and a model M_1 to $y(t-L+1), \ldots, y(t)$.
2. *Change detection:* Decide which hypothesis holds, i.e., the possible presence of double talk and/or a change in the echo path based on a likelihood approach, i.e determine

$$l_i = -2 \log p(y(1), y(2), ..., y(t) | H_i, M_0, M_1), \; i = 0, 1, 2 \tag{28.2}$$

28.3 Parameter Estimation and Detection Algorithm

A sequential detection approach using a sliding window is proposed. The involved quantities are defined below.

$$\text{Data} \quad \underbrace{y(1), y(2), .., y(t-L),} \quad \underbrace{y(t-L+1), .., y(t)}$$

$$\begin{array}{lcc} \text{Model} & M_0 & M_1 \\ \text{Time interval} & T_0 & T_1 \\ \text{RLS quantities} & \theta^0,\ P_0 & \theta^1,\ P_1 \\ \text{Loss function} & V_0 & V_1 \\ \text{Number of data} & n_0 = t - L & n_1 = L \end{array} \tag{28.3}$$

where P_j, $j = 0, 1$ denotes the covariance of the parameter estimate achieved from the RLS algorithm. The loss functions are defined by

$$V_j(\theta) = \sum_{k \in T_j} \left(y(k) - \varphi^T(k)\theta\right)^T \Lambda^{-1}(k) \left(y(k) - \varphi^T(k)\theta\right),\ j = 0, 1 \tag{28.4}$$

28.3.1 Parameter Estimation

The choice of algorithm to estimate the parameter vector θ is of course crucial since the detection method will rely on the estimates and the estimation algorithm highly affects the computational complexity.

Usually, the LMS algorithm is used to update the filter parameters. We choose to apply the RLS method, though. The motivation for using RLS instead of LMS is that the convergence rate is independent of the autocorrelation characteristics and power of the input signal. This is important since speech is highly autocorrelated.

The RLS algorithm minimizes the loss function (28.4) with respect to θ.

28.3.1.1 Estimation of the Active Part of the Impulse Response

Two typical impulse responses of a telephone channel are depicted in Fig. 28.4. Evidently, most of the taps are close to zero. Consider, e.g., the bottom figure. The taps $n_a, n_a + 1, \ldots, n_b$, where $n_a \approx 9$ and $n_b \approx 25$, can be classified as most active. By using a modified version of the low complexity algorithm for fast time-delay estimation suggested in [9], the active part of the impulse response is estimated. The reader is referred to [9] for details. Its complexity is of the order $O(8n)$, where n_l is the length of the considered impulse response. The key idea is to find the active taps of the impulse response using this time-delay estimation algorithm twice. First, the time-delay of the impulse response is estimated, and, thereafter, by reversing the impulse response and applying this time-delay estimation method once more, the last active tap is estimated.

Algorithm 1:

Step 0. t:=1, n:=1. Choose the probability for false alarm, P_{fa}.

Step 1. Run the NLMS algorithm one step.

Step 2. Compute k_2 from

$$erf\left(\sqrt{\frac{k_2}{2}}\right) = (1 - P_{fa})^{1/n} \tag{28.5}$$

Step 3. Estimate $\hat{x}_{1,n}(t)$, using a Kalman filter based on

$$\begin{aligned}
\begin{pmatrix} \hat{x}_1(t+1) \\ \hat{x}_2(t+1) \end{pmatrix} &= A(t) \begin{pmatrix} \hat{x}_1(t) \\ \hat{x}_2(t) \end{pmatrix} + \bar{w}(t) \\
\hat{h}_i(t) &= C(t) \begin{pmatrix} \hat{x}_1(t) \\ \hat{x}_2(t) \end{pmatrix}, \; i = 1, \dots, n \\
A(t) &= \begin{pmatrix} 1 & 0 \\ \frac{\mu}{n_l} & 1 - \frac{\mu}{n_l} \end{pmatrix} \\
C(t) &= (0 \;\; 1) \\
R_1(t) &= \begin{pmatrix} 0 & 0 \\ 0 & \frac{\mu^2}{n_l^2} \left(\frac{\lambda}{\sigma}\right)^2 \end{pmatrix} = E\left\{\bar{w}(t)\bar{w}^T(t)\right\} \\
R_2(t) &= 0 \\
x_0 &= \begin{pmatrix} 0 & \hat{h}_i(-1) \end{pmatrix}^T \\
P_0 &= \begin{pmatrix} E\{h^2_{i,prior}\} & 0 \\ 0 & \frac{\mu}{2n_l} \left(\frac{\lambda}{\sigma}\right)^2 \end{pmatrix}
\end{aligned}$$

where $\hat{x}_1(t)$ is the estimated parameter value and $\hat{x}_2(t)$ is the average estimate of the parameter $\hat{h}_i(t)$ obtained from the NLMS. Furthermore, $E\left\{h^2_{i,prior}\right\}$ is the prior variance of the impulse response parameter values used in the prior P_0.

Step 4. Check if $\hat{x}_{1,n}(t) \in \left[-\sqrt{k_2 P_{11}(t)\mathrm{Var}(u(t))}, \sqrt{k_2 P_{11}(t)\mathrm{Var}(u(t))}\right]$ to decide which of the hypotheses

$$\begin{aligned}
H_0 &: \; \hat{x}_{1,n}(t) \in \mathcal{N}(0, P_{11}(t)) \\
H_1 &: \; \hat{x}_{1,n}(t) \in \mathcal{N}(m, P_{11}(t)), \; m \neq 0.
\end{aligned}$$

holds.

Step 5. If H_1 is accepted, stop the algorithm. The time delay is n.

Step 6. If $n < n_l$, let $n = n + 1$, and iterate steps 2–6.

Step 7. Let $t = t + 1$, and iterate steps 1–7.

28.3.2 Derivation of the Maximum A Posteriori Probabilities

The maximum likelihood approach is a standard technique often used in statistical detection theory. In this chapter, the slightly more general maximum a posteriori approach, where the prior probabilities q_i for each hypothesis can be incorporated, will be considered. Using results in [11], the exact a posteriori probabilities

$$l_i = -2 \log p(H_i | y(1), y(2), ..., y(t)), \; i = 0, 1, 2 \tag{28.6}$$

are derived below.

Assuming that $H_i, i = 0, 1, 2$, is Bernoulli distributed with probability q_i, i.e.,

$$H_i = \begin{cases} \text{does not hold}, & \text{with probability } 1 - q_i \\ \text{holds}, & \text{with probability } q_i \end{cases} \tag{28.7}$$

$\log p(H_i)$ is given by

$$\begin{aligned} \log p(H_i) &= \log\left(q_i^2(1-q_i)^{n_0+n_1-2}\right) \\ &= 2\log(q_i) + (n_0+n_1-2)\log(1-q_i),\ i=0,1,2 \end{aligned} \tag{28.8}$$

To reduce the sensitivity when the input signal u is close to zero, i.e., during silence, an inverse Wishart prior [11, 12] has been used on the variance. Thereby, the algorithm is prevented from alarming for double talk when silence is broken on the input signal.

Consider Model (28.1), where $w \in N(0,\lambda)$. Thus the prior distribution on λ is assumed to be inverse Wishart.

The inverse Wishart distribution has two parameters, m and σ, and is denoted by $W^{-1}(m,\sigma)$. Its probability density function is given by

$$p(\lambda) = \frac{\sigma^{m/2} e^{-\frac{\sigma}{2\lambda}}}{2^{m/2}\Gamma(m/2)\lambda^{(m+2)/2}} \tag{28.9}$$

The expected mean value of λ is given by

$$E(\lambda) = \frac{\sigma}{m-2} \tag{28.10}$$

and the variance is given by

$$Var(\lambda) = \frac{2\sigma^2}{(m-2)^2(m-4)} \tag{28.11}$$

The mean value given in (28.10) is chosen as the noise variance averaged over an interval, and the variance given in (28.11) as the averaged variance of the noise variance.

Result: Consider the signal model (28.1) and the hypotheses given in Section 28.2.2. The a posteriori probabilities are given by

$$\begin{aligned} l_0 &= -2\log p(H_0) + 2\log p(\{y(k)\}_{k=1}^t) + (n_0+n_1)\log(\pi) + 2\log(2) \\ &\quad + D(0) + D(1) + 2\log\Gamma\left(\frac{m}{2}\right) - m\log(\sigma) - 2\log\Gamma\left(\frac{n_0+n_1-2}{2}\right) \\ &\quad + (n_0+n_1-2+m)\log\left(V_0(\theta^0) + V_1(\theta^0) + \sigma\right) \end{aligned} \tag{28.12}$$

$$\begin{aligned} l_1 &= -2\log p(H_1) + 2\log p(\{y(k)\}_{k=1}^t) + (n_0+n_1)\log(\pi) + 2\log(2) \\ &\quad + D(0) + D(1) + 2\log\Gamma\left(\frac{m}{2}\right) - m\log(\sigma) - 2\log\Gamma\left(\frac{n_0+n_1-2}{2}\right) \\ &\quad + (n_0+n_1-2+m)\log\left(V_0(\theta^0) + V_1(\theta^1) + \sigma\right) \end{aligned} \tag{28.13}$$

$$\begin{aligned} l_2 &= -2\log p(H_2) + 2\log p(\{y(k)\}_{k=1}^t) + (n_0+n_1)\log(\pi) + 4\log(2) \\ &\quad + D(0) + (n_0-2+m)\log(V_0(\theta^0)+\sigma) - 2\log\Gamma\left(\frac{n_0-2}{2}\right) \\ &\quad + 2\log\Gamma\left(\frac{m}{2}\right) - m\log(\sigma) + D(1) \\ &\quad + (n_1-2+m)\log(V_1(\theta^0+\sigma)) - 2\log\Gamma\left(\frac{n_1-2}{2}\right) \end{aligned} \tag{28.14}$$

where

$$D(i) = \log\det P^0 - \log\det P_i + \sum_{t=1}^{n_0+n_1} \log\det \Lambda(t) \tag{28.15}$$

□

Remark: By removing terms that are small for large t and L, $t \gg L$ and constants that are equal in the Equations (28.12), (28.13) and (28.14) and using Stirling's formula, the following approximative formulas are achieved.

$$\begin{aligned} l_0 &\approx (n_0+n_1-2+m)\log\left(\frac{V_0(\theta^0)+V_1(\theta^0)+\sigma}{n_0+n_1-4}\right) \\ &\quad +\log\det(P_0^{-1}+P_1^{-1})+2\log(q_0) \end{aligned} \tag{28.16}$$

$$\begin{aligned} l_1 &\approx (n_0+n_1-2+m)\log\left(\frac{V_0(\theta^0)+V_1(\theta^1)+\sigma}{n_0+n_1-4}\right) \\ &\quad -\log\det P_0 - \log\det P_1 + 2\log(q_1) \end{aligned} \tag{28.17}$$

$$\begin{aligned} l_2 &\approx (n_0-2+m)\log\left(\frac{V_0(\theta^0)+\sigma}{n_0-4}\right) + (n_1-2+m)\log\left(\frac{V_1(\theta^0)+\sigma}{n_1-4}\right) \\ &\quad -2\log\det P_0 + 2\log(q_2) \end{aligned} \tag{28.18}$$

where P^0 is the prior covariance matrix on θ.

28.3.3 Stopping Rule

To conclude which hypothesis is valid at a specific instant, the values of the likelihood functions are calculated. The hypothesis corresponding to the least value is assumed to hold.

In real-time applications, the alarm time is the crucial quantity. Of course, it should be as small as possible. The most natural stopping rule is to say that a change has occurred as soon as l_1 or l_2 is smaller than l_0. However, to avoid false alarms, it might be advisable to accept a change hypothesis only if its likelihood has been the smallest for a number of consecutive samples.

In methodological investigations, it might be interesting to estimate the change time as accurately as possible. The idea is to try to find the peak value of the likelihood as a function of time by requiring that $l_1 - l_0$ (or $l_2 - l_0$) has been decreasing for at least half a window length L and then starts to increase, that is, the stopping rule for accepting hypothesis H_i at time $t - L$ is as follows

$$l_i(t-L) < l_k(t-L), \quad k \neq i \tag{28.19}$$

and

$$l_i(t-L-j-1) - l_0(t-L-j-1) < l_i(t-L-j) - l_0(t-L-j) \;\; \forall j \in \{1,2,\ldots,\lfloor L\rfloor/2\} \tag{28.20}$$

28.3.4 Detection Algorithm

The key idea of the proposed algorithm is to utilize the fact that the most active part of a typical impulse response of a telephone channel is only about 4 ms long

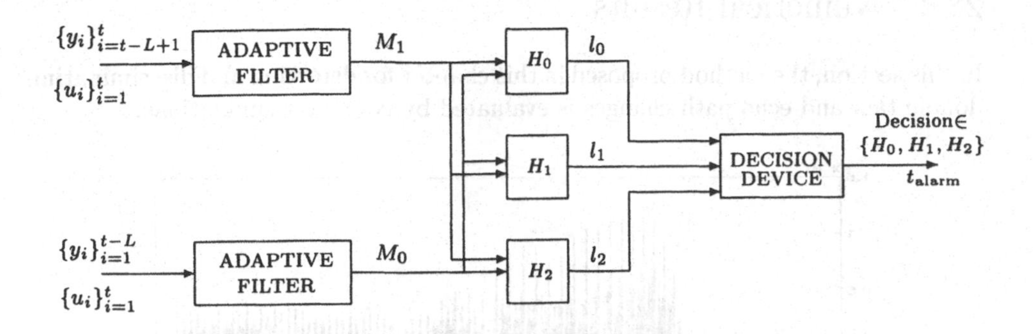

FIGURE 28.2. The model used to describe the detection scheme

(i.e., 32 samples with the standard 8 kHz sampling frequency). By using the time-delay estimation algorithm described in Algorithm 1, the most active region of the impulse response is estimated as described in Section 28.3.1.1.

Now the scheme for detecting and discriminating of double talk and echo path changes in a telephone channel can be formulated.

Algorithm 2:

Step 0. Choose the sliding window length L and the model structure, i.e., n_b, and the sensitivity constant ε used in step 4.

Step 1. If $t < L$, let $t = t + 1$.

Step 2. Using Algorithm 1, estimate the active part of the impulse response based on $\{y_t : t \in T_0\}$, where $T_0 = \{1, 2, \ldots, t - L\}$, i.e., determine n_a^0 and n_b^0.

Step 3. Using Algorithm 1, estimate the active part of the impulse response based on $\{y_t : t \in T_1\}$, where $T_1 = \{t - L + 1, t - L + 2, \ldots, t\}$, i.e., determine n_a^1 and n_b^1.

Step 4. If $|n_a^0 - n_a^1| > \varepsilon$ and/or $|n_b^0 - n_b^1| > \varepsilon$, H_1 holds.

Step 5. Estimate recursively $\theta_{active}^0 = \{h_{n_a^0}, \ldots, h_{n_b^0}\}$ and its covariance matrix P_0 based on $\{y_t : t \in T_0\}$, where $T_0 = \{1, 2, \ldots, t - L\}$.

Step 6. Estimate recursively $\theta_{active}^1 = \{h_{n_a^1}, \ldots, h_{n_b^1}\}$ and its covariance matrix P_1 based on $\{y_t : t \in T_1\}$, where $T_1 = \{t - L + 1, t - L + 2, \ldots, t\}$.

Step 7. Compute the likelihood function l_i corresponding to hypothesis i, $i = 0, 1, 2$.

Step 8. Compare the likelihood functions and use a stopping rule, discussed below, to decide whether H_1 or H_2 has occurred at time $(t - L)$.

Step 9. If the null hypothesis is rejected, restart the algorithm by letting $t = 0$.

Step 10. Let $t = t + 1$, and iterate the steps 1–9.

28.4 Numerical Results

In this section, the method proposed in this chapter for detecting and discriminating double talk and echo path changes is evaluated by computer simulations.

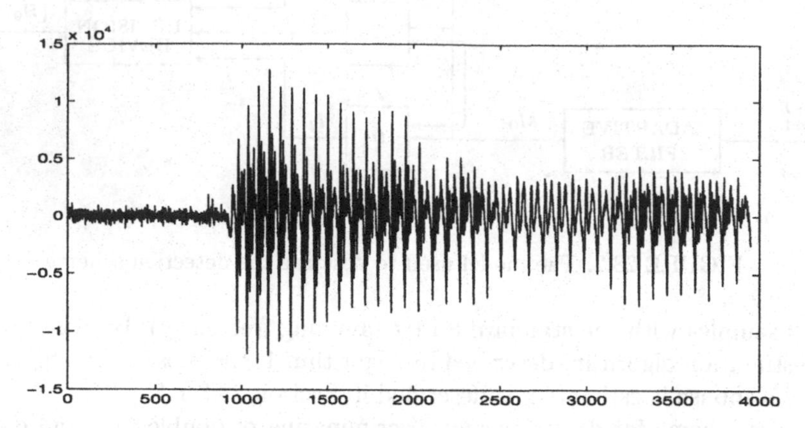

FIGURE 28.3. Input signal used in the simulations

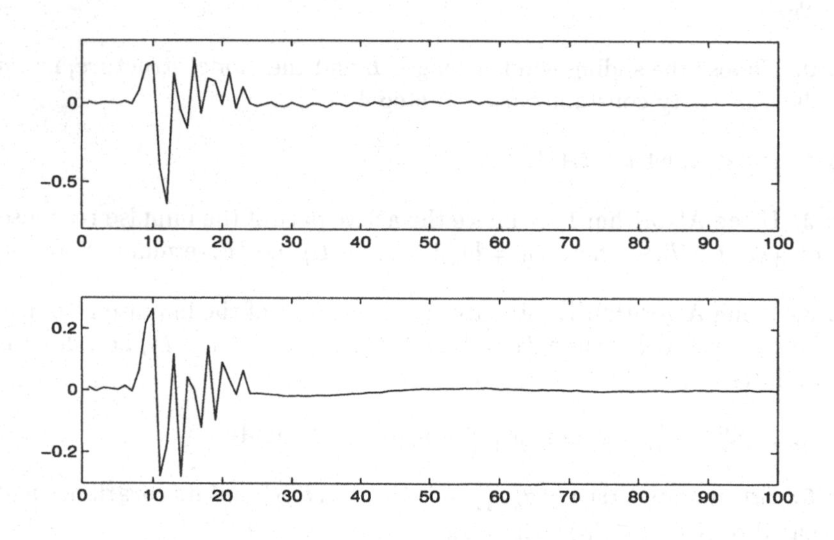

FIGURE 28.4. Impulse responses used in the simulations

Experiment 1: A real speech signal of length 4000 samples was used as the input signal. It is shown in Fig. 28.3. The signal-to-noise ratio was approximately 15 dB. To generate the different hypotheses, the input signal was segmented. A change in the echo path was simulated at sample 1976 by filtering the segments through different impulse responses. The impulse responses used are shown in Fig. 28.4. The window length was chosen to be 75. One hundred noise realizations were performed. The test detected abrupt changes at samples 2025, 2026, 2027, 2028 with 4, 72, 23, and 1% probability, respectively, and were all classified as echo path changes, i.e., the algorithm alarmed for echo path change within the interval [2025, 2028] with 100% probability. The result is shown in Fig. 28.6.

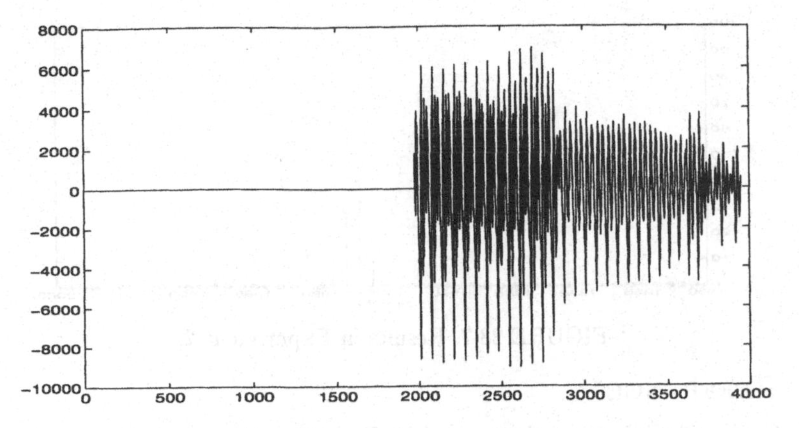

FIGURE 28.5. Double talk signal used in the simulations

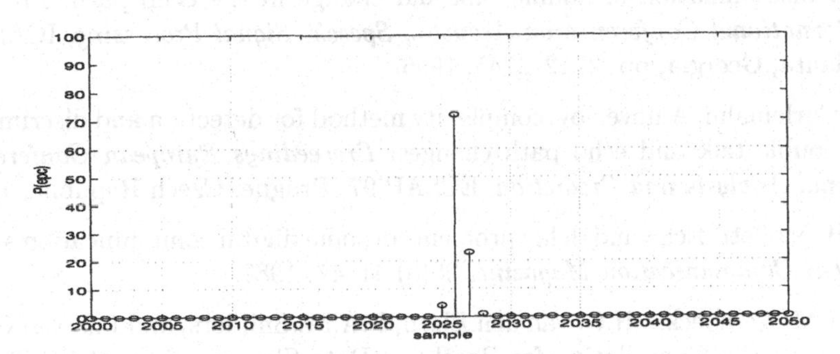

FIGURE 28.6. Results in Experiment 1

Experiment 2: The same speech signal as in Experiment 1 was used as the input signal. To generate the different hypotheses, the input signal was segmented. Double talk was simulated at sample 1976 by filtering both segments through the same impulse response (that shown at the top in Fig. 28.4) and then adding speech to one of the segments. In Fig. 28.5, the double talk signal is depicted. One hundred noise realizations were performed. The window length was chosen to be 50. The test detected abrupt changes at samples 1976, 1977, 1978, 1979, 1980, and 1981 with 10, 40, 29, 8, 0, and 13% probability, respectively, and were all classified as double talk, i.e., the algorithm alarmed for double talk within the interval [1976, 1981] with 100% probability. The result is shown in Fig. 28.7.

28.5 Conclusions

In this chapter, we studied a problem of major importance in communication, detecting and discriminating of double talk and abrupt changes in the echo path. These two phenomena are considered abrupt changes in channel characteristics. Based on model assumptions, the maximum a posteriori probabilities are derived. The proposed algorithm has relatively low complexity and is based on sequential detection theory. It is experimentally verified using real speech as the input signal and impulse responses created from measured real hybrids.

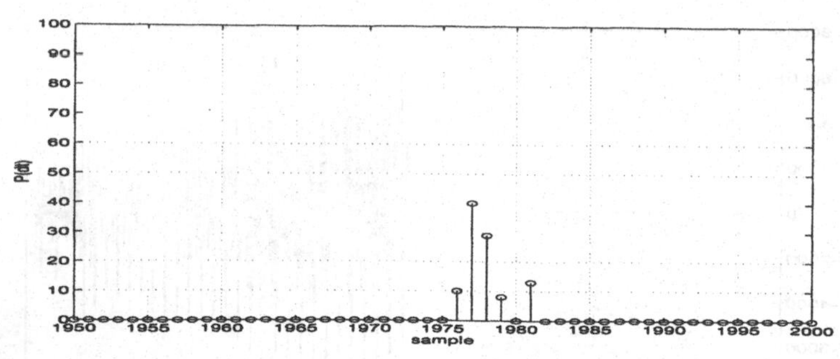

FIGURE 28.7. Results in Experiment 2

28.6 REFERENCES

[1] C. Carlemalm, F. Gustafsson, and B. Wahlberg. On the problem of detection and discrimination of double talk and change in the echo path. *Proceedings International Conference on Acoustic, Speech, Signal Processing*, ICASSP'96, Atlanta, Georgia, pp. 2742–2745, 1996.

[2] C. Carlemalm. A novel low complexity method for detection and discrimination of double talk and echo path changes. *Proceedings European Conference on Signal Analysis and Prediction*, ECSAP'97, Prague, Czech Republic, 1997.

[3] R.H. Moffett. Echo and delay problems in some digital communication systems. *IEEE Communication Magazine*, 25(8):41–47, 1987.

[4] N.A.M. Verhoeckx, H.C. van den Elzen, F.A.M. Snijders, and P.J. van Gerwen. Digital Echo Cancellation for Baseband Data Transmission. *IEEE Trans. on Acoust., Speech and Signal Proc.*, 27(6):768-781, 1979.

[5] D.G. Messerschmidt. Echo cancellation in speech and data transmission. *IEEE Journal on Selected Areas in Communications*, 2:283–297, 1984.

[6] J. Homer. Adaptive echo cancellation in telecommunications. Ph.D. Dissertation. The Australian National University, Department of Systems Engineering, Canberra, Australia, 1994.

[7] D.T.M. Slock. On the convergence behavior of the LMS algorithm. *IEEE Transaction on Signal Processing*, 41(9):2811–2825, 1993.

[8] D.L. Duttweiler. A twelve-channel digital echo canceler. *IEEE Transaction on Communications*, 26:647–653, 1978.

[9] C. Carlemalm, S. Halvarsson, T. Wigren, and B. Wahlberg. Algorithms for Time Delay Estimation Using a Low Complexity Exhaustive Search, submitted to *IEEE Transaction on Automatic Control.*

[10] U. Forssén. Tomlinson filters in the HDSL application. Report no. TRITA–TTT–9111, Royal Inst. of Tech., Stockholm, Sweden, 1991.

[11] F. Gustafsson. Estimation of Discrete Parameters in Linear Systems, Ph.D. Thesis. Linköping University, Department of Electrical Engineering, Linköping, Sweden, 1992.

[12] T.W. Anderson. The non-central Wishart distribution and certain problems of multivariate statistics. *Annals of Mathematical Statistics*, 17:409–431, 1946.

29

A CELP-Based Speech-Model Process

M. Reza Serafat[1]

ABSTRACT

A new Speech-Model Process (SMP) is proposed as a test signal for speech processing applications to avoid several shortcomings (e.g., speaker dependencies) of using natural speech as a test signal. The generation procedure for this SMP, which is based on a modified CELP-structure, is presented. The controlling part of this SMP involves several Markov chains to adapt the time-varying properties of the process to those of natural speech. Furthermore, we show that this signal can also be used as a suitable test signal for telephone speech-applications, if we limit it to the telephone frequency-band.

29.1 Introduction

One of the major problems in testing adaptive transmission systems and in objective quality assessment of speech-coding systems is the choice of a suitable test signal. The test signals should ideally be constructed such that measurement results are reproducible, comparable, and speaker as well as speech material independent. A good solution for this problem can be found by using speech-model processes.

A speech-model process is a stochastic process with characteristics similar to those of natural speech. Here, we would like to draw your attention to the differences between the characteristics of natural telephone speech (300–3400 Hz) and those of natural wide-band speech (0–8 kHz). For example, telephone speech has a symmetric Probability Density Function (PDF). But the PDF of wide-band speech has an asymmetric form. Furthermore, the contour lines of the bivariate PDF of telephone speech are nearly ellipsoidal or circular [1]. This property was the motivation for modelling telephone speech as a Spherically Invariant Random Process (SIRP) [2]. The multivariate PDFs of SIRPs depend on a nonnegative definite quadratic form $s = \underline{x}^T \mathbf{C}_{xx}^{-1} \underline{x}$ of their argument $\underline{x}^T = (x_1, \ldots, x_n)$, $\mathbf{C}_{xx}$ denoting the covariance matrix. Thus, the bivariate PDF of a SIRP exhibits ellipsoidal or circular contour lines, leading to the nomenclature 'spherical invariance'. Therefore, it is desired that a test signal for telephone speech-applications also includes this property.

Several speech-model processes have been proposed and investigated in the past

[1]Institute for Network & System Theory, University Kiel, Germany,
E-Mail: res@techfak.uni-kiel.d400.de

as speaker, independent test signals like, e.g., NSIRP (Nearly Spherically Invariant Random Process) [2], MSIRP (Markov Spherically Invariant Random Process) [3], and MSMP (Markov Speech-Model Process) [4]. But all of these speech-model processes are constructed with specific considerations. Therefore, they include some compromises to fulfil the construction constraints. For example, NSIRP and MSIRP take the special characteristics of telephone speech into account and can be used only in narrow-band speech processing applications. Besides, they involve several compromises to be mathematically tractable in a relatively simple way. MSIRP and MSMP, which generate a speech-model process as a product process, have to use comb filters to introduce a time-varying pitch structure instead of using periodically spaced impulses as excitation of the formant filters during the voiced regions. Otherwise, the probability density function (PDF) of these test signals would suffer from a large approximation error. Therefore, if we consider a spectrogram of these signals, the pitch lines are not very well recognizable.

There is also a standardized speech-model process, namely the ITU-T recommendation P.50 (artificial voice) [5]. But the short-time characteristics and the sequence behaviour of this signal are not well adapted to those of natural speech. For example, the variation of the formant structure is carried out by the uniformly distributed random choice of one of its 17 representative formant filters. Furthermore, the one- and two-dimensional PDFs of artificial voice exhibit strong deviations from those of natural speech.

Each one of the above named test signals has advantages, fortes, and drawbacks. Therefore, none of them can be seen as a general test signal for every application. Sometimes, a wide-band test signal is required, which includes a pitch structure better than that of MSMP, or a test signal is needed with properties similar to those exploited in speech coders. For these cases, we propose the CELP-based speech-model process termed MCSMP (Markov CELP-Based Speech-Model Process).

The design idea of the MCSMP generator is based on using the structure of a communication system which consists of a transmitter, a communication channel, and a receiver. In such a system, the embedded coder in the transmitter encodes the input speech signal and transmits the resulting coded parameters through the communication channel to the receiver. Then the corresponding decoder, implemented in the receiver, can decode the received parameters and produce a synthetic speech signal, which is ideally a copy of the input speech signal. If we substitute the transmitter of such a communication system with a system, which produces the required transmission parameters by applying suitable random generators, we obtain a speech-like random signal as the output of the decoder. The main problem in the designing of the desired generator is the construction of such a parameter controlling system. For such a signal generation procedure, a suitable coding system with a very efficient structure is needed. The CELP-structure exhibits the desired properties. Therefore the MCSMP applies a simplified CELP coder.

This article is organized as follows. In Section 29.2, we give a short description of our modified (simplified) CELP coder, which is used to produce the training material for the implemented trained Markov Chains (MC) in the MCSMP generator. In Part 29.3, after a short discussion about the design constrains of MCSMP generators, a brief description of the generation procedure of MCSMP is be given. Thereafter, we present some measurement results and comparisons with natural wide-band speech. We conclude with some remarks in Section 29.5.

29.2 Basic Structure of the Proposed CELP Coder/Decoder and Preparation of the Training Material for the Markov Chains

Code-Excited Linear-Prediction (CELP) coding has proven a very efficient structure for encoding natural speech with good quality and a relatively small number of transmission parameters. Therefore, we use the CELP-structure to construct a new speech-model process. Figure 29.1 shows the block diagram of the CELP coder.

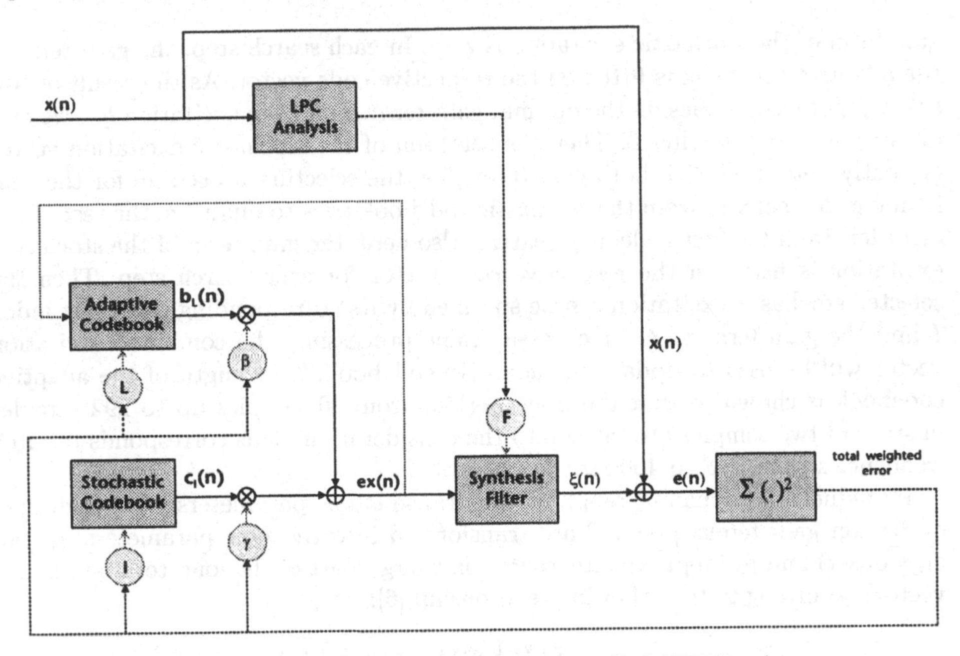

FIGURE 29.1. Block diagram of the applied CELP coder

First of all, for each frame of 20 msec duration, the applied coder specifies a lattice synthesis filter $H(z) = 1/A(z)$ of 16th order, using the Levinson-Durbin algorithm. The filter parameters are updated for each subframe of 5 msec by linear interpolation between the two reflection coefficient sets of neighboring frames. Thus, we obtain four subframes with a length of $N = 80$ samples. A gain term G_q, which represents the quantized average energy per frame, is also coded once per frame and updated for each subframe by geometric interpolation with the gain term of the preceding frame.

The coder/decoder utilize two excitation sources. The first comes from an adaptive codebook, responsible for the pitch structure. The remaining source is a stochastic codebook, containing 128 vectors of length 80. It is responsible for the excitation innovation. Both excitation sources are multiplied by their corresponding gain terms and summed to give the combined excitation sequence $ex(n)$. The selection of the excitations are carried out sequentially, using an analysis-by-synthesis procedure. This procedure attempts to minimize the total weighted error between the input signal and the synthesized speech signal, using a noise-weighting filter $W(z) = \frac{A(z)}{A(z/\lambda)}$. The weighting factor λ is set to 0.8.

First an optimal long-term prediction vector $\underline{b}_L$ is selected, assuming that the

Parameter	Type	Codebook Length
Filter	Vector $(k_1, \cdots, k_{16})$	100
G_q	Scalar	32
G_s	Vector $(g_{s,1}, \cdots, g_{s,4})$	128
P_0	Vector $(p_{0,1}, \cdots, p_{0,4})$	128
Lag	Vector $(L_1, \cdots, L_4)$	128
Index	Vector $(I_1, \cdots, I_4)$	128

TABLE 29.1. Characteristic parameters of each frame

gain term of the stochastic excitation is zero. In each search step, the gain term of the adaptive codebook is fitted on the respective code-vector. As the result of this selecting procedure, beside the optimal gain term β and the excitation $\underline{b}_L$, we also obtain the best lag value L. Then the selection of the stochastic excitation vector is jointly optimized with both gain terms, i.e. the selecting procedure for the best fitting code-vector $\underline{c}_I$ from the stochastic codebook tries to minimize the rest of the error left from the first codebook search. Also here, the gain term of the stochastic excitation is fitted on the respective code-vector for each search step. Then the selected stochastic excitation can be specified by its corresponding codebook index I and the gain term γ. After each subframe processing, the combined excitation vector will be used to update the adaptive codebook. The length of the adaptive codebook is chosen so that the pitch periods from 40 samples up to 292 samples in steps of two samples are taken into the consideration. This corresponds to pitch frequencies from 55.8 to 400 Hz.

To reduce the dynamic range of the transmission parameters, the subframe-excitation gain terms γ and β are transformed into the new parameters g_s (energy offset) and p_0 (approximate excitation energy part of the long-term predictive vector) according to the following relationship [6]:

$$g_s = \frac{\sum_{\mu=0}^{N-1} \beta^2 b_L^2(\mu) + \gamma^2 c_I^2(\mu)}{N G_q \prod_{\nu=1}^{16}(1 - k_\nu^2)} \tag{29.1}$$

$$p_0 = \frac{\sum_{\nu=0}^{N-1} \beta^2 b_L^2(\nu)}{\sum_{\mu=0}^{N-1} \beta^2 b_L^2(\mu) + \gamma^2 c_I^2(\mu)} \tag{29.2}$$

where k_ν is the νth reflection coefficient of the current synthesis filter.

Table 29.1 shows the parameters obtained for each frame. The parameters resulting from subframe processing are arranged with the corresponding vectors. Thus, there are a lag-, an index-, a G_s-, and a P_o-vector including four values of respective parameter per frame. We also obtain a vector including the 16 reflection coefficients and the quantized average energy.

We use the CELP coder described above and encode a speech data set of about 30 minutes to obtain the training material for the implemented Markov chains of the MCSMP generator. Furthermore, we restrict the possible values of the parameters obtained to a limited number of significant states. Otherwise, the Markov chains would obtain too large transition matrices, requiring an even larger set of speech material to calculate transition probabilities. Therefore, we apply vector and scalar quantizers to generate corresponding 'optimal' codebooks for the quantization of the parameters obtained from the CELP coder. The right column of Table 29.1

shows the length of these codebooks. The formant filter codebook is designed by an algorithm described in [7]. The other codebooks are optimized with the LBG-Algorithm, using the euclidean distance measure [9].

29.3 Generation Procedure of MCSMP

As mentioned before, the generation procedure of MCSMP applies the basic structure of a simplified communication system. We substitute the encoder in the transmitter of such a system with a suitable parameter generator, which produces the needed coding parameters randomly for each frame.

The desired parameter generator has to be constructed such that the coding parameters appear in a meaningful manner. Hereby, the natural interparameter relationships and the time behaviour of each parameter have to be taken into account. Therefore, the bivariate histograms of all possible pairs of parameters are calculated and compared with each other. Additionally, for the calculation of the histograms, we distinguished between the voiced and the unvoiced regions, so that we obtained two sets of histograms, one for the voiced and one for the unvoiced case. This is necessary because of the differences between the characteristics of the voiced and the unvoiced sounds related to the different ways of their generation. The voiced sounds are produced by exciting the vocal tract with periodically spaced impulses (glottis impulses), whereas for the unvoiced sounds, the excitation of the vocal tract can be modeled as a white noise with small average power [8]. Furthermore, to search the possible correlations in time for each parameter, the corresponding joint probability densities are calculated.

From these investigations, several interparameter dependencies, and time correlations of the parameters were detected. We found out that there are strong dependencies between the values of the vector of the energy offset $\underline{G}_s$ and of the vector of the adaptive excitation parts $\underline{P}_o$ and the value of the gain term G_q. Besides, there was a considerable correlation in time of the gain terms, the lag-vectors, formant filter indices, and the P_o-vectors. But the index-vector $\underline{I}$ seemed to be independent from other coding parameters, if we neglect the weak correlation between this vector and the gain terms.

As mentioned in the last section, the applied coder/decoder updates the gain term of each subframe, depending on the current and the last value of the frame gain term. Therefore, we introduced a gain vector $(\underline{G}_q(k))^T = (G_q(k-1), G_q(k)$, where the $G_q(k)$ is the gain term of the k-th frame. Thus, we obtained to the sequences of coding parameters in addition a sequence of the gain-vectors $\underline{G}_q$. Also, we applied an 'optimal' vector quantizer with 128 entries to restrict the possible values of the gain vectors to a limited number of significant values and investigate the relationship between the quantized gain vectors and the other coding parameters. Thereby, we observed a considerable dependency between the gain-vectors and the P_o-vectors and a coupling between the index vectors $\underline{I}$ and the gain vectors.

We exploited the observed interparameter dependencies and the time correlations of the parameters by the design of the controlling part of the MCSMP generator. In addition, we took the structure of the MSMP generator [4], especially the realization of the interparameter coupling in its controlling part, into consideration and implemented the MCSMP generator.

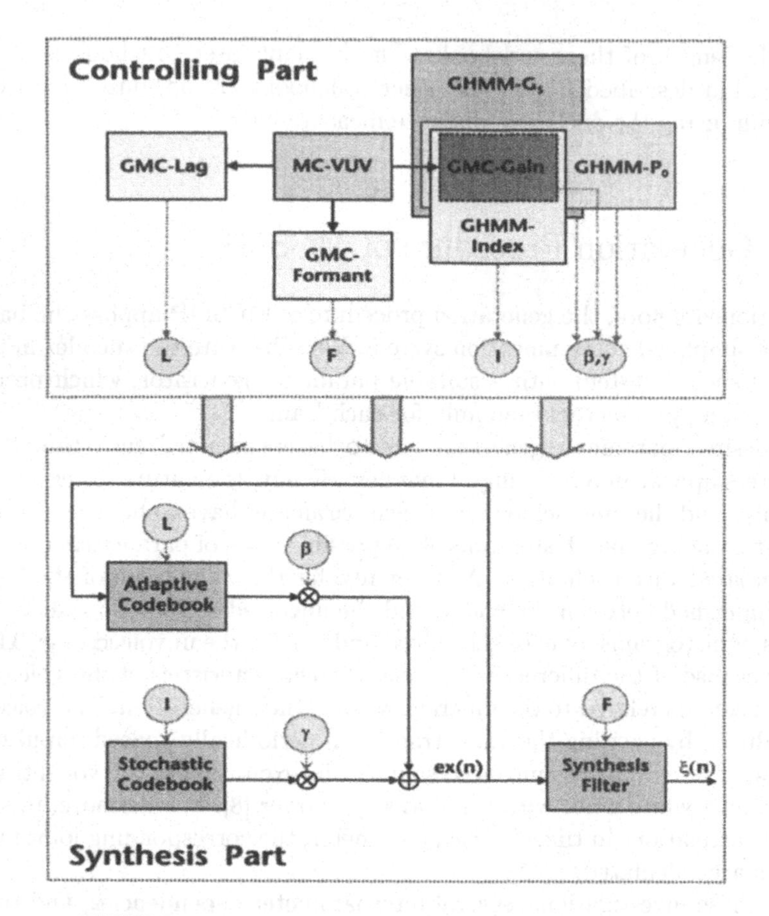

FIGURE 29.2. Block diagram of the MCSMP generator

The block diagram of the MCSMP generator is shown in Fig. 29.2. It consists of two parts, namely, the controlling part and the synthesis part. The controlling part involves several Markov chains and Generalized Hidden Markov Models (GHMMs), which are linked together in a hierarchical structure.

At the top of the hierarchical structure of the controlling parts is the MC-VUV, which decides whether a region of maximally 10 frames is voiced (v) or unvoiced (uv). This Markov chain includes $N_{(\mathrm{vuv})} = 20$ states, which are related in increasing order to the time indices $\{-N_{(\mathrm{vuv})}/2, \ldots, -1,\ 1, \ldots, N_{(\mathrm{vuv})}/2\}$. The states with the negative time indices (the first $N_{(\mathrm{vuv})}/2$ states) correspond to the unvoiced regions and the other states are related to the voiced regions. The transition from one state of this MC to another happens in nonequidistant time points. The absolute value of the time index of each state indicates the duration of staying in the current frame. For example, during the period in the third state of the MC-VUV, which is related to the time index $(-N_{(\mathrm{vuv})}/2 + 2) = -8$, an unvoiced region of length 8 frames will be produced. Furthermore, the MC-VUV produces an binary output signal $y(n)$, which is updated in each frame. This process is defined as follows:

$$y(n) = \begin{cases} 1 & ; \quad \text{voiced frame} \\ 0 & ; \quad \text{unvoiced frame} \end{cases}$$

where n indicates the number of the current frame.

Depending on the decision of the MC-VUV, three Generalized Markov Chains (GMC) are controlled. As a GMC, we define an MC, which can be influenced by an external process. In our case, we use the binary output signal of the MC-VUV $y(n)$ as such an external process. Thus, an Mth-order GMC involves 2^M transition matrices $\mathbf{A}^{(\lambda_1 \cdots \lambda_M)}$; $\lambda_\nu \in \{0,1\}$. The elements of this transition matrices are defined according to the following equation:

$$a_{ij}^{(\lambda_1 \cdots \lambda_M)} = P\,(z(n) = i|\; z(n-1) = j,\; y(n) = \lambda_M, \ldots, y(n-M+1) = \lambda_1)$$

where $P(\cdot)$ represents the probability measure and $z(n)$ denotes the current state of the GMC. Such a GMC uses the M previous states of the external process and the last state of the internal process to choose the current state.

The GMC-Lag chooses a lag-vector out of the lag-codebook with respect to the values of $y(n)$ and $y(n-1)$. The GMC-Formant chooses a formant filter, and the GMC-Gain specifies a suitable gain term from the corresponding codebooks with respect to the value of $y(n)$. GMC-Gain also works as the hidden part of GHMMs, which we use to control the excitation gain terms (β, γ) and to select the excitation code-vector from the stochastic codebook. A GHMM can be described similarly to a GMC. An Mth-order GHMM involves 2^M symbol matrices $\mathbf{B}^{(\lambda_1 \cdots \lambda_M)}$; $\lambda_\nu \in \{0,1\}$, with the elements $b_{ij}^{(\lambda_1 \cdots \lambda_M)}$ defined as follows:

$$b_{ij}^{(\lambda_1 \cdots \lambda_M)} = P\,(s(n) = i|\; z(n) = j,\; y(n) = \lambda_M, \ldots, y(n-M+1) = \lambda_1)$$

where $s(n)$ and $z(n)$ indicate the currently observed symbol and the current state of the hidden part, respectively.

Thus, we introduce a direct coupling between the average energy of the current frame and the combined excitation sequence $ex(n)$. GHMM-$\mathrm{P_0}$ specifies the P_0-vector , GHMM-$\mathrm{G_S}$ chooses a suitable G_s-vector, and GHMM-Index selects one index vector from the corresponding codebooks, depending on the current state of GMC-Gain and the values of $y(n)$ and $y(n-1)$, where each selected vector contains one value for each subframe. The excitation gain terms of each subframe can be calculated according to equations (29.1) and (29.2). Furthermore, we update the formant filter and the average frame energy by interpolating of the related parameters of the neighboring frames. In this way, we obtain a sequence of four values/indices of each parameter. For each subframe, the parameter set $(L_\nu, F_\nu, I_\nu, \beta_\nu, \gamma_\nu); \nu \in \{1, \cdots, 4\}$, where ν indicates the subframe number, will be transmitted to the synthesis part, which can be seen as a CELP-decoder. This decoder produces a sequence of 80 samples of the resulting process ξ, using the parameter set received according to the procedure described in Chapter 29.2.

29.4 Measurement Results

The PDF and the long-term spectrum of natural speech are compared to those of MCSMP. These measurement results show that the characteristics of MCSMP agree well with those of natural speech.

For example, Fig. 29.3.a shows that the spectrum of MCSMP approximates the spectrum of natural speech very closely. We also compare the spectrum of the signal envelope of MCSMP to that of natural speech (Fig. 29.3.b). The deviation between them is below the ITU tolerance limit of ± 5 dB [10].

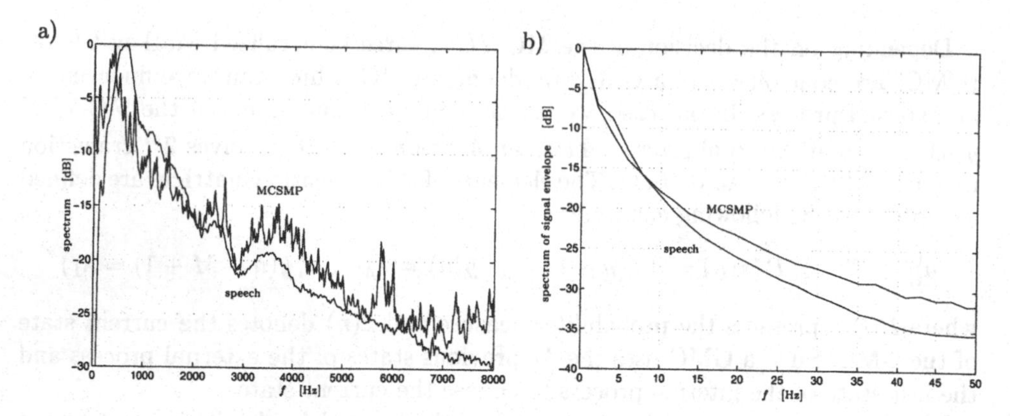

FIGURE 29.3. (a) Comparison of long-term spectrum and (b) spectrum of the signal envelope of MCSMP and natural speech

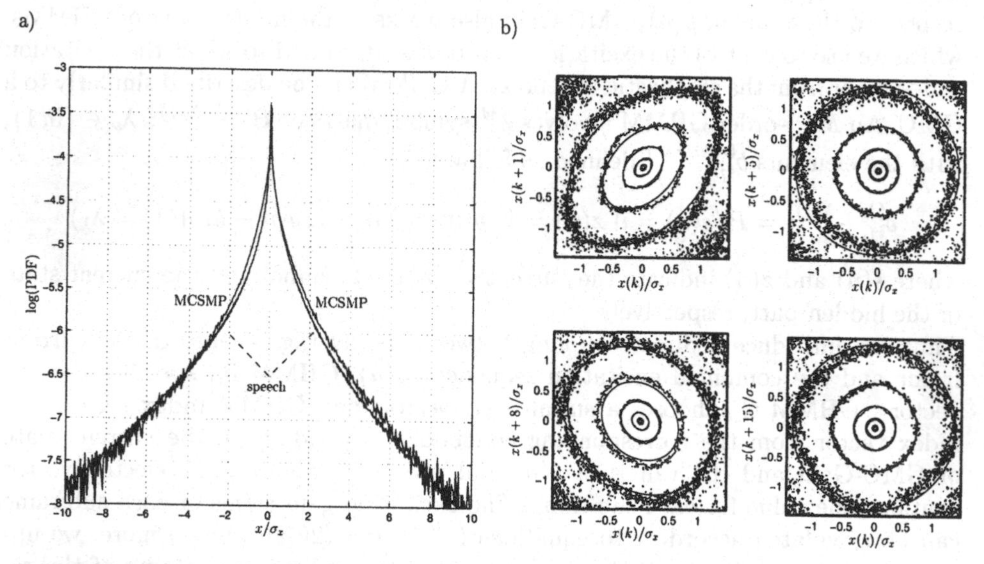

FIGURE 29.4. (a) 1-dimensional PDF of MCSMP and natural speech. (b) Comparison of contour lines of bivariate PDF of band-limited MCSMP and the theoretical curves

The Fig. 29.4.a exhibits the PDF of MCSMP and that of natural speech in good agreement. In addition, we compare the characteristics of band-pass filtered MCSMP with narrow-band speech. For this purpose, the MCSMP signal is filtered with the intermediate reference system P.48 [10] to guarantee dealing with suitably modeled telephone speech. In this case, we measure the contour lines of the bivarate PDF of MCSMP to examine whether it is nearly spherically invariant because, as mentioned before, this feature is one of the important characteristics of natural telephone speech. The comparison of these contour lines with the theoretical curves of a SIRP (Fig. 29.4.b) shows that MCSMP fulfills this requirement very well.

Furthermore, we compare the synthetic index sequence of formants, lag-vectors, and gain-terms, where the indices are used to specify the entries of formant-, lag-, and G_q-codebook respectively, with the corresponding sequence of natural speech by using their PDFs and autocorrelation functions (ACF). Some results of these measurements are presented in Fig. 29.5. The results of these comparisons can be

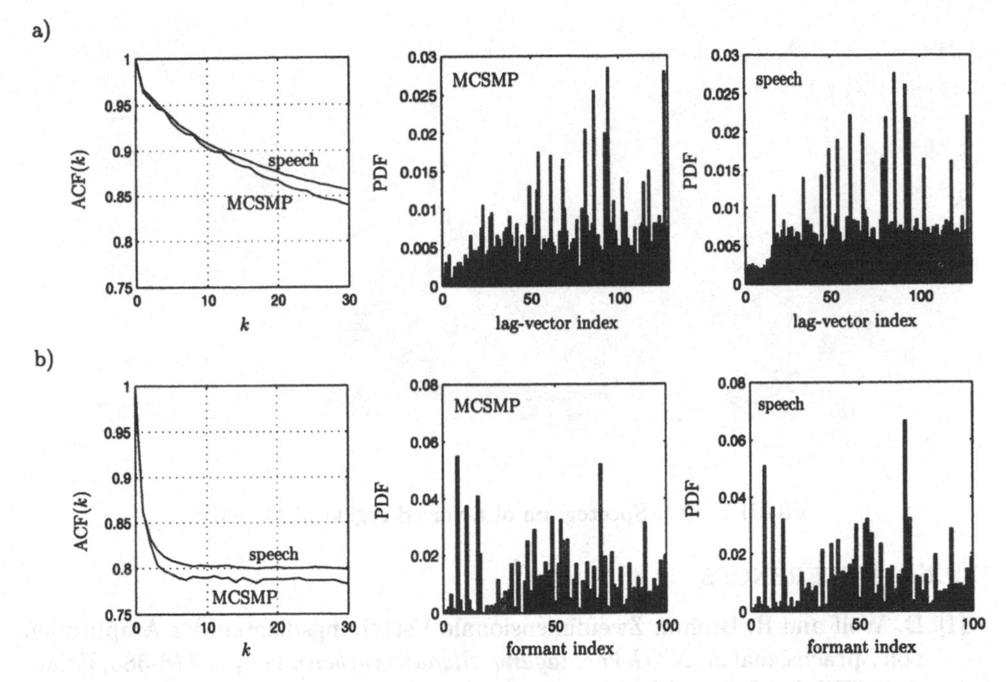

FIGURE 29.5. Autocorrelation function and PDF of *(a)* the lag-vector and *(b)* the formant indices

seen as a measure for adapting the sequence behaviour of the MCSMP to those of natural speech.

As mentioned before, one of the main aims in designing the MCSMP was to construct a test signal with a pitch structure, which is better adapted to the pitch structure of natural speech than that of MSMP. Therefore, we applied a coder with an embedded closed-loop long-term predictor. Figure 29.6 shows the spectogram of a voiced region of MCSMP. The pitch lines and their contours are well recognizable there, especially in the lower frequency band, where the power spectral density of the signal is higher.

29.5 Conclusion

The generation of a novel wide-band speech-model process is shown to be feasible with characteristics similar to those of natural speech. Even after a succeeding band limitation, the characteristics of narrow-band speech are well reproduced. This concept is applicable to generating a speaker-independent test signal, which represents some 'average-speech signals'. Besides, MCSMP renders it possible to produce a test signal with the characteristics of several languages and/or to generate a test signal which includes required special features if needed, like male/female-type speech, weak/strong variation of pitch frequency, or even the characteristics of one specific language. For such purposes, we only have to use varying speech-data sets to train the Markov chains. Furthermore, because of the closed-loop structure of the synthesis part of the MCSMP generator, we achieve good adaptation of the pitch contour of MCSMP to that of natural speech.

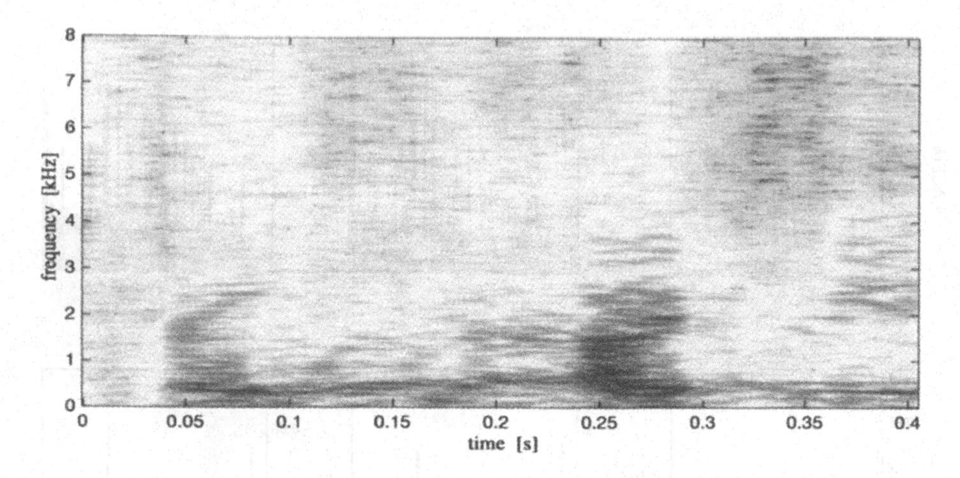

FIGURE 29.6. Spectogram of a voiced region of MCSMP

29.6 References

[1] D. Wolf and H. Brehm. Zweidimensionale Verteilungsdichten der Amplituden von Sprachsignalen. *NTG-Fachtagung 'Signalverarbeitung'*, pp. 378-385, Erlangen, 1973 (in German).

[2] H. Brehm and W. Stammler. Description and Generation of Spherically Invariant Speech-Model Signal. *Signal Processing*, Elsevier Science Publishers B.V., North Holland, 12:119–141, 1987.

[3] U. Halka and U. Heute. Speech-Model Process Controlled by Discrete Markov Chains*Proc. Asilomar Conf. Sig. Syst. Comp.*, pp. 1196–1200, Pacific Grove, U.S.A., 1993.

[4] M.R. Serafat and U. Heute. A Wide-Band Speech-Model Process As A Test Signal. *Proc. VIII European Signal Processing Conference*, Trieste, Italy, Vol. I, pp. 487–490, 1996.

[5] ITU-T Recommendation P.50, Artificial Voices*ITU-T Blue Book IXth Plenary Assembly*, Vol. V, pp. 87–99, 1988.

[6] B.S. Atal, V. Cuperman, and A. Gersho. *Advances in Speech Coding.* Kluwer Academic Publishers, Boston, 1991.

[7] A. Buzo, A. Gray, Jr., R. Gray, and J. Markel. Speech Coding Based Upon Vector Quantization.*IEEE Transactions on Acoustics, Speech, and Signal Processing*, 28:562–574, 1980.

[8] L. R. Rabiner and R. W. Schafer. *Digital Processing of Speech Signals.* Prentice-Hall, Englewood Cliffs, NJ, 1978.

[9] A. Gersho and R. M. Gray. *Vector Quantization and Signal Compression.* Kluwer Academic, Boston, Dordrecht, London, 1992.

[10] ITU-T Recommendation P.48, Specification for an Intermediate Reference System.*ITU-T Bluebook IXth Plenary Assembly*, Vol. V, pp. 81–86, 1988.

30

Vector-Predictive Speech Coding with Quantization Noise Modelling

Søren Vang Andersen[1]
Søren Holdt Jensen[1]
Egon Hansen[1]

ABSTRACT
This chapter studies the modelling of quantization noise in a predictive coding algorithm. The specific algorithm is a low-delay, vector-predictive transform coder for speech signals. The coder uses a Kalman filter with a backward estimated LPC model for the speech signal combined with an additive noise model for the quantization noise. Three different approaches to the quantization noise problem are described and compared. These are: neglecting the quantization noise in the prediction, modelling it as an uncorrelated additive noise source, and modelling it as a noise source correlated with the quantizer input. Simulations indicate that this last approach is the most efficient among those described. Moreover, we study the choice of measurements for the Kalman filter in connection with signal smoothing. In the coding algorithm described, the measurements are obtained by a transform matrix. Simulations indicate that an overlap in the vectors, which are transformed, provides an improvement in coder performance.

30.1 Introduction

The linear predictive coding (LPC) model for a speech signal can predict new speech samples from previous speech samples. This is used in predictive speech coding algorithms. In such algorithms we quantize the prediction error, and less prediction error leads to less quantization noise in the decoded signal. However, in a predictive speech coding algorithm, the prediction is not obtained from previous samples of the speech signal itself but from previous samples of the decoded speech signal contaminated by quantization noise. Therefore a model for the quantization noise along with the LPC model for the speech signal make a better speech prediction and thereby an extra reduction of the quantization noise possible.

The Kalman filter [16] provides a framework in which a linear model for the quantization noise can be combined with the LPC model for the speech signal in a coder configuration. According to Gibson [11], Irwin and O'Neal proposed the Kalman filter for differential pulse code modulation (DPCM) in 1968. From then on several workers have studied coding algorithms with Kalman filtering [9, 10,

[1] Aalborg University, Institute of Electronic Systems, Center for Personkommunikation (CPK), Fredrik Bajers Vej 7, DK-9220 Aalborg, Denmark, E-mails: {sva, shj, eh}@cpk.auc.dk

12, 17, 18]. An extensive study of Kalman filtering algorithms for adaptive DPCM (ADPCM) systems was given in 1980 by Gibson [11]. Recently Crisafulli [5] and Crisafulli, Mills, and Bitmead [6] have proposed signal smoothing in conjunction with Kalman filtering in an ADPCM coder. Also Ramabadran and Sinha [19, 20, 21] proposed a vector-predictive coding algorithm. In their algorithm a Kalman filter provides the vector prediction, and a linear measurement on the vector prediction error is scalar quantized. This linear measurement is optimized for the Kalman filter with signal smoothing [19]. The coding algorithm proposed by Ramabadran and Sinha was further studied by Andersen et al. [3]. Here it was proposed to use a cross-correlated model for the quantization noise. Computational complexity is important in the application of speech coding algorithms; Watkins, Crisafulli, and Bitmead [22] describe how to trade performance versus computational complexity when a Kalman filter is applied for speech coding.

So far most work has been focused on scalar quantization. However, the algorithms generalize directly to vector quantization [1, 2]. This chapter describes the combination of quantization noise modelling and trained vector quantization in a low-delay, vector-predictive transform coder.

30.2 Outline of the Coder

The LPC model for a speech signal is given by[2]

$$X_{t+1} = \boldsymbol{\alpha}_t \begin{bmatrix} X_t & X_{t-1} & \ldots & X_{t-p+1} \end{bmatrix}^T + Y_t \tag{30.1}$$

where t is the time index, p is the model order, $\boldsymbol{\alpha}_t$ is a row vector containing model coefficients, X_t is the process that models the speech signal, and Y_t is a zero-mean white process uncorrelated with other involved processes up to time t. Equation (30.1) may be recast in the following state-space form:

$$\boldsymbol{S}_{t+j} = \boldsymbol{\Theta}_{t+j,t}\boldsymbol{S}_t + \boldsymbol{R}_{t+j,t} \tag{30.2}$$

Here j is the number of samples we advance in the signal for each iteration of the state-space model. The state transition matrix $\boldsymbol{\Theta}_{t+j,t}$ is a square matrix defined by

$$\boldsymbol{\Theta}_{t+j,t} \equiv \begin{cases} \begin{bmatrix} \boldsymbol{\alpha}_t \\ \boldsymbol{I} \quad \boldsymbol{0} \end{bmatrix}, & j = 1 \\ \boldsymbol{\Theta}_{t+j,t+j-1}\boldsymbol{\Theta}_{t+j-1,t}, & j > 1 \end{cases}$$

The state $\boldsymbol{S}_t$ and state excitation $\boldsymbol{R}_{t+j,t}$ are column vectors defined by

$$\boldsymbol{S}_t \equiv \begin{bmatrix} X_t & X_{t-1} & \ldots & X_{t-p+1} \end{bmatrix}^T$$

and

$$\boldsymbol{R}_{t+j,t} \equiv \begin{cases} \begin{bmatrix} Y_t & \boldsymbol{0} \end{bmatrix}^T, & j = 1 \\ \boldsymbol{\Theta}_{t+j,t+j-1}\boldsymbol{R}_{t+j-1,t} + \boldsymbol{R}_{t+j,t+j-1}, & j > 1 \end{cases}$$

[2] Uppercase Arabic letters denote stochastic processes. Corresponding lowercase letters denote their outcomes. $\mathsf{E}[\cdot]$, $\mathsf{Var}[\cdot] = \mathsf{E}\left[(\cdot - \mathsf{E}[\cdot])^2\right]$, and $\mathsf{Corr}[\cdot] = \mathsf{E}\left[(\cdot - \mathsf{E}[\cdot])(\cdot - \mathsf{E}[\cdot])^T\right]$ denote expectation, variance, and autocorrelation, respectively. $\boldsymbol{I}$ denotes an identity matrix. $\boldsymbol{0}$ denotes a zero matrix.

respectively. In the following, we will suppress time indexes to avoid an overcrowded notation. Instead we introduce a delay operator D[·] such that, e.g., (30.2) becomes

$$\boldsymbol{S} = \boldsymbol{\Theta}\mathsf{D}[\boldsymbol{S}] + \boldsymbol{R}$$

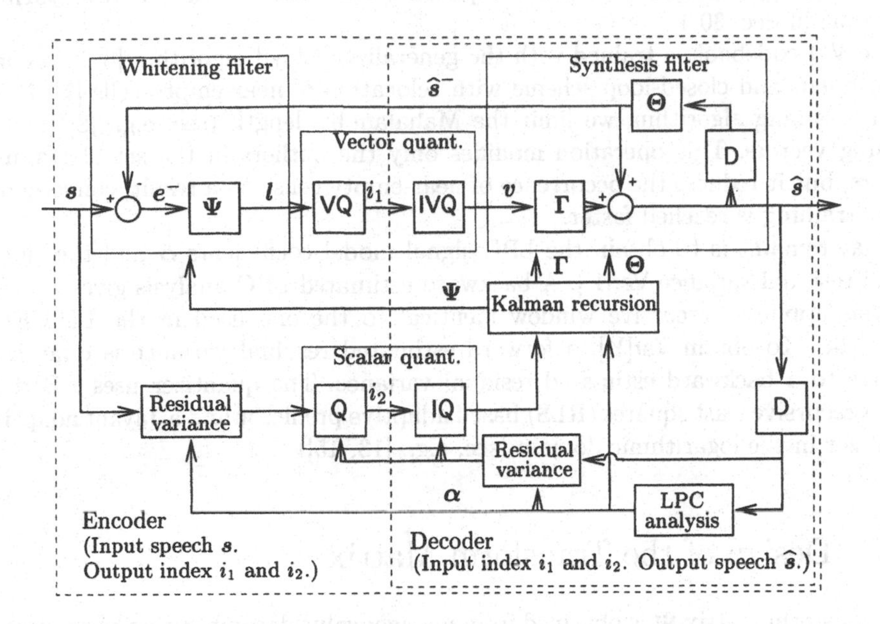

FIGURE 30.1. Block diagram of the encoder. The decoder is equivalent to the indicated subsystem of the encoder

With the state-space formulation (30.2), the coder can be illustrated as shown in Fig. 30.1. The synthesis filter provides a state prediction $\widehat{\boldsymbol{s}}^- = \boldsymbol{\Theta}\mathsf{D}[\widehat{\boldsymbol{s}}]$ from an old state estimate $\mathsf{D}[\widehat{\boldsymbol{s}}]$. Then the whitening filter calculates the prediction error $\boldsymbol{e}^- = \boldsymbol{s} - \widehat{\boldsymbol{s}}^-$ and obtains a linear transform $\boldsymbol{l} = \boldsymbol{\Psi}\boldsymbol{e}^-$. This vector forms the input to the vector quantizer VQ[·]. The quantizer index is transmitted to the decoder, and an inverse quantization of this index results in the innovation $\boldsymbol{v}$. The innovation is the input to the synthesis filter. Pre-multiplying $\boldsymbol{v}$ with the Kalman gain matrix $\boldsymbol{\Gamma}$ gives the term to update the state prediction $\widehat{\boldsymbol{s}}^-$ into the state estimate $\widehat{\boldsymbol{s}}$. The decoded signal samples are obtained from $\widehat{\boldsymbol{s}}$. By the choice of which elements of $\widehat{\boldsymbol{s}}$ are used, we trade off between coding quality and algorithmic delay. If estimates of the most recent signal samples are used, then we introduce a minimum algorithmic delay. On the other hand, if estimates of delayed signal samples are used, then the smoothing ability of the Kalman filter may result in an improved quality of the decoded signal [6, 20].

To run the whitening and synthesis filters, we have to provide the transform matrix $\boldsymbol{\Psi}$ and the Kalman gain matrix $\boldsymbol{\Gamma}$. The calculation of these matrices is based on the Kalman recursion [13]. At this point, we state the following general form:

$$\mathsf{Corr}\left[\boldsymbol{E}^-\right] = \boldsymbol{\Theta}\mathsf{D}[\mathsf{Corr}[\boldsymbol{E}]]\,\boldsymbol{\Theta}^T + \mathsf{Corr}[\boldsymbol{R}] \tag{30.3}$$

$$\boldsymbol{\Gamma} = \mathsf{E}\left[\boldsymbol{E}^-\boldsymbol{V}^T\right]\mathsf{Corr}[\boldsymbol{V}]^{-1} \tag{30.4}$$

$$\mathsf{Corr}[\boldsymbol{E}] = \mathsf{Corr}\left[\boldsymbol{E}^-\right] - \boldsymbol{\Gamma}\mathsf{E}\left[\boldsymbol{V}(\boldsymbol{E}^-)^T\right] \tag{30.5}$$

The transform matrix $\boldsymbol{\Psi}$ is derived from $\mathsf{Corr}[\boldsymbol{E}^-]$, which due to the Kalman recursion is available in both the encoder and the decoder. We return to this topic in Sec. 30.3. The quantities $\mathsf{E}\left[\boldsymbol{E}^-\boldsymbol{V}^T\right]$ and $\mathsf{Corr}[\boldsymbol{V}]$ are needed in the Kalman recursion, and they will depend on how the quantization noise is modeled. We return to this topic in Sec. 30.4.

The VQ codebook is trained with the generalized Lloyd algorithm [8] in a combined open- and closed-loop scheme with relocation of near-empty cells [7]. Prior to the training algorithm, we limit the Mahalanobis length (see, e.g., [8]) of the training vectors. This operation modifies only the outliers in the set of training vectors, but it reduces the occurrence of near-empty cells. As a result, convergence of the training is reached faster.

What remains is to obtain the LPC signal model coefficients $\boldsymbol{\alpha}$ and the signal model residual variance $\mathsf{Var}[Y]$. A backward estimated LPC analysis gives $\boldsymbol{\alpha}$. The analysis applies a recursive window identical to the one used in the LD-CELP coder [14]. To obtain $\mathsf{Var}[Y]$, a forward estimated residual variance is quantized relative to a backward estimated residual variance. The quantizer uses ADPCM with a recursive least squares (RLS) based adaptive predictor and a Jayant adaptive quantizer in the logarithmic domain (see, e.g., [13, 15]).

30.3 Design of the Transform Matrix

The transform matrix $\boldsymbol{\Psi}$ is obtained from an eigenvalue decomposition of a suitable submatrix of $\mathsf{Corr}[\boldsymbol{E}^-]$. To describe $\boldsymbol{\Psi}$, we introduce two new dimensional parameters in addition to p and j. The first dimensional parameter is the dimension m of $\boldsymbol{l}$ and $\boldsymbol{v}$. The second dimensional parameter n indicates the number of elements from $\boldsymbol{e}^-$ that should influence the transformed vector $\boldsymbol{l}$. Now, let the eigenvalue decomposition of the n-by-n upper left submatrix of $\mathsf{Corr}[\boldsymbol{E}^-]$ be given by

$$\mathsf{Corr}\left[\boldsymbol{E}^-\right]_{1..n,1..n} = \boldsymbol{\Omega}\boldsymbol{\Lambda}\boldsymbol{\Omega}^T \tag{30.6}$$

Here $\boldsymbol{\Lambda}$ is a diagonal matrix where the diagonal elements are eigenvalues in decreasing order, and $\boldsymbol{\Omega}$ is an orthogonal matrix where the columns are the corresponding eigenvectors. We choose $\boldsymbol{\Psi}$ to be

$$\boldsymbol{\Psi} \equiv \left[\begin{array}{cc} \boldsymbol{\Xi}^{\frac{1}{2}}\boldsymbol{\Lambda}^{-\frac{1}{2}}_{1..m,1..m}\boldsymbol{\Omega}^T_{1..n,1..m} & \boldsymbol{0} \end{array}\right] \tag{30.7}$$

where $\boldsymbol{\Xi}$ is a time-invariant diagonal matrix. With this $\boldsymbol{\Psi}$, the quantizer input $\boldsymbol{L}$ can be expected to have a time-invariant diagonal autocorrelation matrix equal to $\boldsymbol{\Xi}$. Moreover, we choose $\boldsymbol{\Xi}$ to be a time average of $\boldsymbol{\Lambda}_{1..m,1..m}/\boldsymbol{\Lambda}_{1,1}$, i.e., an average of the eigenvalues normalized relative to the largest eigenvalue. This approach has similarities with a fixed bit allocation for scalar quantizers. Therefore it is not an optimal approach to adaptive quantization. However, we judge it sufficient to convey our ideas about quantization noise modelling.

If $m = n = j$, then the transformed vector $\boldsymbol{l}$ is obtained by a full-rank transformation of the prediction error from the prediction of the j most recent signal samples. Without quantization noise we can exactly reconstruct the original signal from this transform. However, in the presence of quantization noise, $n > j$ will result in a smaller state-estimation error [19]. In this case the transformed vector $\boldsymbol{l}$ will contain information about, not only the prediction error from the prediction of the j most recent samples but also the coding error from the coding of earlier samples. We term this an overlapped transform (OLT). To fully benefit from the OLT, the decoded speech signal should be constructed from elements of the state estimate $\widehat{\boldsymbol{s}}$ delayed by at least $n-j$ samples.

With a limited number of bits available for the VQ, the coding of those elements in $\boldsymbol{l}$ that correspond to the smallest eigenvalues may become very noisy. Choosing $m < n$, we can then reduce the complexity of the coding algorithm with little or no loss in coder performance, compared to the case $m = n$.

30.4 Quantization Noise Models

Define the quantization noise $\boldsymbol{N}$ as

$$\boldsymbol{N} \equiv \boldsymbol{V} - \boldsymbol{L}$$

In the following, we study three different treatments of this noise: neglecting the quantization noise in the prediction (Sec. 30.4.1), modelling it as an uncorrelated additive noise source (Sec. 30.4.2), and modelling it as a noise source correlated with the quantizer input (Sec. 30.4.3).

30.4.1 The Neglected Model (NM)

The simplest case of noise modelling is to neglect $\boldsymbol{N}$ by setting $\mathsf{Corr}[\boldsymbol{N}] = \boldsymbol{0}$. With this model the Kalman equations for the NM become

$$\boldsymbol{\Gamma} = \mathsf{Corr}\left[\boldsymbol{E}^-\right] \boldsymbol{\Psi}^T \left(\boldsymbol{\Psi} \mathsf{Corr}\left[\boldsymbol{E}^-\right] \boldsymbol{\Psi}^T\right)^{-1} \tag{30.8}$$

$$\mathsf{Corr}[\boldsymbol{E}] = (\boldsymbol{I} - \boldsymbol{\Gamma}\boldsymbol{\Psi}) \mathsf{Corr}\left[\boldsymbol{E}^-\right] \tag{30.9}$$

This follows from (30.4) and (30.5), respectively.

We consider the case where $m = n = j$. For the first vector to be coded,

$$\mathsf{D}[\mathsf{Corr}[\boldsymbol{E}]] = \boldsymbol{0} \tag{30.10}$$

Therefore (30.3) reduces to $\mathsf{Corr}\left[\boldsymbol{E}^-\right] = \mathsf{Corr}[\boldsymbol{R}]$, which is a matrix that is non-zero only in the upper left j-by-j submatrix. From (30.6),

$$\mathsf{Corr}\left[\boldsymbol{E}^-\right] = \begin{bmatrix} \boldsymbol{\Omega}\boldsymbol{\Lambda}\boldsymbol{\Omega}^T & \boldsymbol{0} \\ \boldsymbol{0} & \boldsymbol{0} \end{bmatrix} \tag{30.11}$$

Inserting (30.7) and (30.11) in (30.8), we obtain

$$\boldsymbol{\Gamma} = \boldsymbol{\Psi}^T \boldsymbol{\Lambda} \boldsymbol{\Xi}^{-1} \tag{30.12}$$

With this expression for the Kalman gain, the product of the Kalman gain and the transform matrix reduces to

$$\boldsymbol{\Gamma}\boldsymbol{\Psi} = \begin{bmatrix} \boldsymbol{\Omega}\boldsymbol{\Lambda}^{\frac{1}{2}}\boldsymbol{\Xi}^{-\frac{1}{2}} \\ \boldsymbol{0} \end{bmatrix} \begin{bmatrix} \boldsymbol{\Xi}^{\frac{1}{2}}\boldsymbol{\Lambda}^{-\frac{1}{2}}\boldsymbol{\Omega}^T & \boldsymbol{0} \end{bmatrix} = \begin{bmatrix} \boldsymbol{I} & \boldsymbol{0} \\ \boldsymbol{0} & \boldsymbol{0} \end{bmatrix} \tag{30.13}$$

Using (30.11) and (30.13), we simplify (30.9) to

$$\mathsf{Corr}[\boldsymbol{E}] = \begin{bmatrix} \mathbf{0} & \mathbf{0} \\ \mathbf{0} & \boldsymbol{I}_{p-j\times p-j} \end{bmatrix} \begin{bmatrix} \boldsymbol{\Omega\Lambda\Omega}^T & \mathbf{0} \\ \mathbf{0} & \mathbf{0} \end{bmatrix} = \mathbf{0} \tag{30.14}$$

Hence, (30.10) will hold for the following vectors, too, and (30.12) gives the Kalman gain for all vectors. From (30.13) we see that this Kalman gain acts as an inverse to the transform matrix $\boldsymbol{\Psi}$; the algorithm reduces to an LPC predictive transform coder. This extends the result given by Crisafulli for DPCM coding [5] .

30.4.2 The Uncorrelated Model (UCM)

As a first refinement of the noise model, we express the autocorrelation matrix of the quantization noise $\boldsymbol{N}$ relative to the autocorrelation matrix of the quantizer input $\boldsymbol{L}$:

$$\mathsf{Corr}[\boldsymbol{N}] = \boldsymbol{\Delta}\mathsf{Corr}[\boldsymbol{L}] \tag{30.15}$$

Here $\boldsymbol{\Delta}$ is a time-invariant diagonal matrix. The diagonal elements of $\boldsymbol{\Delta}$ are noise factors for the quantization of the elements in $\boldsymbol{l}$. These noise factors are estimated for a given VQ codebook and corresponding training vectors before they are applied in the encoding algorithm. In the UCM we assume that $\boldsymbol{N}$ is zero-mean and uncorrelated with $\boldsymbol{E}^-$. With this model the Kalman equations for the UCM become

$$\boldsymbol{\Gamma} = \mathsf{Corr}\left[\boldsymbol{E}^-\right]\boldsymbol{\Psi}^T\left((\boldsymbol{I}+\boldsymbol{\Delta})\,\boldsymbol{\Psi}\mathsf{Corr}\left[\boldsymbol{E}^-\right]\boldsymbol{\Psi}^T\right)^{-1} \tag{30.16}$$

$$\mathsf{Corr}[\boldsymbol{E}] = (\boldsymbol{I}-\boldsymbol{\Gamma\Psi})\,\mathsf{Corr}\left[\boldsymbol{E}^-\right] \tag{30.17}$$

This follows from (30.4) and (30.5), respectively.

30.4.3 The Cross-Correlated Model (CCM)

Provided that the quantizer has been trained to convergence using the generalized Lloyd algorithm on a representative set of training vectors, the quantizer will meet the centroid condition [8]. This implies that $\boldsymbol{N}$ is zero-mean and

$$\mathsf{E}\left[\boldsymbol{N}\boldsymbol{L}^T\right] = -\mathsf{Corr}[\boldsymbol{N}] \tag{30.18}$$

To include (30.18) in the Kalman filter, we model the quantizer by the following linear system:

$$\boldsymbol{V} = \boldsymbol{L}+\boldsymbol{N} = \boldsymbol{\Upsilon}\boldsymbol{L}+\boldsymbol{Q} \tag{30.19}$$

and we choose $\boldsymbol{\Upsilon}$ such that

$$\mathsf{E}\left[\boldsymbol{Q}\boldsymbol{L}^T\right] = \mathsf{E}\left[\boldsymbol{V}\boldsymbol{L}^T\right] - \boldsymbol{\Upsilon}\mathsf{Corr}[\boldsymbol{L}] = \mathbf{0} \tag{30.20}$$

This implies that

$$\boldsymbol{\Upsilon} = \mathsf{E}\left[\boldsymbol{V}\boldsymbol{L}^T\right]\mathsf{Corr}[\boldsymbol{L}]^{-1} \tag{30.21}$$

Now our task is to derive expressions for $\mathsf{E}\left[\boldsymbol{V}\boldsymbol{L}^T\right]$. From (30.15) and (30.18), it follows that

$$\begin{aligned} \mathsf{E}\left[\boldsymbol{V}\boldsymbol{L}^T\right] &= \mathsf{E}\left[(\boldsymbol{L}+\boldsymbol{N})\boldsymbol{L}^T\right] = \mathsf{Corr}[\boldsymbol{L}] + \mathsf{E}\left[\boldsymbol{N}\boldsymbol{L}^T\right] \\ &= \mathsf{Corr}[\boldsymbol{L}] - \mathsf{Corr}[\boldsymbol{N}] = (\boldsymbol{I}-\boldsymbol{\Delta})\,\mathsf{Corr}[\boldsymbol{L}] \end{aligned} \tag{30.22}$$

When we use the relationship

$$\mathsf{Corr}[\boldsymbol{V}] = \mathsf{Corr}[\boldsymbol{L}] + \mathsf{Corr}[\boldsymbol{N}] + 2\mathsf{E}\left[\boldsymbol{L}\boldsymbol{N}^T\right] = \mathsf{Corr}[\boldsymbol{L}] - \mathsf{Corr}[\boldsymbol{N}]$$

we can rewrite (30.22) as

$$\mathsf{E}\left[\boldsymbol{V}\boldsymbol{L}^T\right] = \mathsf{Corr}[\boldsymbol{V}] \tag{30.23}$$

Before (30.19) is used to derive the Kalman equations for the CCM, we analyze the cross-correlation between the state prediction error $\boldsymbol{E}^-$ and the noise source $\boldsymbol{Q}$. For this analysis, we split $\boldsymbol{E}^-$ into an estimate of $\boldsymbol{E}^-$ from $\boldsymbol{L}$ and the corresponding estimation error:

$$\boldsymbol{E}^- = \boldsymbol{\Phi}\boldsymbol{L} + \boldsymbol{U} \tag{30.24}$$

Here $\boldsymbol{\Phi}$ is chosen such that

$$\mathsf{E}\left[\boldsymbol{U}\boldsymbol{L}^T\right] = \mathsf{E}\left[\boldsymbol{E}^-\boldsymbol{L}^T\right] - \boldsymbol{\Phi}\mathsf{Corr}[\boldsymbol{L}] = 0 \tag{30.25}$$

which implies that

$$\boldsymbol{\Phi} = \mathsf{E}\left[\boldsymbol{E}^-\boldsymbol{L}^T\right]\mathsf{Corr}[\boldsymbol{L}]^{-1} \tag{30.26}$$

To proceed, we assume that $\boldsymbol{E}^-$ is zero-mean Gaussian. The stochastic vectors $\boldsymbol{L}$ and $\boldsymbol{U}$ are both linear functions of $\boldsymbol{E}^-$; if $\boldsymbol{E}^-$ is zero-mean Gaussian, then so are $\boldsymbol{L}$ and $\boldsymbol{U}$. Therefore (30.25) implies independence between $\boldsymbol{L}$ and $\boldsymbol{U}$ [4]. For a given quantizer, $\boldsymbol{Q}$ is a deterministic function of $\boldsymbol{L}$. Hence, $\boldsymbol{Q}$ and $\boldsymbol{U}$ are independent. Because they both are zero-mean, it follows that

$$\mathsf{E}\left[\boldsymbol{U}\boldsymbol{Q}^T\right] = 0 \tag{30.27}$$

From (30.20), (30.24), and (30.27) it follows that

$$\mathsf{E}\left[\boldsymbol{E}^-\boldsymbol{Q}^T\right] = \boldsymbol{\Phi}\mathsf{E}\left[\boldsymbol{L}\boldsymbol{Q}^T\right] + \mathsf{E}\left[\boldsymbol{U}\boldsymbol{Q}^T\right] = 0 \tag{30.28}$$

In conclusion to the extent that $\boldsymbol{E}^-$ can be considered zero-mean Gaussian, the new noise source $\boldsymbol{Q}$ can be considered uncorrelated with $\boldsymbol{E}^-$.

Now we are in a position to find expressions for $\mathsf{E}\left[\boldsymbol{E}^-\boldsymbol{V}^T\right]$. From (30.19), (30.21), and (30.28), it follows that

$$\begin{aligned}\mathsf{E}\left[\boldsymbol{E}^-\boldsymbol{V}^T\right] &= \mathsf{E}\left[\boldsymbol{E}^-\boldsymbol{L}^T\right]\boldsymbol{\Upsilon}^T + \mathsf{E}\left[\boldsymbol{E}^-\boldsymbol{Q}^T\right] \\ &= \mathsf{Corr}\left[\boldsymbol{E}^-\right]\boldsymbol{\Psi}^T\mathsf{Corr}[\boldsymbol{L}]^{-1}\,\mathsf{E}\left[\boldsymbol{L}\boldsymbol{V}^T\right]\end{aligned} \tag{30.29}$$

When we use (30.29) together with (30.22) and (30.23), the Kalman equations for the CCM become

$$\boldsymbol{\Gamma} = \mathsf{Corr}\left[\boldsymbol{E}^-\right]\boldsymbol{\Psi}^T\left(\boldsymbol{\Psi}\mathsf{Corr}\left[\boldsymbol{E}^-\right]\boldsymbol{\Psi}^T\right)^{-1} \tag{30.30}$$

$$\mathsf{Corr}[\boldsymbol{E}] = \left(\boldsymbol{I} - \boldsymbol{\Gamma}\left(\boldsymbol{I} - \boldsymbol{\Delta}\right)\boldsymbol{\Psi}\right)\mathsf{Corr}\left[\boldsymbol{E}^-\right] \tag{30.31}$$

This follows from (30.4) and (30.5), respectively.

Note that with the CCM the noise factor matrix $\boldsymbol{\Delta}$ does not appear explicitly in the calculation of the Kalman gain (30.30). However, it appears implicitly through the update of the autocorrelation matrix for the prediction error in (30.3) and (30.31). The computational complexity for the Kalman recursion with the CCM is not significantly larger than with the UCM. In the UCM, the m-by-m diagonal matrix $(\boldsymbol{I}+\boldsymbol{\Delta})$ must be multiplied by the m-by-m symmetric matrix $\boldsymbol{\Psi}\mathsf{Corr}[\boldsymbol{E}^{-}]\,\boldsymbol{\Psi}^T$ (c.f. (30.16)), whereas in the CCM the m-by-m diagonal matrix $(\boldsymbol{I}-\boldsymbol{\Delta})$ must be multiplied by $\boldsymbol{\Psi}$, and the non-zero part of this matrix has dimension m-by-n.

30.5 Simulation Results

The impact of quantization noise modelling on the performance of the coding algorithm was studied by computer simulations.

30.5.1 Test Configurations

In the realizable coder (RC), described in Sec. 30.2, the influence of quantization noise on the vector prediction is twofold:

1. the predictor state $\widehat{\boldsymbol{s}}$ is contaminated by quantization noise;
2. the predictor coefficients $\boldsymbol{\alpha}$ are obtained by LPC analysis of the decoded signal, which is derived from $\widehat{\boldsymbol{s}}$ and therefore contaminated by the same quantization noise.

The first influence is explicitly modeled by the noise models described in Sec. 30.4; it is relevant to study this influence both in connection with and disconnected from the second influence. For this reason, along with the RC, we simulated a second coder configuration in which the predictor coefficients $\boldsymbol{\alpha}$ are obtained from the noise-free speech signal. We term this the noise-free analysis (NFA). With the NFA the second influence of the quantization noise is not present in the vector prediction.

In a third coder configuration the predictor coefficients $\boldsymbol{\alpha}$ are obtained as in the NFA, and $\boldsymbol{s}$ is used in place of $\widehat{\boldsymbol{s}}$ as the predictor state. We term this the noise-free prediction (NFP). With the NFP none of the influences of the quantization noise are present in the vector prediction. Thus the performance of the NFP provides a useful upper bound for the performance of the NFA and the RC. To be specific, the simulated configurations can be characterized as follows:

- RC: $\widehat{\boldsymbol{s}}^{-} = \boldsymbol{\Theta}\mathsf{D}[\widehat{\boldsymbol{s}}]$ and $\boldsymbol{\alpha}$ is obtained from $\mathsf{D}[\widehat{\boldsymbol{s}}]$;
- NFA: $\widehat{\boldsymbol{s}}^{-} = \boldsymbol{\Theta}\mathsf{D}[\widehat{\boldsymbol{s}}]$ and $\boldsymbol{\alpha}$ is obtained from $\mathsf{D}[\boldsymbol{s}]$;
- NFP: $\widehat{\boldsymbol{s}}^{-} = \boldsymbol{\Theta}\mathsf{D}[\boldsymbol{s}]$ and $\boldsymbol{\alpha}$ is obtained from $\mathsf{D}[\boldsymbol{s}]$.

30.5.2 Parameter Settings

Several coders were simulated. In all coders the LPC signal model had an order of $p = 50$. For every j samples, two bits were allocated to the quantization of $\mathsf{Var}[Y]$ with a predictor order of 10 and the forgetting factor equal to 0.99. The VQ codebook contained 256 vectors. Thus eight bits were allocated to the VQ index. In the coders j was chosen to be 10, 8, 6, and 5. This corresponds to bit rates of 8, 10, 13.3, and 16 kbps, respectively.

The output was taken from the state estimate $\widehat{s}$ at a zero sample delay as well as a j sample delay for all coders. We did this to study the impact of smoothing. The simulations included two different designs of the transform matrix. In the first design, $n = m = j$ was used. In the second design, an overlapped transform (OLT) with $n = 2m = 2j$ was used.

The VQ codebooks were trained on speech data from 15 female and 15 male speakers. Each speaker read a different Danish text. In the training we used the parameters listed below. To obtain satisfactory results for the combination of the RC and the OLT at 8 and 10 kbps it was necessary to modify these parameters, as indicated in the brackets.

- The training ratio was 200 (400). Thus a total of 1707 (3414) vectors from each speaker were used.
- For each speaker, the signal was partitioned into 10 ms segments. For these segments the average power was calculated. Only vectors originating from speech segments with a signal power no more than 10 dB (40 dB) below this average were used in the training.
- The threshold for near-empty cells was set to 5% of the training ratio.
- The Mahalanobis length for training vectors was limited to 4.0 (5.0). With this value, typically less than 2% (1%) of the training vectors were moved.
- The training was closed-loop iterated. Twenty (40) iterations were used here.

These parameters were chosen as a compromise between performance and computational demands.

30.5.3 *Objective Measures*

The performance of the coders was measured on a total of 150 sec. of speech obtained in the same manner as the training data from four female and four male speakers. None of these speakers, nor the texts they read, were represented in the training data. We used the segmental SNR as an objective performance measure. The segmental SNR was calculated from 10 ms segments. Only segments with a signal power no more than 40 dB below the average for the speaker were used in the calculation of the segmental SNR.

The three different approaches to the noise modelling were studied. Fig. 30.2 gives the results. The NFA was used in Fig. 30.2.A. We see that noise modelling generally improves the segmental SNR measure and that the CCM provides better results than the UCM. This remained true when going to the RC, as can be seen in Fig. 30.2.B.

The smoothing and the OLT were studied in connection with the CCM. Fig. 30.3 gives the results. The NFA was used in Fig. 30.3.A. We see that the smoothing and especially the smoothing combined with the OLT significantly improves the performance of the coder with the CCM. We take the performance of the NFP as an upper bound for the performance of the NFA. For the CCM with the smoothing

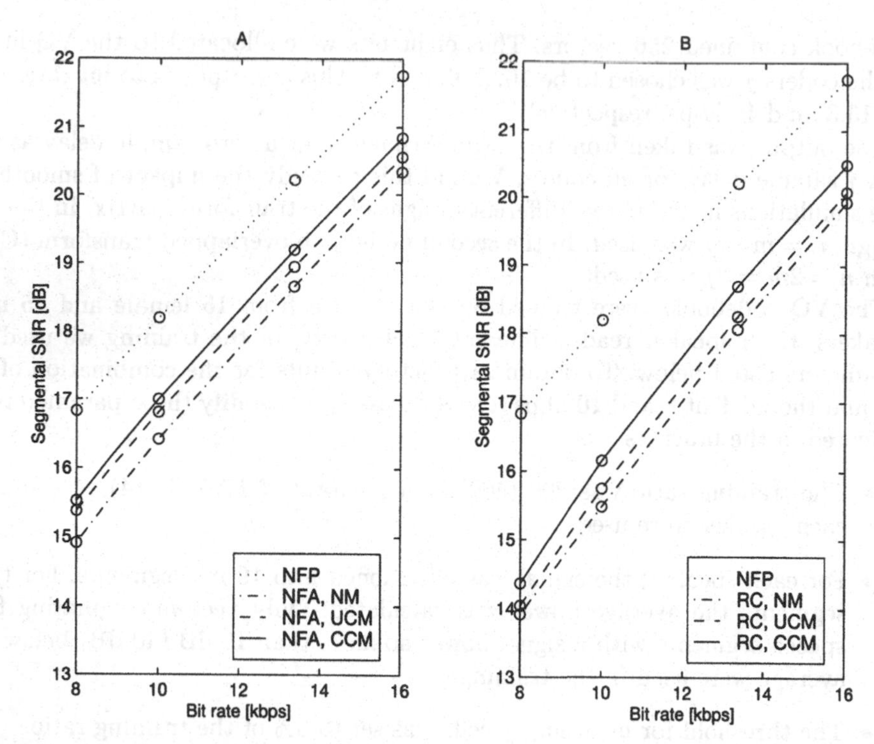

FIGURE 30.2. Segmental SNR for coding with the three noise models at different bit rates. A: Results obtained with the noise free analysis (NFA). B: Results obtained with the realizable coder (RC)

and the OLT, the performance of the NFA is less than 0.5 dB below this upper bound. The smoothing and the OLT still improves the performance of the CCM when going from the NFA to the RC, as can be seen in Fig. 30.3.B. However, the performance of the RC with the CCM, smoothing, and OLT is now significantly lower than the performance of the NFP.

For comparison we used the 16 kbps LD-CELP coder [14] without the perceptual postfilter. This coder had a segmental SNR of 20.0 dB when tested on the same data.

30.5.4 Informal Listening

Informal listening tests were carried out for the RC's. At the high bit rates (13.3 and 16 kbps) the coding noise was perceived as a signal correlated background noise, and at the low bit rates (8 and 10 kbps) as harshness of the speech signal. The algorithm with the CCM, smoothing, and OLT generally provided a better perceptual quality than the NM version. At high bit rates this improvement was due to a reduction of the background noise. At lower bit rates the decoded signal became less harsh. An example of the error signals for the two 10 kbps coders compared can be seen in Fig. 30.4.

An improvement of the CCM over the UCM was noticed. The reduction of background noise and signal harshness was judged the same in the two cases. In the UCM case, however, a noticeable spectral distortion was introduced, resulting in an artificially bass voice in the decoded signal.

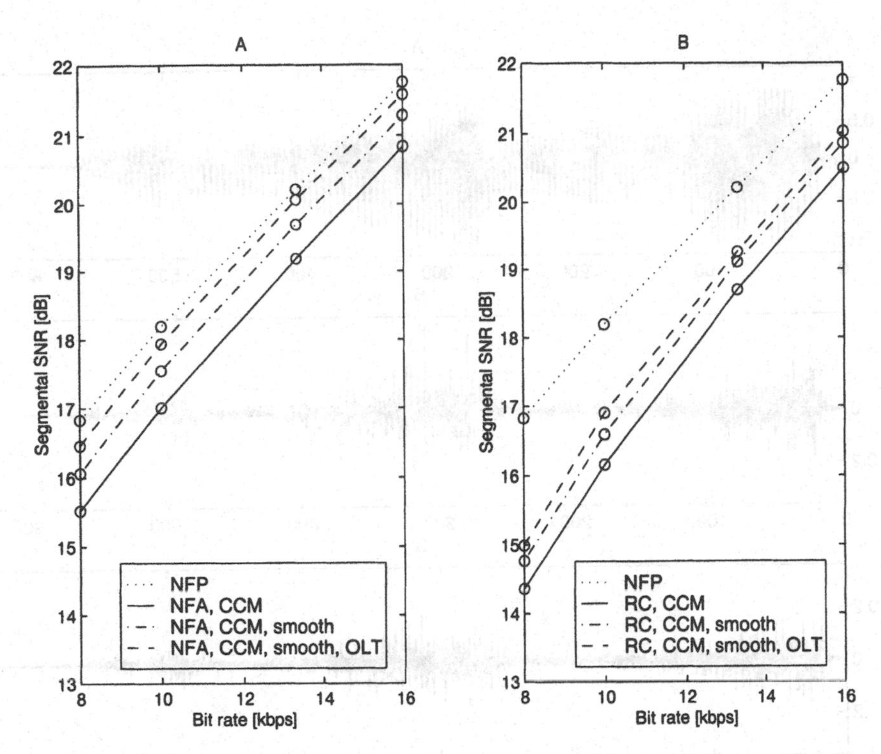

FIGURE 30.3. Segmental SNR for coding with the smoothing and the overlapped transform (OLT) at different bit rates. A: Results obtained with the noise free analysis (NFA). B: Results obtained with the realizable coder (RC)

For female voices the 16 kbps coder with the CCM, smoothing, and OLT was judged perceptually comparable to the LD-CELP coder. The LD-CELP coder introduced less signal distortion whereas our coder introduced less signal correlated background noise. For male voices the LD-CELP coder was judged generally better.

30.6 Conclusions

In this chapter a low-delay, vector-predictive speech coding algorithm has been studied. The algorithm is a vector-predictive transform coder with a backward estimated LPC and a trained vector quantization. It applies a Kalman filter for the vector prediction. This involves explicit modelling of the quantization noise.

The coding algorithm has been simulated at bit rates between 8 and 16 kbps. The performance has been evaluated by segmental SNR and informal listening tests.

Three models for the quantization noise have been described: the neglected noise model (NM), the uncorrelated noise model (UCM), and the cross-correlated noise model (CCM). With the NM we have shown that the coding algorithm reduces to an LPC predictive transform coder. Simulations have shown that the UCM and especially the CCM provide better performance than the NM.

With an appropriate quantization noise model, the Kalman filter provides signal smoothing at the expense of delay but at no extra computational cost. Simulations have shown that smoothing is efficient in improving the performance of the coding algorithm when the CCM is used.

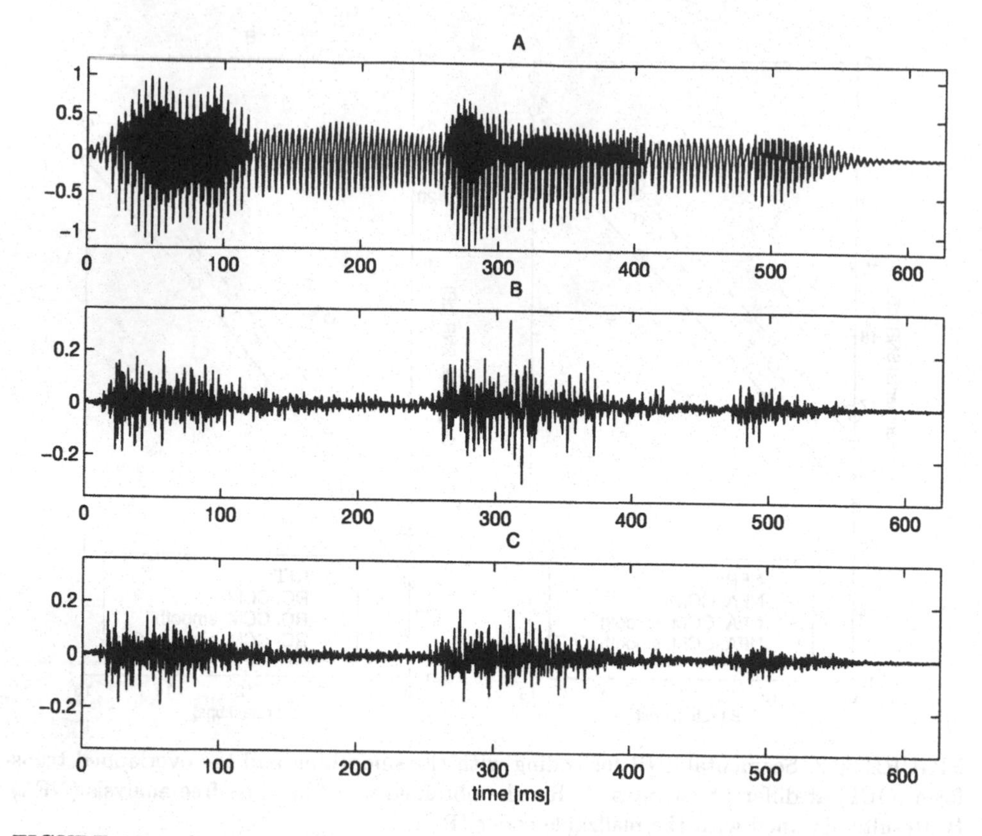

FIGURE 30.4. Coding at 10 kbps with the RC. A: The word "vandvarmer" from the Danish sentence "Jeg har et problem med min vandvarmer" spoken by a female speaker. B: The error signal after coding with the NM. C: The error signal after coding with the CCM, smoothing and overlapped transform (OLT)

The transform matrix can be designed so that the transformed vector contains information about the prediction error and also the quantization error from the quantization of previous prediction errors. Simulations have shown that this approach can further improve the performance of the coding algorithm when used in conjunction with the CCM and smoothing.

An upper bound for the SNR of the algorithms with quantization noise modelling is the SNR of an algorithm in which the quantization noise has been artificially removed from the vector prediction. With a noise-free LPC analysis, we obtained a segmental SNR for the noise modelling approach within 0.5 dB from this upper bound. However, the Kalman filter is sensitive to noise in the LPC analysis. When the quantization noise affects the backward estimated LPC, a drop in segmental SNR of as much as 1.5 dB was observed. With noise in the LPC analysis, the coding algorithm with the CCM and smoothing provides a performance gain of 1.0 to 1.4 dB compared to the coding algorithm with the NM.

This chapter does not propose a final coder. The coding algorithm used has no perceptual weighting, and postfiltering has not been introduced. However, the results presented in this chapter indicate that modelling of the quantization noise can improve the performance of a low-delay, vector-predictive speech coder.

30.7 References

[1] S.V. Andersen, S.H. Jensen, and E. Hansen. Quantization Noise Modeling in Low-Delay Speech Coding. In *IEEE Workshop on Speech Coding for Telecommunications Proceedings*, pp. 65–66, Pocono Manor, PA, 1997.

[2] S.V. Andersen, S.H. Jensen, and E. Hansen. Quantization Noise Modeling in Low-Delay Vector-Predictive Speech Coding. In *Signal Analysis and Prediction I*, pp. 299–302. ICT Press Prague, Czech Republic, 1997.

[3] S.V. Andersen, M. Olesen, S.H. Jensen, and E. Hansen. Innovation Coding with a Cross-Correlated Quantization Noise Model. In *Signal Processing VIII: Theories and Applications*, pp. 1019–1022. Edizioni LINT Trieste, Italy, 1996.

[4] P.J. Brockwell and R.A. Davis. *Time Series: Theory and Methods.* Springer-Verlag, 1991.

[5] S. Crisafulli. Adaptive Speech Coding via Feedback Techniques. Ph.D. Thesis, Department of Systems Engineering, Research School of Physical Sciences and Engineering, The Australian National University, Canberra, Australia, 1992.

[6] S. Crisafulli, J.D. Mills, and R.R. Bitmead. Kalman Filtering Techniques in Speech Coding. In *Proc. IEEE Int. Conf. Acoust., Speech, Signal Processing*, 1, pp. 77–80, San Francisco, California, 1992.

[7] V. Cuperman and A. Gersho. Vector-Predictive Coding of Speech at 16 kbits/s. *IEEE Trans. Comm.*, 33:685–696, 1985.

[8] A. Gersho and R.M. Gray. *Vector Quantization and Signal Compression.* Kluwer Academic Publishers, 1992.

[9] J.D. Gibson. Optimal and Suboptimal Estimation in Differential PCM and Adaptive Coding Systems. In *Proc. IEEE Conf. Decision and Control*, p. 353, Houston, TX, 1975.

[10] J.D. Gibson. Sequential Filtering of Quantization Noise in Differential Encoding Systems. In *Proc. Int. Conf. Cybernetics and Society*, pp. 685–689, Washington, DC, 1976.

[11] J.D. Gibson. Quantization Noise Filtering in ADPCM Systems. *IEEE Trans. Sys., Man, and Cybern.*, 10:529–536, 1980.

[12] J.E. Gunn and A.P. Sage. Speech Data Rate Reduction I: Applicability of Modern Estimation Theory. *IEEE Trans. Aerospace Electron. Syst.*, 9:130–138, 1973.

[13] S. Haykin. *Adaptive Filter Theory*, 3rd edition. Prentice Hall, Engelwood Cliffs, NJ, 1996.

[14] ITU International Telecommunication Union, Geneva. Coding of Speech at 16 kbit/s Using Low-Delay Code Excited Linear Prediction, Recommendation G.728, 1992.

[15] N.S. Jayant. Adaptive Quantization With a One-Word Memory. *The Bell System Technical Journal*, 52:1119–1144, 1973.

[16] R.E. Kalman. A New Approach to Linear Filtering and Prediction Problems. *Trans. ASME, J. Basic Engineering*, 82:35–45, 1960.

[17] J.L. Melsa and R.B. Kolstad. Kalman Filtering of Quantization Error in Digitally Processed Speech. In *Conf. Rec. IEEE Int. Conf. Commun.*, pp. 310–313, Chicago, Illinois, 1977.

[18] G. Pirani and C. Scagliola. Performance Analysis of DPCM Speech Transmission Systems using the Kalman Predictors. In *Conf. Rec. 1976 IEEE Int. Conf. on Acoustics, Speech and Signal Processing*, pp. 262–265, Philadelphia, Pennsylvania, 1976.

[19] T.V. Ramabadran and D. Sinha. On the Selection of Measurements in Least-Squares Estimation. In *Proc. IEEE Int. Conf. on Systems Eng.*, pp. 221–226, Dayton, Ohio, 1989.

[20] T.V. Ramabadran and D. Sinha. Speech Data Compression Through Sparse Coding of Innovations. *IEEE Trans. Speech, Audio Processing*, 2:274–284, 1994.

[21] T.V. Ramabadran and D. Sinha. Speech Coding using Least-Squares Estimation. In B.S. Atal, V. Cuperman, and A. Gersho, eds, *Advances in Speech Coding*, chapter 33, pp. 349–359, Kluwer, 1991.

[22] C.R. Watkins, S. Crisafulli, and R.R. Bitmead. Reduced Complexity Kalman Filtering for Signal Coding. In *Int. Workshop on Intelligent Signal Processing and Communication Systems*, pp. 269–273, Sendai, Japan, 1993.

31

Modelling Signal Dynamics in Voiced Speech Coding

Domingo Docampo[1]
Juan Francisco Fariña[2]

ABSTRACT
We present a non-linear and non-parametric prediction algorithm for modelling the internal dynamics of the speech signal through a state-space geometric description. The proposed model extracts the intrinsic determinism embedded in the voiced speech signal from comparisons among the observed trajectories in state space. The combination of a metric criterion and frequency similarity results in a model which evolves throughout a quasi-periodic trajectory with a pitch period similar to the speech signal under analysis. Results for voiced frames of this model and comparisons with LPC-based methods are promising at medium bit rates.

31.1 Introduction

Time-domain coders differ in the way the receptor generates the excitation to the LPC filter. The quality of the synthetic signal in this type of coders relies on the number of bits used to encode the excitation signal [1]. There are currently some techniques which render acceptable quality within the range from 4.5 to 6.5 Kbps. Reducing the bit-rate without affecting the quality is not an easy task. Kleijn [2] suggests an interpolative algorithm (PWI), based on the quasi-periodic nature of the excitation signal during voiced frames, which allows the reconstruction of the signal from two different sequences delayed in a certain amount of time – typically 20–30 ms.

We are interested in studying whether this quasi-periodic nature of the voiced frames is really caused by an internal determinism of speech dynamics. If that were the case, it would be possible to model the production of voiced speech so that the excitation signal is generated by a dynamic process driven by a trigger sequence. Such an approach was taken in [6], where the speech signal is modeled as the output of a low-dimension determinism generator.

In this paper we propose a non-linear and non-parametric algorithm to model the internal dynamics of the signal through a geometric description of its state space, under the hypothesis that the mechanism for the production of voiced speech is

[1]Departamento de Tecnologías de las Comunicaciones, E.T.S.I. Telecomunicación, Campus Universitario, 36200 Vigo, Spain, E-mail: ddocampo@tsc.uvigo.es
[2]Departamento de Tecnologías de las Comunicaciones, E.P.S.I. Telecomunicación, Univ. Carlos III de Madrid, 28911 Leganés, Madrid, Spain, E-mail: jfarina@tsc.uc3m.es

quasi-deterministic and of low dimension[3]. Our main contribution is the development of a metric which takes into account the non-stationary nature of the speech signal to enable an appropriate neighbor-state selection. It results in a more accurate prediction of speech dynamics which can be exploited to achieve a better compression rate for voiced speech frames.

Section 31.2 is dedicated to introducing qualitative aspects of the Dynamic System Theory. The voiced speech production mechanism is analyzed in this framework in Section 31.3. Next, the coding algorithm which results from this analysis is explained in Section 31.4. Section 31.5 presents the results of applying the algorithm to synthesize actual voiced speech. Finally, Section 31.6 gives the summary and conclusions.

31.2 Dynamic System Model

Modelling and predicting a deterministic dynamic system relies on finding a function F that can mimic, with the desired degree of accuracy, the evolution of the system in state space [5]. We take the state of a deterministic dynamic system $\mathbf{s}_t$ as the necessary information to determine its future evolution: the sequence of states the system goes through in time.

Then our objective is to track the deterministic internal dynamics defined as

$$\mathbf{s}_{t+N\tau} = \mathcal{F}_N(\mathbf{s}_t) \tag{31.1}$$

The problem we deal with can be split in two, well-defined stages. First, we should identify a mapping which transforms the actual state of the system to a vectorial representation (the problem of ***state-space reconstruction***). Then, we need to address the modelling of the dynamic evolution of the system efficiently (the problem of the ***dynamic system approach***).

Reconstruction of the attractor is carried out starting from observations of the system x_n – usually the only information we have. We want to find a one-to-one correspondence (***embedding***) between the attractor states in phase space and the image states built from the observations. The one-to-one requirement is useful since then the future evolution of the system can be completely specified from a point in the reconstructed space. Otherwise, there would be crossed trajectories in some points of the space, in which the prediction would thus be impossible. Besides, a differentiable embedding is desirable, so that the topological structure of the attractor is also preserved. A particular solution is the ***delay coordinate embedding*** which associates each state $\mathbf{s}$ with a vector in the Euclidean space $\mathcal{R}^m$: the state of the system is represented by a vector (***delay coordinate vector***) containing the last m samples taken from the observations:

$$d_t = [x_t, \ldots, x_{t-(m-1)\tau}]^T \tag{31.2}$$

Let S denote a compact finite-dimensional set of states of a system and D a mapping from S to $\mathcal{R}^m$. A well-known result by Takens [4] guarantees, under certain conditions, the existence of an embedding. Provided that m is greater than twice the dimension of S and S is a smooth manifold, then $D(S)$ is also a smooth manifold, and D is a diffeomorphism.

In practice, this transformation requires an appropriate selection of m and τ (the sampling interval) because of the noise present both in the observations and the measurement devices. As in [6], to cope with this problem, we project the state space onto a subspace of lower dimension ($\hat{m}$, $\hat{m} < m$), oriented across the principal directions of the original state space, removing the transverse directions associated with noise components:

$$\begin{aligned} V : \mathcal{R}^m &\longrightarrow \mathcal{R}^{\hat{m}} \\ \boldsymbol{d}_t &\longrightarrow V(\boldsymbol{d}_t) = \boldsymbol{b}_t \end{aligned}$$

We try to capture the dynamic evolution of the system through the attractor using a local, linear, and non-parametric model. Therefore we assume a reasonable degree of smoothness on every predictive region – location of the nearest neighbors to the prediction state. Given some samples from a trajectory in state space, the prediction of its future is based on the observation of the K closest trajectories in the history of the signal. These trajectories enable us to estimate the parameters of the local model to mimic the dynamic of the system in that region of state space.

31.2.1 The Local Model

Let $\boldsymbol{b}_t$ be the state of the system used to make the prediction – what, from now on, shall be called the ***prediction state*** – and F a function that models the dynamics of the system within the interval $(t + N_o\tau, t + (N_o + q - 1)\tau)$, defined as follows:

$$\begin{aligned} F : \mathcal{R}^{\hat{m}} &\longrightarrow \mathcal{R}^q \\ F(\boldsymbol{b}_t) &= [x(t + N_o\tau), \ldots, x(t + (N_o + q - 1)\tau)]^T \end{aligned} \tag{31.3}$$

Note that the range of F is $\mathcal{R}^q$. Rather than forecasting just a state, F predicts N_o samples ahead a complete ***subframe***. NOw the estimation of the local model can be summarized in two stages:

1. First, K nearest neighbors are selected. Then each state is linked to a vector $\boldsymbol{y}_i \in \mathcal{R}^q$ defined as follows:

$$\begin{aligned} \boldsymbol{y}_i &= [x(t_i + N_o\tau), \ldots, x(t_i + (N_o + q - 1)\tau)]^T \\ &= F(\boldsymbol{b}_{t_i}) \qquad i = 1, \ldots, K \end{aligned} \tag{31.4}$$

2. Next, using the selected states, a linear model is built. The estimate given by the model can be written as

$$F(\boldsymbol{b}_t) = L(\boldsymbol{x}) = M^T\boldsymbol{x} + \boldsymbol{N} \tag{31.5}$$

where L is a linear function, $L : \mathcal{R}^p \longrightarrow \mathcal{R}^q$, $M \in \mathcal{R}^p \times \mathcal{R}^q$, $\boldsymbol{N} \in \mathcal{R}^q$, $\boldsymbol{x} \in \mathcal{R}^p$, and p ($p \leq m$) is the dimension of the subspace where the linear approximation takes place.

To settle the dimension (p) of that subspace, we use the following procedure:

1. First, we compute the singular values λ_i of a matrix, whose rows are the vectors $\boldsymbol{b}_i - \boldsymbol{c}$, $\boldsymbol{b}_i$, $\quad i = 1 \ldots K$, are the neighbor states (in $\mathcal{R}^{\hat{m}}$), and $\boldsymbol{c}$ their center of mass.

2. Then we compute the $\max_{i<\hat{m}} (\lambda_i - \lambda_{i+1})$.

3. If the maximum happens at any index $k > 1$, we choose $p = 2$. We settle $p = 1$ otherwise.

Before estimating the parameters of the linear model, the attractor of the system is projected onto a plane or a line. These simple hyperplanes are those that better match (in a least square sense) the geometric distribution of the neighbor states.

The ***bias/variance*** dilemma arises with the number of neighbors selected. Because of the noise present, a small number of neighbors will increase the variance. Increasing their number to diminish the variance, however, will also enlarge the prediction region. Therefore, the bias will be increased. Besides, the optimal solution for each state would be different if the density of known states around were to changed, and calculating that optimum number is not feasible. We employ a more practical strategy consisting of selecting a fixed number of neighbors and introducing a weight for each neighbor to gauge its contribution to the model according to its distance to the predictive state. Those weights are based on the distances d_i to the neighbors and the largest distance of all them, d_{max}:

$$p_i = 1 - \frac{d_i}{d_{max}} \quad i = 1, \ldots, K \tag{31.6}$$

31.3 Dynamic of Voiced Speech Production Mechanism

31.3.1 Speech Dimension

A key step in analyzing the dynamics of a system is to determine the dimension of its attractor. If the attractor dimension is lower than the state-space dimension, it is not unlikely that the system may be governed by a set of deterministic equations. If both dimensions are similar, then either there is no deterministic relationship or it is strongly masked by a noisy component.

A useful tool for analysing dynamic systems is the concept of ***generalized dimensions***. If the system attractor is covered by E-dimensional boxes of side length ϵ, the generalized dimensions are defined as follows:

$$D_q \equiv \frac{1}{q-1} \lim_{\epsilon \to 0} sup \frac{\log \sum_i p_i^q}{\log \epsilon} \tag{31.7}$$

where p_i is the measure of the attractor on box i and the sum is carried over occupied boxes. The generalized dimension is defined for all real values of q. The most interesting is the ***correlation dimension*** ($q = 2$) since its measure turns out to be a lower bound for the other dimensions.

Several measures of the dimension of speech were carried out in [3] and in [6]. The result for voiced speech, a figure between three and five, points to a likely internal determinism of low dimension in the signal. With respect to unvoiced speech the measures were higher, between five and seven. Therefore, the assumption of an internal determinism is less obvious.

31.3.2 Non-Stationarity of the Speech Signal: Surface Multiplicity

Modelling techniques based on the reconstruction of the dynamic system are mainly used in systems satisfying the following properties: the system must be deterministic, of low dimension, autonomous, and quasi-stationary. Under such conditions, the deterministic relationship between each state and its output can be represented as a surface in state space. Therefore, a local linear method approaches this surface by means of a plane-interpolation. If the system does not match any of these conditions, the modelling problem becomes more difficult to overcome since the surface to be reconstructed is not unique.

We will assume (as in [3]) that the production mechanism of voiced speech is autonomous during small intervals of time (*frames*), quasi-deterministic, and of low dimension. However, the mechanism is non-stationary because of both the movement of the articulations in the vocal tract and the variability of the vocal cord vibration (*pitch* frequency). As the next section will show, this non-stationarity is present in the libraries where the neighbor states are looked for, which contain the history of the signal. Because of this non-stationarity, the likeness between the values associated with each state (which determine the smoothness of the surface in the predictive region) does not depend on the distance between them and the prediction state. Trying to estimate a model in such conditions would result in a large bias regardless of the distances.

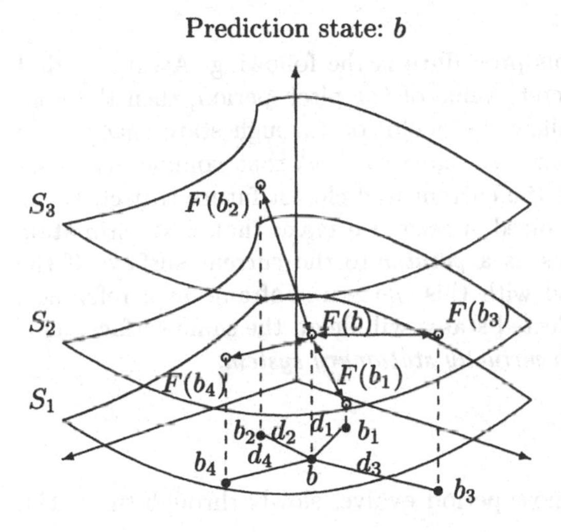

Distance: $d_1 < d_2 < d_4 < d_3$.

Neighbour states: b_1, b_2, b_3, b_4.

$F(b_1) \in$ surface S_1.

$F(b_2) \in$ surface S_3.

$F(b_3), F(b_4)$ and $F(b) \in$ surface S_2.

The metric criterion selects states b_3 and b_4.

The actual prediction is based on a model built using the pairs b_3–$F(b_3)$ and b_4–$F(b_4)$.

FIGURE 31.1. Surface multiplicity

In Fig. 31.1 we show a schematic representation of this problem in three dimensions. If the model is estimated using only two neighbors, a simple selection criterion based on distances between states is not suitable. To overcome this problem, we propose a metric criterion which selects neighbors with associated values lying on the same surface as $F(b)$. Now, the goodness of fit of our prediction depends on the density of the representation within that surface.

By using an LPC filter with coefficients varying from frame to frame, the first cause of non-stationarity is coped with. Then, the non-linear model is applied to the excitation sequence. The general scheme of the coder, as shown in Fig. 31.2, comprises the LPC filter and a ***non-linear generator of the excitation sequence*** (NLG), which reconstructs the appropriate excitation to that filter.

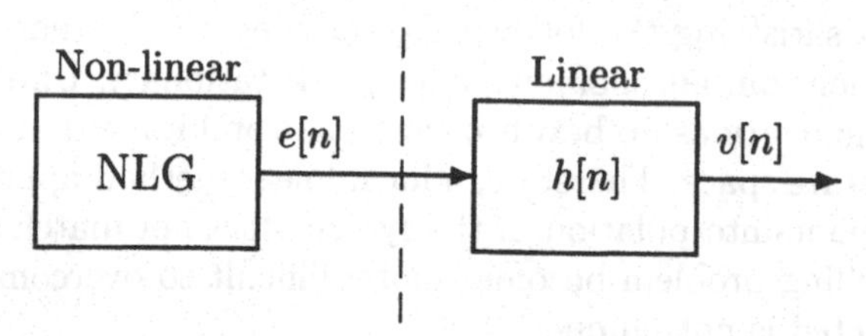

FIGURE 31.2. Coder scheme

As mentioned above, the non-stationarity still present in the variability of the pitch frequency requires the use of a non-traditional metric to select the neighbor states. This selection process takes place as follows:

1. We make a first estimate of the target synthesis subframe, which shall be called from now on the ***pattern subframe***, based on the quasi-periodic nature of voiced speech.
2. Then neighbor state selection is performed according to a perceptual distance between the pattern subframe and those subframes associated with the states in the voiced libraries.

The geometric interpretation of this procedure is the following. Assuming that each surface is associated with a discrete value of the pitch period, then the non-stationarity of the signal does not allow the evolution through state space to fit into a surface but rather into a volume. We have to ***break*** that volume into a set of surfaces and to know in which one the current and closest future trajectory lie. Using the quasi-periodic nature of voiced speech, we make that first estimation of the synthesis subframe. That gives us a ***pointer*** to the current surface. If the distance measurements are calculated with this pattern subframe as a reference, the subframes associated with the selected states will lie on the same surface. This procedure transforms our system in a ***virtually stationary system***.

31.3.3 Pattern Subframe

Let $v(t)$ be a quasi-periodic signal whose period evolves slowly through time. The vector $\boldsymbol{v}(t)$ is defined as follows:

$$\boldsymbol{v}(t) = [v(t), v(t-\tau), \ldots, v(t-(q-1)\tau)]^T \tag{31.8}$$

If the current period is T, a first estimate of that subframe ($\boldsymbol{v}(t)$) could be:

$$\boldsymbol{v}(t) \simeq \boldsymbol{v}(t-T) = \boldsymbol{u}(t) \tag{31.9}$$

where $\boldsymbol{u}(t)$ is the pattern subframe. This first estimate is based on information from the history of the signal. This produces an essentially periodic model along the frame which results in a step evolution of the pitch period. Turning back to Fig. 31.1, the estimated trajectory would lie on the same surface all along the frame. A better solution consists of smoothing this evolution using information from the future [2].

Now, the forecast will be closer to the real trajectory of the signal. According to this, we build the pattern subframe as follows:

$$v(t) \simeq \alpha v(t-T) + (1-\alpha)v(t+kT') = u(t), \quad \alpha = 1 - \frac{T}{T+kT'}$$

Within a coding scheme, information about the future is extracted from the trigger sequence of the next frame.

31.4 Non-Linear Generator of Excitation

As Fig. 31.3 shows, the synthesis of an excitation frame (a sequence of 240 samples) depends primarily on its ***Phonetic Classification***. If the frame is classified as ***unvoiced*** (or ***transition***), the synthesis is carried out using any well-known method like those employed in CELP schemes [1]. For (***voiced*** frames), the process is summarized below:

In the first block (***frame initiation***) the receptor reconstructs an initial set of samples (***trigger sequence***) by the same algorithm used in CELP schemes [1]. The length of the trigger sequence must be greater than both the pitch period and N_0 (the number of samples between the target subframe and the subframe on which the model is based).

The second block (***neighbors selection***) involves the following procedure, which is applied to each subframe – usually a sequence of 30 samples – all across the frame:

1. Selection of the K states (which constitute the neighborhood) with its associated subframes as outlined in 31.4.2. The neighbour states are selected among past states stored in the libraries.
2. Estimation and evaluation of the effect of the local model on the predictive state.
3. Multiplication by a gain factor as outlined in 31.4.3.

31.4.1 State Libraries

There are two kind of libraries: ***fixed*** and ***adaptive***. The first library consists of a fixed set of states with its associated subframes. Its objective is to guarantee a sufficiently dense representation of the system attractor so that the approximation error is reasonably low. However, the size of this library is limited by the available memory, resulting in a trade-off between density of representation and memory capacity. The second library comprises the states taken from the last sequences synthesized in the receptor. its objective is to increase the density representation in the region of the attractor which contains the current system dynamics. Therefore, the adaptive library evolves in parallel with the system.

31.4.2 Metric Criteria for Neighbor States

Suppose we want to synthesize the subframe within the interval $(n, n+q-1)$, which we denote as

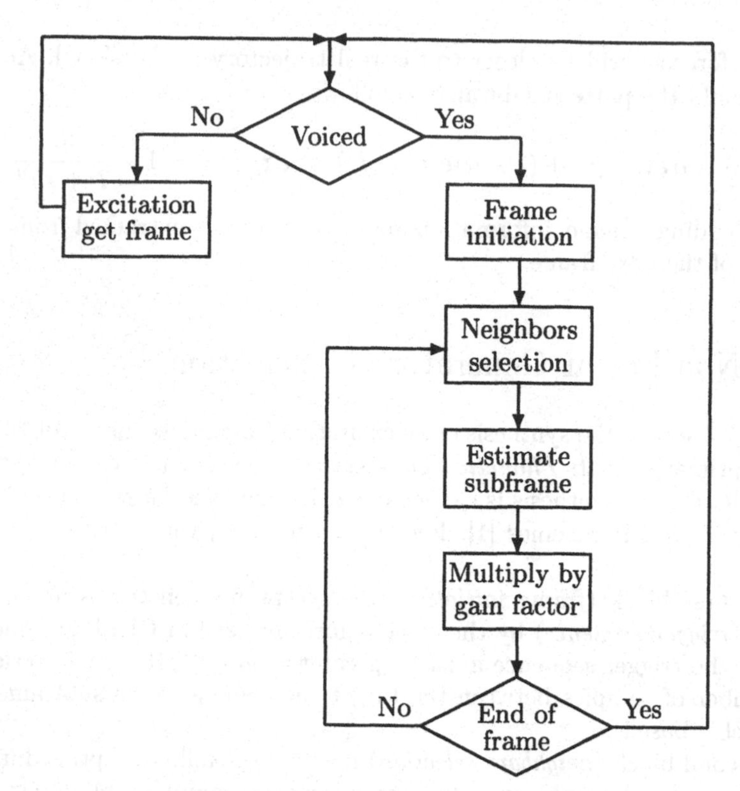

FIGURE 31.3. Non-linear generator scheme

$$\boldsymbol{v} + \boldsymbol{m}_v = G H \boldsymbol{e} + \boldsymbol{m}_v \tag{31.10}$$

where $\boldsymbol{v}$ is the component due to the excitation in that interval; $\boldsymbol{m}_v$ the contribution from the memory of the LPC filter; $\boldsymbol{e}$ is an excitation subframe, H the LPC filter; and G a gain factor. The task of the NLG is to generate the sequence $\boldsymbol{e}$ so that $\boldsymbol{v} + \boldsymbol{m}_v$ reasonably matches the original subframe.

Let $\boldsymbol{u}$ be the pattern subframe and $\boldsymbol{y} \in \mathcal{R}^q, \boldsymbol{y} = H_p(\boldsymbol{u} - \boldsymbol{m}_v)$, where H_p is the *perceptual weighting filter* [1].

Let $\boldsymbol{\xi}_i$ ($i = 1, \ldots, L$) be the subframe associated with the state i, stored in the libraries (L denotes the dimension of the library vectors). This subframe generates the following weighting synthetic speech sequence:

$$\boldsymbol{x}_i = H_p H \boldsymbol{\xi}_i = H_c \boldsymbol{\xi}_i \quad , \quad i = 1, \ldots, L \tag{31.11}$$

Using the gain

$$g_i = \frac{\boldsymbol{y}^T \left(H_c \boldsymbol{\xi}_i\right)}{||H_c \boldsymbol{\xi}_i||^2} \tag{31.12}$$

the neighborhood will be constituted by the K states whose associated sequences minimize the following expression:

$$E_i = ||\boldsymbol{y} - g_i H_c \boldsymbol{\xi}_i||^2 = ||H_p(\boldsymbol{u} - \boldsymbol{m}_v) - g_i H_c \boldsymbol{\xi}_i||^2 \tag{31.13}$$

In practice, speech is scaled during recording or other processing, so neighbors need to be multiplied by p_i:

$$p_i = \frac{\boldsymbol{b}_{n-N_0}^T \cdot \boldsymbol{b}_{n_i}}{||\boldsymbol{b}_{n_i}||^2} \tag{31.14}$$

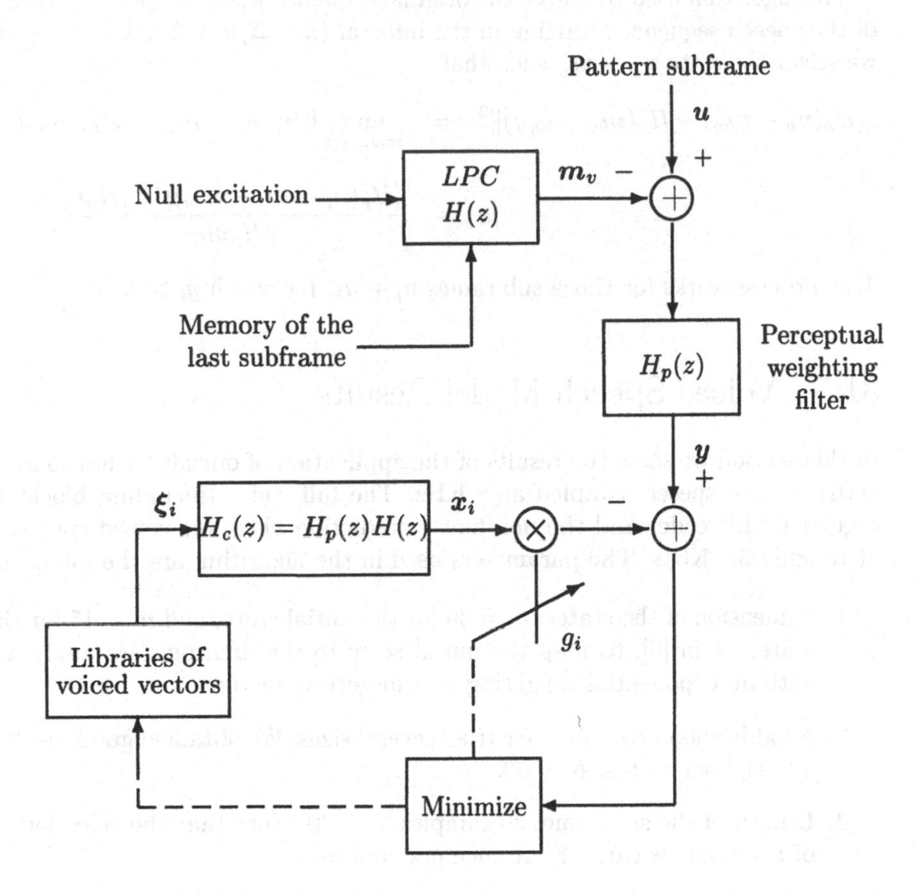

FIGURE 31.4. Metric criterion for neighbor selection

The sequence generated by the NLG is given by

$$e = F(b_{n-N_0}) \tag{31.15}$$

31.4.3 Gain Factor

This procedure, which takes place at the transmitter, calculates the gain factor G, which minimizes the perceptual distance between the original and synthetic subframes. It is carried out in two steps:

1. Search the sample $\eta \in (n - \Delta, n + \Delta)$ so that $v_\eta + m_\eta$ (original subframe) corresponds to $v + m_v$ (synthetic subframe). The underlying reason for this step resides in that we need to know the original subframe that better matches the synthetic one because of the use of rounded pitch values.

2. Obtain the gain factor using the following equation:

$$G = \frac{[H_p(v_\eta + m_\eta - m_v)]^T \cdot [H_p v]}{\|H_p v\|^2} \tag{31.16}$$

The algorithm used to search the original sequence works as follows: given the set of the speech sequences starting in the interval $(n-\Delta, n+\Delta)$, $A = \{\boldsymbol{v}_i + \boldsymbol{m}_i\}_{i=1}^{2\Delta}$, we select the state $\boldsymbol{v}_\eta + \boldsymbol{m}_\eta$ such that

$$\begin{aligned} \|H_p(\boldsymbol{v}_\eta + \boldsymbol{m}_\eta) - H_p(\boldsymbol{m}_v + g_\eta \boldsymbol{v})\|^2 &= \min_{i=0\ldots 2\Delta} \|H_p(\boldsymbol{v}_i + \boldsymbol{m}_i) - H_p(\boldsymbol{m}_v + g_i \boldsymbol{v})\|^2 \\ g_i &= \frac{[H_p(\boldsymbol{v}_i + \boldsymbol{m}_i - \boldsymbol{m}_v)]^T \cdot [H_p \boldsymbol{v}]}{\|H_p \boldsymbol{v}\|^2} \end{aligned} \tag{31.17}$$

This process works for those subframes $\boldsymbol{v}_i + \boldsymbol{m}_i$ for which $g_i > 0$.

31.5 Voiced Speech Model Results

In this section we show the results of the application of our algorithm to modelling actual voiced speech sampled at 8 KHz. The full coder (including blocks from a regular CELP coder and the non-linear prediction block for voiced speech) works at roughly 3.8 Kbps. The parameters used in the algorithm are the following:

1. Dimension of the states: $m = 30$ for the initial state, and $\hat{m} = 15$ for the final state. As in [6], to map the initial state to the final one, we apply an SVD with an exponential weighting as a forgetting factor.
2. Neighborhood size: we have tried several sizes. We obtained good results using 29 neighbor states: $K = 29$.
3. Length of the subframe: 30 samples, $q = 30$. Note that the selection process of neighbors is carried out once per subframe.
4. Prediction depth: 30 samples ahead, $N_0 = 30$.
5. The final local linear model consists of one or two parameters plus a bias term: $p = 1$ or $p = 2$. Therefore, the attractor of the system is projected onto a line or a plane to obtain these parameters.

We have made extensive simulations to test the approximating capabilities of the algorithm. In Fig. 31.5 we show the results of the method on a segment of voiced speech of 0.25 seconds.

The result is that our model fits the trajectory of the original signal during its evolution. The reason lies in the metric used to search the neighbor states since it *forces* the model to evolve through a quasi-periodic trajectory with a pitch similar to the original speech signal. The main problem with it is the lack of synchronism between the original and the synthetic excitation signal around one sample because of the round in the pitch period along the frame.

We have performed a perceptual analysis of the algorithm with speech taken from different speakers (male and female). Although the coder does not match the quality provided by well-established coding standards, we have identified two promising good features of the synthetic signal, namely:

a) the original intonation is kept;

b) unlike crude LPC-based methods, it is possible to recognize the speaker.

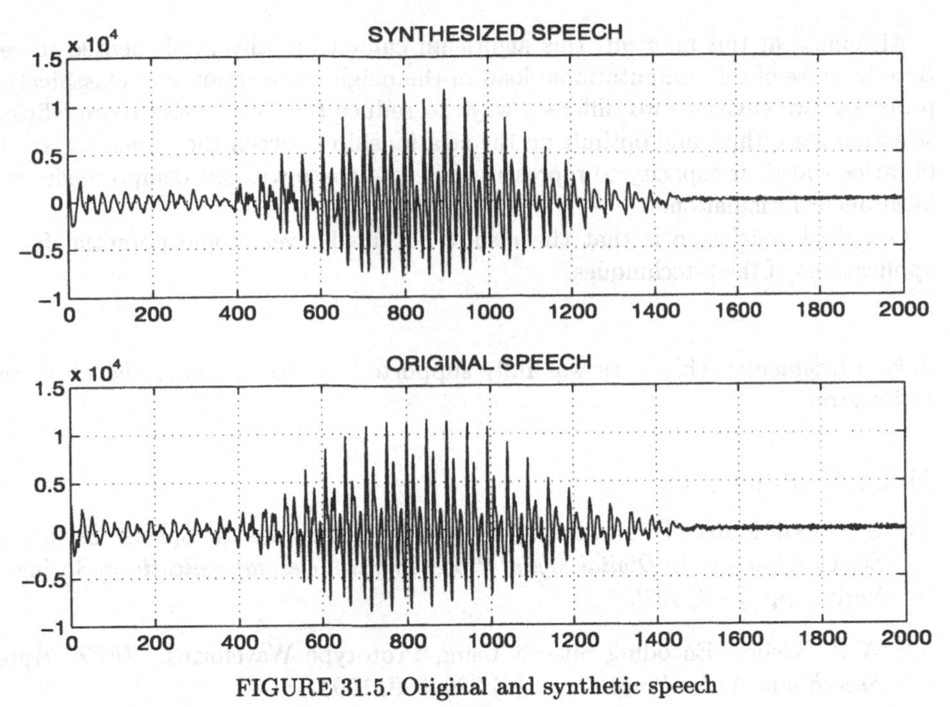

FIGURE 31.5. Original and synthetic speech

Both features contribute to a synthetic signal which is still noisier but sounds more *natural* than that synthesized by LPC-based methods.

The results obtained at these exploratory stages of our work are very encouraging, and suggest that further work should be done in this area.

31.6 Conclusions

This paper describes a non-linear and non-parametric prediction algorithm for modelling the internal dynamics of voiced speech signals. The basic foundations of the coder rely upon Dynamic System Theory, modelling the determinism of a system based on a state-space representation. The main innovation in our algorithm resides in a new metric criterion to select the neighbor states used by the linear model. This criterion makes a first estimate of each subframe based on linear interpolation between the trigger sequence of two consecutive frames, which is used as a pattern subframe. The selection of neighbors is carried out by perceptual distance measurements among the pattern and the respective subframes for each state in the libraries. The final objective is to transform our system into a virtually stationary system before building the linear local model.

Experimental results show that our predictive model (and implicitly the metric used) fits the determinism within the excitation signal. In particular, the model evolves throughout a quasi-periodic trajectory with a pitch frequency similar to the speech signal under analysis. It results in a more accurate prediction of the speech dynamics, which can be exploited to achieve a better compression rate for voiced speech frames. It is worth noting that the trigger sequence is the only one which needs an index of reference, sent from the transmitter, to be looked up in the libraries.

Although, at this moment, this algorithm cannot be fully implemented in real time because of the computational load in the neighbor selection and classification processes, we envision two different ways to reduce this cost: using more efficient selection algorithms and optimizing the relationship between the dimension of the libraries and their capacity to represent the state space without compromising the local linear formulation.

Our final conclusion is that the results obtained reveal some potential future applications of these techniques.

Acknowledgments: This work was fully supported by Nortel under the Multicom 21 Program.

31.7 References

[1] C. García Mateo, D. Docampo. Modelling Techniques for Speech Coding: A Selected Survey. In *Digital Signal Processing in Telecommunications.* Springer-Verlag, pp. 1–43, 1996.

[2] W.B. Kleijn. Encoding Speech Using Prototype Waveforms. *IEEE Trans. Speech and Audio Processing*, 1(4):386–399, 1993.

[3] N. Tishby. A Dynamical Systems Approach to Speech Processing. In *Proc ICASSP'90*, Albuquerque NM, pp. 365–368, 1990.

[4] F. Takens. Detecting Strange Attractors in Fluid Turbulence. In *Dynamical Systems and Turbulence*, Springer-Verlag, 1981.

[5] M. Casdagli. A Dynamical Systems Approach to Input-Ouput Systems. In *Nonlinear Modelling and Forecasting*, Addison-Wesley, Reading MA, pp. 265–281, 1992.

[6] B. Townshend. Nonlinear Prediction of Speech Signals. In *Nonlinear Modelling and Forecasting*, Addison-Wesley, Reading MA, pp. 433–453, 1992.

32

Experimental Study of Speech Recognition in Noisy Environments

Tomáš Kreisinger[1]
Pavel Sovka[1]
Petr Pollák[1]
Jan Uhlíř[1]

ABSTRACT
Achieving reliable performance in a speech recognizer for car telephone applications has been studied intensively for more than a decade. This paper addresses the effects of mismatched conditions and their minimization with respect to the performance of speaker-independent isolated-word recognition in a car-noise environment without considering the Lombard effect. This study is primarily intended to evaluate the dependence of the recognition rate on the signal-to-noise ratio (SNR) of an input signal either without any noise-compensation method or with a noise-compensation or noise-adaptive method and especially to find the appropriate conditions so that an isolated word recognizer can be used in a real car-noise environment. When hidden Markov models (HMMs) are trained on noisy speech with a SNR of 10dB, it is possible to recognize noisy speech with a SNR in the interval from 40 dB to 5 dB with a recognition rate better than 93%. If modified spectral subtraction is used and models are trained on the enhanced speech, the SNR interval increases to 0 dB. If the parallel model combination (PMC) technique is used, there is no need to train models on noisy or enhanced speech. The model adaptation enables recognizing noisy speech with any SNR from 40 to -10 dB with a recognition rate greater than 73% (for a SNR from 40 to 5 dB, the recognition rate is above 93%). In this respect PMC offers great flexibility with better recognition rates than other noise-compensation techniques.

32.1 Introduction

Isolated word recognition is applicable in many systems. We assume a voice control of dialing for mobile telephony. That is why the vocabulary used is limited to 12 words.

The utterances recorded in a stopped car are used as "clean" speech . It means that this speech set is not influenced by the variability of pronunciation caused by stress, tiredness, noise, and other driving conditions (Lombard effect). Speech and noise are recorded under the same acoustic conditions which allows generating noisy

[1]Czech Technical University, Faculty of Electrical Engineering, Technická 2, 166 27 Praha 6, Czech Republic, E-mails: {kreising, sovka, pollak, uhlir}@feld.cvut.cz

speech with various SNRs and therefore the effects of mismatched conditions[2] on the recognition rate can be analyzed.
The dependence of the recognition rate on the SNR of the test signal is studied under the following conditions:

- the training phase is done on a clean speech signal and no compensation technique is used (this case represents the lower performance of a speech recognizer);
- the training phase is done on a noisy speech signal (if SNR used for training and testing is the same, then this case represents the upper performance of a speech recognizer - in figures marked as "matched");
- some noise-compensation or noise-adaptive technique is used.

The goals are as follows:

- to verify the type of signal on which the recognizer should be trained: clean speech, noisy speech, or processed (enhanced) speech; which SNR should be used?
- to test how some chosen technique can adapt the whole recognition system to all possible input SNRs.

To answer these questions, the dependence of the recognition rate on the input SNR is evaluated. As a criterion, the SNR for which the recognition rate does not fall below a certain threshold (for our study, it is set to 93%) is used.
Some spectral subtraction methods were chosen as noise compensation techniques. Wiener filtration (the best noise-compensation technique [11]) was not used because it can be considered a special case of the parallel model combination technique (PMC). Noise-adaptive methods are represented only by PMC in this study which is one of the best methods for noise-resistant HMM recognizers [4, 11].

32.2 HMM Recognizer

32.2.1 Recognizer Construction

The recognition system is a continuous density HMM with eight states and without skips (simple left-right model). For each state in the model, the distribution is represented by an unimodal Gaussian density (no mixtures are used). No streams are used. The feature vector is composed of eight static coefficients, eight delta (or dynamic) coefficients, one energy coefficient, and one delta energy coefficient. The syntax diagram for connected word recognition contains 12 word models connected in parallel (for 12 words used in one utterance) and one model for a silence (pauses) or background noises. The weighting factors (p and s) for the transition from a word model to the silence model were varied to find their optimum for our database.

[2]Mismatched conditions mean that the different noisy conditions are used for the training phase and for the recognition.

32.2.2 Database Used and Processing Parameters

Following are the test conditions: sampling frequency 8 kHz; 16 bits per sample; frame length 20 ms; overlap 10 ms; Hamming window; number of speakers 14; number of utterances 212; number of words in one utterance eight on average; number of utterances for the training part of the database 123; number of utterances in the test part of the database 89.

Noises were recorded under various driving conditions in three types of cars. Noises are divided into three groups: stationary, non-stationary (relatively slow changes), and highly non-stationary noises (very fast changes).

Signals from speech and noise databases were mixed according to the specific SNR and used in training and testing (recognition) phases.

32.2.3 Parametrization

The parametrization used is as follows (the notation according to HTK):

a) the type of parametrization: MEL spectra = MELSPEC, MEL cepstra = MFCC, LPC spectra = LPC, LPC cepstra = LPCEPSTRA. The order for cepstral parameters is set to 20. Final cepstral vectors are then truncated to eight coefficients. The order for spectra is set to eight;

b) the type of feature vector for par=MFCC or LPCEPSTRA:
$[c_1 \dots c_8\ \Delta c_1 \dots \Delta c_8\ E\ \Delta E]$ stands for par_D_E, where E = energy.
$[c_1 \dots c_8\ \Delta c_1 \dots \Delta c_8]$ = par_D, $[c_1 \dots c_8\ E]$ (par_E), $[c_1 \dots c_8]$ = par.

32.2.4 Recognizer Behavior in Noisy Environments

The influence of parametrization is illustrated in Figure 32.1:

- cepstra give higher recognition rates than spectra because of better orthogonalization properties of cepstra;
- for higher SNR the LPC cepstrum is better than the MEL cepstrum because the LPC spectral modeling requires generally less data than DFT spectral modeling. On the other hand LPC is more sensitive to noise disturbances;
- the MEL cepstrum (MFCC) gives the best noise immunity and the MEL spectrum the worst;
- speech recognition rates for matched conditions represent the best performance the recognizer can achieve.

The influence of various types and numbers of parameters on the MFCC feature vector is shown in Figure 32.2. The best performance can be achieved by using static and dynamic coefficients (MFCC_D_E). The better noise immunity excludes energy and delta energy (cases MFCC_D or MFCC). This expected behavior can be explained by using spectra in the linear domain. The changes of the speech spectrum for various SNRs are greater when the energies of speech and noise are used. If the energy is omitted, then spectral changes are less which means that the models trained without using the energy are less sensitive to the disturbing noise.

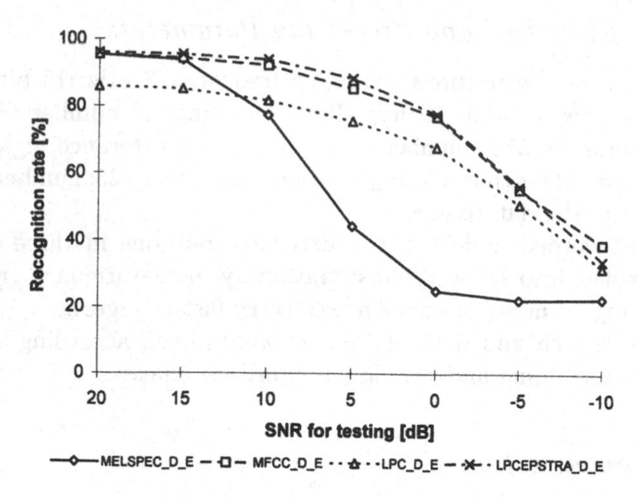

FIGURE 32.1. Recognition rates for various types of parametrization.

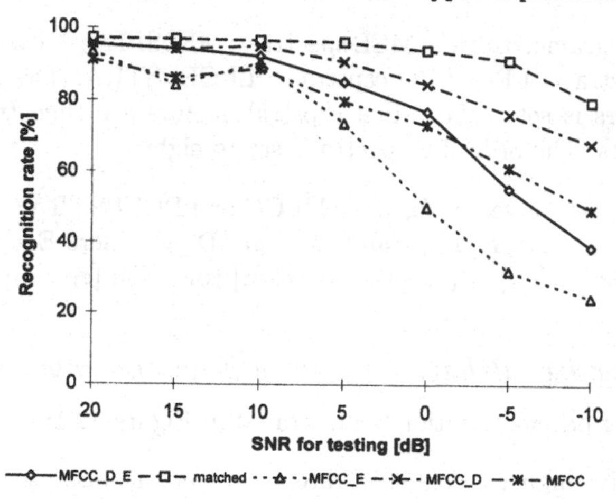

FIGURE 32.2. Recognition rates for various types of MFCC feature vector.

On the other hand, models trained this way do not represent the proper spectra of noisy speech. Therefore the recognition rates are less than when the energy is used.

For further experimentation we chose MFCC_D_E parametrization. The results are shown in Figure 32.3 which contains the table and its map for the training/testing procedure on various SNRs. This figure suggests that if no noise compensation is used then the best choice is to train models on noisy speech as follows (recognition rate should not fall below 93%!):

- training with a SNR of about 10 dB if the expected SNR of the recognition is above 5 dB;
- training with a SNR of about 0 dB if the expected SNR of the recognition is in the range from 10 to 0 dB.

That means that we must use two different sets of HMM for SNRs above 0 dB. These conclusions are valid if the model for pauses is substituted by the model for background noise (not only the models for words but also the model for pauses are modified by training on noisy speech).

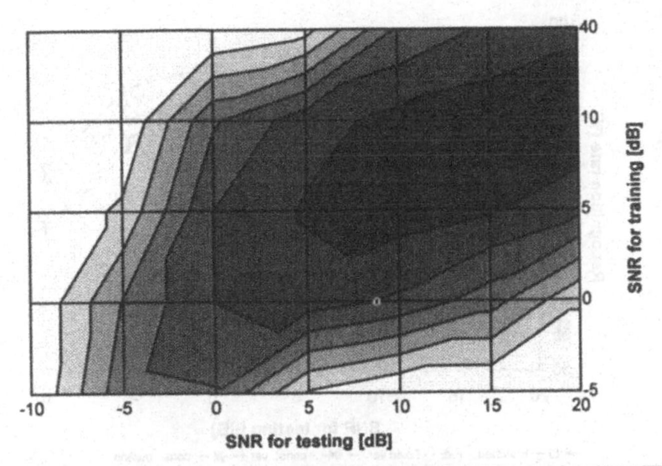

SNR tr. [dB]	SNR Testing [dB]						
	20	15	10	5	0	-5	-10
40	95.47	94.30	92.23	85.23	77.07	55.31	38.86
10	97.80	97.28	96.63	95.21	91.71	78.11	55.57
5	94.56	96.24	97.02	96.24	94.04	86.40	73.83
0	86.27	90.93	93.65	95.08	94.04	90.16	78.63
-5	45.21	62.31	77.98	85.75	91.84	91.32	79.79

FIGURE 32.3. Map of recognition rates for training HMM MFCC on noisy speech.

The diagonal of this table (or this map) represents matched conditions (the models are trained and tested under similar noise conditions with the same SNR). The first row represents the case when the training is done on clean speech and testing on noisy speech. The first case represents the upper performance of a system and the latter case represents the lower performance.

For matched conditions the influence of the various types of training procedures on the recognition rate was also tested (see Figure 32.4). First we trained the HMM on clean speech. Then we retrained these models on noisy speech keeping either variances or means constant which means that we adapted either variances (marked "const-var") or means ("const-means") to the noisy speech signal. It can be seen that omitting the variances adaptation causes a smaller decrease in the recognition rate than omitting the means adaptation. But both cases differ little from the full training (marked as "matched"). Also the differences between normal training and fixed-variance ("fix-var") training seem to be small for our database. That is why we decided to use the PMC technique for means only and not for variances (see Section 32.4).

32.3 Noise Compensation

For the purposes of this study we have focused on spectral subtraction techniques only. Various spectral subtraction techniques have been frequently used for noisy speech preprocessing [1, 2, 6, 7, 8, 9] and they also have proved the efficiency for the speech recognition.

Spectral subtraction algorithms cause speech distortion as the result of noise spectrum variation in time. To evaluate how this speech distortion influences the recogni-

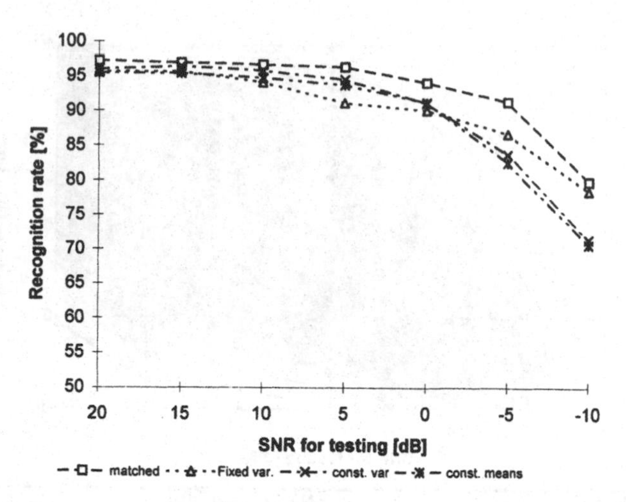

FIGURE 32.4. Various types of training for matched conditions

tion rate, raw spectral subtraction without any additional postprocessing was used. We tested spectral subtraction with half-wave rectification (acronym "so1hw") and twice repeated spectral subtraction with full-wave rectification (acronym "so1ff") [13]. The repetition of spectral subtraction allows comparable noise suppression and speech distortion in the case of half-wave rectification. Since the influence of speech distortion is studied, both algorithms use voice-activity detection according to manually created labels of signal database. The results achieved by simulations are as follows:

- HMM trained on clean speech:
  - for SNRs above 0 dB the recognition rates fall more quickly for the half-wave rectification than for full-wave because of the higher speech distortion (the difference is about 1%);
  - for SNRs below 0 dB half-wave rectification gives better results. The difference increases linearly from 1% to 7% as the SNR decreases from -5% to -10%;
- if the speech distortion is compensated for training on enhanced speech then the recognition rate is increased. This can be seen from the rows of Figure 32.5 showing the table and its map for full-wave rectification "so1ff";
- the comparison of Figures (tables) 32.3 and 32.5 reveals that spectral subtraction gives better results for lower SNRs (last three columns of both tables).

To suppress the speech distortion caused by noise spectrum variation in time and VAD failures, methods other than spectral subtraction must be used. These methods can track noise spectrum variations even during speech activity. Some examples of these methods are RASTA [7], and recently suggested Martin's [10] and Doblinger's [5] (acronyms "martin" and "dobl"). These methods track spectral minima in frequency subbands to estimate the background noise spectrum. Both methods have very good performance in non-stationary environments. The Martin method requires more computational effort than Doblinger.

a)	-5 dB	0 dB	b)	-5 dB	0 dB	c)	-5 dB	0 dB
exten	6.72	4.34	exten	7.45	4.70	exten	7.38	4.70
solff	7.95	8.52	solff	5.72	6.25	solff	4.84	2.69
dobl	5.20	3.58	dobl	6.72	4.10	dobl	6.85	3.25
martin	6.28	4.10	martin	7.58	4.47	martin	7.58	4.44

TABLE 32.1. SSNRE for Contamined Noisy Speech : a) stationary, b) non-stationary, c) highly non-stationary noises with SNR 0 and -5 dB.

We tried to use another approach combining spectral subtraction with an adaptive Wiener filter [15] (called extended spectral subtraction, acronym "exten"). This method uses only one parameter p controlling the smoothing of the background noise spectrum (the optimal value of p seems to be 0.95). The average noise suppression is 6 dB independent of the stationarity of noises; see Table 32.1. As the criterion the segmental signal-to-noise ratio enhancement (SSNRE [dB]) was used:

$$SSNRE = \frac{1}{L}\sum_{i=1}^{L} 10\log\frac{P_N[i]}{P_{NR}[i]} \tag{32.1}$$

where P_N and P_{NR} are the powers of noise and residual noise respectively.

Tables 32.1(a) - 32.1(c) show the performance of the methods discussed for various types of noise from the standpoint of stationarity. By comparing the results in these three tables, it is possible to conclude that full-wave rectified spectral subtraction is better than Martin's or Doblinger's methods or extended spectral subtraction for nearly stationary noises. But for highly non-stationary noises, full-wave rectified spectral subtraction fails whereas the other three methods give noise suppression similar to the preceding case. Figures 32.5 and 32.6 illustrate the differences between spectral subtraction and extended spectral subtraction[3]. The first rows of these figures are shown in Figure 32.7. The differences in recognition rates are smaller than one should expect on the basis of the SSNRE results in Table 32.1 because all three types of noise were used simultaneously[4]. The differences in speech rates become evident especially for lower SNR. Martin's method shows the best noise immunity. The worse result of full-wave rectified spectral subtraction compared to the result of half-wave rectified spectral subtraction for lower SNR is caused by the spectral subtraction repetition which generates echoes in the processed speech. It follows from Figures 32.5 and 32.6 that if the recognition rates are to be greater than 93%, the models should be trained on enhanced speech generated from noisy speech with a SNR of about 10 dB. This approach ensures the proper recognition for noisy speech with a SNR above 0 dB. Training on enhanced speech forces the models to be less sensitive to speech distortion caused by the filtration of noisy speech. Contrary to the case of speech recognition without any noise-compensation preprocessor (see discussion for Figure 32.3), we need not use two different sets of models.

Another very often used noise-compensation method is Wiener filtration. We did not use this method because it can be considered a special case of the PMC (see next chapter).

---

[3]Martin's method gives slightly better results than extended spectral subtraction and Doblinger's method gives worse results.

[4]For highly non-stationary noises only, the differences in the recognition rates are greater.

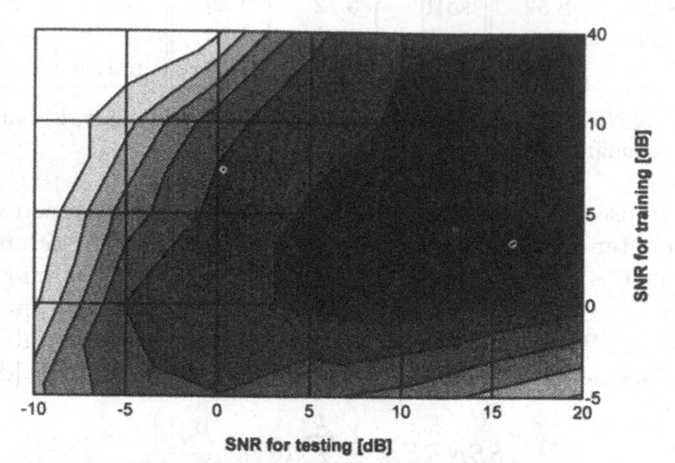

SNR tr. [dB]	SNR Testing [dB]						
	20	15	10	5	0	-5	-10
40	95.98	95.98	95.85	93.52	85.62	76.30	56.35
10	97.41	96.89	96.24	95.21	93.39	87.31	72.93
5	97.67	97.41	97.02	96.24	94.69	90.93	80.44
0	94.56	95.98	97.02	96.24	95.60	94.04	85.62
-5	87.69	89.38	91.32	92.36	93.91	93.26	89.51

FIGURE 32.5. Recognition rates for twice repeated full-wave rectified spectral subtraction.

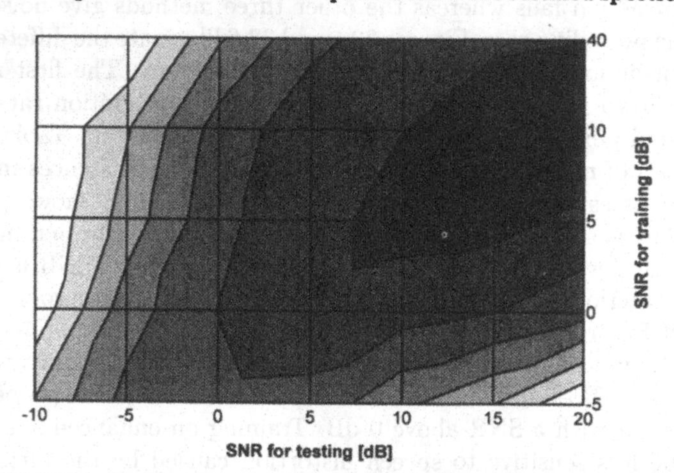

SNR tr. [dB]	SNR Testing [dB]						
	20	15	10	5	0	-5	-10
40	95.98	95.73	94.95	93.52	89.77	82.90	68.78
10	97.28	97.02	95.98	95.21	92.88	87.82	73.58
5	95.47	96.63	96.76	95.34	93.52	88.86	80.83
0	92.62	94.04	94.82	95.34	94.04	91.06	84.59
-5	83.94	88.08	90.41	93.39	93.52	92.62	88.08

FIGURE 32.6. Recognition rates for extended spectral subtraction.

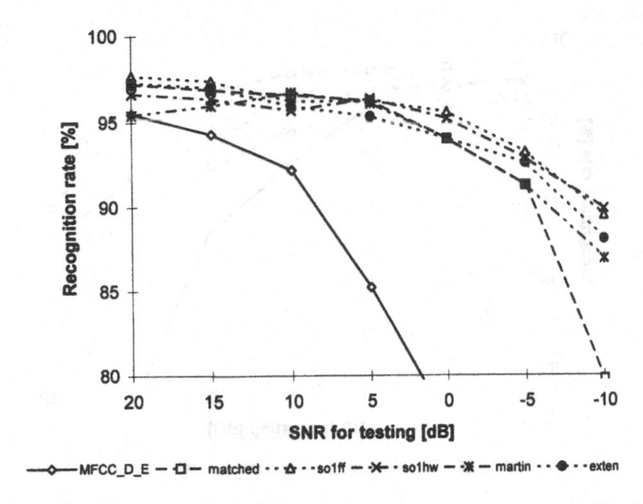

FIGURE 32.7. Comparison of noise-compensation methods.

32.4 Noise-Adaptive Methods

The problem analyzed in the foregoing section is that the important parts of speech may be removed from the signal. This is caused by filtering noisy speech by a noise-compensation method. Noise-adaptive methods are based on the adaptation of HMM parameters (trained on clean speech) to the noise whereas noisy speech is not filtered and therefore not distorted. We decided to use the non-iterative PMC method [12] which modifies state probabilities and not transition ones. Moreover the frame-state alignment is not changed by this method. Since training on noisy speech affects all these three characteristics, the matched conditions represent the upper performance of the speech recognizer.

As mentioned in Section 32.2, we chose the Log Add approximation which adapts only means of HMM and not variances. This is possible because of the small database. When only static parameters are used, the PMC Log Add approximation gives results similar to Wiener filtering results [11]. Wiener filtering can be seen as adapting the static means of HMM models. This approach improves results, but there is an even more efficient PMC Log Normal approximation which can update the means and also the variances of HMM models.

The PMC Log Add approximation for static parameters is described by

$$\hat{\mu} = \mu + g\tilde{\mu} \tag{32.2}$$

where g is a weighting factor and $\tilde{\mu}$, μ, and $\hat{\mu}$ are the means of the models for the noise, clean speech, and the adapted model for noisy speech, respectively. The influence of g on the PMC behaviour was analyzed in [4]. We verified that the best results can be achieved if this factor is set according to the expected SNR during recognition. Contrary to [4] and [11], we used g as the multiplicative factor for the noise means only. We studied two possibilities: to transform $cepstra \longrightarrow log_{spectra} \longrightarrow lin_{spectra}$ and back considering the energy and without the energy. This leads to two different ways of computing the weighting factor g, but results are the same. The first way uses normalized spectra without any information about the signal or noise energy. The latter deals with spectra containing the information about the signal/noise energy.

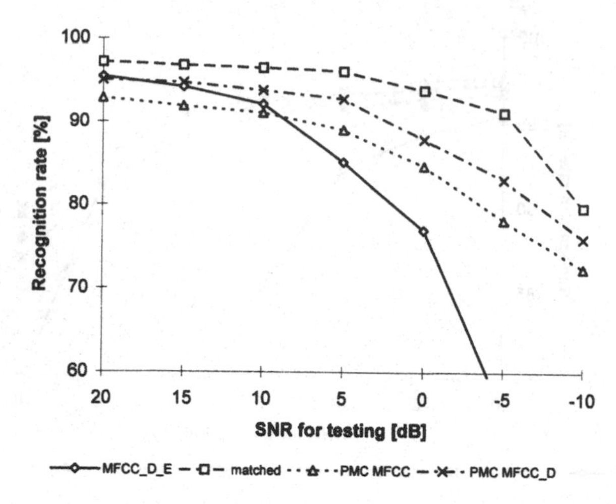

FIGURE 32.8. Performance of PMC for static and dynamic model parameters

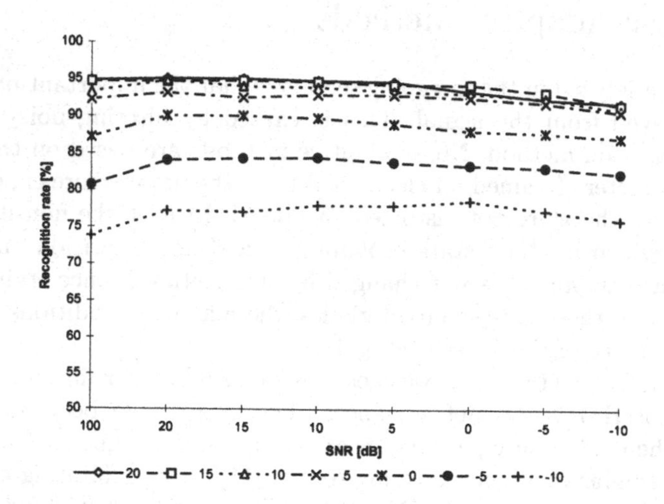

FIGURE 32.9. Influence of PMC weight g for static and dynamic coefficients.

In agreement with [3] and [4], to estimate $\tilde{\mu}$ we used approximately one second of the noisy signal preceding the word being recognized. To reliably detect non-speech segments, a robust real-time VAD [14] should be used.

Results achieved by PMC applied to static only (PMC MFCC) or static and dynamic (PMC MFCC_D) cepstral parameters can be seen in Figure 32.8. If PMC uses both the static and dynamic coefficients simultaneously, the recognition rates are better when only static coefficients are used.

The influence of weighting in the PMC is summarized in Figure 32.9 (the case with normalized spectra) where SNR[dB] represents the expected SNR of noisy speech which equals $20 \log_{10} g$. Each type of line is generated for the given test SNR varying from 20 to -10 dB (see the legend under the figure). The maxima of the recognition rates corresponding to the proper value of g are rather flat. Similar results are also valid for static parameters. It follows from this fact that finding the proper value g (given by the expected SNR) is not a critical problem.

32.5 Conclusions

Models for connected word recognition also contain a model for pauses. If the SNR increases, this pause model becomes inadequate to describe background noise between words, and that is why the recognition rate immediately falls below 95% even when the SNR is relatively high (10 dB). The value of the weighting (penalty) factors p and s must be carefully balanced in this case. This disadvantage can be partially suppressed by training on a noisy speech signal. Training HMM, in noisy conditions similar to those expected in the recognition, forces models to learn features of speech and also features of the noise environment. This approach depends on the SNR of the input signal. Therefore it is necessary to use a noise-compensation method as the preprocessor for noisy speech. Unfortunately, these methods may cancel or distort important parts of speech. This distortion leads to lower recognition rates. The decrease of recognition rates can be partially suppressed by training models on enhanced (distorted) speech. When this idea is used, the models still depend on the SNR although less. The best results can be achieved by using the noise-adaptive method PMC. This method adapts the model parameters leaving the speech uncanceled. Therefore this approach is independent of the SNR. For real-time implementation of PMC, it is necessary to detect non-speech segments and the SNR. Less than one second of non-speech signals is sufficient to estimate noise characteristics.

Acknowledgments: This study was supported within the project COST 249 "Speech recognition over telephone line" and the grant GACR 102/96/k087 "The theory and application of speech communication in Czech".

32.6 References

[1] M. Berouti, R. Schwartz, and J. Makhoul. Enhancement of speech corrupted by acoustic noise. In *Proceedings of ICASSP*, pp. 208–211, 1979.

[2] S.F. Boll. Suppression of acoustic noise in speech using spectral subtraction. *IEEE Trans. on ASSP*, 27(2):113–120, 1979.

[3] M.K. Brendborg and B. Lindberg. Noise robust recognition using feature selective modelling. In *Report of COST 249*, Roma, Italy, 1997.

[4] M.K. Brendborg. Toward noise immune automatic speech recognition using phoneme models. Ph.D. Thesis, Aalborg University-Center for PersonKommunikation, Aalborg, Denmark, 1996.

[5] G. Doblinger. Computationally efficient speech enhancement by spectral minima tracking in subbands. In *Proceedings of EUROSPEECH'95*, Madrid, Spain, pp. 1513–1516, 1995.

[6] Y. Ephraim and D. Malah. Speech enhancement using a minimum mean-square log-spectral amplitude estimator. *IEEE Trans. on ASSP*, 33(6):443–445, 1985.

[7] H. Hermansky and N. Morgan. Rasta processing of speech. *IEEE Trans. on SAP*, 2:578–579, 1994.

[8] G.S. Kang and L.J. Fransen. Quality improvement of LPC-processed noisy speech by using spectral subtraction. *IEEE Trans. on ASSP*, 37(6):939–942, 1989.

[9] P. Lockwood and J. Boudy. Experiments with non-linear spectral subtractor (NNS), hidden markov models, and the projection for robust speech recognition in cars. *Speech Communication*, 11:215–228, 1992.

[10] R. Martin. Spectral subtraction based on minimum statistics. In *Proceedings of EUSIPCO'94*, Edinburgh, Scotland, U.K., pp. 1182–1185, 1994.

[11] B.P. Milner and S.V. Vaseghi. Comparison of some noise-compensation methods for speech recognition in adverse environments. *IEE Proc.-Vis. Image Signal Process.*, 141(5):280–288, 1994.

[12] M.J.F. Gales and S.J.Young. HMM recognition in noise using parallel model combination. In *Proceedings of EUROSPEECH'93*, Berlin, Germany, pp. 837–840, 1993.

[13] P. Pollák, P. Sovka, and J. Uhlíř. Noise suppression system for a car. In *Proceedings of EUROSPEECH'93*, Berlin, pp. 1073–1076, 1993.

[14] P. Sovka and P. Pollák. The study of speech/pause detectors for speech enhancements methods. In *Proceedings of EUROSPEECH'95*, Madrid, Spain, pp. 1575–1578, 1995.

[15] P. Sovka, P. Pollák, and J. Kybic. Extended spectral subtraction. In *Proceedings of EUSIPCO'96*, Trieste, Italy, pp. 963–966, 1996.

33

Non-Linear Parametric Modelling by Means of Self-Exciting Threshold Autoregressive Models

Matthias Arnold[1]
Roland Günther[2]
Herbert Witte[3]

ABSTRACT
This paper gives a short introduction to non-linear parametric modelling by means of self-exciting threshold autoregressive (SETAR) models. After a discussion of the formal definition we summarize established identification techniques for the model parameters. All these methods make use of non-recursive calculations. To overcome this drawback we present here a new recursive procedure for the estimation of the coefficients of those models where the order and the number of thresholds can be chosen arbitrarily. The algorithm was tested with simulated data. The usefulness of such models for the analysis of biological signals was demonstrated with two examples in ECG and EEG analysis.

33.1 Introduction

In the analysis of time series a parametric modelling approach is frequently used. One of the most popular parametric models is the autoregressive (AR) model. In its classical form the AR model is defined by a stochastic difference equation with *constant* coefficients:

$$Y_n = \sum_{i=1}^{p} a_i Y_{n-i} + \varepsilon_n \tag{33.1}$$

From a mathematical point of view this type of model is attractive since AR processes approximate the class of so-called regular processes (purely non-deterministic processes). From a practical point of view AR modelling can be used to characterize the temporal variation of a time series by means of a few parameters or to estimate the spectral density by performing a z-transform of the model parameters. Nevertheless, the application of AR models to real data is limited for at least

[1]Friedrich Schiller University Jena, Medical Faculty, Institute of Medical Statistics, Computer Sciences and Documentation, Jahnstraße 3, 07740 Jena, Germany, E-mail: iia@imsid.uni-jena.de
[2]Faculty of Mathematics and Informatics, Institute of Stochastics, Ernst-Abbe-Platz 1-3, 07740 Jena, E-mail: guenther@minet.uni-jena.de
[3]Institute of Medical Statistics, Computer Sciences and Documentation, Jahnstraße 3, 07740 Jena, E-mail: iew@imsid.uni-jena.de

two reasons: 1. Such models are only suitable for analyzing stationary processes. 2. Their application is restricted to processes generated by linear systems.

Nonstationary processes can be analyzed by means of AR models with *time-varying* coefficients. An efficient way of fitting such models is to use a state-space representation of AR models with stochastic coefficients. The coefficients can be estimated recursively by means of Kalman filtering [9, 1].

The observation that linear models frequently fail to describe important properties of time series from many fields led to the development of several non-linear alternatives. Their superiority in comparison with the classical linear model could be demonstrated in various applications, especially in the field of econometrics [19], and in the analysis of some typical examples of time series which have been proved to be non-linear (Wolf's sunspot series, Canadian lynx data [20, 16, 5]).

At present, research interest is focused on the investigation of four particular non-linear models. The bilinear model was introduced by Ruberti et. al. [14] and extensively investigated in [6] and [13]. Tong and Lim [17, 18] proposed the threshold model and Haggan and Ozaki [8] the exponential autoregressive model. Additionally, Priestley [12] developed a generalized class of non-linear models, the so- called state-dependent models, which includes each of the above-mentioned models as a special case. These models can also be regarded as AR (or ARMA) models with *time-dependend* coefficients.

Despite the general nature of state-dependent models there are factors which motivate the use of SETAR models in practical applications:

- A wide class of non-linear time series which includes the exponential autoregressive and the invertible bilinear models can be approximated by threshold models [11].

- Compared with other non-linear models SETAR models are more easily interpretable and tractable.

- Patterns which are characteristic of non-linear systems such as oscillations with amplitude- dependent frequencies, (asymmetric) limit cycles, jump resonances and synchronization phenomena can be generated by means of SETAR models [18].

- Many systems from the real world show saturation characteristics which can be modeled by the introduction of thresholds.

Nevertheless, only little attention has been focused on SETAR models in practical applications up to now. This seems to be due to the difficulties in identification of the structural model parameters (threshold variables, delay parameter) and to the lack of suitable procedures for the estimation of the SETAR-coefficients. Procedures available (e.g. [20]) are non-recursive and are not suitable for the processing of long time series.

In Section 33.2 the definition of SETAR models is given and some notations are introduced. After a short summary of known identification techniques (Section 33.3) we introduce a simple and fast recursive stochastic gradient procedure for the estimation of the coefficients of SETAR models (Section 33.4). In Section 33.5 the results of performance tests on the basis of simulated data are given. Furthermore, as a first application in biosignal analysis, the modelling of ECG and a special type of EEG pattern is discussed.

33.2 Definition and Notations

A SETAR model is given by a collection of AR models and a corresponding number of thresholds which define a partition of the real axes. At each instant one AR model is chosen to be active and is assumed to generate the corresponding observation. The active regime is determined by detecting the interval which covers the observation at a defined time point in the past. This definition can be formalized as follows:

Let

$$\begin{aligned} \mathbf{R} &= \bigcup_{j=1}^{l} R_j \\ R_j &= (r_{j-1}, r_j], \quad -\infty = r_0 < r_1 < \ldots < r_l = \infty \end{aligned}$$

be a disjunctive decomposition of the real axis and d, p, p^*, $p_1, p_2, \ldots, p_l$ positive integers with $p = \max\{p_1, p_2, \ldots, p_l\}$ and $p^* = \max\{p, d\}$. The process $(Y_n)_{n>0}$ with

$$Y_n = \sum_{j=1}^{l} \left[Y_{n-d}^{(j)} \varepsilon_n^{(j)} - Y_{n-d}^{(j)} a_0^{(j)} - \sum_{i=1}^{p_j} Y_{n-d}^{(j)} a_i^{(j)} Y_{n-i} \right], \tag{33.2}$$

where

$$Y_n^{(j)} = \mathbf{I}_{R_j}(Y_n) \quad (\mathbf{I} \text{ indicator function}),$$

and $(\varepsilon_n^{(j)})_{n>0}$ are white noise sequences, is called a SETAR$(l, p_1, p_2, \ldots, p_l)$ process with delay parameter d. For the sake of simplicity we will use the following abbreviations:

$$\begin{aligned} \mathbf{a}^{(j)} &= [a_1^{(j)}, a_2^{(j)}, \ldots, a_p^{(j)}]^T \\ \underline{Y}_{n-1} &= (Y_{n-1}, Y_{n-2}, \ldots, Y_{n-p})^T \end{aligned}$$

where $a_{p_j+1}^{(j)} = \ldots = a_p^{(j)} = 0$. In the following we deal with the simplified model

$$Y_n + \sum_{j=1}^{l} Y_{n-d}^{(j)} \langle \mathbf{a}^{(j)}, \underline{Y}_{n-1} \rangle = \varepsilon_n. \tag{33.3}$$

At time point n the process is in regime j_n, where the sequence $(j_n)_{n>p^*}$ is defined by means of the condition

$$Y_{n-d}^{(j_n)} = 1.$$

Figure 33.1 illustrates signal generation in SETAR models and the meaning of the model parameters.

33.3 Identification Techniques

Identification of SETAR models involves estimation of, first, the structural parameters, i.e. AR orders, thresholds, and delay, and second, estimation of AR coefficients of each regime (SETAR coefficients). Known estimation procedures use classical approaches such as minimization of least squared prediction errors or maximization of conditional density functions (Bayes estimation).

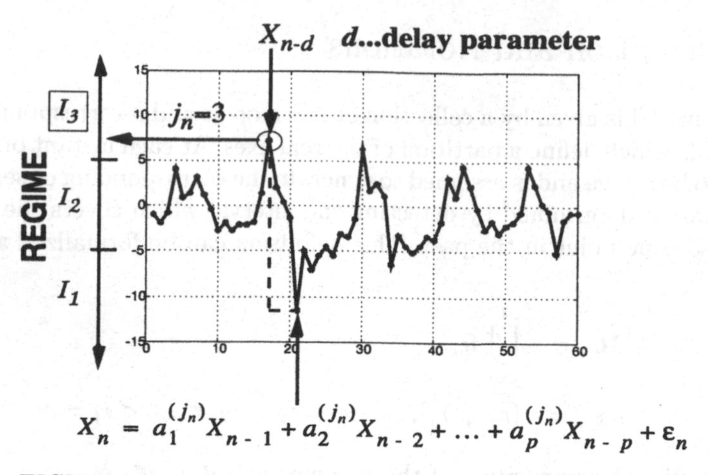

FIGURE 33.1. Schema of signal generation in SETAR models

Tsay [20] developed a simple 4-step procedure which includes a test of threshold nonlinearity, determination of thresholds by visual inspection of various scatterplots, and calculation of the SETAR coefficients. The basic facts of this procedure are outlined in Table 33.1.

The derivation of the a posteriori distribution of the model parameters [5] was an important step towards automatic determination of the structural parameters. However, the search for parameters which would maximize this density is only practical for strongly restricted parameter spaces (number of thresholds, admissable delay parameters).

Chen and Lee [3] make use of conditional marginal distributions for each parameter to approximate a posteriori marginal distributions by means of the iterative Gibbs sampler. The maxima of these empirical marginal distributions determine the value of the corresponding parameter. Though this technique offers many computational advantages it has only been developed for the case of two regimes.

All these procedures are essentially based on the technique of so-called arranged autoregressions which assumes knowledge of the entire data set prior to the estimation. This prevents recursive implementations. Recursive techniques, however, are desirable especially in the treatment of long time series. In the following section we present a recursive procedure for the estimation of SETAR coefficients.

33.4 Recursive Parameter Estimation

Obviously, if $l = 1$ then (33.3) defines a linear AR-process. In this case the following stochastic gradient procedure can be used to estimate the AR-coefficients:

$$\begin{aligned} \hat{\mathbf{a}}(n) &= \hat{\mathbf{a}}^{(1)}(n) = 0; \quad n \leq p \\ e_n &= Y_n + \langle \hat{\mathbf{a}}(n-1), \underline{Y}_{n-1} \rangle \\ \hat{\mathbf{a}}(n) &= \hat{\mathbf{a}}(n-1) - g_{n-1} e_n \underline{Y}_{n-1} \end{aligned}$$

where the so-called control sequence g_n is given by either

$$g_n = s_n^{-1}, s_n = \begin{cases} 1 & ; \ n < p \\ s_{n-1} + Y_n^2 & ; \ n \geq p \end{cases}$$

Step	Remarks
Choose an initial AR order p and a set K of possible values for the delay parameter d.	Use the partial autocorrelation function or an information criterion (i.e. AIC) for order determination.
Fit the arranged autoregressions for fixed p and each $d \in K$ by means of the RLS algorithm. Calculate test function $\hat{F}(d,p)$.	Choose d as delay parameter which maximizes $\hat{F}(\cdot,p)$.
Identify the thresholds by inspection of scatterplots of the threshold variable Y_{n-d} against standardized prediction errors or the course of estimated AR coefficients from step 2.	The distribution of these quantities will be biased by model changes at the location of thresholds. Consequently abrupt changes in the distribution of the scatterplots point to possible thresholds.
Refine AR order and thresholds for each regime if the model accuracy is unsatisfactory.	Use linear autoregressive techniques (see step 1) for the local linear models.

TABLE 33.1. Outline of the 4-step procedure, developed by Tsay, for identification of SETAR models.

(see [7]) or

$$g'_n = a/s'_n, s'_n = \begin{cases} 1 & ;\ n < p \\ s'_{n-1} + \|Y_n\|^2 & ;\ n \geq p \end{cases}$$

with $a \in (0,1]$ (see [2]). In [15] this scheme was used to construct a similar recursive procedure for the estimation of SETAR(2,1,1) models. Here we generalize this procedure for SETAR models with arbitrary order and number of thresholds. Individual control sequences for each regime are used. The gradient descent step for the coefficient vector of regime j and the update of the corresponding control sequence are performed if and only if the process resides in that regime:

$$\begin{aligned} \hat{\mathbf{a}}^{(j)}(n) &= 0; \quad n \leq p^* \\ e_n &= Y_n + \langle \hat{\mathbf{a}}^{(j_n)}(n-1), \underline{Y}_{n-1} \rangle \\ \hat{\mathbf{a}}^{(j)}(n) &= \hat{\mathbf{a}}^{(j)}(n-1) - g^{(j)}_{n-1} e_n Y^{(j)}_{n-d} \underline{Y}_{n-1} \end{aligned}$$

where

$$g^{(j)}_n = a/s^{(j)}_n, s^{(j)}_n = \begin{cases} 1 & ; n < p^* \\ s^{(j)}_{n-1} + Y^{(j)}_{n-d+1} \|\underline{Y}_n\|^2 & ; n \geq p^* \end{cases}$$

33.5 Results

33.5.1 Simulations

In this section some simulation results are shown. The proposed algorithm was tested with random samples of the following SETAR(3,4,4,4)-process with threshold values -1 and 1:

$$Y_n = \begin{cases} (.5, .3, .05, -.1) & \underline{Y}_{n-1} + \varepsilon_n;\ Y_{n-d} \leq -1 \\ (1.5, -.5, .1, -1.2) & \underline{Y}_{n-1} + \varepsilon_n;\ -1 < Y_{n-d} < 1 \\ (.4, -.1, -.03, .1) & \underline{Y}_{n-1} + \varepsilon_n;\ Y_{n-d} \geq 1 \end{cases} \tag{33.4}$$

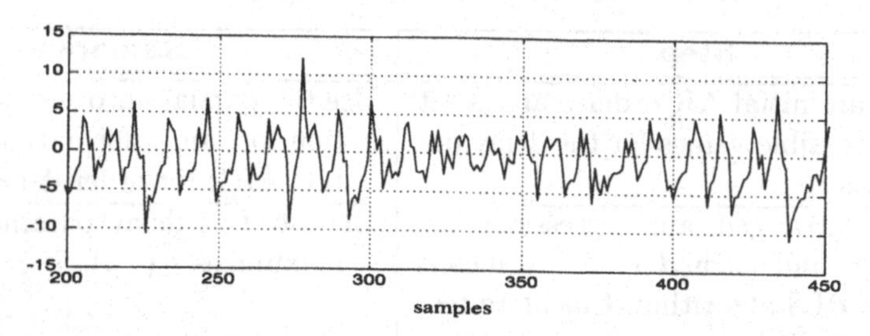

FIGURE 33.2. Part of a sample of the process (33.4)

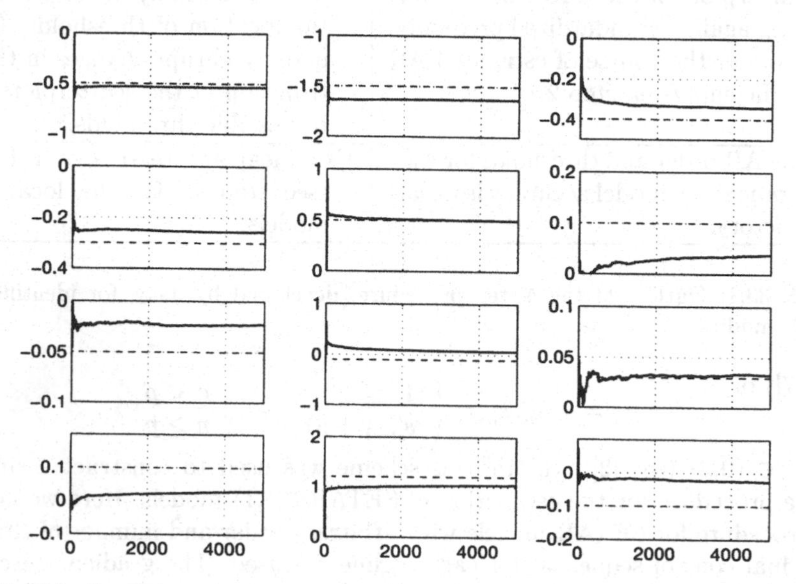

FIGURE 33.3. Coefficient estimates of a particular sample (length 5000) of the process (3). Coefficients of regime j are given in column j

We investigated the time courses of the estimates with respect to single trajectories of length 5000 and statistics with respect to samples of trajectories (number of samples $N = 500$). To measure the accuracy of the estimates we used the absolute deviation of the individual coefficients from the true values and the quadratic norm of the error vector $\|\mathbf{a} - \hat{\mathbf{a}}(n)\|^2$ where $\mathbf{a}$ denotes the vector of all coefficients and $\hat{\mathbf{a}}^k(n)$ its estimate for sample k at time point n. From samples of trajectories we computed the bias

$$\overline{\hat{\mathbf{a}}(n)} = \left\| \frac{\sum_{k=1}^{N} \hat{\mathbf{a}}^k(n)}{N} - \mathbf{a} \right\|^2$$

and the empirical variance

$$\hat{\delta}^2_{\hat{\mathbf{a}}(n)} = \frac{\sum_{k=1}^{N} \|\hat{\mathbf{a}}^k(n) - \overline{\hat{\mathbf{a}}(n)}\|^2}{N-1}$$

The estimation results are given in Figs. 33.2 - 33.4. It can be observed that the estimates approximate the true values rather quickly, but there are differences between the individual coefficients with respect to accuracy. This seems to be due to the fast decrease in the control sequences. An approach to solving this problem is the introduction of *relaxed* control sequences which result from a slight modification of $s_n^{(j)}$:

$$\tilde{s}_n^{(j)} = \begin{cases} 1 & ; n \leq p^* \\ \tilde{s}_{n-1}^{(j)} + Y_{n-d+1}^{(j)}(\max\{\alpha^{(j)} s_{n-1}^{(j)}, 1\} + \|\underline{Y}_n\|^2 - \tilde{s}_{n-1}^{(j)}) & ; n > p^* \end{cases}$$

Figs. 33.5 - 33.7 demonstrate the significant improvement in accuracy which was obtained by the use of relaxed control sequences ($\alpha^{(j)} = 0.1$).

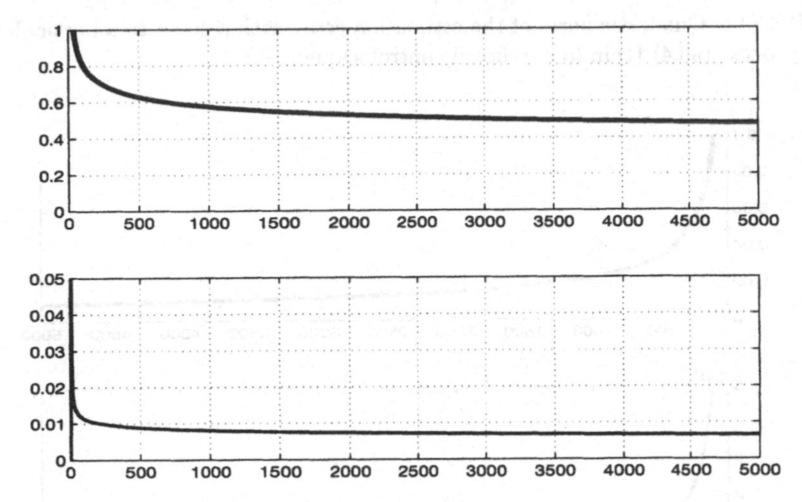

FIGURE 33.4. Bias (bold) and variance of the estimates with respect to 500 samples of the process (33.4)

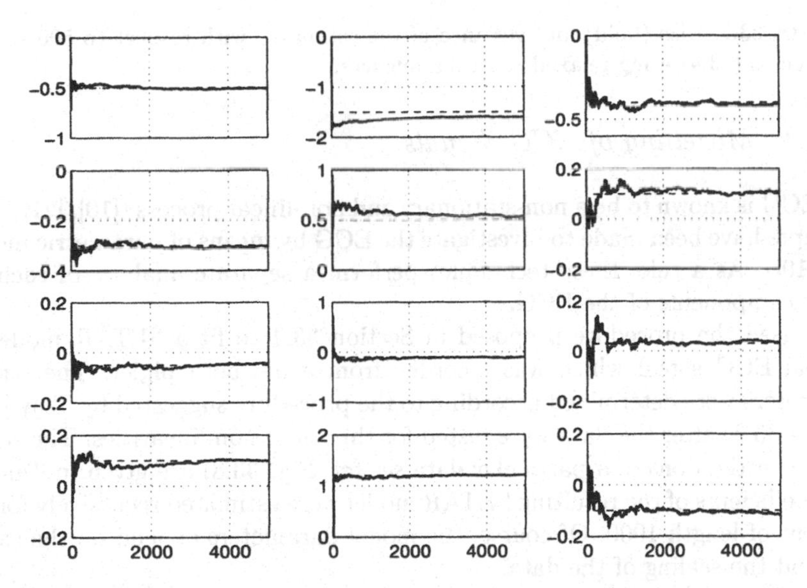

FIGURE 33.5. Coefficient estimates of a particular sample of the process (33.4) using relaxed control sequences. Coefficients of regime j are given in column j.

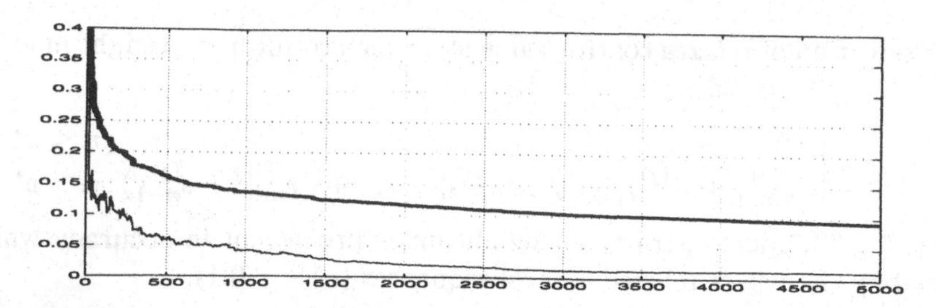

FIGURE 33.6. Quadratic norm of the estimation error with respect to a particular sample of the process (33.4) (thin line: relaxed control sequence).

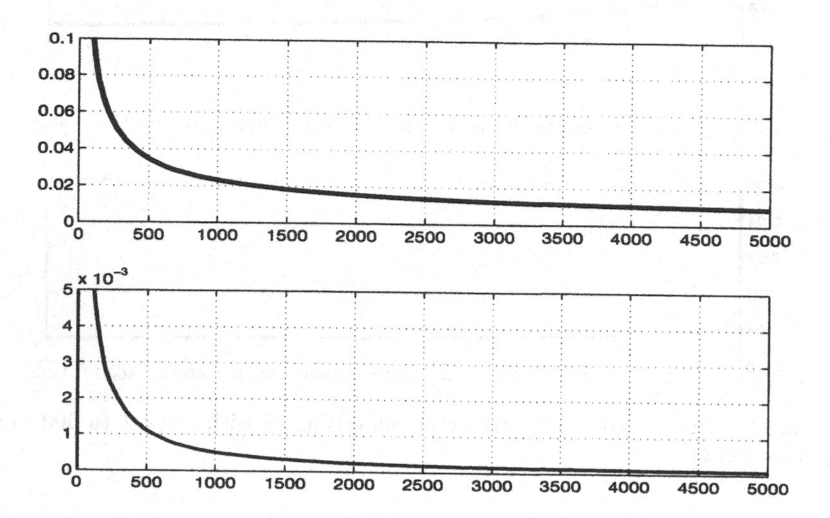

FIGURE 33.7. Bias (bold) and variance of the estimates with respect to 500 samples of the process (33.4) using relaxed control sequences.

33.5.2 Modelling of ECG Signals

The ECG is known to be a non-stationary and non-linear process ([10], [21]). Several attempts have been made to investigate the ECG by means of parametric modelling (see [10]). As a rule, these techniques perform a separate analysis of each of the major components of the ECG.

We used the procedure proposed in Section 33.4 to fit a SETAR model to an original ECG signal which was recorded from a newborn piglet. The structural parameters were determined according to the procedure suggested by Tsay [20] (see Section 33.3) after the data were tested for threshold non-linearities. The results of these investigations of a particular data set (cf. Fig. 33.8) are given in Table 33.2. The coefficients of the resulting SETAR model were estimated recursively for a data segment of length 1000. Of course, the model parameters depend on the sampling rate and the scaling of the data.

Impulse responses of the resulting non-linear system were investigated in order to explore how well the model captures the characteristics of the ECG. We found that the response to a particular impulse (amplitude=1, length=4) has a shape very similar to the waveform of the analyzed ECG (see Fig. 33.8). This first result motivates the use of SETAR models for parametric modelling of the ECG.

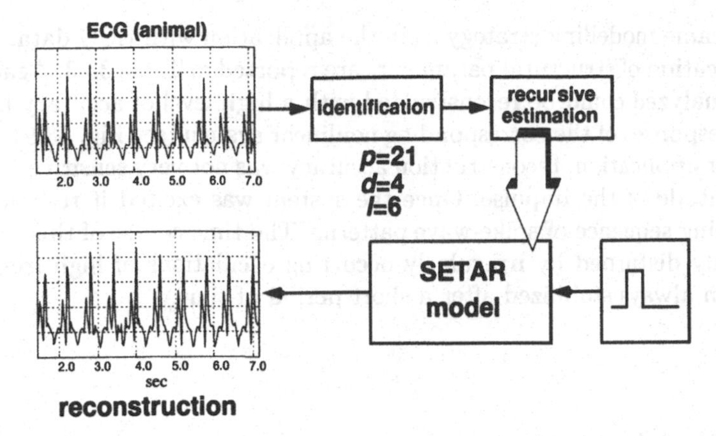

FIGURE 33.8. Modelling of ECG signals (newborn piglet) by means of a SETAR model. After identification of the model parameters the ECG was reconstructed by an impulse response (amplitude of the impulse = 1, length = 4) of the corresponding nonlinear system

Parameter	estimated values
p-value of F-statistics	> 0.99
AR-order	21
Delay parameter	4
Number of thresholds	5

TABLE 33.2. Analysis of ECG data: result of the non-linearity test, parameters of the SETAR model used

33.5.3 Modelling of Spike-Wave Patterns in the EEG

A second application deals with parametric modelling of spike–wave–discharges (SWD) in EEG recordings. These patterns occur during childhood absence epilepsy. The EEG of children suffering from this neurological disorder is characterized by sporadic emergence of clear rhythmic oscillations of high amplitude where a spike of high frequency is followed by a slow wave of lower frequency. Such periods typically have a duration of about 10 seconds.

The basic mechanisms of such disorders are still unclear. One approach to gaining insight into the neural processes which generate such patterns is to investigate assumed models of the processes involved. Neural oscillations are frequently modeled by sets of interacting differential equations [4].

Parametric modelling of SWD by means of SETAR models leads to more tractable systems and seems to be promising for nonlinear analysis of such EEG patterns. We

Parameter	estimated values
p-value of F-statistics	> 0.99
AR-order	20
Delay parameter	2
Number of thresholds	4

TABLE 33.3. Analysis of EEG data: result of the non-linearity test, parameters of the SETAR model used

used the same modelling strategy as in the application with ECG data. The results of identification of structural parameters are reported in Table 33.3. Again the EEG pattern analyzed could be reconstructed with a high level of accuracy by means of impulse responses of the corresponding nonlinear system (see Fig. 33.9). In contrast to the first application, reconstruction accuracy was not very sensitive to the length and amplitude of the impulse. Once the system was excited it responded with a never-ending sequence of spike-wave patterns. The time course of this response was temporarily disturbed by irregularly occurring oscillations of high frequency, but the system always stabilized after a short period of time.

33.6 Conclusion

We have introduced an efficient method for recursive parameter estimation in SETAR models. This approach is especially useful for high speed analysis of long time series. The reliability of the estimates was demonstrated by means of simulation results. As a first application in biosignal analysis we used this procedure for parametric modelling of ECG signals and patterns of SWD in EEG recordings. After estimation of the model parameters by means of the proposed recursive procedure these patterns could be reconstructed as impulse responses of the corresponding nonlinear systems. It can be concluded that the dynamics of such physiological recordings can be well described by means of SETAR models.

These first results are promising and suggest the use of SETAR models as additional tool for nonlinear analysis in applications with physiological signals. The modelling procedure can be interpreted as a decomposition of the underlying complicated system (model) in a set of several simple, interacting (alternatingly active) subsystems (submodels). These subsystems can, for instance, be related to excitatory and inhibitory physiological processes.

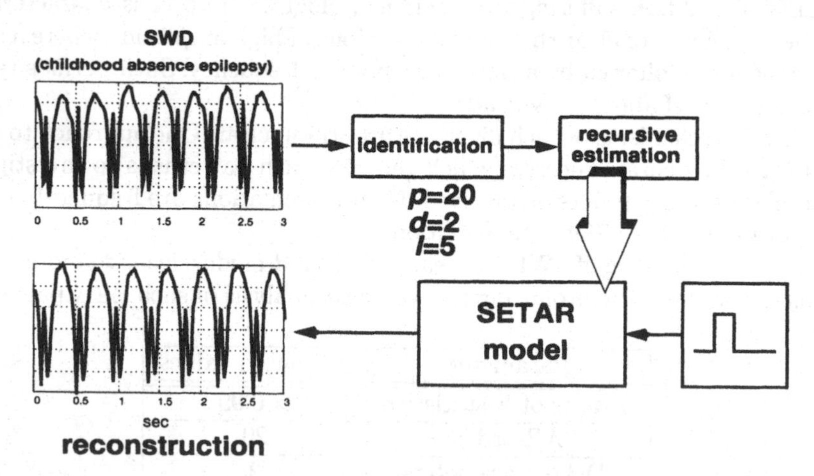

FIGURE 33.9. Modelling of SWD in an EEG recording by means of a SETAR model. After identification of the model parameters the EEG segment was reconstructed by an impulse response of the corresponding nonlinear system

Acknowledgments: This work has been supported by the German Ministry for Education and Research (project *Clinical Oriented Neurosciences* 01 ZZ 9602).

We are grateful to Dr. R. Bauer, Institute of Pathophysiology, Friedrich Schiller University Jena (Germany), and Dr. M. Feucht, Universitätsklinik für Neuropsychiatrie des Kindes- und Jugendalters, Vienna (Austria) who provided the ECG and EEG recordings, respectively.

33.7 References

[1] M. Arnold, W. Miltner, H. Witte, R. Bauer, and C. Braun. Adaptive AR modeling of non-stationary time series by means of kalman filtering. *IEEE Trans. Biomed. Eng.*, accepted.

[2] H.H. Chen and L. Guo. *Identification and Stochastic Adaptive Control.* Birkhäuser, Boston, 1991.

[3] R. Chen. Threshold variable selection in open-loop threshold autoregressive models. *J. Time Series Anal.*, 16(5):461, 1995.

[4] W.J. Freeman. Simulation of chaotic EEG patterns with a dynamic model of the olfactory system. *Biol. Cybern.*, 56:139-150, 1987.

[5] J. Geweke and N. Terui. Bayesian threshold autoregressive models for nonlinear time series. *J. Time Series Anal.*, 14(5):441-454, 1993.

[6] C.W.J. Granger and A.P. Anderson. *An Introduction to Bilinear Time Series Models.* Vandenhoeck and Ruprecht, Göttingen, 1978.

[7] R. Günther. On the convergence of some adaptive estimation procedures,. *Math. Operationsforsch. u. Statist.*, 14:535-550, 1983.

[8] V. Haggan and T. Ozaki. Modeling non-linear random vibrations using an amplitude-dependent autoregressive time series model. *Biometrika*, 68:189-196, 1981.

[9] S. Haykin. *Adaptive Filter Theory.* Prentice Hall, Englewood Cliffs, New Jersey, 1986.

[10] S. Mukhopadhyay and P. Sircar. Parametric modelling of ECG signal. *Med. Biol. Eng. & Comput.*, 34:142-148, 1996.

[11] J.D. Petruccelli. On the approximation of time series by threshold autoregressive models. *The Indian Journal of Statistics*, 54:106-113, 1992.

[12] M.B. Priestley. State-dependent models: a general approach to non-linear time series analysis. *J. Time Series Anal.*, 1(1):47-71, 1980.

[13] T. Subba Rao and M.M. Gabr. *An Introduction to Bispectral Analysis and Bilinear Time Series Models.* Springer, Berlin, 1984.

[14] A. Ruberti, A. Isidori, and P. d'Allessandro. *Theory of Bilinear Dynamical Systems.* Springer, Berlin, 1972.

[15] K. Schmidt. Adaptive Parameterschätzung und Vorhersage im SETAR(2;1,1)-Modell. Master's thesis, Friedrich Schiller University, Jena, Germany, 1992.

[16] B.Y. Thanoon. A threshold model with piece-wise non-linear dynamics for sunspot series. *J. Univ. Kuwait (Sci.)*, pp. 19–29, 1990.

[17] H. Tong. On a threshold model. In C.H. Chan, editor, *Pattern Recognition and Signal Processing.* Sijthoff and Noordhoff, Amsterdam, 1978.

[18] H. Tong and K.S. Lim. Threshold autoregression, limit cycles and cyclical data (with discussion). *J. R. Statist. Soc.*, 42:245-292, 1980.

[19] H. Tong and I. Yeung. On tests for self-exciting threshold autoregressive-type non-linearity in partially observed time series. *Appl. Statist.*, 40(1):43-62, 1991.

[20] R.S. Tsay. Testing and modeling threshold autoregressive processes. *J. Am. Statist. Assoc.*, 84(405):231-240, 1989.

[21] Q. Xue, Y.H. Hu, and W.J. Tompkins. Neural-network-based adaptive matched filtering for QRS detection. *IEEE Trans. Biomed. Eng.*, 39(4):317-329, 1992.

34

Lyapunov Exponents of Simulated and Measured Quasi-Periodic Flows

Maja Bračič[1]
Aneta Stefanovska[1]

ABSTRACT
Lyapunov exponents are calculated from numerically simulated and measured signals. The existing algorithms for their estimation have free parameters and their impact and the impact of noise are analysed on quasi-periodic simulated signals. Furthermore, calculation of the exponents of the blood flow signal, measured on a healthy subject, is presented. Two typical patterns of both the global and the local Lyapunov exponents are obtained over a wide range of parameter values. Either we have four paired and one zero exponent, or five paired exponents. This may indicate the deterministic and almost conservative nature of the system governing blood circulation on the time-scale of minutes.

34.1 Introduction

The Lyapunov or characteristic exponents measure the rate of convergence of the nearby trajectories of a dynamic system in phase space. They are one of the most meaningful characterisations of a non-linear dynamic system. Using Lyapunov exponents, one can distinguish among fixed points and periodic, quasi-periodic or chaotic motions.

Consider a dynamic system of dimension d, defined by equations

$$\dot{\mathbf{x}}(t) = \mathbf{f}(\mathbf{x}(t)) \quad \mathbf{x}(0) = \mathbf{x}_0,\ \mathbf{x} \in \Re^d \tag{34.1}$$

where $\mathbf{x}(t)$ is a state vector and the flow $\mathbf{f}$ is some generally non-linear function. The movement of the state vector $\mathbf{x}(t)$ in the phase space results in a system trajectory. After transients, the trajectories, generated by different initial conditions $\mathbf{x}_0$ settle near an attractor [6]. The time evolution of a small perturbation to a trajectory is governed by a linearised equation in the tangent space

$$\delta\dot{\mathbf{x}}(t) = D\mathbf{f}(\mathbf{x}(t))\delta\mathbf{x}(t) \quad \delta\mathbf{x}(0) = \delta\mathbf{x}_0 \tag{34.2}$$

where $D\mathbf{f}$ is the Jacobi matrix of the flow $\mathbf{f}$. At the end of the last century, A.M. Lyapunov introduced a measure of average contraction of the perturbation to a given trajectory as

$$\lambda(\delta\mathbf{x}_0) = \lim_{t\to\infty} \frac{1}{t} \log\left(\frac{\|\delta\mathbf{x}(t)|_{\delta\mathbf{x}_0}\|}{\|\delta\mathbf{x}_0\|}\right) \tag{34.3}$$

[1] University of Ljubljana, Faculty of Electrical Engineering, Tržaška 25, 1000 Ljubljana, Slovenia, E-mails: {maja, aneta}@osc.fe.uni-lj.si

today known as the Lyapunov exponents. He proved that $\lambda = \lambda(\delta\mathbf{x}_0)$ is finite for every nonzero solution $\delta\mathbf{x}(t)$ of Eq. (34.2). Using d linearly independent initial conditions $\delta\mathbf{x}_{01} \ldots \delta\mathbf{x}_{0d}$, one obtains a fundamental system of solutions $\lambda_1 \geq \lambda_2 \ldots \geq \lambda_d$. For ergodic systems the set of λ_i does not depend on the initial condition $\delta\mathbf{x}_{0i}$, and so the λ_i are global properties of the attractor [10].

The procedure for determining the Lyapunov exponents can be considered a generalisation of linear stability analysis for unsteady situations [8]. If any λ_i is positive, small perturbations will grow exponentially. If all λ_i are negative, any perturbation will decrease, and the attractor is a stable fixed point. A stable periodic state has one zero exponent which corresponds to a perturbation tangent to the limit cycle, and all other exponents are negative. A quasi-periodic system with k incommensurate frequencies has k zero exponents, all others are negative. Every attractor of a smooth dynamic system, given by Eq. (34.1), has at least one zero Lyapunov exponent corresponding to a perturbation tangent to the trajectory. The exponents of a Hamiltonian system come in conjugate pairs, consisting of a positive and a negative exponent of the same magnitude, and two of them are zero [6].

For systems whose equations of motion are known explicitly, there is a straightforward method of computing all Lyapunov exponents [3]. This method cannot be directly applied to experimental data. There are two approaches to estimating the exponents from measured signals. By the first, introduced by [12, 15, 18], two nearby points in the phase space are followed and only the largest exponent is evaluated. The second approach, introduced by [6, 14], is based on estimating the Jacobians of the map. The great advantage of this method compared to the trajectory tracing method is that one can deal with arbitrary vectors in tangent space whereas the observed data points are used only to approximate local flow. Thus, we can calculate all exponents (including negative ones) as long as the approximation of the flow is adequate [14].

In the past decade Lyapunov exponents have been widely used to characterise chaotic data. We have used them to analyse human blood flow dynamics within non-linear system theory. On the time-scale of minutes, the dynamic of the peripheral blood flow is shown to be quasi-periodic [16]. Therefore, the methods for estimating all Lyapunov exponents were first tested on simulated quasi-periodic signals. Since their exponents are known analytically, the influence of algorithm-free parameters can be studied. We briefly present the algorithm in the next section. Guidelines for parameter settings are summarised in the third section. In the fourth section, the Lyapunov exponents of the blood flow signal are reported, and then some conclusions are drawn in the last section.

34.2 Estimating Exponents from Scalar Signals

In experiments one typically observes only one or at best a few components of the state vector $\mathbf{x}(t)$ in discrete moments $t = nt_s$, where t_s is a sampling time. In the case of scalar observations, $s(n) = s(t_0 + nt_s)$, where t_0 is an initial time. The calculation of Lyapunov exponents, however, is performed on the attractor in phase space. The attractor can be reconstructed from the signal by the method of delay coordinates, introduced by [11]. Using d scalar values of the signal, vectors

$$\mathbf{x}(n) = [s(n), s(n+T), \ldots, s(n+(d-1)T)] \tag{34.4}$$

are constructed in d-dimensional embedding space. The time lag T is some integer. If d is chosen large enough, the invariants of the attractor reconstructed by Eq. (34.4) and those of the original attractor will be the same [17].

The evolution of the system in the thus embedded state space is given by a map

$$\mathbf{x}(n+z) = \mathbf{f}(\mathbf{x}(n)) \tag{34.5}$$

where z is some positive integer. The Jacobi based methods determine the Lyapunov exponents by tracing the evolution of small perturbations to a chosen (fiducial) trajectory. After S steps the initial perturbation will grow or shrink to

$$\begin{aligned}\delta\mathbf{x}(n+Sz) &= D\mathbf{f}(\mathbf{x}(n+(S-1)z))\ldots D\mathbf{f}(\mathbf{x}(n+z))\cdot D\mathbf{f}(\mathbf{x}(n))\cdot\delta\mathbf{x}(n) = \\ &= \mathbf{Y}(\mathbf{x}(n),S)\delta\mathbf{x}(n)\end{aligned} \tag{34.6}$$

The logarithms of the eigenvalues of the positive and symmetric matrix

$$\Lambda_x = \lim_{S\to\infty}\left[\mathbf{Y}(\mathbf{x}(n),S)\mathbf{Y}(\mathbf{x}(n),S)^T\right]^{1/2S} \tag{34.7}$$

are the global Lyapunov exponents associated with the chosen trajectory [10]. These eigenvalues are independent of $\mathbf{x}(n)$ for almost all $\mathbf{x}(n)$ on the attractor.

When dealing with measured signals, one has no knowledge about the equations of motion and the Jacobi matrix. Therefore, the dynamic has to be approximated. The mapping of $\mathbf{x}(n)$ to $\mathbf{x}(n+z)$ can be approximated as

$$f_l(\mathbf{x}(n)) = \sum_{j=1}^{N_o} c_{lj}\chi_j(\mathbf{x}(n)) \tag{34.8}$$

where f_l denotes the lth component of the vector map $\mathbf{f}$ [2] and $\chi_j(\mathbf{x}(n))$ is a set of basis functions. The map may be approximated either in original or in tangent space. Initially, linear approximations of a tangent map were used [6], and the basis functions were hence set to $\chi_j(\mathbf{x}(n)) = \delta\mathbf{x}_j(n) = \mathbf{x}(n) - \mathbf{x}(p(j))$, where $\mathbf{x}(p(j))$ is the jth neighbour of $\mathbf{x}(n)$. Lately, this method has been generalised to higher order polynomial approximations [5]. An approximation of original flow using radial basis functions $\chi_j(\mathbf{x}(n)) = \sqrt{r^2 + ||\delta\mathbf{x}_j(n)||}$ with stiffness parameter r, was also proposed [9].

The coefficients c_{lj} are determined by the least squares fit. In this way, the residuals for mapping a set of N_b neighbours of $\mathbf{x}(n)$ to a set of neighbours of $\mathbf{x}(n+z)$ are minimised. At least N_O neighbours are needed to determine c_{lj} uniquely, and thus interpolate the flow. Usually, N_b greater than N_O is used to achieve flow approximation.

From the linear part of the flow in tangent space the matrix $\mathbf{Y}(\mathbf{x}(n),S)$ can be constructed and the eigenvalues of Λ_x computed. Numerically, the computation of $\mathbf{Y}(\mathbf{x}(n),S)$ fails for large S because the column vectors in $\mathbf{Y}(\mathbf{x}(n),S)$ converge extremely rapidly to the subspace of the tangent space with the largest expansion rate. The eigenvalues of Λ_x can, however, be evaluated using recursive QR decomposition, proposed [6]. At each step the decomposition

$$D\mathbf{f}(\mathbf{x}(n+jz))\mathbf{Q}^{j-1} = \mathbf{Q}^j\mathbf{R}^j \quad \mathbf{Q}^0 = \mathbf{I} \tag{34.9}$$

is performed. Then the Lyapunov exponents $\lambda_1 \ldots \lambda_d$ are obtained as

$$\lambda_i = \frac{1}{Szt_s}\sum_{j=1}^{S}\ln[R_{ii}^j] \tag{34.10}$$

The global Lyapunov exponents bear no information of the local behaviour on the attractor. Therefore, in [1] local Lyapunov exponents are introduced as the logarithms of the eigenvalues of the matrix (34.7) for finite S. These values may vary significantly over the attractor, but their means converge toward the global exponents.

A brief summary of the procedure for estimating Lyapunov exponents from scalar signals reveals the free parameters of the algorithm:

- The attractor is reconstructed from N points of measured signal by the method of delay coordinates. The embedding dimension d and embedding time $\tau = Tt_s$ must be set.
- A starting point for the fiducial trajectory is selected. N_b neighbouring points are sought. The tangent vectors are evolved for the chosen evolutionary time $t_{evol} = zt_s$ and the local mapping is approximated by a set of basis functions.
- The fiducial trajectory is followed for a chosen observation time St_{evol} and the Lyapunov exponents are estimated from the approximated Jacobi matrices according to Eq. (34.9) and (34.10).

The results obtained depend strongly on the algorithmic parameters, and no analytical criteria for parameter settings are available. Therefore, we have studied the effect of each parameter numerically. Calculations were done on different test signals with known Lyapunov exponents [4]. In the following section, we briefly present our experience with quasi-periodic signals.

34.3 A Quasi-Periodic Example

The impact of all algorithmic parameters and the influence of noise was analysed on quasi-periodic signals with up to five incommensurate frequencies $\omega_1 = 1\,\mathrm{s}^{-1}$, $\omega_2 = (\sqrt{5}-1)/2\,\mathrm{s}^{-1}$, $\omega_3 = \sqrt{7}-\sqrt{5}\,\mathrm{s}^{-1}$, $\omega_4 = (\sqrt{11}-\sqrt{7})/3\,\mathrm{s}^{-1}$, $\omega_5 = (\sqrt{17}-\sqrt{11})/7\,\mathrm{s}^{-1}$.

Convergence Rates

The zero exponents which characterise quasi-periodic systems can be obtained only if an infinite number of steps along fiducial trajectory are taken. If the local map is known exactly, the estimate of the Lyapunov exponent approaches the correct value as $1/St_{evol}$ [8]. Moreover, the convergence is slowed down (Figure 34.1) if noise is added to the signal and an error in the map approximation is generated. If the signal-to-noise ratio is too low, the exponents no longer decrease toward zero but have a finite negative or positive value. The inherent dynamics of the system is reflected on the convergence of exponents and, therefore, the observation time should exceed some periods of the slowest rhythm in the signal.

Attractor Reconstruction

The number of calculated exponents is equal to the dimension of the reconstructed phase space d. If d is too small, the attractor cannot unfold and the calculated exponents are erroneous. On the other hand, if d is too large, the reconstructed attractor is contained in a submanifold of dimension $m < d$, and $(d - m)$ spurious exponents unrelated to the dynamics of the system are obtained. Under time reversal, true exponents change sign, whereas the spurious exponents do not [2].

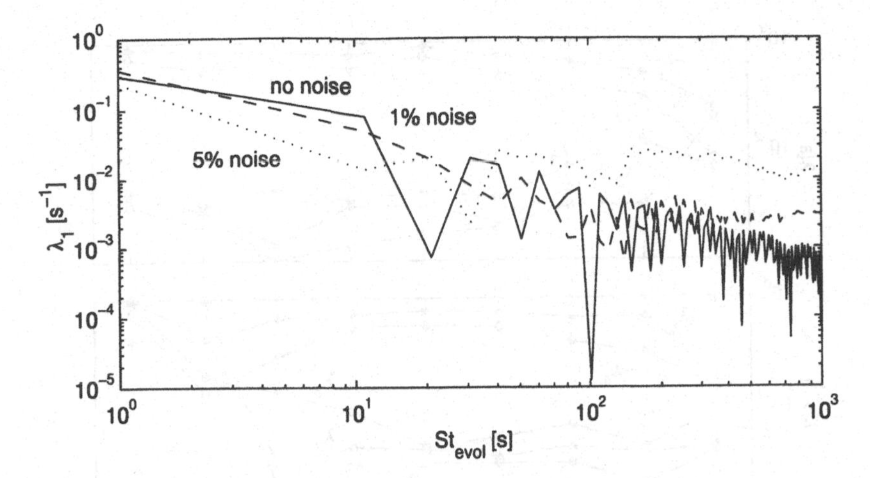

FIGURE 34.1. The convergence of the first exponent of quasi-periodic flow with two incommensurate frequencies with different levels of additive noise. $N = 30000$, $t_s = 0.1$ s, $d = 4$, $\tau = 1.5$ s, $t_{evol} = 1$ s, $S = 1000$, $N_b/N_0 = 8$. Quadratic functions were used

Several methods for determining the embedding dimension d and embedding time τ were proposed [2]. We have used the average mutual information [2] and the average displacement method [13] for a first approximation of the embedding time τ, and the false nearest neighbours method [2] for the dimension.

The choice of the embedding time τ is not critical for quasi-periodic systems, although it should neither be very small nor close to a periodicity of the system. The embedding dimension $(k+1)$ is obtained for quasi-periodic system with k frequencies by the false nearest neighbours method. However, $d \geq 2k$ is needed to obtain k zero exponents. The other $(d-k)$ are negative and spurious.

Different recommendations regarding the necessary number of data-points for correct attractor reconstruction can be found [2]. Figure 34.2a shows that for larger N the convergence of vanishing exponents $(\lambda_1 \ldots \lambda_3)$ is improved and the ratio between the order of magnitude of vanishing and spurious (λ_4 and λ_5) exponents is increased. However, the number of points should be gained by longer measurement time rather than by increasing the sampling frequency. One can see in Figure 34.2b that oversampling may result in uncorrect estimation of the Lyapunov exponents. It is more important how many times a particular part of the state space is revisited by the trajectory than the number of points itself.

Approximation of Local Mapping

To approximate the local mapping at a chosen point, N_b vectors in the neighbourhood of this point are considered. In noisy signals, large N_b/N_o should be used. But at the same time, the maximum distance of neighbouring points ε_{max} must be within reasonable limits compared to the attractor size r_A, $\varepsilon_{max}/r_A < 0.2$, to allow for an adequate approximation. For quasi-periodic signals without noise $N_b/N_0 \geq 2$ ensures the $1/t$ convergence toward zero, whereas $N_b/N_o \geq 8$ is needed if 1% noise is added to the signal and $N_b/N_o \geq 64$ if 5% noise is added.

Different basis function were tested on the quasi-periodic example. The use of linear basis functions leads to $2k$ instead of k zero exponents. Moreover, in this case one cannot distinguish between periodic and quasi-periodic behaviour. Quadratic

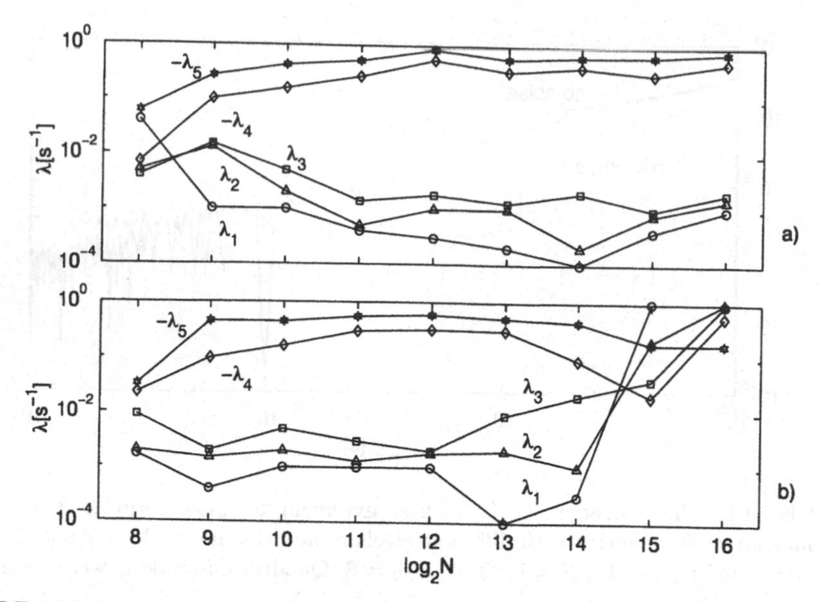

FIGURE 34.2. Exponents of a quasi-periodic flow with three frequencies for a different number of points, used for attractor reconstruction. In figure a) the sampling time is constant $t_s = 0.5\,\mathrm{s}$ and the measurement time is $t = Nt_s$. In figure b) the measurement time is fixed at $t = 512\,\mathrm{s}$ and the sampling time changes as $t_s = t_{sm}t/N$, $t_{sm} = 2\,\mathrm{s}$. $\tau = 2\,\mathrm{s}$, $d = 7$, $t_{evol} = 1\,\mathrm{s}$, $S = 500$. Quadratic basis functions are used, $N_b/N_0 = 2$

and radial basis functions perform equally well on noise-free signals, whereas radial basis functions are less sensitive to noise. For higher order polynomials, the number of basis functions grows very rapidly with the order of the polynom, and they cannot be used if the noise in the signal dictates large N_b/N_o.

The vanishing exponents of quasi-periodic flow do not depend strongly on the evolutionary time t_{evol}, but the spurious ones decrease with increasing t_{evol}. Since the method of delay coordinates introduces certain symmetries into the reconstructed attractor, $t_{evol} = l\tau$, $l = 1 \ldots (d-1)$ should be avoided [7].

34.4 Lyapunov Exponents of the Blood Flow Signal

To gain an insight into the dynamics of the complex non-linear system of blood flow regulation, Lyapunov exponents were calculated from the signal of peripheral blood flow. The signal was measured on a healthy male subject using a laser Doppler flow-meter (PeriFlux 3, Perimed, Sweden). This non-invasive technique allows continuous recording of the blood flow. The signal was collected for 11 minutes and sampled by 50 Hz. Figure 34.3 presents the signal in (a) time and (b) frequency domains.

The power spectrum of the signal reveals five characteristic frequency peaks in the range from 0.01 Hz to 1 Hz. We hypothesise that five almost periodic systems contribute to the dynamics of the observed blood flow. Because of mutual couplings among these systems, the characteristic frequencies vary in time [16].

The signal contains the instrumental noise and the noise resulting from movements of the subject during the recording. Filtering the signal in the phase space [2] can affect the inherent dynamics when little is known about the system. Therefore no dynamic filtering was used. A low-pass moving average filter was used to remove the trend from the signal. Since we are reconstructing the dynamics within a one minute cycle of the blood through the body, a window length of 100 s was chosen. To avoid errors due to oversampling, the signal was also smoothed using a 0.1 s long window.

The Lyapunov exponents were calculated from the $N = 27000$ points of the signal. The fiducial trajectory was followed for at most six periods of the slowest rhythm. Within this time, all exponents approach finite values. However, one may expect that the values obtained are increased by the noise.

The embedding time was estimated by the average mutual information and average displacement methods. The first local minimum of mutual information was found at $\tau = 0.3\,\mathrm{s}$. For embedding dimension two, the same value was obtained by the geometrically based average displacement criterion, whereas it gave $\tau \approx 0.08\,\mathrm{s}$ for $d \geq 8$.

Based on the power spectrum and guidelines obtained on quasi-periodic signals, the embedding dimension $d \geq 10$ should be used. The false nearest neighbours method suggested $d > 5$.

The choice of evolutionary time depends on the choice of embedding time. The calculations with different t_{evol} revealed that after it exceeds a few sampling times, all exponents decrease monotonically, except for the resonance-like phenomena at $l\tau$, but none of them changes sign. $t_{evol} = 0.2\,\mathrm{s}$ was taken.

Quadratic, higher polynomial, or radial basis functions may be used to approximate the local mapping. However, for polynomial basis functions N_o grows rapidly with the increasing order of the polynom and the embedding dimension. Since noise is present, a large N_b/N_o is recommended. Consequently, the size of the neighbourhood becomes too large compared to the attractor size to obtain a reliable approximation. To reduce the effect of noise and avoid approximation errors due to large ε_{max}/r_A, radial basis functions with $N_o = 100$ were used, and N_b/N_o equalled two.

Figure 34.4 presents the exponents of the blood flow signal in different embedding

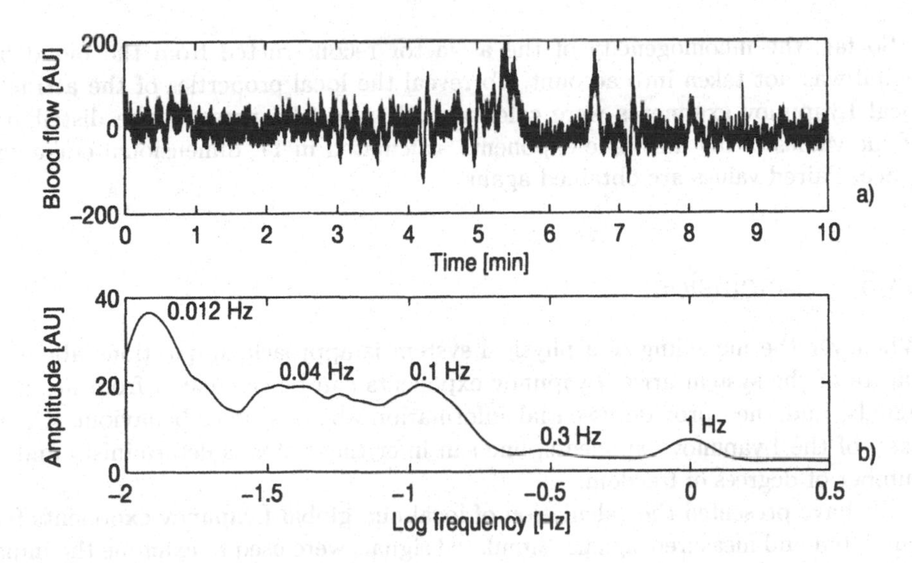

FIGURE 34.3. The blood flow signal measured from a healthy young male subject in the (a) time and (b) frequency domains

dimensions. In low dimensions, the exponents vary from one dimension to the other, but as the dimension reaches 10, two patterns can be observed: either four paired and one zero or five paired exponents are calculated. In the latter case, one pair equals zero within the calculation error. Among the exponents of the reversed signal, four or five pairs are observed again, and the other exponents are negative. Only the latter can be identified as spurious.

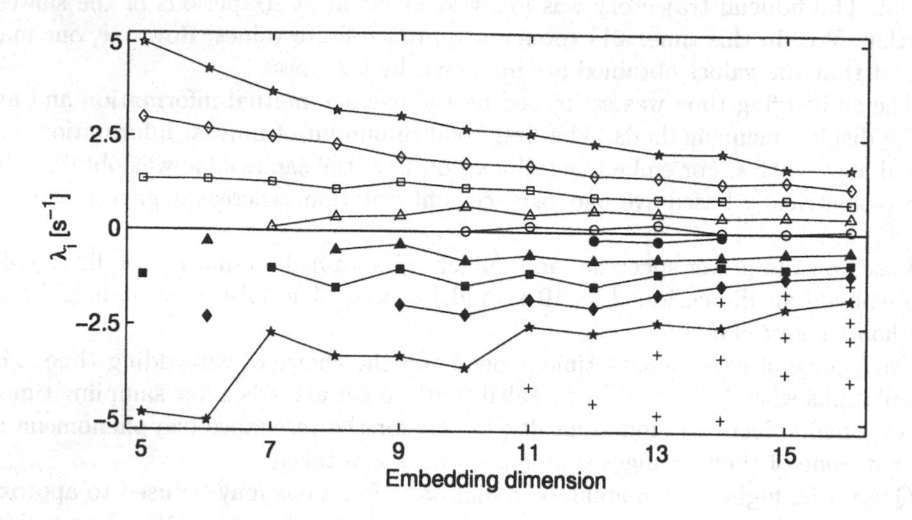

FIGURE 34.4. Exponents of the blood flow signal of a healthy person for different embedding dimensions. The exponents appear in pairs. Paired exponents have markers of the same shape, the positive exponent has an empty marker, and the negative exponent has a filled marker. Spurious exponents are marked by crosses. The calculation error is approximately $0.1\,s^{-1}$. $N = 27000$, $t_s = 0.02\,s$, $\tau = 0.08\,s$, $t_{evol} = 0.2\,s$, $S = 1000$, $N_b/N_o = 2$. Radial basis functions were used

So far, the inhomogeneity of the attractor reconstructed from the blood flow signal was not taken into account. To reveal the local properties of the attractor, local Lyapunov exponents were calculated. Figure 34.5 presents the distribution of the values of the first nine exponents, calculated in 11, dimensional embedding space. Paired values are obtained again.

34.5 Conclusions

Whenever the modelling of a physical system is approached, questions about the nature of the system arise. Lyapunov exponents can be calculated from measured signals, and they provide essential information about system behaviour. On the basis of the Lyapunov exponents, one can infer the system's determinism and the number of degrees of freedom.

We have presented the calculation of local and global Lyapunov exponents from simulated and measured signals. Simulated signals were used to examine the impact of algorithmic parameters on the results. Although a plateau of relatively stable values of Lyapunov exponents can be found in the parameter space, it is our view

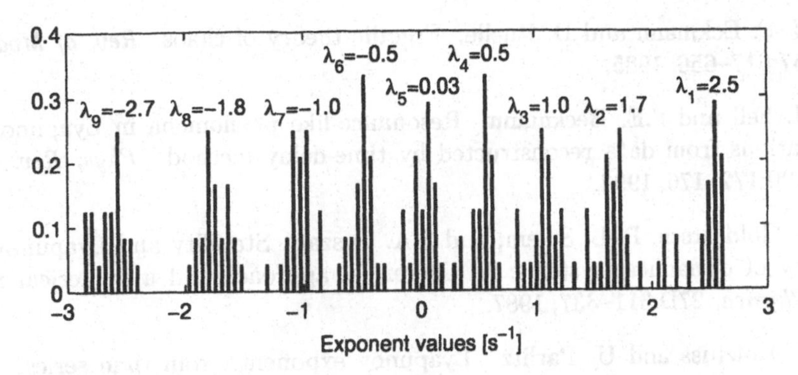

FIGURE 34.5. The distribution of local exponents of the blood flow signal of a healthy person, calculated from 50 different starting points. $N = 27000$, $t_s = 0.02\,\text{s}$, $\tau = 0.08\,\text{s}$, $d = 11$, $t_{evol} = 0.2\,\text{s}$, $S = 400$, $N_b/N_o = 2$. Radial basis functions were used

that conclusions should not be drawn from the exact values of exponents but rather from their patterns.

Over an interval of embedding dimension d, both the global and the local Lyapunov exponents of the blood flow signal are relatively stable with two patterns: four paired and one zero or five paired exponents. The other exponents are negative and spurious. Similar results were obtained for other subjects and with signals measured by the instrument of another producer [4].

The appearance of an exponent that equals zero within the calculation error shows that the blood flow dynamics are governed by a deterministic system. Moreover, blood flow dynamics are governed by a finite number of subsystems, most likely five.

Acknowledgments: The authors gratefully acknowledge the support of the Slovenian Ministry of Science and Technology.

34.6 References

[1] H.D.I. Abarbanel, R. Brown, and M.B. Kennel. Variation of Lyapunov exponent on a strange attractor. *J. Nonlinear Sci.*, 1:175–199, 1991.

[2] H.D.I. Abarbanel, R. Brown, J.J. Sidorowich, and L.Sh. Tsimiring. Analysis of observed chaotic data in physical systems. *Rev. of Mod. Phys.*, 65:1331–1392, 1993.

[3] G. Bennetin, L. Galgani, A. Giorgilli, and J.M. Strelcyn. Lyapunov characteristic exponents for smooth dynamical systems and for Hamiltonian systems: A method for computing all of them. *Meccanica*, 15:9–30, 1980.

[4] M. Bračič and A. Stefanovska. Local and global Lyapunov exponents of blood flow. To be published in *Open Systems and Information Dynamics* in 1997.

[5] R. Brown, P. Bryant, and H.D.I. Abarbanel. Computing Lyapunov spectrum of a dynamical system from observed time series. *Phys. Rev. A*, 43:2787–2806, 1991.

[6] J.-P. Eckmann and D. Ruelle. Ergodic theory of chaos. *Rev. of Mod. Phys.*, 57:617–656, 1985.

[7] J. Fell and P.E. Beckmann. Resonance-like phenomena in Lyapunov calculations from data reconstructed by time-delay method. *Phys. Rev. Lett. A*, 190:172–176, 1994.

[8] I. Goldhirsch, P.-L. Sulem, and S.A. Orszag. Stability and Lyapunov stability of dynamical systems: A differential approach and a numerical method. *Physica*, 27D:311–337, 1987.

[9] J. Holzfuss and U. Parlitz. Lyapunov exponents from time series. *Lecture Notes in Mathematics*, 1486:263–270, 1990.

[10] V.I. Oseledec. A multiplicative ergodic theorem. Lyapunov characteristic numbers for dynamical systems. *Trans. Moscow Math. Soc.*, 19:197–231,1968.

[11] N. Packard, J. Crutchfield, D. Farmer, and R. Shaw. Geometry from time series. *Phys. Rev. Lett.*, 45:712–716, 1980.

[12] M.T. Rosenstein, J.J. Collins, and C.J. De Luca. A practical method for calculating the largest Lyapunov exponent from small data sets. *Physica D*, 65:117–134, 1993.

[13] M.T. Rosenstein, J.J. Collins, and C.J. DeLuca. Reconstruction expansion as a geometry-based framework for choosing proper delay times. *Physica D*, 73:82–98, 1994.

[14] M. Sano and Y. Sawada. Measurement of the Lyapunov spectrum from a chaotic time series. *Phys. Rev. Lett.*, 55:1082–1085, 1985.

[15] S. Sato, M. Sano, and Y. Sawada. Practical methods of measuring the generalised dimension and the largest Lyapunov exponent in high dimensional chaotic systems. *Prog. Theor. Phys.*, 77:1–5, 1987.

[16] A. Stefanovska and P. Krošelj. Correlation integral and frequency analysis of cardiovascular functions. To be published in *Open Systems and Information Dynamics* in 1997.

[17] F. Takens. Detecting strange attractors in turbulence. *Lecture Notes in Mathematics*, 898:366–381, 1980.

[18] A. Wolf, J.B. Swift, H. Swinney, and J.A. Vastano. Determining Lyapunov exponents from a time series. *Physica*, 16D:285–317, 1985.

35

Rank Order Based Decomposition and Classification of Heart Rate Signals

Aleksej Makarov[1]
Gilles Thonet[2]

ABSTRACT
A new technique for decomposing composite signals into trend and cyclic components is addressed. Contrary to other techniques, it does not require any a priori knowledge about the functional form or distribution of the signal components. The algorithm is applied to extract cyclic components of heart rate time series retrieved from implanted defibrillators. It is shown that after the decomposition of these signals, better classification can be performed between healthy subjects and patients suffering from ventricular tachyarrhythmias.

35.1 Introduction

Composite signals are widespread in various fields, such as natural sciences and biomedical engineering. A *composite signal* is defined as the sum of a fluctuating baseline (called the *trend*) and one or more *cyclic components*, which are assumed to be stochastic, narrow-band, and zero-mean. Their spectral limits are expected to be time-varying, which means that they have an evolving cycle width and a changing shape.

The methods used up to now for the analysis of such signals are based on a structural model, such as the Box and Jenkins SARIMA[3] model [2], a Bayesian dynamic linear model (DLM), or a composite parametric regression model [13]. Alternatively, a spectral splitting of the signal into fixed frequency bands can be used, in which the trend is assumed to be contained in a low frequency band, and the cyclic components in higher frequency bands.

There are cases where the trend may be subject to abrupt and large changes and where the amplitude, shape and period of the cyclic components may vary. In this context, we refer to the trend as the *acyclic component* [10]. The conventional analytical methods mentioned above cannot be applied in such cases for the following

[1]Swiss Federal Institute of Technology, Micro-Processor and Interface Laboratory, 1015 Lausanne, Switzerland, E-mail: Aleksej.Makarov@epfl.ch

[2]Swiss Federal Institute of Technology, Signal Processing Laboratory, 1015 Lausanne, Switzerland, E-mail: Gilles.Thonet@epfl.ch

[3]SARIMA: Seasonal Autoregressive Integrated Moving Average.

reasons. First, the SARIMA, DLM and parametric decomposition models require an a priori knowledge of the cyclic component period. This parameter can not be obtained by classical techniques (such as the short-time Fourier transform) if the spectral content of the acyclic component buries the spectral peak corresponding to the cyclic one. On the other hand, the spectral splitting can also be ineffective since neither the acyclic nor the cyclic components are exactly contained within the fixed subband limits.

A new method for the decomposition of signals possessing an acyclic component and a fixed number of cyclic components is introduced in Sec. 35.2. The average cycle width is estimated by counting repetitive features of cycles called ***characteristic points*** (Sec. 35.2.1 and 35.2.2). Then, the computed cycle width is used for the decomposition step. An iterative procedure is derived in Sec. 35.2.4 for extracting several cyclic components. An example of application to a synthetic signal is presented in Sec. 35.3.

Eventually, the proposed method is applied to the analysis of heart rate time series before the onset of ventricular tachyarrhythmias. Significant features are extracted on the second cyclic component of these signals to obtain better risk stratification of persons susceptible of overcoming such arrhythmias.

35.2 Description of the Method

35.2.1 Detection of Characteristic Points

A cycle can be represented as a pattern of non-zero and zero slopes. For example, a pure sinusoidal cycle contains one positive and one negative slope, with zero slopes in between, corresponding to the cycle extrema. Similarly, cyclic signals of other functional forms contain zero-slope points, which obviously correspond to the possible extrema. Cyclic time series are seldom smooth but mostly noisy, implying that it is more difficult to find their extrema. However, one can extract consistent monotonic microtrends within a cycle and find the instants when these microtrends change their monotonic behaviour, as shown in Fig. 35.1. Such instants will be called ***characteristic points.***

There is usually an approximately constant number of characteristic points per cycle, and their mutual distances fluctuate around integer multiples of cycle widths. A way to detect a consistent microslope within a cycle is to use ***order statistics envelopes*** [7, 8, 9]. Assuming that impulsive noise deteriorates the signal only sparsely, a special case of order statistics envelopes can be used, namely the ***extrema envelopes.*** Consider $\Delta\mathcal{M}_t$, the difference between successive maxima,

$$\Delta\mathcal{M}_t = \max_{t-M+1\leq i\leq t}\{x(i)\} - \max_{t-M\leq i\leq t-1}\{x(i)\} \tag{35.1}$$

and $\Delta\mathcal{W}_t$, the difference between successive minima,

$$\Delta\mathcal{W}_t = \min_{t-M+1\leq i\leq t}\{x(i)\} - \min_{t-M\leq i\leq t-1}\{x(i)\} \tag{35.2}$$

Then, a consistent microtrend is detected at time t using the product $\Delta\mathcal{M}_t\Delta\mathcal{W}_t$:

$$\begin{aligned}\Sigma_t &= h(\Delta\mathcal{M}_t\Delta\mathcal{W}_t) \\ &= \begin{cases} 1 & \text{for an increasing or decreasing trend} \\ 0 & \text{for no trend} \end{cases}\end{aligned} \tag{35.3}$$

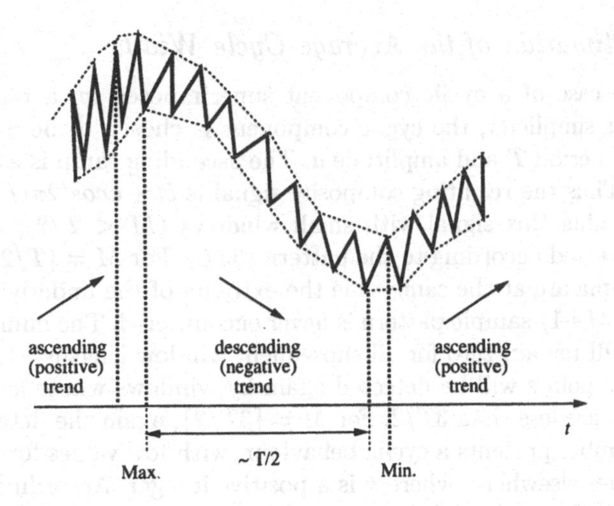

FIGURE 35.1. Characteristic points are instants at which consistent monotonic microtrends within a noisy cycle change the signs of their slopes. They are denoted by "Max" and "Min"

where h stands for the unit step function with the convention $h(0) = 0$. A characteristic point of a cycle is detected at time t whenever the sequence $\{\Sigma_{t-1}\Sigma_t\}$ equals either $\{01\}$ or $\{10\}$, corresponding to a change in the microtrend monotonic behaviour.

From (35.1), (35.2), and (35.3), it is easily seen that a non-zero slope is detected by observing the extrema changes of an M-sample sliding window. The necessary and sufficient condition to detect an increasing microtrend is that the new signal sample $x(t)$ is larger than the previous maximum $\max_{t-M\leq i\leq t-1}\{x(i)\}$, so that

$$x(t) = \max_{t-M\leq i\leq t}\{x(i)\} \tag{35.4}$$

On the other hand, for the product $\Delta\mathcal{M}_t\Delta\mathcal{W}_t$ to be larger than zero, $\Delta\mathcal{W}_t$ also has to be larger than zero. This holds if $\min_{t-M\leq i\leq t-1}\{x(i)\}$ is the global minimum of the (M+1)-sample window $\{x(t-M), x(t-M+1), ..., x(t)\}$, entailing that

$$x(t-M) = \min_{t-M\leq i\leq t}\{x(i)\} \tag{35.5}$$

From (35.4) and (35.5), it is easily noticed that for an ascending microtrend, the detection according to (35.3) is equivalent to the detection of the following (M+1)-sample pattern:

$$\underbrace{\overbrace{x(t-M)}^{\min\{x\}_{t-1}} \quad x(t-M+1) \quad \ldots\ldots\ldots \quad x(N)}_{\text{previous window } \{x\}_{t-1}} \quad \overbrace{x(t)}^{\max\{x\}_t} \tag{35.6}$$

In the case of a decreasing microtrend, the order statistics in (35.4) and (35.5), as well as in the pattern (35.6), must be replaced by their opposites.

35.2.2 Estimation of the Average Cycle Width

Consider the case of a cyclic component superimposed on a ramp, as shown in Fig. 35.2. For simplicity, the cyclic component is chosen to be a strictly periodic sinusoid with period T and amplitude a. The ascending ramp is assumed to have a slope $b > 0$. Thus the resulting composite signal is $bt + acos(2\pi t/T)$.

When scanning this signal with small windows ($M < T/2$), all characteristic points are detected according to the pattern (35.6). For $M = \lceil T/2 \rceil$, the M-sample window extrema are at the same time the extrema of the underlying cycle. Consequently, the $(M+1)$-sample pattern is never encountered. The number of characteristic points will remain zero for all subsequent window lengths M, until $M = \lfloor T \rfloor$. Characteristic points will be detected again by windows whose length exceeds the period T, but are less than $3T/2$. For $M = \lceil 3T/2 \rceil$, again the detection will fall to zero. This number presents a cyclic behaviour, with low values for $M \in (kT/2, kT)$ and high values elsewhere, where k is a positive integer. Accordingly, T is also the cycle width of the number of characteristic points.

In Fig. 35.2, the $(M + 1)$-sample patterns allow the detection of cycle extrema for windows spreading from points A to B, C to D, and C to E. On the contrary, windows CF and CG never meet the pattern (35.6). For such windows, the product $\Delta\mathcal{M}_t\Delta\mathcal{W}_t$ is always zero. Hence no characteristic point can be detected. Similar behaviour can be derived for decreasing trends with slope $b < 0$.

RESONANCE OF THE WINDOW LENGTHS WITH THE PERIOD OF A CYCLIC COMPONENT

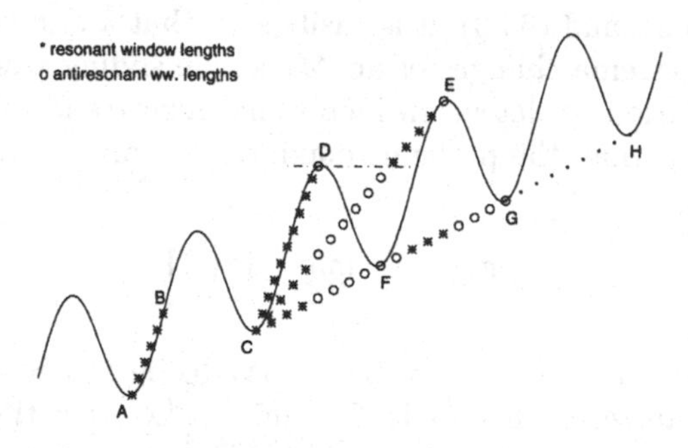

FIGURE 35.2. Detection of characteristic points in a sinusoid superimposed on a ramp. The window lengths which allow the detection are denoted by asterisks, whereas those which miss the detection are marked with circles

By using a set of window lengths ranging from M_{min} to $M_{max} \gg T$, the number of detected characteristic points over an interval of N samples, where $N > M_{max}$, presents cyclic behaviour, as shown in Fig. 35.3. The cycle width T can be computed as the distance between two neighbouring peaks.

The proposed method, particularly designed for non-stationary signals, fails to estimate the average cycle length in two cases.

1. A noiseless periodicity on a constant baseline. In this case, characteristic points are detected only with window sizes $M < T/2$. However, if the number of characteristic points per cycle is known in advance, as well as the fact that the trend slope is $b = 0$, the period can be computed from the detected number of characteristic points.

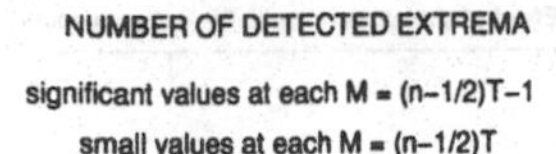

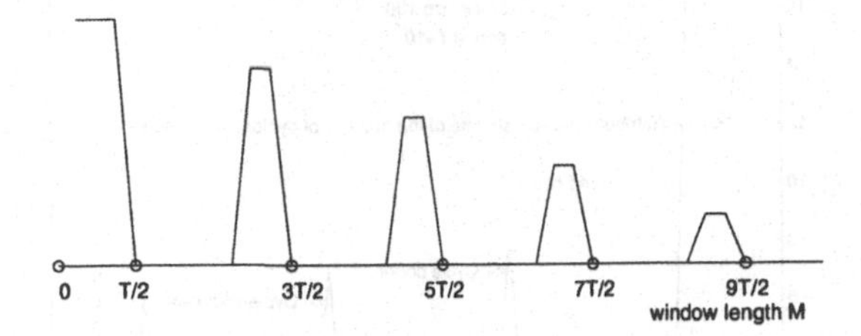

FIGURE 35.3. Number of characteristic points with respect to window length

2. A trend with slope $|b| > 2\pi a/T$, where a is the amplitude of a sinusoidal signal. Similarly, for non-sinusoidal signals, very steep trends may cause the product $\Delta\mathcal{M}_t \Delta\mathcal{W}_t$ to be always different from zero. Hence no characteristic point can be detected. Fortunately, such steep trends occur only sparsely in most of the natural signals. So only few characteristic points may miss within the interval of N samples, which does not degrade the estimate of the average cycle width.

35.2.3 Effect of Noise

Before estimating the average cycle width of a signal, its composite nature should be checked. Given the fact that the opposite of a composite signal is a stationary process (a constant level corrupted by white noise), this last case will be chosen as the null hypothesis.

Let us find the detection probability of characteristic points under null hypothesis, or false alarm probability. From pattern (35.6), the probability that (35.4) is satisfied for independent continuously distributed noise is $1/(M+1)$. For simplicity, it is assumed that two equal extrema never occur within the same $(M+1)$-sample window. Similarly, the probability that (35.5) is satisfied under the condition that (35.4) holds is $1/M$. Since the signal is independent, the probability of having the pattern (35.6) under the null hypothesis is thus $1/((M+1)M)$. Then, the probability of detecting a non-zero slope trend is given by

$$P_T = \frac{2}{(M+1)M} \tag{35.7}$$

Hence, the characteristic points are detected in stationary noise with probability

$$P_c = P_T(1 - P_T) \tag{35.8}$$

independently of the noise distribution. The number of characteristic points in a data block of N samples therefore is given by

$$N_F = P_c(N - M - 1) \tag{35.9}$$

The evolution of N_F with respect to the window length M is shown in Fig. 35.4 with the dashed line. Thus a white noise test can be performed by computing the

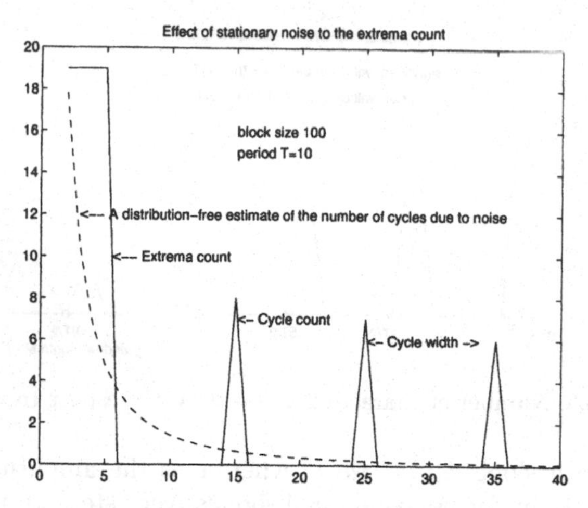

FIGURE 35.4. Number of detected characteristic points with respect to the window length

χ^2 statistics and comparing the theoretical curve N_F with the actually detected number of characteristic points for corresponding window lengths.

If the null hypothesis is rejected, the acyclic component can be assumed to be significant ($|b| > 0$). And if the cyclic component exists, its average cycle width T_{AV} can be estimated. Several cycle width estimates T_i can be computed between successive peaks of characteristic points. Then the average cycle width is calculated not as the mean of previous estimates, but as the median to reduce the significant noise effect at lower window lengths. Therefore comparing the cycle width estimates T_i with the resulting value T_{AV} is suggested, again with the χ^2 statistics, so as to confirm the cyclicity of the signal. If the cyclicity hypothesis is rejected, the fluctuations around the trend level are assumed to be random and independent.

In non-stationary cases, the false alarm rate is highly dependent on the distribution of acyclic, cyclic, and uncorrelated components [10].

35.2.4 Decomposition of Composite Signals

Once the cycle width is estimated, the trend and cyclic components can be separated by more classical methods [3]. Assuming small trend changes over a single cycle width, the trend (or acyclic component) $\mu(t)$ can be reconstructed as an irregularly sampled sequence with a sampling period corresponding to the estimated cycle width.

Assuming that cycles are zero-mean, the trend value $\mu(K)$ for the observed K^{th} cycle is the block average over the cycle width T_K:

$$\mu(K) = \frac{1}{2\lfloor \frac{T_K}{2} \rfloor + 1} \sum_{t=K-\lfloor \frac{T_K}{2} \rfloor}^{K+\lfloor \frac{T_K}{2} \rfloor} x(t) \tag{35.10}$$

The cycle index K in (35.10) can be expressed at time t as the integer upper bound of the ratio of the time t and the sum of all previous cycle widths T_J, $J \leq K$:

$$K = \lceil \frac{t}{\sum_{J=1}^{K} T_J} \rceil \tag{35.11}$$

If T_K is an odd integer, the quantization error introduced in (35.10) is zero. Since the cycle width is estimated for an N-sample block, the cycle widths computed for successive cycles inside the same block will be equal.

Then the trend values $\mu(t)$ can be interpolated between irregularly sampled points $(K, \mu(K))$ to be defined for all time samples. The easiest way to do it is to use straight-line interpolation.

Finally, the cyclic component $\phi(t)$ is computed as

$$\phi(t) = x(t) - \mu(t) \tag{35.12}$$

In the case of several cyclic components, the above procedure can be applied iteratively. For instance, if two cyclic components $\phi_1(t)$ and $\phi_2(t)$ are present in the composite signal $x(t)$ and if their cycle widths differ enough, the following procedure can be applied.

1. Segment the signal $x(t)$ into N-sample blocks of data, $N \gg T_{max}$.
2. Estimate the average cycle width of $\phi_1(t)$ for each block of $x(t)$. It is assumed that $\phi_1(t)$ has shorter cycle widths than $\phi_2(t)$.
3. Extract the acyclic component $\mu_1(t)$ by averaging the signal $x(t)$ and interpolating over cycle widths.
4. Compute $\phi_1(t) = x(t) - \mu_1(t)$.
5. Estimate the average cycle width of $\phi_2(t)$ for each block of $\mu_1(t)$.
6. Extract the acyclic component $\mu(t)$ by averaging the signal $\mu_1(t)$ and interpolating over cycle widths.
7. Compute $\phi_2(t) = \mu_1(t) - \mu(t)$.

Extension to more than two cyclic components is straightforward.

35.3 Application to a Synthetic Signal

This section addresses an example of an application to a synthetic signal. Fig. 35.5 displays a composite signal obtained by superimposing a sinusoid of amplitude $a = 1$ and period $T = 10$ over a trend consisting of a rectangular impulse and the lower half of an ellipse. The signal length is $N = 200$ samples.

By using window sizes ranging from $M = 1$ to $M = N$, the evolution of the number of characteristic points with respect to M can be computed, as displayed in the upper part of Fig. 35.6. The white noise hypothesis is rejected by the χ^2 statistics comparing the numbers of characteristic points obtained with respect to the expected values under null hypothesis, as shown in the lower part of Fig. 35.6. Thus the signal can be considered possibly composite, and the median distance between peaks of the characteristic point curve is computed, giving an average cycle width estimate of 10 samples. This estimate is accepted by the χ^2 cyclicity test (described in Sec. 35.2.3) with very high probability, which confirms the composite signal hypothesis.

The extracted components are shown in the middle and the lower parts of Fig. 35.7. As expected, the largest errors occur around steepest trends.

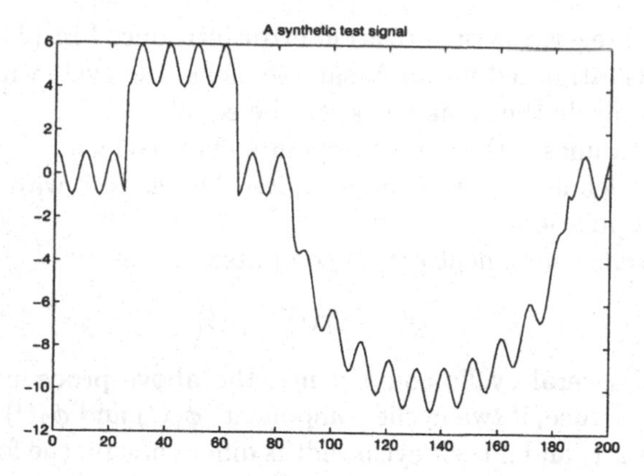

FIGURE 35.5. Signal presenting two abrupt changes, a continuously varying trend and a superimposed sinusoid

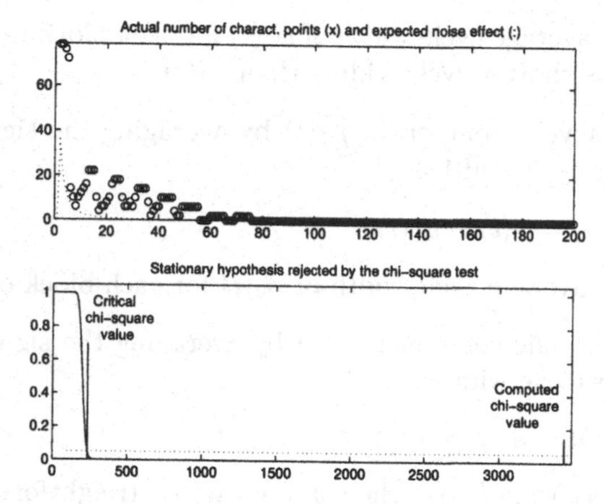

FIGURE 35.6. Top diagram: Number of detected characteristic points with respect to window length (the dotted line represents the expected number of characteristic points under null hypothesis). Bottom diagram: Rejection of the null hypothesis

35.4 Application to Heart Rate Signals

The decomposition scheme is applied to heart rate (HR) signals to detect subjects at increased risk of overcoming a sudden cardiac death (SCD) in the immediate or in the more distant future.

After a brief introduction to the main underlying physiological concepts, the database of HR signals is presented. Then, the new decomposition method is used to extract features in cyclic components so as to distinguish better between healthy subjects and patients at risk.

35.4.1 Heart Rate Analysis

Cardiovascular diseases are one of the most common cause of mortality in a large number of countries. In particular, SCD hits between 0.1 and 0.2 % of the global population in Western countries. SCD is usually the result of an abnormally rapid heart rhythm whose origin takes place in the heart ventricles, leading to *ventricular*

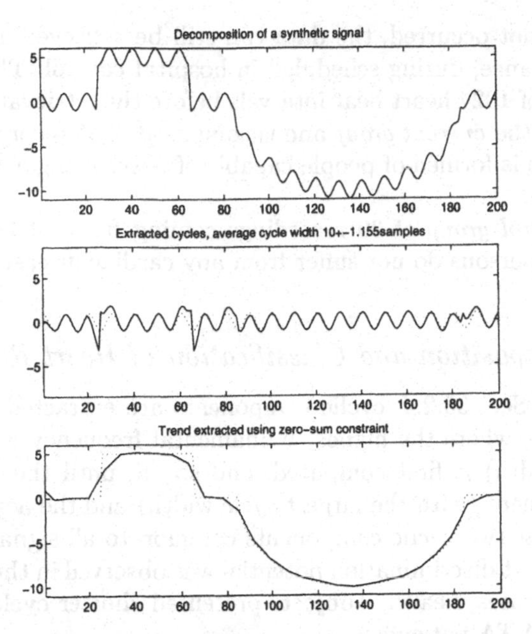

FIGURE 35.7. Original signal (top) and extracted cyclic and acyclic components (middle and bottom). The dotted line represents original components

tachyarrhythmia (VTA). This term includes tachycardias and, in the worse cases, fibrillations. The autonomous nervous system is expected to be one of the main factor leading to the onset of VTA. It contains two components, namely, the *sympathetic system* (which increases the HR) and the *parasympathetic* or *vagal* system (which decreases the HR). A recent study [12] has pointed out the importance of the sympatho-vagal balance in VTA initiation.

For years numerous works have been devoted to the analysis of heart rate variability (HRV). The HR signal is defined as the temporal evolution of the time interval between two successive R peaks (or beats). It is now well established that the study of HRV yields some noninvasive insights into the interaction between the autonomous nervous system and the cardiovascular system [1]. Accordingly, a better understanding of the phenomena underlying VTA is expected from HRV study.

Previous works have attempted to find some particular features in the HR dynamics so as to distinguish healthy persons from patients suffering from VTA. Preliminary results were obtained by [6] which showed that people with higher cardiac mortality have lower 24-hour variability. Then, frequency-domain analysis has been applied [11], but results are still confusing and difficult to use in clinical practice.

35.4.2 Description of the Database

Healthy subjects are compared with patients suffering from recurrent VTA. The data concerning ill patients were retrieved from implanted Medtronic defibrillators provided with a memory capable of storing 1024 RR intervals. The recording is stopped as soon as a VTA is detected, in order to store the last 1024 RR intervals preceding its onset. Thus the first group will be called *VTA group* and contains 41 recordings including 33 tachycardia and eight fibrillations.

If a VTA has not occurred, the data can still be retrieved at any time of interrogation, for instance, during scheduled in-hospital control. The data collected by this way consist of 1024 heart beat intervals before the retrieval time. This group of patients is called the ***current group*** and is composed of 35 recordings. Consequently, the current group is formed of people capable of overcoming a VTA, but not in the immediate future.

Finally, a ***control group*** of 39 recordings coming from 12 healthy subjects was gathered. These persons do not suffer from any cardiac disease.

35.4.3 Decomposition and Classification of Heart Rate Signals

As mentioned in Sec. 35.2.4, cyclic components are extracted according to an iterative procedure, where the highest fundamental frequency component (with the shortest cycle width) is first computed, and so on, until the lowest fundamental frequency component (with the largest cycle width) and the acyclic component are obtained. The first two cyclic components common to all signals were retained for analysis. The widest discrimination potential was observed in the second cyclic component. In most cases, healthy subjects presented shorter cycle widths and larger amplitudes than VTA patients.

35.4.3.1 Physiological Interpretation of Cyclic Components

The first cyclic component of a HR signal, extracted as the highest discrete frequency constituent, can be associated with fluctuations due to vagal activity [1], responsible for slowing down the HR. Past works have clearly pointed out the influence of the parasympathetic system on HR fluctuations due to respiration (***respiratory sinus arrhythmia*** [4]).

The second cyclic component contains lower frequencies, generally assumed to be covered by both vagal and sympathetic activities. Increased sympathetic activity speeds up the HR and is expected to trigger VTA. The response of the HR to this influence is much slower than to the vagal one [1]. Consequently, a frequency shift of the second cyclic component toward lower frequencies (larger cycle widths) can be interpreted as an increase in sympathetic and a decrease in vagal activity.

Eventually, the trend or acyclic component is believed to be influenced mostly by the sympathetic part of the autonomous nervous system. Nevertheless, a vagal influence can also be present in this component. The regulatory effect of the vagal activity is expected to be more obvious in the control group, producing larger variations for healthy subjects than for VTA and current groups.

35.4.3.2 Classification Features

After visual inspection, it clearly appeared that the second cyclic component had different forms for control and VTA subjects. These differences were quantified by two nonconventional indicators, namely, the ***range*** (a measure of amplitude over a given interval) and the ***higher order crossings*** (a measure of dominant frequency over a given interval) of the second cyclic component.

The use of range may be justified by medical studies showing that patients with heart failure have diminished power at all frequencies larger than 0.02 Hz with respect to healthy persons [1].

The higher order crossings of a signal are defined as the number of crossings through zero of the signal filtered by a combination of high-pass and low-pass filters [5]. The usual filter choice for the prefiltering step is $(1 + z^{-1})^i(1 - z^{-1})^j$, which is equivalent to i successive backward summations followed by j successive backward differences of the analysed signal. The differences can remove polynomial trends of order $p \leq j$, whereas the summations smooth the signal and suppress random noise. The effect of such a filtering is that almost all cycles in the second cyclic component are smoothed and de-trended.

The need for de-noising arises from the observation that breathing cycles are not the only high-frequency cycles that can occur in the first component. Occasionally, two distinct cycle subseries occur simultaneously at high frequencies. Then the slower cycle is interpreted as the second cyclic component. It was found that the smoothing exponent $i = 10$ removes remnants of these high-frequency vagal fluctuations in the second cyclic component.

Finally, on some fixed-length interval, it is sometimes impossible to distinguish between cycles and trend. By differentiating a cyclic component superimposed on a trend, the number of crossings of the zero level is closer to the double of the number of cycles (two crossings per cycle are assumed). Hence, the trend remnants are removed by a single differentiation of the second cyclic component ($j = 1$).

35.4.3.3 Results

Examples of extracted components for healthy and VTA subjects are presented in Figs. 35.8 and 35.9. They illustrate the typical behaviour of heart rhythm in healthy and VTA subjects. In practice, the large variability among patients makes the classification more difficult.

Fig. 35.10 displays the feature space for healthy subjects (control group) and patients at risk (current and VTA groups). The separation into two classes is correct for about 90% of cases. Note that the class boundaries (represented by the dotted line) are derived empirically.

In Fig. 35.11, ill subjects (VTA and current groups) are tried to be classified. The patients, who recently suffered from VTA but who are in baseline condition at time of investigation (current group), have cycle widths and amplitudes that

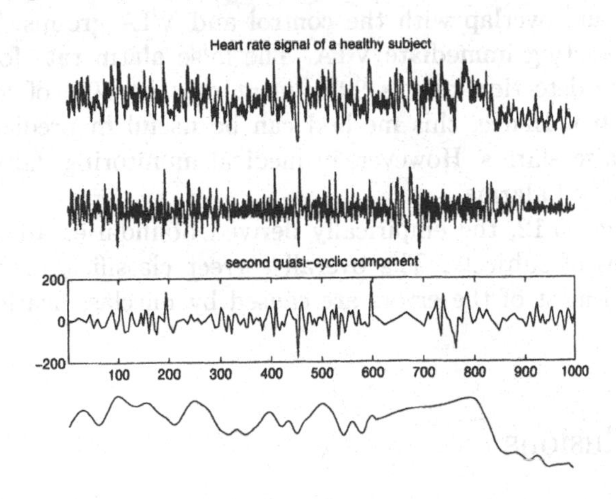

FIGURE 35.8. From top to bottom: Original signal, first cyclic component, second cyclic component, and second acyclic component of a healthy subject

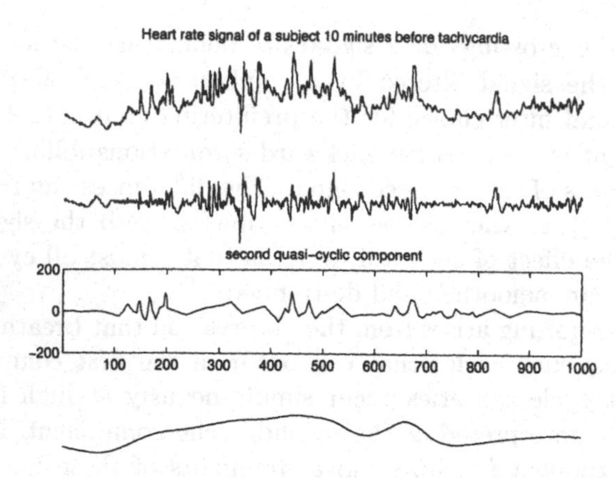

FIGURE 35.9. From top to bottom: Original signal, first cyclic component, second cyclic component and second acyclic component of a VTA patient

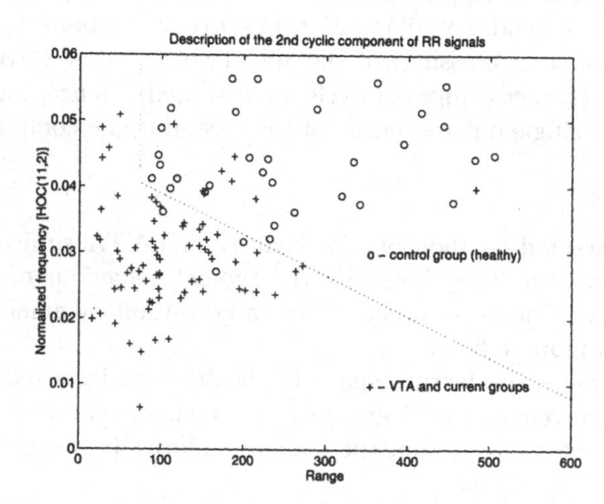

FIGURE 35.10. Classification between healthy subjects (control) and patients at risk (current and VTA)

are in between and overlap with the control and VTA groups. This reduces the efficiency of predicting immediate VTA. The false alarm rate for this prediction test is 32%. The detection rate is 90%, meaning that 10% of forthcoming VTA were not predicted. Hence, this method can be useful in predicting VTA but is susceptible to false alarms. However, in medical monitoring, false alarms are less harmful than missed alarms.

Finally, in Fig. 35.12, the empirically derived boundaries are used to separate the three classes of subjects. The overall correct classification rate is 80% (see Table 35.1), and most of the errors are caused by misclassification of the current group.

35.5 Conclusions

This chapter has addressed an emerging technique for the decomposition of composite signals into a trend or acyclic component and one or more cyclic components. It

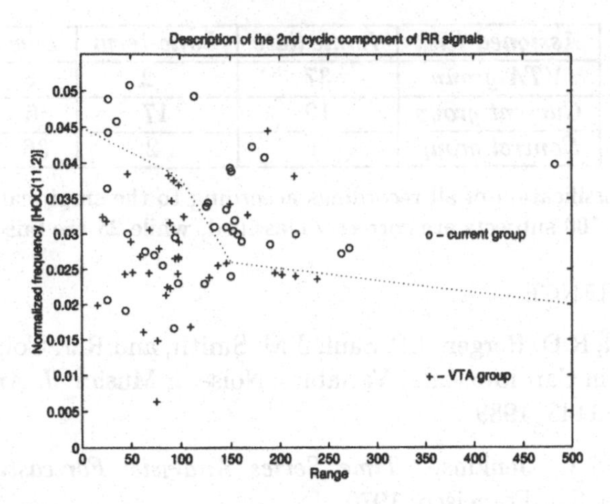

FIGURE 35.11. Classification between patients at immediate risk of overcoming a VTA (VTA) and patients who recently suffered from VTA (current)

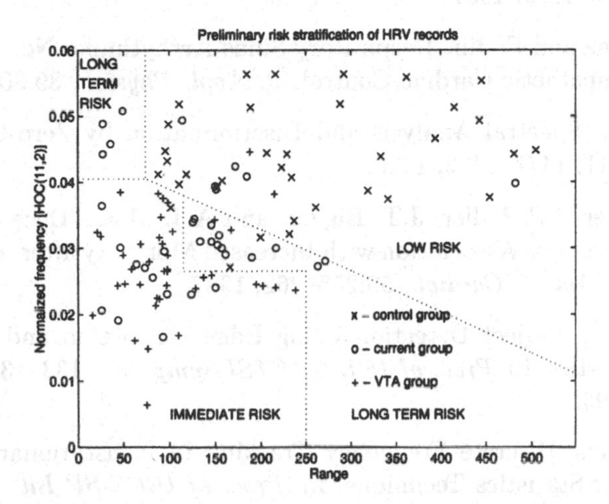

FIGURE 35.12. Global feature space split into three parts, with assigned short-term VTA risk, long-term VTA risk, and low VTA risk

has been shown that the new method is efficient for a wide range of non-stationary signals and requires a small amount of prior information about the signal structure.

The relevance of this method for heart rate variability analysis has been enhanced due to a direct physiological interpretation of the cyclic components. Three groups of subjects have been considered: healthy subjects; persons suffering from ventricular tachyarrhythmias but who are under baseline condition during recording; and patients undergoing an arrhythmia at the end of the recording. A classification method has been introduced by considering cycle widths and amplitudes of the second cyclic component. Indeed, a higher number of zero-crossings of the filtered second cyclic component can be expected for healthy subjects with respect to subjects at risk. Moreover, the second cyclic component of patients under baseline condition has shown features that are between the two other groups.

These results pave the way for new processings of heart rate signals, where their composite nature is taken into account and related to physiological observations.

Assigned Risk	*Immediate*	*Long-term*	*Low*
VTA group	37	2	2
Current group	12	17	6
Control group	1	2	36

TABLE 35.1. Classification of all recordings according to the empirical boundaries previously mentioned. 100 subjects are correctly classified, while 25 are misclassified

35.6 References

[1] M.L. Appel, R.D. Berger, J.P. Saul, J.M. Smith, and R.J. Cohen. Beat to Beat Variability in Cardiovascular Variables: Noise or Music? *J. Am. Coll. Cardiol.*, 14(5):1139–1148, 1989.

[2] G. Box and G. Jenkins. *Time Series Analysis: Forecasting and Control.* Holden-Day, San Francisco, 1970.

[3] P.J. Brockwell and R.A. Davis. *Time Series: Theory and Methods.* Springer-Verlag, New York, 1991.

[4] P.G. Katona and F. Jih. Respiratory Sinus Arrhythmia: Noninvasive Measure of Parasympathetic Cardiac Control. *J. Appl. Physiol.*, 39:801–805, 1975.

[5] B. Kedem. Spectral Analysis and Discrimination by Zero-Crossings. *Proc. IEEE*, 74(11):1477–1492, 1986.

[6] R.E. Kleiger, J.P. Miller, J.T. Bigger, and A.J. Moss. Decreased Heart Rate Variability and its Association with Increased Mortality after Acute Myocardial Infarction. *Am. J. Cardiol.*, 59:258–262, 1987.

[7] A. Makarov. Object Detection Using Edge Extraction and Joint Nonlinear Order Statistics. In *Proc. of 15th GRETSI Symp.*, pp. 133–136, Juan-les-Pins, France, 1995.

[8] A. Makarov. Discrete Frequency Tracking in Nonstationary Signals Using Joint Order Statistics Technique. In *Proc. of IEEE-SP Int. Symp. on Time-Frequency and Time-Scale Analysis*, pp. 441–444, Paris, France, 1996.

[9] A. Makarov. Periodicity Retrieval from Nonstationary Signals. *Signal Processing VIII*, pp. 1949–1952, Trieste, Italy, 1996.

[10] A. Makarov. Nonparametric Data Qualification: Detection and Separation of Cyclic and Acyclic Changes. Ph.D. Thesis, No 1715, Swiss Federal Institute of Technology, Lausanne, Switzerland, 1997.

[11] M. Muzi and T.J. Ebert. Quantification of Heart Rate Variability with Power Spectral Analysis. *Cur. Opinion Anaesthesiol.*, 6:3–17, 1993.

[12] G. Thonet, E. Pruvot, P. Celka, R. Vetter, and J.-M. Vesin. Subband Decomposition of Heart Rate Signals Prior to Ventricular Tachyarrhythmia Onset. *Med. & Biol. Eng. & Comp.*, 35, *World Congress on Med. Physics and Biomed. Eng.*, Nice, France, 1997.

[13] M. West and J. Harrison. *Bayesian Forecasting and Dynamic Models.* Springer-Verlag, 1989.

Printed in the United States
By Bookmasters